Roger Nkambou, Jacqueline Bourdeau, and Riichiro Mizoguchi (Eds.)

Advances in Intelligent Tutoring Systems

Studies in Computational Intelligence, Volume 308

Editor-in-Chief
Prof. Janusz Kacprzyk
Systems Research Institute
Polish Academy of Sciences
ul. Newelska 6
01-447 Warsaw
Poland
E-mail: kacprzyk@ibspan.waw.pl

Further volumes of this series can be found on our homepage: springer.com

Vol. 285. Roberto Cipolla, Sebastiano Battiato, and Giovanni Maria Farinella (Eds.)
Computer Vision, 2010
ISBN 978-3-642-12847-9

Vol. 286. Zeev Volkovich, Alexander Bolshoy, Valery Kirzhner, and Zeev Barzily
Genome Clustering, 2010
ISBN 978-3-642-12951-3

Vol. 287. Dan Schonfeld, Caifeng Shan, Dacheng Tao, and Liang Wang (Eds.)
Video Search and Mining, 2010
ISBN 978-3-642-12899-8

Vol. 288. I-Hsien Ting, Hui-Ju Wu, Tien-Hwa Ho (Eds.)
Mining and Analyzing Social Networks, 2010
ISBN 978-3-642-13421-0

Vol. 289. Anne Håkansson, Ronald Hartung, and Ngoc Thanh Nguyen (Eds.)
Agent and Multi-agent Technology for Internet and Enterprise Systems, 2010
ISBN 978-3-642-13525-5

Vol. 290. Weiliang Xu and John Bronlund
Mastication Robots, 2010
ISBN 978-3-540-93902-3

Vol. 291. Shimon Whiteson
Adaptive Representations for Reinforcement Learning, 2010
ISBN 978-3-642-13931-4

Vol. 292. Fabrice Guillet, Gilbert Ritschard, Henri Briand, Djamel A. Zighed (Eds.)
Advances in Knowledge Discovery and Management, 2010
ISBN 978-3-642-00579-4

Vol. 293. Anthony Brabazon, Michael O'Neill, and Dietmar Maringer (Eds.)
Natural Computing in Computational Finance, 2010
ISBN 978-3-642-13949-9

Vol. 294. Manuel F.M. Barros, Jorge M.C. Guilherme, and Nuno C.G. Horta
Analog Circuits and Systems Optimization based on Evolutionary Computation Techniques, 2010
ISBN 978-3-642-12345-0

Vol. 295. Roger Lee (Ed.)
Software Engineering, Artificial Intelligence, Networking and Parallel/Distributed Computing, 2010
ISBN 978-3-642-13264-3

Vol. 296. Roger Lee (Ed.)
Software Engineering Research, Management and Applications, 2010
ISBN 978-3-642-13272-8

Vol. 297. Tania Tronco (Ed.)
New Network Architectures, 2010
ISBN 978-3-642-13246-9

Vol. 298. Adam Wierzbicki
Trust and Fairness in Open, Distributed Systems, 2010
ISBN 978-3-642-13450-0

Vol. 299. Vassil Sgurev, Mincho Hadjiski, and Janusz Kacprzyk (Eds.)
Intelligent Systems: From Theory to Practice, 2010
ISBN 978-3-642-13427-2

Vol. 300. Baoding Liu (Ed.)
Uncertainty Theory, 2010
ISBN 978-3-642-13958-1

Vol. 301. Giuliano Armano, Marco de Gemmis, Giovanni Semeraro, and Eloisa Vargiu (Eds.)
Intelligent Information Access, 2010
ISBN 978-3-642-13999-4

Vol. 302. Bijaya Ketan Panigrahi, Ajith Abraham, and Swagatam Das (Eds.)
Computational Intelligence in Power Engineering, 2010
ISBN 978-3-642-14012-9

Vol. 303. Joachim Diederich, Cengiz Gunay, and James M. Hogan
Recruitment Learning, 2010
ISBN 978-3-642-14027-3

Vol. 304. Anthony Finn and Lakhmi C. Jain (Eds.)
Innovations in Defence Support Systems, 2010
ISBN 978-3-642-14083-9

Vol. 305. Stefania Montani and Lakhmi C. Jain (Eds.)
Successful Case-based Reasoning Applications, 2010
ISBN 978-3-642-14077-8

Vol. 306. Tru Hoang Cao
Conceptual Graphs and Fuzzy Logic, 2010
ISBN 978-3-642-14086-0

Vol. 307. Anupam Shukla, Ritu Tiwari, and Rahul Kala
Towards Hybrid and Adaptive Computing, 2010
ISBN 978-3-642-14343-4

Vol. 308. Roger Nkambou, Jacqueline Bourdeau, and Riichiro Mizoguchi (Eds.)
Advances in Intelligent Tutoring Systems, 2010
ISBN 978-3-642-14362-5

Roger Nkambou, Jacqueline Bourdeau, and
Riichiro Mizoguchi (Eds.)

Advances in Intelligent Tutoring Systems

Springer

Prof. Roger Nkambou
Knowledge Management Laboratory
Computer Science Department
University of Quebec at Montreal
201, avenue du Président-Kennedy
Montréal (Québec) H2X 3Y7
Canada
E-mail: nkambou.roger@uqam.ca

Prof. Riichirio Mizoguchi
Institute of Scientific and
Industrial Research
Osaka University
8-1 Mihogaoka
Ibaraki, Osaka, 567-0047
Japan
E-mail: miz@ei.sanken.osaka-u.ac.jp

Prof. Jacqueline Bourdeau
Télé-université, UQAM
100, rue Sherbrooke Ouest
Montréal (Québec) H2X 3P2
Canada
E-mail: bourdeau@licef.teluq.uqam.ca

ISBN 978-3-642-14362-5 e-ISBN 978-3-642-14363-2

DOI 10.1007/978-3-642-14363-2

Studies in Computational Intelligence ISSN 1860-949X

Library of Congress Control Number: 2010930137

Typeset & Cover Design: Scientific Publishing Services Pvt. Ltd., Chennai, India.

Printed on acid-free paper

9 8 7 6 5 4 3 2 1

springer.com

Acknowledgements

We are grateful to all contributors, for their willingness to work with us on this book which focuses both on basic concepts of Intelligent Tutoring Systems as well as on current developments in research on Artificial Intelligence in Education (AIED). Writing this book was particularly challenging, due to the interdisciplinary nature of AIED research. We were however fortunate to work with a very good team of contributors. We would like to thank all of them.

We would also like to thank Prof. Gordon McCalla from the University of Saskatchewan, Canada, for his very articulate, kind and inspiring Foreword.

Special thanks go to our Springer contact in Heidelberg, Germany, Dr. Thomas Ditzinger for his assistance and editorial guidance during the entire editing process.

Many thanks go to Othalia Larue and Natacha Vandereruch for their efficient and generous help with the final editing of this book. We also gratefully acknowledge the contribution of Carol Harris, Judith Campbell and Mary Linonge in the English revision of several chapters in this book.

This book is dedicated to our families for allowing us to spend so much time on this project, especially, Joelle Laurence, Arielle, Presley, Audrey Camille, Marthe-Sarah and Roger Jr. for their patience and faithful support of the first editor, who spent literally all holidays working on this book instead of spending quality time with them.

Foreword

May the Forcing Functions be with You: The Stimulating World of AIED and ITS Research

It is my pleasure to write the foreword for Advances in Intelligent Tutoring Systems. This collection, with contributions from leading researchers in the field of artificial intelligence in education (AIED), constitutes an overview of the many challenging research problems that must be solved in order to build a truly intelligent tutoring system (ITS). The book not only describes some of the approaches and techniques that have been explored to meet these challenges, but also some of the systems that have actually been built and deployed in this effort. As discussed in the Introduction (Chapter 1), the terms "AIED" and "ITS" are often used interchangeably, and there is a large overlap in the researchers devoted to exploring this common field. In this foreword, I will use the term "AIED" to refer to the research area, and the term "ITS" to refer to the particular kind of system that AIED researchers build.

It has often been said that AIED is "AI-complete" in that to produce a tutoring system as sophisticated and effective as a human tutor requires solving the entire gamut of artificial intelligence research (AI) problems. In fact, AIED is really even broader than that: it draws from a wider range of computer science than just AI, including human-computer interaction, data mining, the semantic web, multi-agent systems, and information science; and AIED also draws from a wide range of social sciences, including cognitive science, psychology, anthropology, sociology, linguistics, and, of course, education.

Fortunately, working in an educational context also provides useful constraints that allow progress to be made that would otherwise be impossible in such a broad area. From a social science perspective, AIED researchers are focussed on learning and knowing, not all of cognition, and on particular domains to be learned that are often fairly well understood. Most aspects of human behaviour have a projection into AIED, but this projection can be managed more tractably than open human interaction. From a computer science perspective, AIED researchers are typically highly applied, building actual systems, so are satisfied with good results rather than provable theories. Further, system requirements can be limited in highly useful ways by the educational goals. Thus, knowledge representation of the domain is needed, but for most educational domains this knowledge is already at least somewhat explicitly codified, so the problem is usually more tractable than the general knowledge representation problem. Managing the interaction between the tutoring system and the learner is critical, as in any interactive system, but in an intelligent tutoring system the interaction can be informed by pedagogical strategies that are both educationally well tested and also constrain the learner in ways that are not resented as they would be in general human-computer interaction. Moreover, full natural language interaction isn't usually needed: most educational domains have notation or specific vocabulary that either allow limited natural language understanding or an end-run around natural language altogether. A key to making tutoring "intelligent" is individualizing the interactions between a learner and the system, and this is done through a "learner model" (aka a "student model"). In educational applications it is easier than in many other applications to get and maintain a user model: demographic information, student marks, and other performance data are all readily available, and students follow learning paths that are often predictable, thus helping to anticipate changes to the learner model and to diagnose learner understandings and misunderstandings.

While these constraints allow AIED researchers to make progress, they also often set up interesting angles or foci on research problems that are also being explored outside of AIED. From a social science perspective, the focus on learning means that how people grow and revise their knowledge and perspectives becomes fundamental. The need for personalization, but the importance of also having a learning community, means interesting issues in the relationship of individuals to groups can be looked at. And so on. In computer science, AIED forces systems to deal fundamentally with change: an ITS's very goal is to stimulate change in the learner. Knowledge representation has to have cognitive fidelity not just logical purity – both conceptions and misconceptions have to be represented and inconsistency has to be dealt with as an unalterable fact of life. Natural language interaction has to be concerned with performance phenomena not just competence, and must go well beyond syntax to deal with the semantics and pragmatics issues necessary to actually have the system understand the learner. Diagnosis has to be cognitively plausible and track standard learning paths followed by actual human learners. These "forcing functions", as John Seely Brown discussed in an invited talk at the first ITS conference, Brown (1988), actually drive AIED towards solutions that may actually be more useful and scalable for general systems than many existing computer science paradigms that tend to get captured by technological or formal fetish, or get too narrow in their goals.

AIED is not standing still. There are big shifts in how humans are using information technology, and AIED is shifting right alongside. People are now involved in a huge amount of on-line activity, with each other (through social media) and with information (through the web). Intelligent tutoring systems can take advantage of these shifts by deploying new pedagogical strategies that, for example, make use of the web as an information source for the learners, use social media for interaction among the learners, deploy recommender system techniques to find appropriate information or to find suitable helpers for learners facing an impasse, incorporate intelligent agents as companions to help guide a learner through the vast repository of on-line information and media, etc. Vast quantities of data, real time and fine-grained, can be captured from user interactions in this new space, and new data mining and statistical algorithms lead to the possibility of making sense of learner performance by analyzing this interaction data. The key, however, to being able to leverage this rich data source is to understand the user's goals and his or her broader context, Vassileva et al (2001), McCalla (2004). Otherwise, there are just too many statistical patterns lurking in the data that swamp any possibility of distinguishing the relevant from the irrelevant. AIED allows such context capture more readily than other areas of interactive systems research, since the aim of learning something new usually makes both the learner's goals and broader contextual elements more explicit. Often learners have specific learning goals, are working on known tasks, are encountering known misconceptions, are at a specific point in exploring the domain, etc. Knowing these things, an ITS can much more readily decide what a given pattern of learner behaviour means "in context". And, the ITS can react appropriately, now choosing from a significantly expanded set of pedagogical strategies ranging from "traditional" tutoring approaches (eg. teaching and coaching), through to various just-in-time learning strategies (eg. finding a web resource, suggesting a peer helper, offering the learner an opportunity to reflect on the state of their understanding by showing them part of their learner model, etc.) Once again, AIED has the right forcing functions for exploring the highly interactive, information rich, and socially networked modern world enabled by information and communications technology.

It is likely that AIED research will continue to be a leader in exploring how advanced techniques from computer and social science can be deployed to support human interactions in and with a constantly evolving "cyberspace". Certainly, the constraints that learning imposes will continue to allow ITSs to reason more deeply about their users than other interactive systems without such constraints. Moreover, learning itself as a main goal for people will, if anything, be of increasing importance, as revolutionary change impacts most areas of human life and forces everybody into a race to keep abreast of new developments in most fields of human activity. This is probably why a recent trend to study "life long learning" issues seems to be taking hold in AIED. Two recent workshops have studied issues raised by the general goal of building life long learning companions, Lane (2008), and the specific implications of life long user modelling, Kay and Kummerfeld (2009).

While it is too early to say whether this new trend will have lasting traction, the contours of the research activities that will be pursued are emerging. Learner modelling will be central, but the learner model is likely to be computed as needed from a vast amount of (often contradictory) coarse and fine grained data about a learner that is constantly accumulating. This means that data mining will be of increasing centrality to AIED – the growth of an explicit educational data mining community, EDM (2010), suggests that this trend is already firmly ensconced. Recommender system technology (also a newly hot emerging field of its own with the recent creation of its own conference series, Amatriain and Torrens (2010)) is also likely to be important, as people need help in sorting through the vast amount of information now available on the web to find the subset relevant to their needs. Social networking will also be crucial as people seek other people to help them in specific situations and as groups of learners form virtual learning communities to support each other on an on-going basis (the area of Computer Supported Collaborative Learning, Stahl et al (2006), which emerged from the HCI community, will thus be a key part of the greater AIED enterprise). A unifying pursuit for this next generation of AIED research could be the development of learning companions, at a person's side for life, helping them fulfill their individual learning needs "just in time" through access to the vast array of resources and people available in an increasingly ICT-saturated world. Sort of Tak-Wai Chan's pioneering "The Prince" learning companion, Chan and Baskin (1990), meets novelist Neal Stephenson's "Primer" at the side of Nell, a little girl who is the main protagonist in the novel Diamond Age, Stephenson (1995). This thus unites AIED's future with its roots in the one-on-one tutoring paradigm: after all isn't the best tutor a companion to the learner, a wise and sensitive supporter of that individual's particular learning needs?

And, finally, it is possible to carry out a reality check on the credibility of my prognostications in this Preface. I wrote a paper entitled "The Fragmentation of Culture, Learning, Teaching, and Technology" that appeared in a special issue of the AIED Journal kicking off the new millennium in 2000, McCalla (2000). In this paper, I speculated on what issues AIED researchers would be exploring in 2010. You might want to compare what I said back then and what we are actually doing these days. I am too afraid to do so myself!

The papers in this volume explore some of the issues I have discussed, and also many other interesting issues, that arise when building an ITS. The papers also illustrate many of the clever insights AIED researchers have provided into these issues, insights that should be generally useful beyond AIED. The papers not only look backward at successes already achieved, but also forward to new terrains being explored by AIED. Enjoy reading this volume, a "just-in-time" contribution to an area of research on the cusp of many major trends in both social science and computer science as the information revolution accelerates.

Gordon McCalla
University of Saskatchewan

References

Amatriain X, Torrens M (2010) Chairs, 4th ACM Conference on Recommender Systems. Barcelona, Spain. http://recsys.acm.org/2010/

Brown JS (1988) Invited Talk at 1st International Conference on Intelligent Tutoring Systems, Montréal. Paper appeared as "Toward a New Epistemology for Learning". In: Frasson C and Gauthier G (eds) Intelligent Tutoring Systems: At the Crossroad of Artificial Intelligence and Education. Ablex, Norwood, NJ.

Chan T-W, Baskin AB (1990) "Learning Companion Systems". In : Frasson C and Gauthier G (eds) Intelligent Tutoring Systems: At the Crossroad of Artificial Intelligence and Education. Ablex, Norwood, NJ.

EDM (2010). Educational Data Mining community homepage. http://www.educationaldatamining.org/

Kay J, Kummerfeld R (2009) Chairs, Workshop on Lifelong User Modelling, 1st and 17th Conference on User Modeling, Adaptation, and Personalization. Trento, Italy. http://www.cs.usyd.edu.au/~judy/CFP/UMAP09_llum.html

Lane C (2008) Chair, ICT Workshop on Intelligent Lifelong Learning Companions, Institute for Creative Technologies. Marina del Rey, CA. http://projects.ict.usc.edu/companion/abstracts.html

McCalla GI (2000) The Fragmentation of Culture, Learning, Teaching and Technology: Implications for the Artificial Intelligence in Education Research Agenda in 2010, International Journal of Artificial Intelligence in Education. 11(2):177-196

McCalla GI (2004) The Ecological Approach to the Design of E-Learning Environments: Purpose-based Capture and Use of Information about Learners. Journal of Interactive Media in Education (JIME), Special Issue on The Educational Semantic Web, T. Anderson and D. Whitelock (Guest Eds.) (1), http://www-jime.open.ac.uk/2004/1

Stahl G, Koschmann T, Suthers D (2006) Computer-Supported Collaborative Learning: An Historical Perspective. In : Sawyer RK (ed) Cambridge handbook of the learning sciences. Cambridge University Press, UK.

Stephenson N (1995) The Diamond Age: Or, a Young Lady's Illustrated Primer. Bantam, Spectra.

Vassileva J, McCalla GI, Greer JE (2001) Multi-Agent Multi-User Modeling in I-Help, User Modeling and User-Adapted Interaction Journal. 3(1):1-31

Contents

Part II: Tutor Modeling and Learning Environments

Chapter 1
Introduction: What Are Intelligent Tutoring Systems, and Why This Book?

Roger Nkambou[1], Jacqueline Bourdeau[2], and Riichiro Mizoguchi[3]

[1] Université du Québec à Montréal, 201 Du Président-Kennedy Avenue, PK 4150, Montréal, QC, H2X 3Y7, Canada
[2] Télé-université, Université du Québec à Montréal, 100 Sherbrooke Street W., Montréal, QC, H3T 1J3, Canada
[3] The Institute of Scientific and Industrial Research, Osaka University, 8-1 Mihogaoka, Ibaraki, Osaka, 567-0047, Japan
Nkambou.roger@uqam.ca, bourdeau@teluq.uqam.ca, miz@ei.sanken.osaka-u.ac.jp

Abstract. This introductory chapter opens the doors to the field of Intelligent Tutoring Systems (ITS) and ITS research. A historical perspective provides insight into the genesis of the field, which is a prerequisite for understanding the recent advances presented in the book. Challenges specific to the field are introduced, and the community and its dynamics are described. The chapter ends with a presentation of the book's contents and organization.

1.1 Why This book?

The idea for this book on Intelligent Tutoring Systems (ITS) was sparked by the success of the ITS'08 international conference. The number of presentations and their quality bore witness to the vitality and maturity of the field, and the enthusiasm of the participants held out a promise of sustainability and innovative research. Long life to ITS research!

"Not ANOTHER book on ITS!" Actually, this field is not as rich in books as it is in journals and conference proceedings. The first, simply entitled Intelligent Tutoring Systems, was edited by Sleeman and Brown (1982), who coined the term. It was soon followed by Wenger's Artificial Intelligence and Tutoring Systems (1987), which established what would become the so-called "traditional ITS architecture" with its four components: domain, student, tutor and user interface. Recently, Woolf's book *Building Intelligent Interactive Tutors* (2008) offered an authoritative encyclopedia for anyone desiring an initiation into ITS building.

So why have we written this book, and for whom? Our intention is twofold: to provide a basic understanding of the field and its evolution, and to highlight recent advances and work in progress. Novices and experts alike should find it useful; at least that is our hope. Our intention is to reach two main categories of readers: ITS experts and experts-to-be (graduate students); and non-experts who want to gain

R. Nkambou et al. (Eds.): Advances in Intelligent Tutoring Systems, SCI 308, pp. 1–12.
springerlink.com

some understanding of the main ideas and facts in the field of ITS. The book is divided into five parts. The introductory chapters to these parts, which summarize foundations, developments, strengths and weaknesses in each of the areas covered, are addressed to all readers. For those who want more in-depth knowledge, we give the floor to researchers who present their work, their results, and their view of what the future holds. It is our hope that all readers will find the book informative and thought-provoking.

The next sections contain a summary of the origin of ITS, its foundations and goals, its architecture, its success stories and challenges, the people who have been contributing to the field, and the community that has formed over the years, in a close relationship with the Artificial Intelligence and Education (AIED) Society. The chapter ends with a presentation of the book's contents and organization.

1.2 Intelligent Tutoring Systems and Their Architecture

All ITSs share the same goal: to provide tutorial services that support learning. That being said, they show a vast variety of ways to conceptualize, design and develop these services. Efforts in this direction first began in the 1960s and '70s, with the development of what was called Intelligent Computer-Assisted Learning (ICAI) by Carbonell (1970). The term "Intelligent Tutoring Systems" was coined by Sleeman and Brown in their volume of the same title (1982). The first ITS conference (1988) provided an opportunity to share and consolidate ideas, and it evolved into a biannual conference with full proceedings. Several research labs would dedicate their work to ITS, raising considerable funds and deploying their systems in real settings. These achievements are summarized in the following section.

1.2.1 A Growing Field

In 1990, 20 years after its birth, the field had passed its infancy and could reflect on the first generation of contributions, as evidenced by two publications that appeared in that year: an overview of the field (Nwana 1990) and a position paper (Self 1990). In 2008-2009, a generation later, further evolution is described in two volumes: one that reflects the latest developments (Woolf et al. 2008) and one that provides a broad, in-depth survey of the ITS field (Woolf, 2008). These landmark publications will guide our description of a field that has been growing for three generations.

To understand the first generation, from 1970 to 1990, it is essential to be aware of the climate of the time and the motivations underlying the field's emergence. Artificial Intelligence (AI) was in full bloom and seeking applications for its advanced techniques in both computer and cognitive science. Computer-Assisted Instruction (CAI) was a mature and promising technology with a target market. The educational system was looking for solutions to overcome its limitations in order to deal with large groups in schools. Visionary scientists and pioneers imagined that merging AI with CAI could yield solutions to improve school instruction (Carbonell 1970). In 1984, Bloom published an article demonstrating that individual tutoring is twice as effective as group teaching. AI researchers saw a solid foundation upon which they could create intelligent systems that would provide effective tutoring for every student, tailored to her needs and pace of learning. Nwana has reviewed the systems produced by this generation and counted 43 (Nwana 1990). The dream at that time was that children would "have access to what Philip of Macedon's son had as royal prerogative: the personal services of a tutor as well informed as Aristotle" (Suppes, quoted in Nwana 1990).

In 1990, Self published an article in the first volume of the *Journal of Artificial Intelligence in Education (AIED)*, after a provocative talk at the 4th International Conference on Artificial Intelligence in Education, in which he called for theoretical foundations for ITS (Self 1990). In his article, Self (who was to become the first president of the AIED Society in 1993) claimed that in order to attain the status of a scientific field, ITS research needs scientific foundations. In his view, pursuing the goal of *ITS as human teacher* was overstretching an analogy and taking a wrong direction. He suggested viewing ITS as an engineering design field, which should equip itself with the theories, methods and techniques appropriate for a design field.

Some twenty years later, the field of ITS shows signs of vitality and self-confidence. Has it answered Self's call? There can be no doubt that it has. The AIED journal, the biannual AIED conference and the now biannual ITS conference all feature high-level theoretical and technical papers, presenting implementations in schools and other settings, and bold, innovative developments (Woolf et al. 2008; Woolf 2008). These results are reflected in the chapters of this book.

1.2.2 ITS Architectures

Why should architecture be a central issue? In 1990, talking of the three-component architecture (domain, student, tutoring), Self wrote: "an unquestioning adoption of the traditional trinity model will cause problems in ITS implementation and will restrict the scope of ITS research . . . the main concern is that the traditional trinity imposes a philosophy which is not an appropriate basis from which to develop a theory of ITS" (Self 1990). This criticism warrants special attention. What Self called the trinity is also known as the four-component architecture, where the fourth element is the user interface (Fig. 1.1).

What is the semantics of this architecture? Basically, it divides the system into four components where knowledge and reasoning are needed, leaving the integration problem open. In actuality, the systems generally show an emphasis (in both computational and control terms) on one component over the others. What can we expect to find in each component?

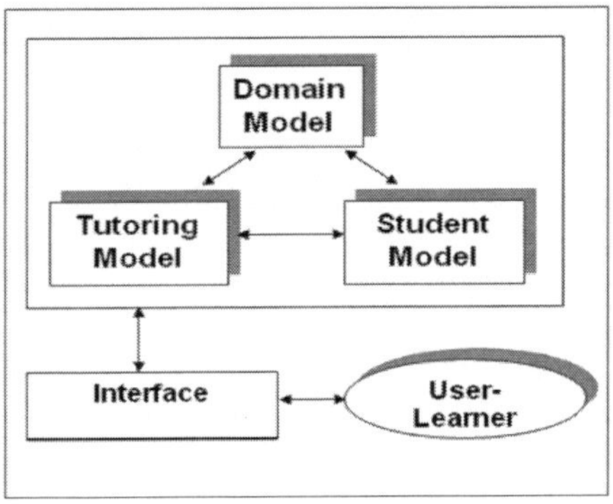

Fig. 1.1 The four-component architecture

The ***domain model*** (also called expert knowledge) contains the concepts, rules, and problem-solving strategies of the domain to be learned. It can fulfill several roles: as a source of expert knowledge, a standard for evaluating the student's performance or for detecting errors, etc. It is sometimes organized into a curriculum, a structure including all the knowledge elements linked together according to pedagogical sequences. Each knowledge unit can be more or less detailed and the curriculum can be practically organized in a dynamic model, according to various structures such as hierarchies, semantic networks, frames, ontology and production rules. The crucial problems concern the capability to reason with the model and gradually adapt the explanation of the reasoning to the learner.

The ***student model*** **is** the core component of an ITS. Ideally, it should contain as much knowledge as possible about the student's cognitive and affective states and their evolution as the learning process advances. The student model is usually viewed as a dynamic model that implements several functions. Wenger (1987) assigned three main functions to the student model: 1) it must gather explicit and implicit (inferred) data from and about the learner; 2) it must use these data to create a representation of the student's knowledge and learning process; and 3) it must account for the data by performing some type of diagnosis, both of the state of the student's knowledge and in terms of selecting optimal pedagogical strategies for presenting subsequent domain information to the student. In the same vein, Self (1988) identified six major roles for the student model: 1) Corrective: to help eradicate bugs in the student's knowledge; 2) Elaborative: to help correct 'incomplete' student knowledge; 3) Strategic: to help initiate significant changes in the tutorial strategy other than the tactical decisions of 1 and 2 above; 3) Diagnostic: to help diagnose bugs in the student's knowledge; 5) Predictive: to help determine the student's likely response to tutorial actions; 6) Evaluative: to help assess the student or the ITS. These functions and roles have been both expanded and diversified in the years since then.

The ***tutoring model*** receives input from the domain and student models and makes decisions about tutoring strategies and actions. Based on principled knowledge, it must make such decisions as whether or not to intervene, and if so, when and how. Content and delivery planning are also part of the tutoring model's functions.

Tutoring decisions would ideally be reflected in different forms of interaction with the student: socratic dialogs, hints, feedback from the system, etc. More generally, student/tutor interactions usually occur through the learning interface, also known as the communication or ***interface component***. This component gives access to the domain knowledge elements through multiple forms of learning environment, including simulations, hypermedia, micro-worlds, etc.

In his survey paper, Nwana reviewed other architectures that have been proposed or adopted (1990). In his discussion, he underlines the strong link between architecture and paradigm (what he calls a philosophy) and explains that differing tutoring philosophies emphasize different components of the learning process: domain, student or tutor. The architectural design of an ITS reflects this emphasis, and this leads to a variety of architectures, none of which, individually, can support all tutoring strategies (Nwana 1990).

Twenty years later, research teams report a wide range of architectures, and they may be unaware of this link with a paradigm, or claim that there is no explicit link. The themes selected for special issues of the *AIED* journal or for categories in the ITS and AIED conferences indicate a wide spectrum of topics, and architecture is not a main concern (Woolf et al. 2008; Woolf 2008). The present book addresses most of these topics and integrates them in the introductory chapter of each part.

One important aspect is that, beside this classic component view of ITSs, these systems offer a number of services. The implementation of some of these services usually transcends the individual components. For instance, the diagnostic service is sometimes considered to be part of the student model, but many other researchers see it as a function of the tutor component.

In the research perspective, a number of advanced ideas and techniques have been explored (open learner model, virtual reality agents, machine learning techniques, data mining techniques, etc.) and will be presented in the chapters of this book.

1.3 ITS and AIED: The Boundaries and the Community

ITS and AIED have developed simultaneously as two complementary fields involving what is in fact one large community. The term ITS is more restrictive than AIED, but the field is inclusive, and open to all AIED-related themes. ITS conferences are organized every other year, alternating with the AIED conference. Both publish their proceedings, and the AIED community is organized as a society, with its journal. Does AIED offer a choice venue for reflection about ITS research? The answer is yes, but reflection and discussions also occur at ITS events. A common topic of discussion is the high level of difficulty of ITS in design and computational terms.

ITS research is an interdisciplinary field, welcoming people from various disciplines, including computer science, psychology, the learning sciences and instructional technology, among others. This interdisciplinarity is both a challenge and a richness. Many ITS researchers study both computer science and cognitive science in order to achieve deep integration. Some have difficulty bridging the gap between bodies of knowledge and cultures from two or more disciplines; others prefer not to address the problem. Some stay and some leave; that's life! C'est la vie!

The future will see the field of ITS continue to embrace new technologies and adapt itself to new generations of researchers and graduate students. ITS will continue to spawn new domains, as it has the fields of qualitative reasoning, hypermedia systems, and educational data mining.

1.4 Organization and Content of This Book

In its structure, this book follows the traditional component architecture, for a shared understanding of the ITS domain. It is composed of five parts. The first three deal with modeling the domain, the tutor and the student, respectively, and the fourth looks at ITS construction. Each of these four parts begins with an introductory chapter that provides an overview of the issue, followed by a number of chapters that deal with specific approaches for dealing with that issue. The last part covers topics related to social, cultural and privacy issues.

1.4.1 Part 1: Modeling the Domain

Part 1 looks at how domain knowledge is represented in ITS. There are five chapters in this part.

Chapter 2, by Nkambou, presents an overview of domain acquisition issues in ITS since the field's emergence. The acquisition and representation of a domain knowledge model is a difficult problem that has been the subject of numerous research efforts in both the AI and AIED fields. The chapter surveys the methods and techniques used for this purpose. It begins by presenting and discussing the epistemological issue associated with domain knowledge engineering. Several knowledge representation languages are then briefly presented, considering their expressiveness, inferential power, cognitive plausibility and pedagogical emphasis. The chapter ends by establishing links with the subsequent chapters in this part of the book.

Chapter 3, by Aleven et al., is about rule-based cognitive modeling for ITS. Rule-based cognitive models play many roles in ITS development. They help in understanding student thinking and problem solving, guide many aspects of the tutor design, and can function as the "smarts" of the system. ITSs using rule-based cognitive models have been demonstrated to be successful in improving student learning in a range of learning domains. A rule-based model used in the Geometry Cognitive Tutor illustrates how, when modeling novice problem-solving knowledge in a complex domain, cognitive fidelity and ease of engineering are two important concerns shaping a model. More recently, rule-based modeling has been used effectively to provide tutoring at the metacognitive level. Much on-going work is aimed at making models easier to develop: for example, by creating efficient and effective authoring tools, or by using machine learning to infer models from author-provided examples or student log data.

Chapter 4, by Mitrovic, describes the constraint-based modeling approach. Stellan Ohlsson proposed constraint-based modeling (CBM) in 1992 as a way to overcome some problems in student modeling. Since then, the approach has been extended and used in numerous intelligent tutoring systems, which authors refer to as constraint-based tutors. CBM is now an established methodology for modeling instructional domains and for representing students' domain knowledge and higher-level skills. Authoring support for constraint-based tutors is now available, as well as mature, well-tested deployment environments. The chapter presents CBM, its foundations and extensions, and various types of instructional domains it has been applied to, and concludes with avenues for future research.

Chapter 5, by Founier-Viger et al., presents methods for coping with what have been termed "ill-defined domains": domains where classical approaches for building tutoring systems are not applicable or do not work well. The chapter provides an updated overview of the problem, and solutions for building intelligent tutoring systems for these domains. In its presentation, it considers three complementary and important perspectives: the characteristics of ill-defined domains, the approaches for supporting tutoring services in these domains, and suitable teaching models. Throughout the chapter, numerous examples are given to illustrate the discussion.

Part 1 ends with Chapter 6, by Zouaq and Nkambou, which gives more details about the ontology-based approach to domain modeling. With the advent of the Semantic Web, the field of domain ontology engineering has gained increasingly in importance. This innovative field may have a big impact on computer-based education and will certainly contribute to its development. The chapter presents a survey on domain ontology engineering and, especially, domain ontology learning, with a particular focus on automatic methods for ontology learning. After summarizing the state of the art in natural language processing techniques and statistical and machine learning techniques for ontology extraction, the chapter explains how intelligent tutoring systems may benefit from this engineering, and talks about the challenges the field is facing.

1.4.2 Part 2: Modeling the Tutor

Part 2 describes tutor modeling approaches and different ways learning environments behave.

Chapter 7, by Bourdeau and Grandbastien, is the introductory chapter for this part. It provides the reader with an overview of tutor modeling in ITS research. Starting with the origin of tutor modeling and a characterization of the tutoring process, it proposes a general definition of tutoring and a description of tutoring functions, variables, and interactions. The Interaction Hypothesis is presented and discussed. This discussion of tutor modeling is followed by the development of the tutorial component of an ITS, and its evaluation. New challenges, such as integrating the emotional states of the learner, are described, and perspectives for opening up the tutor model or providing it with social intelligence are presented.

In Chapter 8, Dubois et al. offer a thorough analysis of the way tutoring decisions are made in cognitive tutors. The chapter describes how AI techniques are utilized in artificial tutoring systems to reach decisions on when and how to intervene. Particular attention is given to the path of "natural" AI for tutoring systems, using human cognition as a model for artificial general intelligence. One tutoring agent built over a cognitive architecture, Conscious Tutoring System (CTS), illustrates this direction. The chapter concludes with a brief look at what may be the future for artificial tutoring systems: biologically-inspired cognitive architectures.

In Chapter 9, Olney, Graesser and Person present their work on conversational interaction in human tutoring and their attempts to build intelligent tutoring systems to simulate this interaction. They address the strategies, actions and dialogue of novice tutors and describe how novice tutoring has been implemented in an ITS called Autotutor, with learning gains comparable to those seen with novice human tutors. They also describe how they have recently extended their investigation to highly accomplished expert human tutors. Their goal is to understand what it is about accomplished, expert human tutors that produces outstanding learning gains. The chapter elaborates on the contrast between novice and expert, both in terms of human tutoring and in the ITS components required to mimic the interaction of novice and expert human tutors.

Chapter 10, by Woolf, describes the automatic recognition of and response to human emotion found in intelligent tutors. The chapter describes how tutors can recognize student emotion with more than 80% accuracy relative to student self-reports, using wireless sensors that provide data about posture, movement, grip tension, facially expressed mental states and arousal. Pedagogical agents have been used that provide emotional or motivational feedback. Students using such agents increase their math value and show improved self-concept and mastery orientation, with females reporting more confidence and less frustration. Low-achieving students—one-third of whom have learning disabilities—report higher affective needs than their higher-achieving peers. After interacting with affective pedagogical agents, low-achieving students improve their affective outcomes and report reduced frustration and anxiety.

Chapter 11, by Mizoguchi et al., is about the use of ontology in modeling the tutoring process. The chapter calls for the extensive modeling of tutoring knowledge by providing in-depth declarative knowledge and higher-level reasoning capabilities. It describes the achievements of a ten-year research project aimed at developing an ontology of education (OMNIBUS), which offers a common framework to be shared by all ITSs, regardless of paradigm or technology. It also describes some recent advances in ontology-based learning design (SMARTIES), and provides access to educational strategies inspired by educational theories or practice.

This part ends with Chapter 12, by vanJoolingen et al., which presents an approach for modeling knowledge in simulation-based inquiry learning. The approach requires a model of the domain that is executable, as well as a model of the learners' knowledge about the domain. An intermediate level is formed by models of the domain that are created by students, as is done in modeling environments. An approach is presented for generating student-created models from drawings.

This approach requires drawing segmentation, shape recognition and model generation, which are performed based on density-based clustering and elementary shape recognition combined with a shape ontology and model fragment composition, respectively. The final result is an executable model that can be used to generate simulation outcomes based on learners' conceptions. The role of such a system is discussed, especially with respect to the diagnosis of misconceptions and the generation of tutoring interventions based on confronting learners with the consequences of their conceptions.

1.4.3 Part 3: Modeling the Student

This part presents an overview of techniques used for student modeling. Specific emphasis is given to some important approaches for building the student model: the data mining approach, the open learner modeling approach and the Bayesian network approach.

Chapter 13, by Woolf, is the introductory chapter of this part. It describes how to build student models for intelligent tutors and indicates how knowledge is represented, updated, and used to improve tutor performance. It provides examples of how to represent domain content and describes evaluation methodologies. Several future scenarios for student models are discussed, including using the student model: 1) to support assessment for both formative issues (the degree to which the student has learned how to learn – for the purpose of improving learning capacity and effectiveness) and summative considerations (what is learned – for purposes of accountability and promotion); 2) to track when and how skills were learned and what pedagogies worked best for each learner; and 3) to include information on the cultural preferences of learners and their personal interests, learning goals, and personal characteristics. Ultimately, student model servers will separate student models from tutors and will be a part of wide area networks, serving more than one application instance at a time.

Chapter 14, by Conati, presents some techniques and issues involved in building probabilistic student models based on Bayesian networks and their extensions. It describes the pros and cons of this approach, and presents examples from two existing systems: an ITS (Andes) and an educational game (PrimeClimb).

Chapter 15, by Bull and Kay, describes the range of purposes that open learner models can serve, illustrating these with diverse examples of the ways they have been made available in several research systems. This is followed by discussion of the closely related issues of openness and learner control and approaches that have been explored for supporting learning by making the learner model available to people other than the learner. The chapter provides a foundation for understanding the range of ways in which open learner models have already been used to support learning, as well as directions yet to be explored.

In Chapter 16, Baker summarizes key data-mining methods that have supported student-modeling efforts, also discussing the specific constructs that have been modeled with the use of educational data mining. He also discusses the relative advantages of educational data mining compared to knowledge engineering, and key directions in which research is needed in order for educational data-mining research to reach its full potential.

This part ends with Chapter 17, by Frasson and Chalfoun. They point out the role of emotions in learning, discussing the different types and models of emotions that have been considered to date. They also address an important issue concerning the different means used to detect emotions. They describe recent approaches which measure brain activity using the electroencephalogram (EEG) in order to determine the learner's mental state, how long this state can exist, and what event can change it. The authors also present the main components of an emotionally intelligent tutoring system capable of recognizing, interpreting and influencing someone's emotions. In conclusion, they look at future perspectives and new directions in the area of emotional learning.

1.4.4 Part 4: Building Intelligent Tutoring Systems

In this part of the book, the issue of building tutoring systems is examined and possible solutions are discussed. A sample ITS is described to give the reader a close-up of how ITSs are used in real life.

Chapter 18, by Nkambou, Bourdeau and Psyché, is the introductory chapter to this part, and provides an overview of methods and tools for building ITSs. In this chapter, the challenge of building or authoring an ITS is addressed, along with problems that have arisen and been dealt with, solutions that have been tested, and evaluation methodologies. The chapter begins by clarifying what is involved in building an ITS. It then positions this challenge in the context of ITS research. The authors conclude with a series of open questions, and an introduction to the other chapters in this part of the book.

Chapter 19, by Paquette et al., presents the ASTUS authoring tool and evaluates its strengths and weaknesses compared to the well-known Cognitive Tutors Authoring Tools (CTAT). In the context of a multi-column subtraction problem, the chapter compares the two tools' ability to handle a comprehensive model of a well-defined domain. The model integrates various known procedural errors taken from the literature to see how each framework deals with complex situations where remedial help is needed. Examples of such situations include handling ambiguous steps, deducing errors from multiple steps, giving pedagogical feedback on composite errors, recognizing off-path steps and producing rich feedback on the tutor's interface. Selected scenarios in the subtraction domain are presented to illustrate that ASTUS can display more sophisticated behavior in these situations than CTAT. ASTUS achieves this by relying on an examinable hierarchical knowledge representation system and a domain-independent MVC-based approach to build the tutor's interface.

Chapter 20, by Razzaq and Heffernan, presents ASSISTment, an open content authoring tool. The difficulty of designing, conducting, and analyzing experiments means that there is often a dearth of empirical data to support or refute ideas. Designing and conducting a simple randomized controlled experiment to compare two different ways of teaching requires a great deal of effort by a teacher or a researcher. The difficulty of conducting such experiments, and then later analyzing the results, may be why so few randomized, controlled experiments are conducted in education. One of the goals of the ASSISTment System is to alleviate some of these difficulties. The chapter describes web-based tools that allow researchers to easily design, build and compare different ways to teach children. These tools can administer randomized controlled experiments to large numbers of students.

Chapter 21, by VanLehn et al., presents one of the most popular ITSs: Andes. The Andes physics tutoring system demonstrates that student learning can be significantly increased merely by upgrading the homework problem-solving support. Five years of experimentation at the United States Naval Academy indicates that Andes significantly improves student learning compared to doing the same homework with paper and pencil. Andes' key feature appears to be the grain size of interaction. The chapter describes Andes' behavior, the system design and implementation, evaluations of pedagogical effectiveness, and recent progress on dissemination/scale-up.

1.4.5 Part 5: Social, Cultural and Privacy Issues

This part presents some important issues currently surfacing in the ITS community. In Chapter 22, Tchounikine, Rummel and McLaren discuss how recent advances in the field of computer-supported collaborative learning (CSCL) have created the opportunity for new synergies between CSCL and ITS research. Three "hot" CSCL research topics are used as examples: analyzing individual and group interactions, providing students with adaptive intelligent support, and providing students with adaptive technological means.

Chapter 23, by Aimeur and Hage, is about privacy issues. The chapter describes the security and privacy-protection approaches taken by ITS and e-learning systems generally, and discusses the challenges they still face, as well as the particular challenges raised by Web 2.0.

The last chapter in the book, Chapter 24, by Blanchard and Ogan, deals with cultural issues. It examines what it means to adapt intelligent tutoring systems for users with diverse cultural backgrounds, and how intelligent tutoring systems can be used to support instruction that takes culture into consideration. The authors discuss the major research issues involved in modifying ITS to support these efforts. To provide insight into the current landscape of the field, they briefly outline some recent research achievements and highlight significant current and future issues raised by the integration of cultural concerns and educational technology.

References

Carbonell, J.: AI in CAI: An Artificial Intelligence Approach to Computer-Assisted Instruction. IEEE Tr. On Man-Machine Systems 11, 190–202 (1970)

Nwana, H.: Intelligent Tutoring Systems: An Overview. Artificial Intelligence Review 4, 251–277 (1990)

Self, J.A.: Student models: What use are they. In: Ercoli, P., Lewis, R. (eds.) Artificial Intelligence Tools in Education. North Holland, Amsterdam (1988)

Self, J.: Theoretical Foundations for Intelligent Tutoring Systems. JAIED 1(4), 3–14 (1990)

Sleeman, D.H., Brown, J.S. (eds.): Intelligent Tutoring Systems. Academic Press, New York (1982)

Wenger, E.: Artificial Intelligence and Tutoring Systems: Computational and Cognitive Approaches to the Communication of Knowledge. Morgan Kaufmann Publishers Inc., Los Altos (1987)

Woolf, B., Aimeur, E., Nkambou, R., Lajoie, S. (eds.): ITS 2008. LNCS, vol. 5091. Springer, Heidelberg (2008)

Woolf, B.: Building Intelligent Interactive Tutors: Student-centered strategies for revolutionizing e-learning, p. 480. Morgan Kaufmann, Burlington (2008)

Part I
Domain Modeling

Chapter 2
Modeling the Domain: An Introduction to the Expert Module

Roger Nkambou

University of Québec at Montréal, 201 Du Président-Kennedy Avenue, PK 4150, Montréal, QC, H2X 3Y7, Canada
Nkambou.roger@uqam.ca

Abstract. Acquiring and representing a domain knowledge model is a challenging problem that has been the subject of much research in the fields of both AI and AIED. This part of the book provides an overview of possible methods and techniques that are used for that purpose. This introductory chapter first presents and discusses the epistemological issue associated with domain knowledge engineering. Second, it briefly presents several knowledge representation languages while considering their expressivity, inferential power, cognitive plausibility and pedagogical emphasis. Lastly, the chapter ends with a presentation of the subsequent chapters in this part of the book.

2.1 Introduction

The purpose of intelligent tutoring systems (ITSs) is to enable learners to acquire knowledge and develop skills in a specific domain. To provide such tutoring services effectively, these systems must be equipped with an explicit representation of the domain knowledge that is the subject of the learning activity. It must also be equipped with the mechanisms by which the representation can be used by the system for reasoning in order to solve problems in the domain.

Acquiring and representing a domain knowledge model is a difficult problem that has been the subject of numerous Artificial Intelligence research projects since research in this field began (Clancey 1985; Brachman and Levesque 2004; Russell and Norvig 2009). Knowledge-based systems and expert systems, in particular, must explicitly represent the knowledge and inferences associated with the expertise in this domain.

Intelligent tutoring systems must also possess a domain-specific expert module that is able to generate and resolve domain problems and provide access to such knowledge in order to facilitate the dissemination (Wenger 1987; Woolf 2008) and acquisition of this knowledge by learners. Hence, developing an explicit model of domain knowledge with sound reasoning mechanisms is an important issue in the field of ITS research. The expert module of an ITS should provide the basis for interpreting learner actions (Corbett et al. 1997). It is therefore important

R. Nkambou et al. (Eds.): Advances in Intelligent Tutoring Systems, SCI 308, pp. 15–32.
springerlink.com

to consider not only the nature and value of the domain knowledge, but also the formalisms used to represent and apply it.

Many solutions have been put forward in an attempt to provide an explicit representation of domain expertise. The solutions presented have been drawn from fields such as philosophy, psychology, AI and education sciences. Philosophy, psychology and education sciences tend to focus on knowledge from an epistemological perspective which is essential for its treatment (Piaget 1997; Gagné 1985; Winograd and Flores 1986; Merrill and Twitchell 1994), whereas AI and cognitive sciences provide solutions that facilitate the expression and computational implementation of knowledge (Anderson 1996; Collins and Quillian 1969; Minsky 1975; Sowa 1984). From the perspective of AIED, it is important to incorporate the above-mentioned approaches so as to effectively meet the requirements associated with the development of a rich knowledge model and the inferential mechanisms associated with ITSs.

The goal of this chapter is to provide a brief overview of the means available for developing the expert module of ITSs. First, we present the epistemological perspective by addressing its importance in the domain knowledge modeling process for the purposes of learning. Next, we explore various languages that are available to represent knowledge and are frequently used to develop domain models for ITSs. We focus on the expressiveness, inferential power and cognitive plausibility of these formalisms and provide examples of the ITSs in which they have been used. We also present two examples of languages which have a strong pedagogical emphasis. We conclude the chapter with an introduction to subsequent chapters in this part of the book, each of which deals with a specific approach to domain modeling.

2.2 The Epistemological Perspective of Domain Knowledge

Epistemology refers to what we know and how we know it. The term "epistemology" is used here according to the definition given by Piaget (1997).Epistemology involves posing fundamental questions about the nature of knowledge (gnoseological considerations), the construction of knowledge (methodological considerations), and the value and validity of knowledge. In our opinion, these considerations are a prerequisite for formalizing knowledge pertaining to a specific domain. In this light, the epistemological perspective is of prime importance when characterizing the nature of knowledge and the inference mechanisms at play in a given domain. This perspective also makes it possible to question several aspects of knowledge: the production modes, the foundations underlying the knowledge in question and the production dynamics. The research carried out by Ramoni et al. (1992) clearly shows that this epistemological perspective is taken into consideration when developing knowledge-based systems. Classical knowledge engineering methodologies involve two distinct levels: the epistemological level and the computational level. The first is the level on which the epistemological analysis is carried out while taking into account the constraints derived from the conceptual structure of the domain knowledge, patterns of inference and tasks to be executed. The second level is the one on which the methods and formalisms must be adopted to formalize these elements. At the epistemological level, the ontology and inference model of a knowledge-based system must be defined. Ontology represents the conceptual model of domain knowledge which focuses on the nature, the properties and the constraints that govern the existence of knowledge. The inference model is the conceptual representation of the nature of the inference structure required to solve a problem or to execute a task by managing the ontology.

When designing ITSs, the nature of the knowledge to be taught or learned should be considered. An important aspect of epistemology consists in defining the nature of the knowledge involved in the learning process. The best known system of classification in the fields of AI and psychology divides domain knowledge into two types: declarative and procedural knowledge. However, more elaborate classification systems exist: Bloom (1975) and Gagne (I985) were among the first educational psychologists to develop clear classifications of knowledge and skills. They assert that different types of knowledge require different types of teaching or instructional methods. Other knowledge-typing schemes were later developed which are more strongly based on modern cognitive theory and are more operational and concrete for the purposes of computational representation. For example, Merrill's Component Display Theory (Merrill 1991) organizes knowledge in a matrix with content type (e.g., fact, concept or procedure) on one axis and performance level (e.g., remember, apply and create) on the other. This matrix scheme is more expressive and intuitive than hierarchical representations such as the one proposed by Gagné. However, epistemology not only involves distinguishing knowledge types, it also includes the quality and role knowledge types play in problem-solving situations. In this sense, Kyllonen and Shute (1988) proposed a multidimensional model which is more complex and which distinguishes knowledge types by means of a hierarchy based on cognitive complexity. This model organizes the knowledge types in relation to the level of the learner's autonomy and the processing speed needed to perform the task. More recently and in a similar manner to Kyllonen and Shute, De Jong and Ferguson-Hessler (1996) stated that, although absolute classification is important, a more pragmatic typology of knowledge should take into account the context in which the knowledge is used. They proposed a "knowledge-in-use" perspective which leads to an epistemological analysis of knowledge that renders ontological types and quality of knowledge.

Although the epistemological perspective of domain knowledge is of some significance in the field of education (as shown in the preceding paragraphs), it rarely surfaces in relation to ITSs and, as a result, has not received any special attention. This may be attributed to the fact that most ITSs have focused on procedural domains having limited scope, which target only the concepts and, more importantly, the tasks that such domains require. However, in the area of learning sciences, there has been a growing interest in the role of epistemology in teaching and learning. Studies suggest that the specific beliefs that teachers have about the nature of knowledge and learning influence their decisions regarding curriculum, pedagogy, and assessment (Schraw and Olafson 2008; Peters and Gray 2006). Moreover, as indicated above, the epistemological perspective of domain knowledge has clearly been a part of the established methodology in the field of knowledge-based systems for several years. Indeed, knowledge-based system methodology generally includes a step involving an epistemological analysis which focuses on the conceptual features of the system's two main components (knowledge about the domain and knowledge about the inference procedures needed to solve a problem) (Ramoni et al. 1992). The epistemological analysis provides a framework containing knowledge-structuring primitives which represent types of concepts, types of knowledge sources, types of structural relationships, such as inheritance, and types of problem-solving strategies (Brachman 1979; Hickman et al. 1989). The expert module of ITSs is similar to a knowledge-based system; accordingly, it should apply the same methodological principle in regard to knowledge representation. Since our goal is not to suggest new practices in the field of ITSs, we will end our discussion of epistemological considerations here. In the next section, we will present methods that allow for the computational implementation of knowledge representation and reasoning.

2.3 Computational Perspective

Epistemological analysis may help in specifying the constraints that are derived from the conceptual structure of the domain knowledge, the patterns of inference, and the tasks to be executed. However, in order to realize a computational model in a domain, certain methods and formalisms must be adopted. In this section, we describe some relevant approaches for the implementation of the expert module in ITSs.

2.3.1 A Historical View of Domain Modules in ITSs

From a historical point of view, three types of models have been identified in connection with ITSs: the black box models, the glass box models and the cognitive models.

2.3.1.1 The Black Box Models

A black box model is a completely inexplicit representation providing only the final results (Nwana 1990). It fundamentally describes problem states differently than those described by the student. A classic example of a black box model system is SOPHIE I (Brown and Burton 1974), a tutoring system for electronic troubleshooting that uses its expert system to evaluate the measurements made by students when troubleshooting a circuit. The expert system comprises a simulated troubleshooting model based on sets of equations. The tutor makes decisions by solving these equations. It can recommend optimal actions for each problem-solving context, but it is up to the student to construct a description of the problem-solving context and his/her rationale for the appropriate action.

2.3.1.2 The Glass Box Models

A glass box model is an intermediate model that reasons in terms of the same domain constructs as the human expert. However, the model reasons with a different control structure than the human expert. In this model, each reasoning step can be inspected. A classic example of a glass box model system is GUIDON (Clancey 1982), a tutoring system for medical diagnosis. This system was built around MYCIN, an expert system for the treatment of bacterial infections. MYCIN consists of several hundred "if-then" rules that probabilistically relate disease states to diagnoses. These rules reference the same symptoms and states that doctors employ in reasoning, but with a radically different control structure, i.e., an exhaustive backward search. During learning sessions, GUIDON compares the student's questions to those which MYCIN would have asked and critiques him/her on this basis.

2.3.1.3 The Cognitive Models

A cognitive model seeks to match representation formalisms and inference mechanisms with human cognition. One of the very important early findings in intelligent tutoring research was the importance of the cognitive fidelity of the domain knowledge module. Cognitive approaches aim to develop a cognitive model of the domain knowledge that mimics the way knowledge is represented in the human mind in order to make ITSs respond to problem-solving situations as the student would (Corbett et al. 1997). This approach, in contrast to the other approaches, has as objective to support cognitively plausible reasoning. In brief, it aims to apply the same style of representation to encode knowledge as that used by the learner.. A positive approach consists of building a system which uses a cognitive architecture such as Adaptive Control of Thought (ACT-R) (Anderson 1996). ACT-R is a theory for simulating and understanding human cognition. ACT-R architecture allows a system to capture in great detail the way humans perceive, think about, and act on the world. Several ITSs have been built using ACT-R (or ACT*, in its early version) production rules, including Algebra Tutor (Singley et al. 1989), Geometry Tutor (Koedinger and Anderson 1990) and LISP Tutor (Corbett and Anderson 1992). It is also generally accepted that tutors representing procedural domain knowledge based on a cognitive analysis of human skill acquisition are also cognitively oriented (Beck et al. 1996). Sherlock (Lesgold et al. 1992) is a good example of such a tutoring system.

2.3.1.4 The Need for Representation Languages

Whether the selected model is a black box model, a glass box model, or a cognitive model, careful consideration must be given to the means (languages) to be used in representing and using knowledge. These means are numerous and the representation thereby obtained may be symbolic (formal or informal), connectionist, or hybrid. As a result, it is not easy for system developers to choose a language. Their selection should take four important points into consideration: the expressivity of the language, the inference capacity of the language, the cognitive plausibility of the language (in terms of language representation, as well as reasoning) and the pedagogical orientation of the language (i.e., the manner in which the specific learning context is considered).

The expressivity of an encoding language is a measure of the range of constructs that can be used to formally, flexibly, explicitly and accurately describe the components of the domain. However, there is a compromise that has to be made between expressivity (what you can say) and complexity (whether the language is computable in real time).

Inference capacity is generally rooted in a formal semantic system and based on an inference procedure. The semantics of a language refers to the fact that the message unambiguously means what it is supposed to mean. For example, consider the language construct "A subconcept of B": Does this mean that all instances of A are also instances of B, or parts of B, or special kinds of B? Clearly defined and well-understood semantics are essential for sound inference procedures.

Whatever language is chosen, the expert module must guarantee mutual understanding between the actions of the system and those of the learner. In fact, if the student does not understand the system's instruction or if the system cannot interpret the student's behaviour in terms of its own view of the knowledge, tutoring can be compromised (Wenger 1987).

In the following sections, we present a few examples of such languages.

2.3.2 General-Purpose Languages

In artificial intelligence, various languages and notations have been proposed for representing knowledge. These are typically based on logic and mathematics and have easily parsed grammars and well-defined semantics to ease machine processing. A logic generally consists of a syntax, a semantic system and a proof theory. The syntax defines a formal language for the logic, the semantic system specifies the meanings of properly formed expressions, and the proof theory provides a purely formal specification of the notion of correct inferences. In the following sections, we present a brief description of some common general-purpose knowledge representation languages. Further details can be found in classic AI books, such as those by Luger (2009) and Russell & Norvig (2009). Readers can also find in these books other non-classic formalisms, such as fuzzy logic, probabilistic approaches, and connectionist approaches. In general, these languages have no pedagogical emphasis, i.e., they do not make any assumptions regarding the pedagogical application of the knowledge that they represent.

2.3.2.1 Production Rules

Production rules is one of the most popular and widely used knowledge representation language. Early expert systems used production rules as their main knowledge representation language. One such example is MYCIN.

A production rule system consists of three components: a working memory, a rule base and an interpreter. The working memory contains the information that the system has acquired concerning the problem. The rule base contains information that applies to all the problems that the system may be asked to solve. The interpreter solves the control problem, i.e., it determines which rule to execute on each selection-execute cycle.

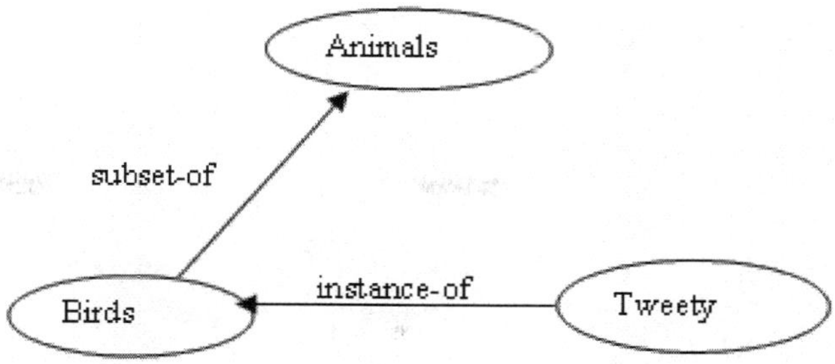

Fig. 2.1 A Simple Example of a Semantic Network

As a knowledge representation language, production rules have many advantages, including their natural expression, modularity, restricted syntax and sound logic basis for making inferences equivalent to First-order Logic (FOL). Production rules are cognitively plausible in terms of representation structures as well as reasoning mechanisms (Anderson 1982). Most cognitive tutors (see Chapter 3) encode the actions of the expert problem-solver as production rules and attempt to determine which rules the student is having difficulty applying.

2.3.2.2 Semantic Networks

An important feature of human memory is the high number of connections or associations between the pieces of information it contains. Semantic networks are one of the knowledge representation languages based on this capacity.

The basic idea of a semantic network representation is very simple: There are two types of primitives: nodes and links or arcs. Nodes, on the one hand, correspond to objects or classes of objects in the world. Links or arcs, on the other hand, are unidirectional connections between nodes that correspond to relationships between these objects. Figure 2.1 shows an example of a semantic net. The basic inference mechanism consists in following inheritant and instance links. To determine whether an object, represented by node A, is a member of a set represented by node B, every link extending upwards from A (is-a and instance link) must be traced to see whether it intersects node B. In order to determine the value of certain properties of an object represented by node A, every link extending upwards from A must be followed (as above) until it intersects a node possessing this property (function link). Many other inferences procedure was proposed including path-based and node-based inferences proposed by Shapiro (1978).

Semantic networks, which correspond to human memory, are cognitively plausible at the structural (representation) level, but not for their reasoning mechanism. In 1969, Collins & Quillian conducted a series of studies to test the psychological plausibility of semantic networks as models for both the organization of memory and human inferencing.

The domain knowledge of SCHOLAR (Carbonell 1970) is that of South America geography. This domain knowledge model is represented using a semantic network whose nodes instantiate geographical objects and concepts. Statements such as “Tell me more about Brazil” simply invoke a retrieval of facts stored in the semantic network. However, the strength of this representation schema lies in its ability to answer questions for which answers are not stored. For example, it is not necessary to store in the semantic network that “Lima is in South America,” provided that the program which interprets the network can make the relevant inference. In other words, the program must know about the attributes concerned, “location” and “capital,” and, in particular, that if X is the capital of Y and Y is located in Z, then X is in Z: This is the rule of inference.

The semantic network form of knowledge representation is especially suitable for describing the taxonomic structure of categories for domain objects and for expressing general statements about the domain of interest. Inheritance and other relationships between such categories can be represented in and derived from subsumptive hierarchies. In contrast, semantic networks are not ideal for representing concrete individuals or data values, such as numbers or strings. Another major problem with semantic networks is the lack of clear semantics for the various network representations (despite the word “semantic”) (Sharples et al. 1989). For example, Figure 2.1 can be interpreted as a representation of a specific bird named Tweety, or it can be interpreted as a representation of some relationship between Tweety, birds and animals.

Many variants of semantic networks have been used in ITS applications, including concept/topic maps (Albert and Steiner 2005; Murray 1998; Garshol 2004; Kumar 2006) and conceptual graphs (Sowa 1984). The latter is presented in the next paragraph.

2.3.2.3 Conceptual Graphs

This formalism (Sowa 1984) is based on semantic networks but is directly linked to the language of first-order predicate logic from which it takes its semantics.

A simple conceptual graph is a bipartite (not necessarily connected) graph composed of concept nodes that represent entities, attributes, states or events, and relation nodes that describe the relationships between these concepts. Each concept node *c* of a graph *G* is labelled by a pair *(type(c), ref(c)),* in which *ref(c)* is either the generic marker * corresponding to the existential quantification or an individual marker corresponding to an identifier. *M* is the set of all individual markers. Each relation node *r* of a graph *G* is labelled by a relation type, *type(r),* and associated with a signature identifying the constraints on the types of concepts that may be linked to its arcs in a graph. For example, the conceptual graph given in Figure 2.2 represents the information: “the experiment *E1* carries out an interaction *I1* between *Nisin* and *Listeria Scott A* in *skim milk* and the result is *reduction.*”

Concept types (respectively relation types of the same arity) create a set T_C (resp. T_R) partially ordered by a generalization/specialization relationship $\leq_c$ (resp. $\leq_r$). (T_C, T_R, M) is a lattice defining the *support* upon which conceptual graphs are constructed. A support thus represents the ontological knowledge that provides the ground vocabulary on which the knowledge base is built.

The semantics of conceptual graphs rely on the translation of both conceptual graphs and their support into first-order logic formulas. For instance, the "kind-of" relationships between types in the support are translated into logical implications.

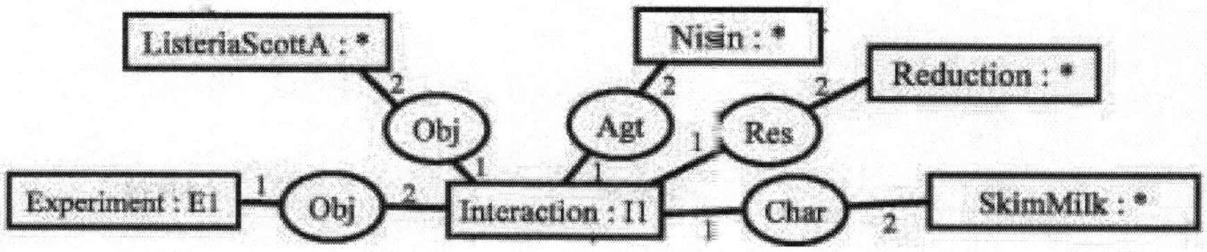

Fig. 2.2 A Sample Conceptual Graph (Adapted from Dibie-Barthélemya, Haemmerléa and Salvatc 2006).

In addition, each conceptual graph has a logical interpretation which is a first-order logic formula, in which each generic marker is associated with a distinct variable, each individual marker with a constant, each concept type with a unary predicate applied to its marker, and each relation type with a predicate applied to the markers of the concept vertices it links. The formula associated with the conceptual graph is then the existential closure of the conjunction of all atoms. For instance, the logical interpretation of the conceptual graph represented in Figure 2.2 is the following:

$$\exists x \exists y \exists z \exists t\ (\text{ListeriaScottA}(x) \wedge \text{Nisin}(y) \wedge \textit{Experiment}(E1) \wedge \text{Interaction}(I1)$$
$$\wedge\ \text{Reduction}(z) \wedge \text{SkimMilk}(t) \wedge \text{Obj}(I1,x) \wedge \text{Agt}(I1,y) \wedge \text{Obj}(E1,I1)$$
$$\wedge\ \text{Res}(I1,z) \wedge \text{Char}(I1,t)\ ).$$

The main inference procedure relies on a subsumptive relationship between two conceptual graphs. This relationship is equivalent to the logical implication between the two logical formulas corresponding to these graphs. In addition, a key operation known as projection makes it possible to compute subsumptive relationships between graphs. Reasoning with conceptual graphs is based on the projection which is complete with respect to logical deduction. However, finding a projection between two graphs is an NP-complete problem.

Many extensions of conceptual graphs have been proposed for the purpose of extending conceptual graphs expressivity with sound semantics. Some of these extensions have focused on concept descriptions, a process equivalent to those of conceptual graphs and description logics (Coupey and Faron 1998; Delteil and Faron 2002). For instance, Delteil & Faron (2002) proposed a graph-based concept description language called GDL, which involves a thorough decision-making procedure. It can be used to represent concept descriptions with complex graph patterns as well as negation and disjunction, which leads to an expressive language. GDL combines features of both conceptual graphs and description logics (see subsequent sections).

Conceptual graphs are used in some ITSs. For instance, they were used in STyLE-OLM and in HYLITE+ (Bontcheva and Dimitrova 2004). STyLE-OLM is an interactive learner modeling system that extracts extended models of the learner's cognition. HYLITE+ is a natural language generation system that creates adaptive Web pages based on a learner model. CBITS, a case-based ITS, is another example of a tutoring system that uses conceptual graphs to represent cases (Fung and Kemp 1999).

2.3.2.4 Frame-Based

Frame-based systems (Minsky 1975) are based on the notion of frames or classes which represent collections of instances (the concepts underlying ontology). Each frame has an associated collection of slots or attributes which can be filled by values or other frames. In particular, frames can have a "kind-of" slot which allows for the assertion of a frame taxonomy. This hierarchy may then be used for the inheritance of slots, thereby allowing for a sparse representation. In addition to frames representing concepts, a frame-based representation may also contain instance frames, which represent particular instances.

An example of the frame-based model is Open Knowledge Base Connectivity (OKBC), which defines an API for accessing knowledge representation systems. OKBC also defines most of the concepts found in frame-based systems, object databases and relational databases. The OKBC API (Chaudhri et al. 1998) is defined in a language-independent manner, and implementations exist for Common Lisp, Java and C.

Frames generally provide quite a rich set of language constructs but impose very restrictive constraints on how they can be combined or used to define a class. Moreover, they only support the definition of primitive concepts, and taxonomy must be hand-crafted.

COCA (Major and Reichgelt 1992) is a shell for building ITSs in which domain knowledge is represented using the frame approach with a number of user-defined attributes and attribute values. Attribute values may be primitive data types (e.g., text strings), procedures to be run, or pointers to other frames.

2.3.2.5 Ontology and Description Logics

Ontology is a formal specification of a domain and includes a definition of concepts and the relationships among them. An ontological knowledge base is composed of two parts: a terminology box (TBox), which contains terminology axioms (concepts and role descriptions), and an assertions box (ABox), containing individuals (concept and role instances). The Semantic Web community has developed a formal language for ontology implementation called Web Ontology Language (OWL).

Frames and ontology semantics are suitable for building on Description Logics (DLs). In fact, DLs (Baader et al. 2007) may be seen as a logical reformulation of frames and ontologies and may provide them with a rigorous and strong basis for reasoning. Indeed, DLs provide representation and reasoning languages with precise semantics. They also limit language expressiveness so that they can guarantee tractable inference. For instance, W3C's OWL-DL is based on *SHOIN* expressive description logic which is known to be decidable (Horrocks and Sattler 2007). A major characteristic of a DL is that concepts are defined in terms of descriptions using other roles and concepts. In this way, the model is built up from small pieces in a descriptive way rather than through the assertion of hierarchies. The DL supplies a number of reasoning services. Main inference (reasoning) tasks include subsumption, classification and satisfiability. Subsumption aims to verify whether a concept is a subset of another by comparing their definitions. Classification verifies whether an instance belongs to a concept. Satisfiability involves the consistency of a concept definition. Satisfiability is accomplished by verifying whether the membership criteria are logically satisfiable. It is the most important reasoning function as subsumption may be redefined as a satisfiability problem. The well-known tableau algorithm (Baader and Sattler 2001) is the decision procedure that implements satisfiability. These reasoning services may subsequently be made available to applications that make use of the knowledge represented in the ontology or in the frames.

There may be some drawbacks with more expressive DLs, which make them difficult to use in real world applications. These include the intractability of their satisfiability or subsumption algorithms, and the increase in the computational complexity of reasoning. However, some research results show that efficient and practical implementations of relatively expressive languages are feasible despite their theoretical complexity (Horrocks 1998; Horrocks and Sattler 2007). As DLs have clear semantics, it is possible to use all of the knowledge encapsulated in the ontology to determine whether the ontology is consistent and complete.

Ontology language is purely declarative and not overly expressive for the description of procedural knowledge. However, knowledge generally goes beyond the description of what exists in the world; it also links goals to actions (Newell 1982). In that sense, knowledge has a strongly procedural aspect. This is also true in the context of ITSs, in which learning of procedural knowledge is the main focus. Accordingly, to build an ITS that teaches a procedural task, one must not only specify the experts' conceptualizations of the domain, but also clarify how problem-solving will ideally occur. In fact, to be able to follow the learner's reasoning and to provide relevant suggestions and explanations, such ITSs must have knowledge of the task that is both robust and explicable.

There is no doubt that ontology is a good tool for representing propositional knowledge and providing shared domain descriptions for various purposes. However, it is not enough. If we wish to include problem-solving tasks for which we have enough knowledge to predict that certain solution strategies will be particularly appropriate, then we need to make the strategies explicit. We need to represent problem-solving methods that could form the basis for the procedural components of the domain knowledge base, thereby making the system more understandable and traceable (Brewster and O'Hara 2007). Task ontology may help achieve this. However, the notion of task ontology should be applied prudently. In fact, ontology should only be used declaratively to specify conceptual building blocks that provide the foundation on which the knowledge base is built. In this sense, the role of ontology should be strictly limited to specifying hidden conceptualization and not be used as a tool for the in-depth definition of procedural knowledge and rules. Such a definition should be built on top of the domain ontology. As a result, task ontology aims to provide relevant terminology that is both necessary and sufficient for building problem-solving models in a given domain (Mizoguchi et al. 1995). The problem-solving model (e.g., task decomposition in terms of goals, sub-tasks and control-flow) needs this task ontology to relate the problem-solving steps to the relevant knowledge in the domain ontology. This idea is consistent with the current practice in semantic web, in which there is a clear separation between the ontology layer and the rule layer. Semantic Web Rule Language (SWRL) is a standard language that extends OWL to include Horn-like rules, making it possible to define procedural knowledge linked to the domain ontology as shown in Chi (2010).

In conclusion, there is no standard means of correctly integrating procedural knowledge and linking it to the declarative domain knowledge. What is important is the fact that these two levels of knowledge should be "loosely coupled" to make computation easier (as in ACT-R).

2.3.3 Pedagogically-Oriented Languages

Several other languages with a significant pedagogical emphasis may be used to represent domain knowledge in a less formal manner. These languages are generally associated with a precise characterization of the knowledge as well as the pedagogical functions that enable learning. We present two examples below.

MOT (Paquette 2008) is one example of such a language. It provides the user with four basic types of knowledge units in graphical form: facts, concepts, procedures and principles (see Figure 2.3).

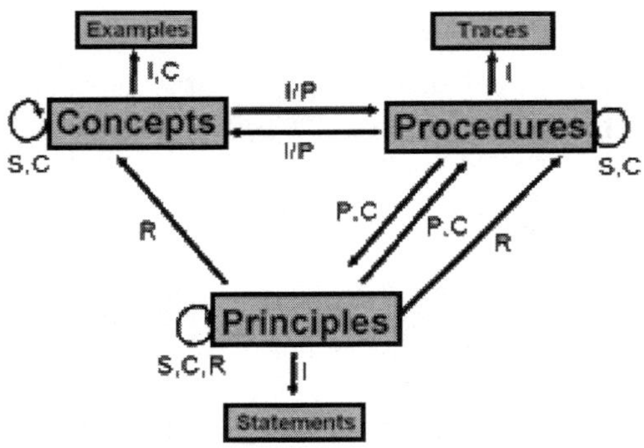

Fig. 2.3 Types of Knowledge in MOT

Knowledge units are implemented as schemas and are connected using six types of links constrained by specific grammar rules. Two examples of these rules are: 1) All abstract knowledge entities (concepts, procedures and principles) may be related by an instantiation link to a set of facts representing individuals called respectively examples, traces and statements. 2) All abstract knowledge entities (concepts, procedures and principles) may be specialized or generalized to other abstract knowledge by means of specialization links. Knowledge units connected by means of links form the domain knowledge model which can be exported into OWL to "become" a formal ontology model. Domain knowledge produced with MOT was used mainly as a semantic basis for learning object semantic referencing.

CREAM language (Nkambou et al. 2001) is another example of a pedagogically- oriented language. It allows for the creation and organization of curriculum elements in three interconnected perspectives: domain, pedagogy and didactic. The domain perspective is represented by a graphical structure of semantic knowledge based on Gagne's taxonomy (1985) and connected by means of semantic links. The pedagogical perspective organizes learning objectives on the basis of Bloom's taxonomy and pre-requisite links. The didactic perspective defines the model of learning resources that supports learning activities. These three perspectives are connected in a complex bi-partite graph that serves as the basis for making tutoring decisions related to content planning and learning resource recommendations. Figure 2.4 shows the basic components in CREAM.

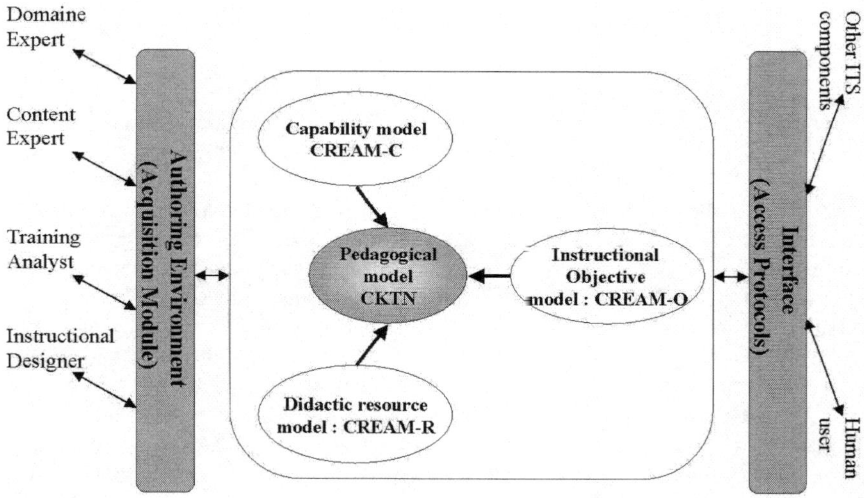

Fig. 2.4 Basic Components in CREAM

The limits of these pedagogically-oriented languages can be attributed to their reduced expressivity as well as their strong pedagogical orientation. The meta-model of the epistemological level is determined in advance (for MOT, see Figure 2.3) with no possibility of modification or extension via other types of objects; conversely, general languages (Conceptual Graphs and ontology) do provide the possibility of determining the epistemological level, e.g., concept and relation types. Moreover, pedagogically-oriented languages that are not based on a precise semantic system are relatively informal in nature. In the case of MOT, even though exporting to an ontological language is possible, it is clear that its limited expressivity is insufficient to describe complex problems (such as defining concepts based on primitive concepts). In addition, the use of unrestrained graphic language does not ensure that the resulting axioms are logically sound and semantically valid. Accordingly, the inference procedure cannot make any guarantees (neither completeness nor completion within a reasonable time). Note that for the two languages cited as examples, cognitive plausibility is not an issue and has no bearing.

For these reasons, it is difficult to see how such languages could be adequate for use in an open ITS. Their relatively meagre semantics do not give rise to any logical foundation on the basis of which inferences may be made regarding the constructed models. However, they may prove to be very effective in certain very specific contexts. For instance, in the case of MOT, the resulting structure serves as a valuable "semantic" base which is available to developers for the purpose of learning design or learning object semantic annotation. In addition, a knowledge model like CREAM can support effective content planning in order to meet specific training needs (Nkambou et al. 1998).

2.4 Conclusion

Although numerous solutions exist to represent domain knowledge and the reasoning mechanisms that operate them, there is no consensus on what can be considered an ideal approach. Even if such a consensus may be difficult to achieve, current approaches need to be analyzed using well-defined criteria. Such analyses could advance ITS developers' knowledge and make their choices easier. In this introductory chapter, we first pointed out the importance of determining the nature and value of domain knowledge and subsequently described some existing solutions for representing the expert module of an ITS in light of four criteria: expressivity, inferential power, cognitive plausibility and pedagogical considerations. We also stated that the AIED community should determine relevant criteria to be considered in selecting a representation language for a given tutoring context.

Furthermore, and in this era of Semantic Web and Web 2.0, ontology can easily be defended as the formalism of choice for several reasons: 1) it offers several levels of expressiveness that can be adapted to needs and inferential constraints; 2) it allows for the integration of multiple views and enables interoperability; and 3) it is built on well-defined semantics resulting from description logic, which provides for sound reasoning mechanisms. However, ontology remains essentially a declarative approach used in the creation of the domain semantic memory. For a procedural learning domain, such semantic memory has to be supplemented by operational and procedural knowledge that refers to its elements. Therefore, another language is needed to build this procedural level on top of the ontology level.

The next three chapters in this part explain specific techniques for representing the expert module. The following two chapters describe two approaches in the field of AIED that are gaining in popularity. First, Chapter 3 presents the "cognitive-tutor" approach which is based on the need for balance between the domain structure and a theory of cognition (in this case, ACT-R). Second, Chapter 4 describes the CBM approach which is based on the need to focus not on an explicit representation of all elements of a domain, but on a definition of a set of principles as constraints that govern the field. One can question the validity of these two approaches in ill-defined domains. Chapter 5 discusses this case and presents appropriate solutions. Finally, given the importance of the ontology-based approach,

Chapter 6 describes how ontology can serve as domain knowledge. Chapter 6 also focuses on how such domain ontology can be automatically learned (extracted) from textual learning resources with a very limited solicitation of domain experts.

The current chapter does not address the problem of the acquisition of domain knowledge. Indeed, regardless of the nature of knowledge, the approach or the representation language, the knowledge that the system uses must be acquired. Such knowledge is usually held by human experts. It can also be found in knowledge sources such as documents. Knowledge acquisition is a major concern in the field of AI and several solutions have been proposed, including automatic extraction approaches using machine-learning or data-mining techniques. The other chapters in this part describe how this problem is addressed in the context of ITSs. Specific tools are generally provided for facilitating the domain knowledge acquisition process. For example, CTAT (cf. Chapter 3) provides tools to facilitate the creation of domain knowledge in the form of production rules. ASPIRE (cf. Chapter 4) provides tools that simplify the implementation of constraints in a given domain. Moreover, automatic learning tools are sometimes used for the acquisition of domain ontology (cf. Chapter 6) or procedural knowledge (cf. Chapter 5). Part 4 of this book describes other examples of authoring tools to facilitate the development of ITSs and their expert modules.

References

Albert, D., Steiner, C.M.: Representing domain knowledge by concept caps: How to validate them. In: Okamoto, T., Albert, D., Honda, T., Hesse, F.W. (eds.) The 2nd Joint Workshop of Cognition and Learning through Media-Communication for Advanced e-Learning, Tokyo (2005)

Anderson, J.R.: Acquisition of cognitive skill. Psychological Review 98(4), 369–406 (1982)

Anderson, J.R.: The architecture of cognition. Lawrence Erlbaum Associates Inc., NJ (1996)

Baader, F., Sattler, U.: An overview of Tableau algorithms for description logics. Studia Logica: An International Journal for Symbolic Logic 69(1), 5–40 (2001)

Baader, F., Horrocks, I., Sattler, U.: Description Logics. In: Harmelen, F., Lifschitz, V., Porter, B. (eds.) Handbook of Knowledge Representation. Elsevier, Amsterdam (2007)

Beck, J., Mia, S., Haugsjaa, E.: Applications of AI in education. Crossroads, Special issue on artificial intelligence 3(1), 11–15 (1996)

Bloom, B.S.: Taxonomy of Educational Objectives: the Classification of Educational Goals. McKay D, New York (1975)

Bontcheva, K., Dimitrova, V.: Examining the use of conceptual graphs in adaptive web-based systems that aid terminology learning. International Journal on Artificial Intelligence Tools 13(2), 299–331 (2004)

Brachman, R.: On the Epistemological Status of Semantic Networks. In: Findler, N.V. (ed.) Associative Networks: Representation and Use of Knowledge by Computers. Academic Press, New York (1979)

Brachman, R.J., Levesque, H.J.: Knowledge Representation and Reasoning. Morgan Kaufmann, San Francisco (2004) ISBN 978-1-55860-932-7

Brewster, C., O'Hara, K.: Knowledge representation with ontologies: Present challenges - Future possibilities. International Journal of Man-Machine Studies 65(7), 563–568 (2007)

Brown, J.S., Burton, R.R.: SOPHIE: a pragmatic use of artificial intelligence in CAI. In: Proceedings of the 1974 annual ACM conference, vol. 2, pp. 571–579 (1974)

Carbonell, J.: AI in CAI: An artificial intelligence approach to computer- assisted instruction. IEEE Transactions on Man-Machine Systems 11, 190–202 (1970)

Chaudhri, V.K., Farquhar, A., Fikes, R., Karp, P.D., Rice, J.P.: OKBC: A Programmatic Foundation for Knowledge Base Interoperability. In: Proceedings of AAAI, pp. 600–607. AAAI Press, Menlo Park (1998)

Chi, Y.-L.: Rule-based ontological knowledge base for monitoring partners across supply networks. Expert Systems with Applications 37, 1400–1407 (2010)

Clancey, B.: Acquiring, representing and evaluating a competence model of diagnostioc strategy. Technical Report: CS-TR-85-1067, Stanford University (1985)

Clancey, W.J.: Overview of GUIDON. Journal of Computer-Based Instruction 10(1-2), 8–15 (1982)

Collins, A.M., Quillian, R.: Retrieval time from semantic memory. Journal of Verbal Learning and Verbal Behavior 8, 240–247 (1969)

Corbett, A.T., Anderson, J.R.: The LISP intelligent tutoring system: Research in skill acquisition. In: Larkin, J., Chabay, R., Scheftic, C. (eds.) Computer Assisted Instruc-tion and Intelligent Tutoring Systems: Establishing Communication and Collaboration. Erlbaum, Hillsdale (1992)

Corbett, A.T., Koedinger, K.R., Anderson, J.R.: Inteligent Tutoring Systems. In: Helander, T.K., Landauer, P. (eds.) Handbook of Human-Computer Interaction. Elsevier Science, Amsterdam (1997)

Coupey, P., Faron, C.: Towards Correspondences Between Conceptual Graphs and Description Logics. In: Mugnier, M.-L., Chein, M. (eds.) ICCS 1998. LNCS (LNAI), vol. 1453, p. 165. Springer, Heidelberg (1998)

De Jong, T., Ferguson-Hessler, M.G.M.: Types and qualities of knowledge. Educational Psychologist 31(2), 105–113 (1996)

Delteil, A., Faron, C.: A Graph-Based Knowledge Representation Language for Concept Description. In: ECAI 2002, pp. 297–301 (2002)

Fung, P.-W., Kemp, R.H.: On developing case-based tutorial systems with conceptual graphs. In: Tepfenhart, W.M. (ed.) ICCS 1999. LNCS, vol. 1640, pp. 214–229. Springer, Heidelberg (1999)

Gagné, R.M.: The Conditions of Learning, 4th edn. Holt Rinehart and Winston, New York (1985)

Garshol, L.M.: Metadata? Thesauri? Taxonomies? Topic Maps! Making Sense of it all. Journal of Information Science 30(4), 378–391 (2004)

Hickman, F.R., Killin, J.L., Land, L., Mulhall, T., Porter, D., Taylor, R.: Analysis for Knowledgebased Systems. In: A Practical Guide to the KADS Methodology. Ellis Horwood, New York (1989)

Horrocks, I.: Using an Expressive Description Logic: FaCT or Fiction? In: Proc. of the 6th Int. Conf. on Principles of Knowledge Representation and Reasoning (KR 1998), pp. 636–647 (1998)

Horrocks, I., Sattler, U.: A Tableau Decision Procedure for SHOIQ. J. of Automated Reasoning 39(3), 249–276 (2007)

Koedinger, K.R., Anderson, J.R.: Abstract planning and perceptual chunks: Elements of expertise in geometry. Cognitive Science 14, 511–550 (1990)

Kumar, A.: Using Enhanced Concept Map for Student Modeling in a Model-Based Programming Tutor. In: Proc. of 19th International FLAIRS Conference on Artificial Intelligence, Melbourne Beach, FL (2006)

Kyllonen, P.C., Shute, V.J.: A Taxonomy of Learning Skills. Brooks Air Force Base, TX, AFHRL Report No. TP-87-39 (1988)

Lesgold, A., Lajoie, S., Bunzo, M., Eggan, G.: SHERLOCK: A coached practice environment for an electronics troubleshooting job. In: Larkin, J., Chabay, R. (eds.) Computer assisted instruction and intelligent tutoring systems: Shared issues and complementary approaches, pp. 201–238. Lawrence Erlbaum Associates, Hillsdale (1992)

Luger, G.F.: Artificial Intelligence: Structures and Strategies for Complex Problem Solving, 6th edn. Addison-Wesley Pearson Education, Boston (2009)

Major, N., Reichgelt, H.: COCA: A shell for intelligent tu-toring systems. In: Frasson, C., McCalla, G.I., Gauthier, G. (eds.) ITS 1992. LNCS, vol. 608, pp. 523–530. Springer, Heidelberg (1992)

Merrill, D., Twitchell, D.: Instructional Design theory. Educational Technology Publication Inc., Englewood Cliff (1994)

Merrill, M.D.: An Introduction to Instructional Transaction Theory. Educational Technology 31, 45–53 (1991)

Minsky, M.: A Framework for Representing Knowledge. In: Winston, P.H. (ed.) The Psychology of Computer Vision. McGraw-Hill, New York (1975)

Mizoguchi, R., Vanwelkenhuysen, J., Ikeda, M.: Task Ontology for Reuse of Problem Solving Knowledge. In: Mars, N.J.I. (ed.) Towards Very Large Knowledge Bases. IOS Press, Amsterdam (1995)

Murray, T.: Authoring Knowledge Based Tutors: Tools for Content, Instructional Strategy, Student Model, and Interface Design. J. of the Learning Sciences 7(1), 5–64 (1998)

Newell, A.: The Knowledge Level. Artificial Intelligence 18(1), 87–127 (1982)

Nkambou, R., Frasson, C., Gauthier, G.: A new approach to ITS-curriculum and course authoring: the authoring environment. Computers and Education 31(2), 105–130 (1998)

Nkambou, R., Frasson, C., Gauthier, G., Rouane, K.: An Authoring Model and Tools for Knowledge Engineering in Intelligent Tutoring Systems. Journal of Interactive Learning Research 12(4), 323–357 (2001)

Nwana, H.: Intelligent Tutoring Systems: an overview. Artificial Intelligence Review 4, 251–277 (1990)

Paquette, G.: Graphical Ontology Modeling Language for Learning Environments. Technology, Instruction., Cognition and Learning 5, 133–168 (2008)

Peters, J.M., Gray, A.: Epistemological perspectives in research on teaching and learning science. Educational Research 5(1), 1–27 (2006)

Piaget, J.: The Principles of Genetic Epistemology. Routledge, London (1997)

Ramoni, M.F., Stefanelli, M., Magnani, L., Barosi, G : An epistemological framework for medical knowledge-based systems. IEEE Trans. Syst. Man Cybern. 22, 1–14 (1992)

Russell, S.J., Norvig, P.: Artificial Intelligence: A Modern Approach, 3rd edn. Prentice Hall, Upper Saddle River (2009)

Schraw, G.J., Olafson, L.: Assessing Tecahers' Epistemological and Ontological World Views. In: Khine, M.S. (ed.) Knowing, Knowledge and Beliefs, pp. 24–44. Springer, Heidelberg (2008)

Shapiro, S.C.: Path-based and node-based inference in semantic networks. In: Proceedings of the 1978 workshop on Theoretical issues in natural language processing. ACM-SIGART, pp. 219–225 (1978)

Sharples, M., Hogg, D., Hutchison, C., Torrance, S., Young, D.: Computers and Thought: A practical Introduction to Artificial Intelligence. MIT Press, Cambridge (1989)

Singley, M.K., Anderson, J.R., Gevins, J.S., Hoffman, D.: The algebra word problem tutor. Artificial Intelligence and Education, 267–275 (1989)

Sowa, J.F.: Conceptual Structures: Information Processing in Mind and Machine. Addison-Wesley, Reading (1984)

Sowa, J.F.: Knowledge Representation: Logical, Philosophical, and Computational Foundations. Brooks Cole Publishing Co., Pacific Grove (2000)

Tennyson, R., Rasch, M.: Linking cognitive learning theory to instructional prescriptions. Instructional Science 17, 369–385 (1988)

Wenger, E.: Artificial Intelligence and Tutoring Systems: Computational and Cognitive Approaches to the Communication of Knowledge. Morgan Kaufmann Publishers, Inc., Los Altos (1987)

Woolf, B.: Building Intelligent Interactive Tutors: Student-centered strategies for revolutionizing e-learning. Morgan Kaufmann, San Francisco (2008)

Winograd, T., Fernando, F.: Understanding Computers and Cognition: A New Foundation for Design. Ablex, Norwood (1986)

Chapter 3
Rule-Based Cognitive Modeling for Intelligent Tutoring Systems

Vincent Aleven

Human-Computer Interaction Institute, Carnegie Mellon University, Pittsburgh, PA, USA
aleven@cs.cmu.edu

Abstract. Rule-based cognitive models serve many roles in intelligent tutoring systems (ITS) development. They help understand student thinking and problem solving, help guide many aspects of the design of a tutor, and can function as the "smarts" of a system. Cognitive Tutors using rule-based cognitive models have been proven to be successful in improving student learning in a range of learning domain. The chapter focuses on key practical aspects of model development for this type of tutors and describes two models in significant detail. First, a simple rule-based model built for fraction addition, created with the Cognitive Tutor Authoring Tools, illustrates the importance of a model's *flexibility* and its *cognitive fidelity*. It also illustrates the model-tracing algorithm in greater detail than many previous publications. Second, a rule-based model used in the Geometry Cognitive Tutor illustrates how ease of engineering is a second important concern shaping a model used in a large-scale tutor. Although cognitive fidelity and ease of engineering are sometimes at odds, overall the model used in the Geometry Cognitive Tutor meets both concerns to a significant degree. On-going work in educational data mining may lead to novel techniques for improving the cognitive fidelity of models and thereby the effectiveness of tutors.

3.1 Introduction

Cognitive modeling has long been an integral part of ITS development. Cognitive modeling is the activity of producing a detailed and precise description of the knowledge involved in student performance in a given task domain, including strategies, problem-solving principles, and knowledge of how to apply problem-solving principles in the context of specific problems. We do not mean to restrict the term "cognitive model" only to models that are executable on a computer, although executable models are the focus of the current chapter. Rather, any precise and detailed description of human knowledge is a cognitive model.

Cognitive models are useful in many ways in ITS development. They summarize the results of analysis of data on student thinking, which often precedes system design and implementation. A cognitive model can also serve as a detailed specification of the competencies (or skills) targeted by an ITS, and as such, can

R. Nkambou et al. (Eds.): Advances in Intelligent Tutoring Systems, SCI 308, pp. 33–62.
springerlink.com

guide many aspects of the design of the ITS. A deep and detailed understanding of the competencies being targeted in any instructional design effort is likely to lead to better instruction (Clark et al. 2007). Further, when a cognitive model is implemented in a language that is executable on a computer, it can function as the "smarts" of the ITS driving the tutoring.

Two types of cognitive models used frequently in ITS are rule-based models (Anderson et al. 1995; Crowley and Medvedeva 2006; Koedinger and Aleven 2007; VanLehn et al. 2005) and constraint-based models (Mitrovic et al. 2001). Whereas rule-based models capture the knowledge involved in generating solutions step-by-step, constraint-based models express the requirements that all solutions should satisfy. Both types of models have been used in successful real-world ITS. For each type of model, mature and efficient authoring tools exist (Aleven et al. 2006; Mitrovic et al. 2009). The current chapter focuses on the models used in Cognitive Tutors, a widely used type of ITS (Anderson et al. 1995; Koedinger et al. 1997; Koedinger et al. 2006). Tutors of this type use a rule-based model, essentially a simulation of student thinking that solves problems in the same way that students are learning to do. The tutor interprets student performance and tracks student learning in terms of the knowledge components defined in the cognitive model. Cognitive Tutors have been shown in many scientific studies to improve student learning in high-school and middle-school mathematics courses, and at the time of this writing, are being used in the US by about 500,000 students annually.

A key concern when developing cognitive models is the degree to which a model faithfully mimics details of human thinking and problem solving. Cognitive scientists have long used rule-based models as a tool to study human thinking and problem solving (Anderson 1993; Newell and Simon 1972). Their models aim to reproduce human thinking and reasoning in significant detail. Often, they take great care to ensure that their models observe properties and constraints of the human cognitive architecture. Outside of basic cognitive science, accurately modeling details of human cognition and problem solving is important in tutor development. We find it helpful to distinguish two main requirements. First, a model used in a tutor must be *flexible* in the sense that it covers the sometimes wide variety in students' solution paths within the given task domain, as well as the different order of steps within each path. This kind of flexibility ensures that the tutor can follow along with students as they solve problems, regardless of how they go about solving them. Second, it is important that a model partitions the problem-solving knowledge within the given task domain in accordance with psychological reality. We use the term *cognitive fidelity* to denote this kind of correspondence with human cognition. As discussed further below, a model with high cognitive fidelity leads to a tutor that has a more accurate student model and is better able to adapt its instruction to individual students. To achieve flexibility and cognitive fidelity, it helps to perform *cognitive task analysis* as an integral part of model development. This term denotes a broad array of methods and techniques that cognitive scientists use to understand the knowledge, skills, and strategies involved in skilled performance in a given task domain, as well as the preconceptions, prior knowledge, and the sometimes surprising strategies with which novices approach a task (Koedinger and Nathan 2004). Although cognitive task analysis and cognitive modeling tend to be (and should be) closely intertwined in ITS development (Baker et al. 2007), the current chapter focuses on cognitive modeling only.

A third main concern in the development of cognitive models is *ease of engineering*. ITS have long been difficult to build. It has been estimated, based on the experience in real-world projects, that it takes about 200-300 hours of highly-skilled labor to produce one hour of instruction with an ITS (Murray 2003; Woolf and Cunningham 1987). Some approaches to building ITS, such as example-tracing tutors (Aleven et al. 2009) and constraint-based tutors (Mitrovic 2001), improve upon these development times. Rule-based systems, too, have become easier to build due to improved authoring tools (Aleven et al. 2006) and remain a popular option (Corbett et al. 2010; Roll et al. 2010). Nonetheless, building tutors remains a significant undertaking. In creating tutors with rule-based cognitive models, a significant amount of development time is devoted to creating the model itself. It may come as no surprise that ITS developers carefully engineer models so as to reduce development time. Further, being real-world software systems, ITS must heed such software engineering considerations as modularity, ease of maintenance, and scalability. Thus, the models built for real-world ITS reflect engineering concerns, not just flexibility and cognitive fidelity. Sometimes, these aspects can go hand in hand, but at other times, they conflict and must be traded off against each other, especially when creating large-scale systems.

We start with a brief description of Cognitive Tutors and the way they use rule-based models to provide tutoring. We describe two examples of cognitive models used in Cognitive Tutors. The first example describes a model for a relatively simple cognitive skill, fraction addition, and emphasizes flexibility and cognitive fidelity. The second example illustrates how the cognitive model of a real-world tutor, the Geometry Cognitive Tutor, is (arguably) a happy marriage between flexibility, cognitive fidelity, and ease of engineering. Although we have tried to make the chapter self-contained, some knowledge of ITS and some knowledge of production rule systems or rule-based programming languages is helpful. Although many excellent descriptions of model tracing and Cognitive Tutors exist (Anderson 1993; Anderson et al. 1995; Koedinger and Aleven 2007; Koedinger et al. 1997; Koedinger and Corbett 2006; Ritter et al. 2007), the current chapter focuses in greater detail than many previous articles on the requirements and pragmatics of authoring a model for use in a Cognitive Tutor.

3.2 Cognitive Tutors

Before describing examples of rule-based models used in ITS, we briefly describe Cognitive Tutors (Koedinger and Corbett 2006). Like many ITS, Cognitive Tutors provide step-by-step guidance as a student learns a complex problem-solving skill through practice (VanLehn 2006). They typically provide the following forms of support: (a) a rich problem-solving environment that is designed to make "thinking visible," for example, by prompting for intermediate reasoning steps, (b) feedback on the correctness of these steps, not just the final solution to a problem; often, multiple solution approaches are possible, (c) error-specific feedback messages triggered by commonly occurring errors, (d) context-sensitive next-step hints, usually made available at the student's request, and (e) individualized problem selection, based on a detailed assessment of each individual student's problem-solving skill ("cognitive mastery learning" (Corbett et al. 2010)). A Cognitive Tutor for geometry problem solving is shown in Figure 3.1. This tutor, a precursor of which was developed in our lab, is part of a full-year geometry course, in which students work with the tutor about 40% of the classroom time.

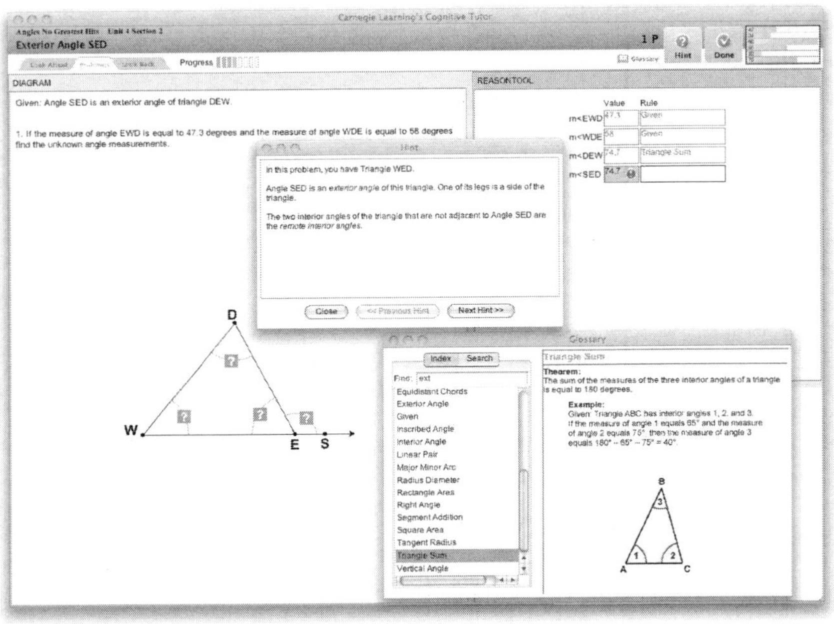

Fig. 3.1 A Cognitive Tutor for geometry problem solving

A number of studies have shown that curricula involving Cognitive Tutors lead to greater learning by students than the standard commercial math curricula used as comparison curricula (Koedinger and Aleven 2007; Koedinger et al. 1997). Cognitive Tutors also have been successful commercially. A Pittsburgh-based company called Carnegie Learning, Inc., sells middle-school and high-school mathematics curricula of which Cognitive Tutors are integrated parts. At the time of this writing, about 500,000 students in the US use Cognitive Tutors as part of their regular mathematics instruction.

Cognitive Tutors are grounded in ACT-R, a comprehensive theory of cognition and learning (Anderson 1993; Anderson and Lebiere 1998). A key tenet of this theory is that procedural knowledge, the kind of knowledge that is exercised in skilled task performance, is strengthened primarily through practice. ACT-R stipulates further that production rules are a convenient formalism for describing the basic components of procedural knowledge. Each production rule associates particular actions with the conditions under which these actions are appropriate. The actions can be mental actions (e.g., a calculation in the head, or a decision which part of the problem to focus on next) or physical actions (e.g., writing down an intermediate step in problem solving). Production rules are commonly written in IF-THEN format, with the "THEN part" or "right-hand side" describing actions and the "IF-part" or "left-hand side" describing conditions under which these actions are appropriate.

Reflecting the roots of this technology in the ACT-R theory, each Cognitive Tutor has a rule-based cognitive model as its central component. The models used in Cognitive Tutors are simulations of student thinking that can solve problems in the multiple ways that students can (or are learning to). The models can also be viewed as detailed specifications of the skills targeted in the instruction with the tutor. Cognitive Tutors use their cognitive models to provide tutoring by means of an algorithm called *model tracing* (Anderson et al. 1995). The tutor assesses and interprets students' solution steps by comparing what the student does in any given situation against what the model might do in the same situation. The basic idea is straightforward: After each student attempt at solving a step in a tutor problem, the tutor runs its model to generate all possible next actions that the model can produce. The student action is deemed correct only if it is among the multiple actions that the model can take next. If the student action is not among the potential next steps, it is deemed incorrect. When a student action is deemed correct, the production rules that generate the matching action serve as an interpretation of the thinking process by which the student arrived at this step. This detailed information enables the system to track, over time, how well an individual student masters these skills. A popular method for doing so is a Bayesian algorithm called *knowledge tracing* (Corbett and Anderson 1995; Corbett et al. 2000), which provides an estimate of the probability that the student masters each targeted skill. Some cognitive models contain rules that are represent incorrect behavior, enabling the tutor, through the same model-tracing process, to recognize commonly occurring errors and provide error-specific feedback.

3.3 A Simple Example of a Cognitive Model

Our first example is a simple rule-based cognitive model for fraction addition. This example emphasizes flexibility and cognitive fidelity; engineering concerns are in the background with a small model such as this. We also illustrate the model-tracing algorithm in the current section. The fraction addition model is not in use in a real-world tutor, although it is a realistic model for the targeted skill. The model is part of a demo tutor that comes with the Cognitive Tutor Authoring Tools (CTAT) (Aleven et al. 2006 and 2009); see Figure 3.2 for a view of this tutor as it is being authored with CTAT. These tools support the development of tutors with rule-based cognitive models in the Jess language, a commercial production rule language (Friedman-Hill 2003). They also support the development of a different type of tutors, example-tracing tutors (Aleven et al. 2009).

The model captures a simple strategy in which the goal of solving a fraction addition problem breaks down into three subgoals converting the fractions so they

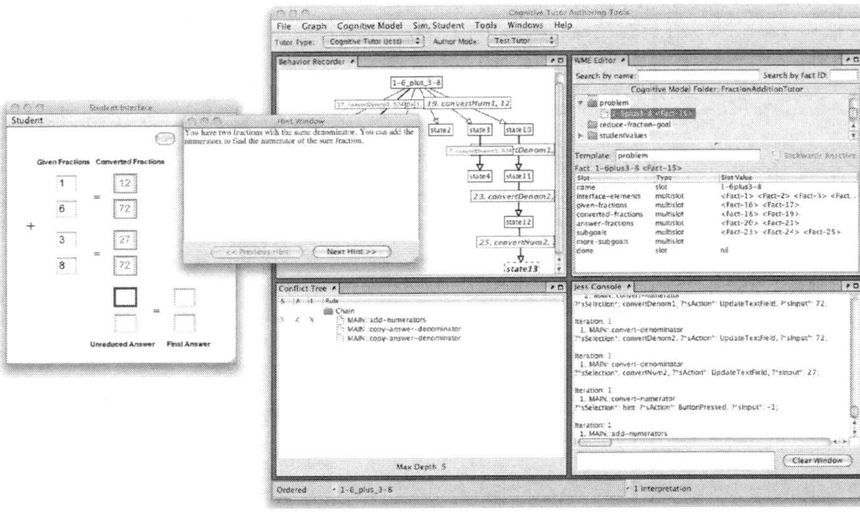

Fig. 3.2 Demo tutor for fraction addition being built with CTAT

share a common denominator, adding the converted fractions, and (if necessary) reducing the resulting fraction to the lowest possible denominator. The converted fractions and the unreduced sum are written out as intermediate steps, as illustrated in the window on the left in Figure 3.2. In order to be useful for tutoring, the model must capture all valid ways of solving a problem. For example, when adding 1/6 and 3/8, the model is capable of generating all possible options for the common denominator, including 24, 48, and 624. The model must be flexible in a different way as well: It must (and does) generate the steps required to solve a fraction addition problem in any reasonable order, not just in a single fixed order. Where the task naturally imposes an order on the steps, on the other hand, the model captures that order (so the tutor enforces the order). For example, the model works on the main subgoals (converting the fractions, adding, reducing) in sequence, since each subgoal depends on the previous. But the tutor should not impose any constraints on the order of steps within each of these subgoals. For example, when converting the two fractions, there is no reason to require that students enter the two numerators and the two denominators in any particular order. Therefore the model is able to generate these observable actions (four in total) in any order.

Now let us look at the model. The main components of a production rule model are its working memory and production rules. In general, working memory comprises a set of data structures, designed specifically for the given task domain, that represent the main entities in a problem and the current state of problem solving. These structures are called "facts," "chunks," or "working memory elements" depending on the production rule language being used. The fraction addition model was implemented in Jess so we will use the Jess-specific term "facts." Each fact has "attributes" or "slots" that contain values, which can be other facts, atomic

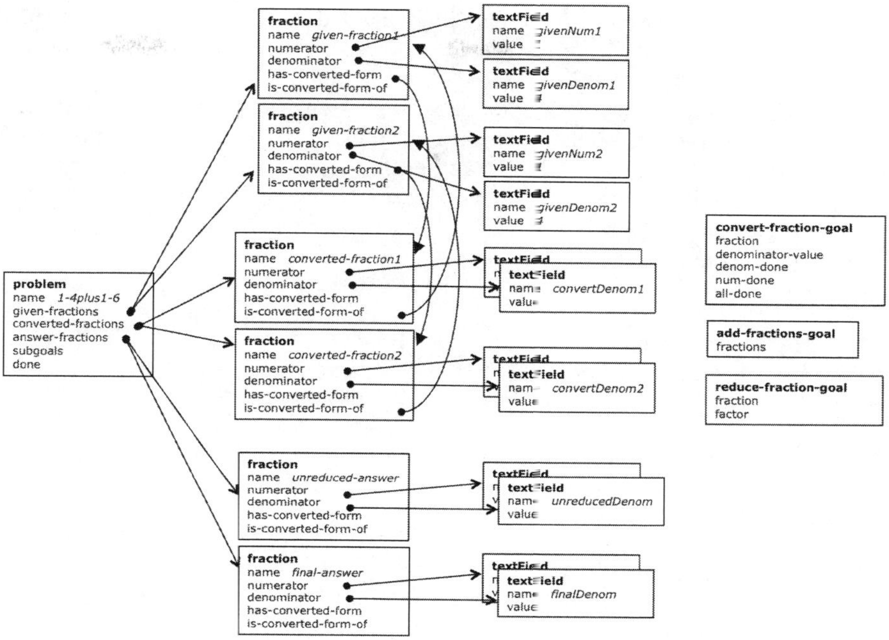

Fig. 3.3 Working memory representation of the fraction addition model

values such as numbers or strings, or lists of such values. The working memory representation for a given fraction addition problem (see Figure 3.3) contains a fact with key information about the problem itself as well as six additional facts representing six fractions: the two given fractions, the converted versions of these fractions, the unreduced sum, and the final answer (i.e., the reduced sum). Each fraction fact has slots for the numerator and denominator. These slots contain further facts that represent the text fields in the tutor interface in which the student enters the numerator and denominator of the various fractions (shown on the left in Figure 3.2), including the values that these text fields may contain. The presence of these facts in working memory reflects a common practice when building production rule models for Cognitive Tutors: working memory typically includes a representation of the interface, however rudimentary. The purpose of doing so is illustrated below, when we discuss the tutor's model-tracing algorithm.

In addition to the types of facts described so far, working memory contains facts representing the three types of subgoals mentioned above: to convert a fraction to a different denominator, to add two fractions with the same denominator, and to reduce a fraction to a lower denominator. These working memory facts can be viewed as representing subgoals that the student holds in her head. At the outset of a problem, only the given fractions are stored in working memory (as in Figure 3.3). The subgoals are generated by the rules in the course of solving a fraction addition problem.

Now let us consider the rules that implement the simple strategy described above, that of solving the three subgoals in order. We start with the first subgoal, that of converting the two given fractions. As mentioned, the model captures multiple possible choices for the common denominator. One such choice (the standard strategy often taught in American schools) is to use the least common multiple of the two denominators. This solution approach is captured in the following rule (the first line indicates the name of the rule; capitalized, italicized font indicates variables that are bound to problem-specific values when the rule is used in a given problem):

DETERMINE-LCD
IF there are no subgoals
and *D1* is the denominator of the first given fraction
and *D2* is the denominator of the second given fraction
THEN
Set *LCD* to the least common denominator of *D1* and D2
Add subgoals to convert the fractions to denominator *LCD*
Add a subgoal to add the two converted fractions

Due to space limitations, we do not present the Jess versions of these rules. The interested reader could download CTAT from http://ctat.pact.cs.cm.edu, install it, and explore the demo model. The DETERMINE-LCD rule models a first thinking step in solving a fraction addition problem. The conditions of this rule, that there are two given fractions and no subgoals, are met at the beginning of each problem. On its right-hand side, the rule sets a number of subgoals, meaning that it creates facts in working memory that represent these subgoals. In this way, the model models unobservable mental actions, namely, a student's planning, in their head, of the things they need to do in order to solve the problem. The DETERMINE-LCD rule does not actually model how to determine the least common multiple of two numbers – instead it uses a function call on the right-hand side of the rule. It is assumed that the student has learned how to determine the least common denominator prior to learning fraction addition, and therefore detailed modeling of that skill is not necessary.

In addition to the DETERMINE-LCD rule, the model contains three rules that capture alternative solution approaches and one rule that captures a common error. A rule named DETERMINE-PRODUCT uses the product of the two denominators as the common denominator, rather than their least common multiple, but is otherwise the same as DETERMINE-LCD. A second rule named DETERMINE-COMMON-MULTIPLE uses *any* common multiple. Although this rule is the same "in spirit" as the previous two, its implementation is different, due to the fact that it must generate *any* value that is a common multiple of the denominators. Rather than generate a random value for the common denominator (which would almost never be the value that the student actually used, meaning that such a rule would be useless for model tracing), the rule always "happens" to use the value that the student is actually using, provided it is a common multiple of the two denominators. One might say the rule guesses right as to what the student is doing. (It is able to do so because the actual student value is made available to the model by the tutor architecture. The demo model that comes with CTAT (version 2.9 and later) provides further detail.) Third, when the denominators of the two fractions are the same, no conversion is necessary. A rule called SAME-DENOMINATORS applies in this particular situation. Finally, the model captures a rule that models the common error of adding the denominators rather than computing a common multiple. This rule, called BUG-DETERMINE-SUM, is the same as DETERMINE-LCD and DETERMINE-PRODUCT, except that it sets the common denominator to the sum of the two denominators. This rule is marked as representing incorrect behavior, simply by including the word "BUG" in the name of the rule. The model contains two more rules corresponding to the subgoal to convert the fractions. These rules take care of converting the numerator and of writing out the converted fractions, once a common denominator has been determined.

CONVERT-DENOMINATOR
IF there is a subgoal to convert fraction *F* so that the denominator is *D*
And the denominator for the converted fraction has not been entered yet
THEN
Write *D* as the denominator of the converted fraction
And make a note that the denominator has been taken care of

CONVERT-NUMERATOR
IF there is a subgoal to convert fraction *F* so that the denominator is *D*
And the numerator for the converted fraction has not been entered yet
And the (original) numerator of *F* is *NUM*
And the (original) denominator of *F* is *DENOM*
THEN
Write *NUM* * (*D* / *DENOM*) as the numerator of the converted fraction
And make a note that the numerator has been taken care of

The conditions of these two rules require that a common denominator has been determined and a "convert-fraction" subgoal has been created. That is, these rules are not activated (i.e., do not match working memory) until one of the "DETERMINE-..." rules described above has been executed. On their right-hand-side, the rules model physical actions ("Write ..."), namely, the action of entering the value for the denominator or numerator into the relevant text field in the tutor interface. These actions are modeled as modifications of facts in working memory that correspond to the relevant interface component, although that fact is hard to glean this from the pseudo code. The reader may want to look at the actual Jess code. In addition to the rules described so far, the model contains a number of other rules related to the other main subgoals, but we do not have the space to describe them.

Let us now look at how the model-tracing algorithm uses the model to provide tutoring (Anderson et al. 1995 and 1991). This algorithm is specific to ITS; it is not used in standard production rule systems. As discussed below, model tracing is somewhat similar to, but also very different from the standard conflict-resolution method used in production rule systems. Simply put, model tracing is the process of figuring out, for any given student action in the tutor interface, what sequence of rule activations (if any) produces the same action. Therefore, just as standard conflict resolution, model tracing is about choosing from among possible rule activations. (A "rule activation" is the combination of a production rule and a set of bindings for its variables. Rule activations are created when a rule is matched against working memory. Different production rule languages use different terminology for this notion. "Rule activation" is a Jess term.) For example, at the outset of our example problem 1/6 + 3/8, a student might, as her first action, enter "24" as the denominator of the second converted fraction. (Note that this denominator is the least common multiple of the two denominators 6 and 8.) The fraction addition model can generate this action, starting with the initial working memory representation, by executing an activation of rule DETERMINE-LCD followed by one of CONVERT-DENOMINATOR, both described above. As an alternative first action, an (adventurous) student might enter "624" as the numerator of the first converted fraction. The model can also generate this action, namely, by executing activations of DETERMINE-COMMON-MULTIPLE and CONVERT-NUMERATOR. Third, a students might (erroneously) enter "14" as the numerator of one of the converted fractions. The model can generate this action by executing activations of BUG-DETERMINE-SUM and CONVERT-DENOMINATOR. Finally, a fourth student might enter "9" as the denominator of the first converted fraction. The model cannot generate this action. No sequence of rule activations produces this action.

As mentioned, the purpose of model tracing is to assess student actions and to interpret them in terms of underlying knowledge components. The first two of our example actions can be "modeled" by rules in the model that represent correct behavior. Therefore, they are deemed correct by the model tracer (and tutor). The production rules that generate the observable action provide an interpretation of the student action in terms of underlying skills. Interestingly, the different actions are modeled by different rules; in other words, they are interpreted as involving different skills. The third student action is modeled also, but one of the rules that is involved represents incorrect behavior. This action is therefore considered to be incorrect. The tutor displays an error message associated with the rule. ("It looks like you are adding the denominators, but you need a common multiple.") Since the model cannot generate the fourth action, the tutor will mark it to be incorrect.

To figure out what sequence of rule activations will produce a given action, the model-tracing algorithm must explore *all* solution paths that the model can generate. Since it is not possible to know in advance what the result of executing a rule activation will be, without actually executing it, the model tracer must in fact execute rule activations as part of this exploration process. Thus, the algorithm (as implemented in CTAT) does a depth-first search over the space of rule activations, expanding each sequence of rule activations up to the point where it results in an observable action. The algorithm fires rule activations to examine their consequences (i.e., the changes made to working memory). When it backtracks, changes made to working memory are undone. The search stops when an action is found that matches the student action, or when all sequences of rule activations (up to a certain depth) have been tried. The space of rule activations searched by the algorithm can be displayed graphically as a "Conflict Tree." CTAT provides an

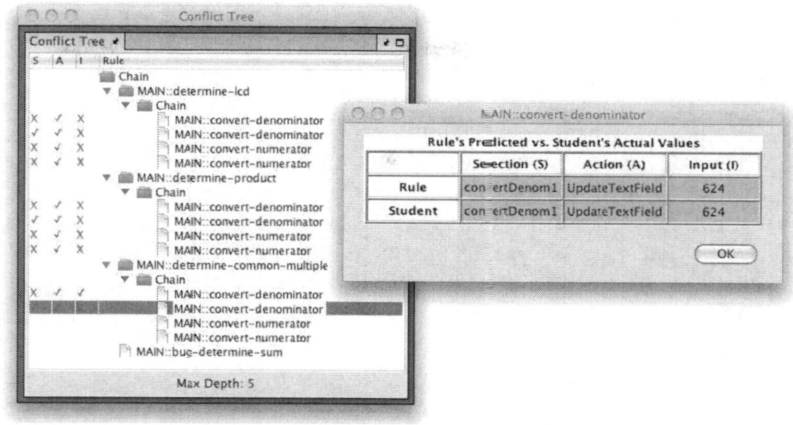

Fig. 3.4 A Conflict Tree displayed by CTAT

essential debugging tool for this purpose (Aleven et al. 2006), shown in Figure 3.4. As illustrated in this figure, at the outset of our fractions problem 1/6 + 3/8, four different rules can be activated, DETERMINE-LCD, DETERMINE-PRODUCT, DETERMINE-COMMON-MULTIPLE, and BUG-DETERMINE-SUM. Each of these activations can be followed by four different rule activations, two each for CONVERT-DENOMINATOR and CONVERT-NUMERATOR. (The Conflict Tree does not contain activations of these rules following BUG-DETERMINE-SUM, because the search stops before that part of the search space is reached.) Appropriately, the Conflict Tree does not show any activations of the SAME-DENOMINATORS rule; this rule's condition, that the two denominators are the same, is not met.

Whenever the model tracer, in the process of building/exploring the Conflict Tree, encounters a rule activation that generates an observable action, it compares that action to the student action being evaluated. To make this comparison possible, the student action and the model actions are encoded in a shared language. Specifically, they are encoded as selection-action-input triples, the components of which represent, respectively, the name of the GUI component (e.g., convertDenom1), the specific action on this component (e.g., UpdateTextField), and the value (e.g., "624"). The result of each comparison is displayed on the left side of the Conflict Tree tool (see Figure 3.4), in three columns that represent the three components – selection, action, and input. The many rows in the Conflict Tree that have "×"s indicate model actions that do not match the student's action, reflecting the fact that the model captures many solution paths that this particular student did not actually take. The highlighted row shows a match between student action and model action. The match is indicated by three check marks ("√"), although the highlighting makes the check marks difficult to see. The pop-up window shows the details of the comparison of the student action and the model action, information that is often helpful to a model author.

The requirement that observable actions are described as selection-action-input triples brings us back to the common practice, described above, that aspects of the tutor interface are represented in working memory. The main purpose of doing so is to give the rules access, in a flexible way, to the names of the GUI components, so they can reference them on their right-hand side, when encoding observable actions as selection-action-input triples. In order for this approach to be effective, the facts in working memory that represent interface elements must be linked to the rest of the problem representation in a way that makes their "semantic" role in the problem clear. The pseudo code for the rules shown above does not fully describe the details; the interested reader is referred to the demo model that comes with CTAT.

At a technical level, one may view the model-tracing algorithm as a new control strategy that replaces (or augments) the typical match-select-act control loop of standard production systems such as Jess (as used outside of CTAT). Rather than letting the model follow its own preferred solution path, as is done in the standard conflict resolution strategy used in standard production rule systems, the model-tracing algorithm "forces" the model to follow the student's solution path. At the same time, the student has to stay within the paths captured in the model, although she does need not follow just the model's *preferred* path. More specifically, model tracing differs in two main ways from the conflict resolution methods found in most "standard" production rule systems. First, in the typical match-select-act cycle that controls a standard production system, the choice of the next rule activation to be fired depends only on properties of the match of the rule's *condition* part with working memory (e.g., the specificity of the rule's conditions, the recency of the match, or the time of last update of the matched working memory elements). In model tracing, on the other hand, the selection is based also on the *observable actions* that a rule activation generates (as part of its *action* part). A second difference is that the model tracer, as it searches for actions the model might generate, searches for (and selects for execution) *sequences* of multiple rule activations, rather than just for *single* rule activations. Put differently, standard production systems do conflict resolution over (a set of) single rule activations, whereas the model-tracing algorithm does conflict resolution over (a set of) chains of rule activations.

Returning to one of our main themes, let us review how the requirements of flexibility and cognitive fidelity have helped shape the fraction addition model. First, consider the ways in which the model is *flexible* – it generates solutions with different common denominators and with different step order. That is, it faithfully mimics the expected variability of students' problem-solving processes in terms of observable actions. This flexibility is motivated by pedagogical considerations. It is necessary if the tutor is to respond (with positive feedback) to *all* students' reasonable solution approaches and not to impose arbitrary restrictions on student problem solving that make the student's learning task harder than necessary.

Second, the model arguably has strong *cognitive fidelity*. Key strategies such as using the least common multiple of the two given denominators, the product of the two denominators, or any other common multiple of the two denominators are modeled by separate rules, even though they could have been modeled more easily by a single rule. By modeling them as separate rules, the model reflects the conjecture that they represent different skills that a student learns separately, as opposed to being different surface manifestations of the same underlying student knowledge (e.g., special cases of a single unified strategy). These two possibilities lead to markedly different predictions about learning and transfer. If skills are learned separately, then practice with one does not have an effect on the other. On the other hand, if seemingly different skills reflect a common strategy, then practice with one should lead to better performance on the other. Since we do not know of a "grand unified denominator strategy" that unifies the different approaches, we make the assumption that the different strategies indeed reflect separate cognitive skills that are learned separately. To the extent that this assumption is correct, the current model (with separate skills) can be said to have greater cognitive fidelity than a model that models the strategies with a single rule.[1] It partitions the knowledge in a way that reflects the psychological reality of student thinking and learning, and presumably leads to accurate transfer predictions.

Incidentally, the model does not contain a separate rule for the situation where one denominator is a multiple of the other. In this situation, the larger of the two denominators could serve as the common denominator, and there is no need to compute the product or to try to compute the least common denominator. It is quite possible that this strategy is learned separately as well. If so, then a model that modeled this strategy with a separate rule would have greater cognitive fidelity than the current model.

Why does such detailed attention to cognitive fidelity matter? Cognitive fidelity helps in making sure that a student's successes and errors in problem solving are attributed to the appropriate skills or are blamed on the appropriate lack of skills. In turn, appropriate attribution of successes and errors helps the tutor maintain an accurate student model through knowledge tracing. In turn, a more accurate student model may lead to better task selection decisions by the tutor, since (in Cognitive Tutors and many other ITS), these decisions are informed by the student model (Corbett et al. 2000). Therefore, greater cognitive fidelity may result in more effective and/or more efficient student learning. So far, this prediction has not been tested empirically, although this very question is an active area of research. We return to this point in the final section of the chapter.

3.4 Cognitive Modeling for a Real-World Tutor: The Geometry Cognitive Tutor

In this section, we present a case study of cognitive modeling for a real-world intelligent tutoring system, namely, the model of the first version of the Geometry Cognitive Tutor described above, developed in our lab in the mid and late 1990s. This tutor is part of a full-year geometry curriculum, which has been shown to

[1] Ideally, the assumptions being made about how students solve fraction addition problems would be backed up by data about student thinking gathered through cognitive task analysis, or by results of mining student-tutor interaction data. We return to this point in the final section of the chapter.

improve student learning, relative to a comparison curriculum (Koedinger et al. 2000). The current Geometry Cognitive Tutor, which derives from the one described in this section, is in use in approximately 1,000 schools in the US (as of May 2010). The case study illustrates that flexibility and cognitive fidelity, emphasized in the fraction addition example, are not the only concerns when creating a model for a large real-world tutor. In a larger effort such as the Geometry Cognitive Tutor, engineering concerns inevitably come to the forefront, due to resource limitations. Arguably, the model for the Geometry Cognitive Tutor represents a happy marriage between flexibility, cognitive fidelity, and engineering. Although the description in this chapter ignores many details of the model, it highlights the model's essential structure and the major design decisions that helped shape it.

The model we describe is the third in a line of cognitive models for geometry developed for use in Cognitive Tutors. The first one was a model for geometry proof, developed by John Anderson and colleagues, and used in the original Geometry Proof Tutor which in lab studies was shown to be highly effective in improving student learning (Anderson 1993; Anderson et al. 1985). The second was developed by Ken Koedinger and captures the knowledge structures that enable experts to plan and generate proof outlines without doing the extensive search that characterizes novices' proof-generation efforts (Koedinger and Anderson 1990; 1993). The third model in this line (and the one on which we focus in the current chapter) was developed and then re-implemented by Ken Koedinger and colleagues, including the chapter author (Aleven and Koedinger 2002). Unlike the fraction addition model, which was implemented using the CTAT authoring tools in the Jess production rule language, the geometry model was implemented in the Tertle production rule language, which was created in our lab and is geared specifically towards model tracing (Anderson and Pelletier 1991). We focus on the Angles unit, one of six units that (at the time) made up the tutor's curriculum. This unit deals with the geometric properties of angles and covers 17 theorems. Other units dealt with Area, the Pythagorean Theorem, Similar Triangles, Quadrilaterals, and Circles. Units dealing with Right Triangle Trigonometry, Transformations, and 3D Geometry were added later. The curriculum was re-structured later, and the current tutor has 41 units. The model described here was used in the Angles, Circles, and Quadrilaterals units, and comprises approximately 665 rules. As mentioned, this model covers only part of the tutor's curriculum.

The tutor focuses on elementary geometry problem solving, not on advanced expert strategies and not on proof. This priority was driven by national curricular guidelines (NCTM 1991), which at the time emphasized problem solving over proof. More specifically, the main instructional objective for the tutor is to help students understand and apply a wide range of geometry theorems and formulas in multi-step problems of (what experts would consider) low to moderate complexity. Examples of such problems, selected from the tutor's Angles unit, are shown in Figures 3.5 and 3.6. Within this unit, the two problems shown in Figure 3.5 are at the lower end of complexity and the problem shown in Figure 3.6 is the single most complex problem. As such, it is not representative of the problems that students solve with the tutor, but it is included here to illustrate the model's capabilities.

Given: Line LT intersects Line HI in Point G.

If the measure of Angle LGH is 82°, find the measures of Angles IGT and TGH.

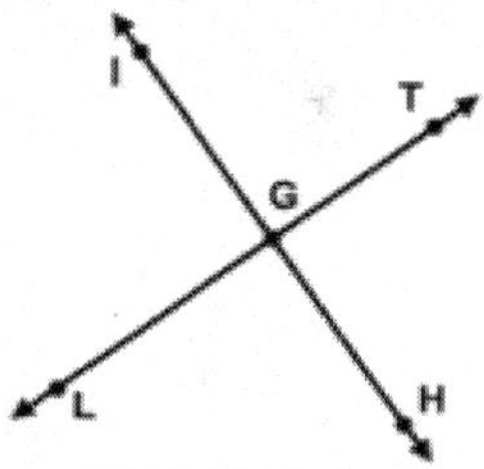

m∠LGH	82	Given
m∠IGT	82	Vertical Angles
m∠TGH	98	Linear Pair

Given: Angle SED is an exterior angle of triangle DEW.

If the measure of angle EWD is 52° and the measure of angle WDE is 59°, find the unknown angle measures.

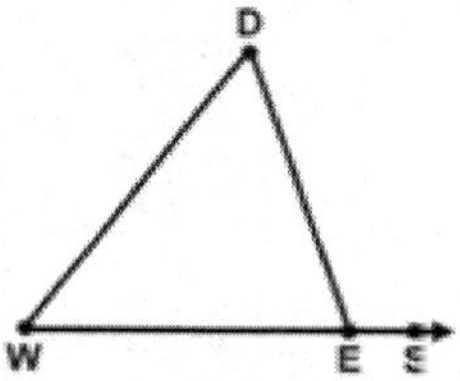

m∠EWD	52	Given
m∠WDE	59	Given
m∠DEW	69	Triangle Sum
m∠SED	111	Triangle Exterior Angle

Fig. 3.5 Examples of geometry problems that the geometry cognitive model is designed to solve. Student input is shown in hand-writing font

The geometry problems used in the tutor typically can be solved in multiple ways, using different theorems. The order of the steps may vary as well. For example, the first of the two problems in Figure 3.5 can be solved (as shown in Figure 3.5) by first finding m∠IGT using the *Vertical Angles* theorem[2], and then finding m TGH by applying the *Linear Pair* theorem[3] . It could also be solved by doing these two steps in the reverse order. Or it could be solved by applying

[2] The *Vertical Angles* theorem states that opposite angles formed at the intersection point of two lines are congruent.

[3] The *Linear Pair* theorem states that adjacent angles formed by two intersecting lines are supplementary, meaning that the sum of their measures equals 180°.

Given: Segment IS // segment CN. Angle ALN is a right angle. Segment YL bisects Angle RYS (so that m∠RYT = m∠TYS).

If the measure of angle IYR is 72° and the measure of angle ATL is 84°, find the measure of Angle TAL.

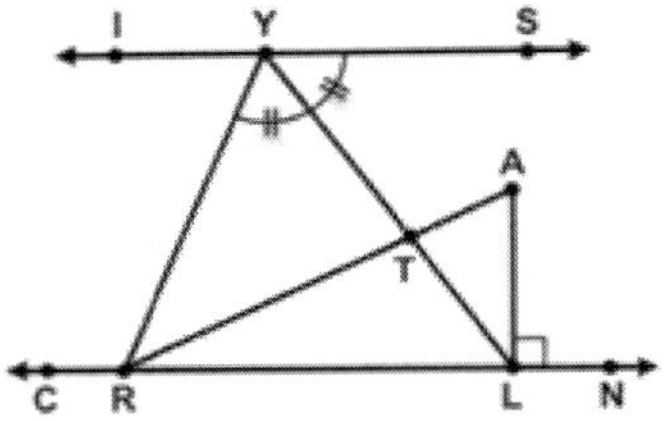

m∠IYR	72	Given
m∠ATL	84	Given
m∠RYS	108	Linear Pair
m∠RYT	54	Angle Bisection
m∠CRY	108	Supplementary Interior Angles
m∠YTR	84	Vertical Angles
m∠TRY	42	Triangle Sum
m∠RTL	96	Linear Pair
m∠TRL	30	Linear Trio
m∠RLT	54	Triangle Sum
m∠TLA	36	Complementary Angles
m∠TAL	60	Triangle Sum

Fig. 3.6 A highly complex tutor problem

Linear Pair to find m TGH, and then applying *Linear Pair* again to find m IGT. Likewise, the second problem can be solved as shown in Figure 3.5, by using the *Triangle Exterior Angle* theorem[4] to infer the measure of ∠SED (an exterior angle of triangle DEW) from the two remote interior angles (∠EDW and ∠DWE). Alternatively, the student could solve the problem without using this theorem, by applying *Triangle Sum*[5] to find ∠DEW and then *Linear Pair* to find m∠SED. The complex problem shown in Figure 3.6 also allows for multiple solution paths; we have not counted them. In order for the tutor to be "flexible," its cognitive model must (and does) capture all these solution variants.

In order to use geometry theorems in the types of problems illustrated in Figures 3.5 and 6, students must learn to recognize the applicability conditions of these theorems as well as to derive quantitative relations from (application of) these theorems. This understanding includes the visual meaning of geometric concepts referenced by each theorem (e.g., "adjacent angles" or "transversal" or "isosceles triangle"). In particular, they must be able to recognize, in diagrams, instances of these concepts, so as to recognize when each theorem applies (and does not apply). For example, to understand the *Triangle Exterior Angle* theorem, students must learn concepts such as the exterior angle of a triangle, or its remote interior angles, so that they can recognize instances of these concepts in actual diagrams and can recognize when this theorem applies (e.g., the second problem in Figure 3.5). When the theorem applies, they must infer an appropriate quantitative relation (i.e., that the measure of the exterior angle is equal to the sum of the measures of the remote interior angles). The key skills targeted in the tutor's angles unit are the ability to apply the 17 theorems in the manner just described. That

[4] The *Triangle Exterior Angle* theorem states that the measure of an exterior angle of a triangle (i.e., an angle between one side of a triangle and the extension of another side of that triangle) is equal to the sum of the two remote interior angles of the triangle.

[5] The *Triangle Sum* theorem states that the sum of the measures of the interior angles of a triangle equals 180°.

is, the tutor interprets student performance and tracks student learning largely at the level of applying individual theorems.

It was decided early on during the design process that the tutor interface would prompt the student for each step in the process of solving a geometry problem, as can be seen in Figures 3.1 (which is a later version of the same tutor), 5, and 6. That is, for each problem in the tutor, an empty table is given at the outset, with the labels for the angle measures visible. The student completes the problem by filling out in the table, for each step, a value (typically, the measure of an angle) and a justification for the value by citing the name of a theorem. (The value of having students provide justifications for their steps was confirmed by subsequent research (Aleven and Koedinger 2002).) Typically, the steps need not be done strictly in the order in which they appear in the table. Rather, the step order must observe the logical dependencies among the steps, meaning that any given step can be completed only when the logically-prior steps have been completed. Typically, the order of the quantities in the table observes the logical dependencies (and so going in the order of the table usually works), but other orders may be possible as well. The purpose of listing the steps in advance, rather than letting the student figure out what the steps are was to minimize the cognitive load associated with searching for solution paths in diagrams. This kind of search (and the concomitant cognitive load) was thought to hinder the acquisition of problem-solving knowledge. This design was loosely inspired by work by Sweller, van Merriënboer, and Paas (Sweller et al. 1998), who showed that completion problems and goal-free problems lead to superior learning during the early stages of skill acquisition, compared to problem solving. Prompting for steps significantly reduces the need for students to search the diagram in the process of solving a problem. Although experts are adept at searching a diagram in a strategic manner, it was thought that strategic diagram need not be in the foreground in the Geometry Cognitive Tutor. The kinds of highly complex problems in which sophisticated search strategies would pay off were deemed to outside of the scope of the instruction.

These key design decisions had a significant impact on the cognitive model. The model captures a problem-solving approach in which the student, in order to generate a problem-solving action (i.e., the next step in a given problem), takes the following four mental steps:

1. *Focus on the next quantity to solve (e.g., by looking at the table).* For example, consider a student who works on the first problem in Figure 3.5, at the point after she has filled out the first row the table (m∠LGH equals 82° as given in the problem statement). Looking at the table, the student might decide to focus on m∠TGH as the next quantity to solve; let us call this quantity the "desired quantity." (Note that this quantity is listed in the third row of the table. The student skips one of the quantities in the table.)
2. *Identify a quantitative relation between the desired quantity and other quantities in the problem, justified by a geometry theorem. Find a way of deriving the value of the desired quantity.* For example, searching the diagram, our student might recognize that ∠LGH and ∠TGH form a linear pair and she might derive, by application of the *Linear Pair*

theorem, the quantitative relation m∠LGH + m∠TGH = 180°. Applying her knowledge of arithmetic (or algebra), our student might determine that m∠IGT can be found once m∠LGH is known,

3. *Check if a sufficient number of quantities in the selected quantitative relation are known to derive the value of the desired quantity.* For example, our student might ascertain (by looking at the table) that the value of m∠LGH (which we call "pre-requisite quantity") is known.
4. *Derive the desired quantity's value from the quantitative relation using the values of the other known quantities.* For example, our student might conclude that m∠ LGH equals 98° and enter this value into the table.

Let us look at the key components of the model in more detail, starting with the organization of working memory. As mentioned, in general, working memory contains a representation of a problem's structure as well as of its evolving solution state. The geometry model's working memory contains two main types of elements. (Following the Tertle production rule language, we employ the term "working memory elements" rather than "facts.") First, working memory lists the key quantities in the given problem, that is, the quantities whose value is given or whose value the student is asked to find. In the tutor unit dealing with angles, the quantities are angle measures. In other units, we encounter arc measures, segment measures, circumference measures, area measures, etc. In addition to the key quantities, working memory contains elements representing quantitative relations among these quantities. These working memory elements specify, for each, the type of quantitative relation, the quantities involved, and the geometry theorem whose application gives rise to (and justifies) the relation. For example, the key quantities in the first problem illustrated in Figure 3.5 are the measures of the three angles LGH, TGH, and IGT. There are three quantitative relations relating these quantities:

- m∠TGH + m∠IGT = 180°, justified by *Linear Pair*
- m∠LGH + m∠TGH = 180°, justified by *Linear Pair*
- m∠LGH = m∠IGT, justified by *Vertical Angles*

Similarly, in the second problem, there are four key quantities and three relations, one justified by application of *Triangle Sum,* one justified by *Linear Pair,* and one by *Triangle Exterior Angle.* The third problem involves 12 key quantities and 15 relations between these quantities.

To model quantitative relations in working memory, we created a hierarchy of relation types, a small part of which is shown in Figure 3.7. This hierarchy reflects both quantitative aspects and geometric aspects of geometry problem solving. It is intended to capture how, over time, geometry problem-solving knowledge might be organized in the student's mind. At higher levels in this hierarchy, relation types are differentiated based on the number of quantities they involve and on how these quantities are quantitatively related. At the lowest level, subtypes are differentiated on the geometry theorem that justifies the relation. The actual relation instances in working memory are instances of the lowest-level subtypes. For

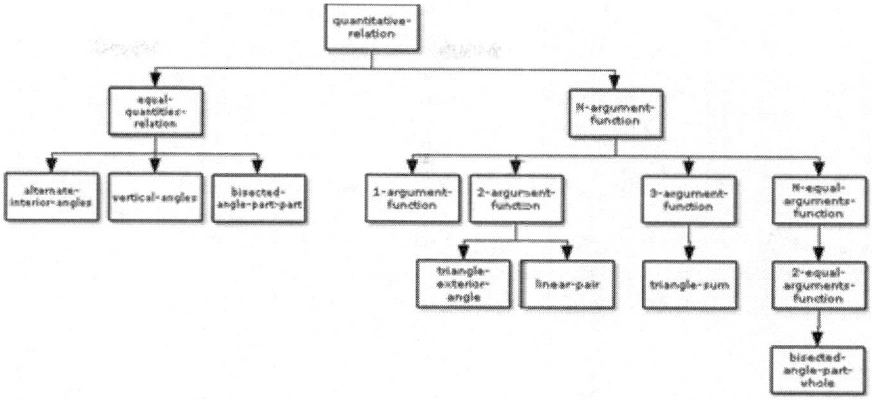

Fig. 3.7 Excerpt from the model's hierarchy of quantitative relation types

example, quantitative relations that state that two quantities are equal are modeled as "equal-quantities-relation." The subtypes of this relation differ with respect to the geometry theorem that justifies the relation, such as the *Vertical Angles* and *Alternate Interior Angles* theorems. As another example, relations that state that some function of certain quantities (usually, their sum) equals another quantity or a constant are modeled as "N-argument-function." The (leaf) subtypes of this type are (again) differentiated based on the geometry theorem that gives rise to the relation, such as *Triangle Exterior Angle* or *Linear Pair.*

The quantitative relations for any given problem are stored at the start of the problem and therefore a complete set – for the given problem – must be provided by the problem author. It may be clear that these quantitative relations contain information about the *solution* of the problem and not just problem definition information. That is, quantitative relations are normally derived in the course of solving a geometry problem; they are not given at the outset. Thus, the model can be said to *pre-store* certain solution information, a common technique for engineering cognitive models that can make them easier to build, as discussed further below. As a final comment about the organization of working memory, we note that it does not contain a symbolic representation of the diagram. Given the decision that diagram search is not a main instructional objective, such a representation was unnecessary. As a result of this decision, the model was simplified considerably, and the initial working memory configuration that must be provided for each problem was reduced.

Turning to the second key component of the geometry cognitive model, the production rules, a large set of rules implements the basic four-step problem-solving strategy described above, and implements the key skills involved in these steps. Roughly speaking, each of the mental steps in the four-step strategy is modeled by one or more production rules. The first mental step in generating a problem-solving action (i.e., a single step in a tutor problem) is to decide which quantity to focus on next (i.e., the desired quantity). This mental step is captured by a single rule that models a student's checking the table in the tutor interface to find a quantity (*any* quantity) whose value has not yet been determined:

IF Q is one of the key quantities in the problem
And the value of Q has not been determined yet
THEN
Set a goal to determine the value of Q (i.e., focus on Q as the desired quantity)

Importantly, this rule generates a match with working memory (i.e., a rule activation) for each quantity whose value has not been determined yet, which is crucial if the model is to generate all acceptable next actions. As discussed, modeling all solution paths is a major precondition for models used for tutoring (by model tracing). Although in this first mental step, any key quantity whose value has not been determined yet can be chosen, the later mental steps impose additional restrictions. That is, not all choices for the desired quantity made in this first step will "survive" the later steps. As a final comment about this rule, there are other possible ways, besides looking at the table, in which a student might decide on the next desired quantity. For example, a student might search the diagram for an angle whose measure can be determined next. Such diagrammatic reasoning is not modeled, however, as discussed above.

The second mental step has two parts, namely, (a) to identify a quantitative relation that can be applied to find the desired quantity, justified by a geometry theorem, and (b) to determine a set of "pre-requisite quantities" from which the value of the desired quantity can be derived. For example, assume that the model is working on the first problem in Figure 3.5, at a point where it has determined that $m\angle LGH = 82°$. (This value is given in the problem statement.) If in mental step 1, the model selects $m\angle TGH$ as the desired quantity, then in mental step 2 (the current step), it may infer that the desired quantity can be derived from the quantitative relation $m\angle LGH + m\angle TGH = 180°$, justified by *Linear Pair*, once $m\angle LGH$ and the constant 180° are known (i.e., the latter two quantities are the pre-requisite quantities). Presumably, the student would recognize the applicability of *Linear Pair* by searching the diagram. The cognitive model does not capture this kind of diagrammatic reasoning, however. Instead, it looks for a quantitative relation in working memory that involves the desired quantity – as mentioned, all quantitative relations for the given problem are pre-stored. Having found a relation from which the desired quantity can be derived, the model determines the pre-requisite quantities through quantitative reasoning. To model this kind of reasoning, the model contains a separate rule (or rules) for each quantitative relation defined at the intermediate levels of the hierarchy. For example, application of *Linear Pair* (as well as application of *Angle Addition, Complementary Angles,* etc.) yields a "two-argument function" (e.g., $m\angle LGH + m\angle TGH = 180°$). This kind of relation asserts that a function of two quantities, called "input quantities," is equal to a third quantity or constant, called "output quantity." The relation may be applied either forwards or backwards. In our running example, the relation must be applied backward to derive the desired quantity:

IF there is a goal to find the value of quantity *Q*
And *R* is a two-argument-function
And *Q* is one of the input quantities of *R*
And *Q1* is the other input quantity of *R*
And *Q2* is the output quantity of *R*
THEN
Set a goal to apply *R* backwards with *Q1* and *Q2* as pre-requisite quantities

This rule and the many other rules like it model the quantitative reasoning involved in geometry problem solving. The use of the hierarchy of quantitative relations is a significant advantage in defining these rules (which was the main reason for creating it in the first place). Because the rules are defined at the intermediate levels of the hierarchy, we need far fewer of them than if rules were defined separately for each geometry theorem, which are found at the lowest level of the hierarchy. Importantly, the set of rules as a whole is capable of identifying all ways of deriving the value of the desired quantity from the quantitative relations in working memory, which is crucial if the model is to generate a complete set of solution paths.[6]

The model also contains rules that "determine" the geometric justification of any given quantitative relation. All these rules do is "look up" the geometric justification pre-stored with the quantitative relation, which as mentioned is encoded in the relation type – recall that the leaf nodes of the hierarchy are differentiated based on the geometry theorems. There is one rule exactly like the following for every relation type at the bottom of the types hierarchy:

IF there is a goal to apply *R*
And *R* is of type *Linear Pair*
THEN
(do nothing)

It may seem odd to have a rule with no action part, and it may seem odd to derive (as this rule does) a geometric justification from a quantitative relation. In normal problem solving, of course, each quantitative relation is justified by the application of a geometry theorem, not the other way around. It is the pre-storing of information, that makes it possible to (conveniently) turn things upside down. Further, the reason why we need these rules without an action part is because they signal which theorem is being applied. As mentioned, the educational objective of the tutor is for students to learn to apply a relatively large set of targeted geometry theorems. In order for the tutor to track student learning at the level of individual geometry theorems, the model must have a separate rule that represents application of each individual theorem. That is exactly what these action-less rules are.

The third mental step in generating a problem-solving action is to check that the values of the pre-requisite quantities have been determined already. Continuing our example, having decided that m∠LGH and the constant 180° are pre-requisite

[6] It is of course equally crucial that, for the given problem and the given set of key quantities, *all* quantitative relations have been pre-stored in working memory.

quantities for deriving the value of the desired quantity m∠TGH, the model ascertains that both pre-requisite quantities are known. Constants are always known (by definition), and a value for m∠LGH has already been entered into the table. The following rule (which is slightly simpler than the actual rule) models a student who checks that pre-requisite quantities have been found by looking them up in the table in the tutor interface:[7]

IF there is a goal to apply relation R in manner *M* to find desired quantity *Q*
And Q1 and *Q2* are pre-requisite quantities
And *Q1* is either a constant, or its value has been determined
And *Q2* is either a constant, or its value has been determined
THEN
Set a goal to use the values of *Q1* and *Q2* to apply relation *R* in manner *M*

Incidentally, although this rule imposes a strong constraint on the order in which the key quantities in the problem can be found, it does not necessarily lead to a single linear order, as discussed above.

The fourth and final mental step is to determine the desired quantity's value from those of the pre-requisite quantities. The student must apply the quantitative relation that has been selected to derive the desired quantity's value. The model, by contrast, relies on having values pre-stored with the quantities, another example of the common engineering technique of pre-storing solution information, discussed below.

If *Q* is the desired quantity
And its value can be derived from pre-requisite quantities whose value is known
And *V* is the pre-stored value for *Q*
THEN
Answer *V* (i.e., enter *V* into the table as value for *Q)*
Mark *Q* as done

This description highlights the model's main components and structure. Many details are left out, and the model is much larger than may be evident from the current description, but we hope to have illustrated how the model is suitable for tutoring.

Discussion of the Geometry Model

The Geometry Cognitive Tutor is designed to help students learn to apply a wide range of geometry theorems in a step-by-step problem-solving approach. The

[7] The model does not model "thinking ahead," meaning, solving a few key quantities in one's head and then entering the last one (and get tutor feedback on it) before entering the logically-prior ones (e.g., the pre-requisite quantities). As the tutor was designed to minimize cognitive load for the student, it seemed reasonable to try to steer the student toward a "low load" strategy in which the student always records logically-prior steps in the table (which reduces cognitive load because it obviates the need for remembering the steps).

main educational objective is for students to come to understand and recognize the applicability conditions of the theorems, including the visual meaning of geometric concepts referenced by each theorem (e.g., "adjacent angles" or "transversal"). A second objective is for students to learn to derive and manipulate the quantitative relations that follow from application of these theorems. As discussed, the Geometry Cognitive Tutor does not reify diagram search, meaning that it does not make explicit or communicate *how* the student should go about searching a diagram in the course of problem solving (e.g., to find the next possible step, or a theorem to apply). The tutor's step-by-step approach to problem solving reduces the need for diagram search, as it was thought that the extraneous cognitive load induced by this kind of search would get in the way of the tutor's main instructional objectives, to obtain a solid understanding of the theorems targeted in the instruction.

Although we have glossed over many details, this case study illustrates our main themes related to cognitive modeling for ITS. First, the model is fully flexible. Within the targeted set of geometry theorems, the model accounts for all different ways of solving the targeted problems, which enables the tutor to follow along with students, regardless of which solution path they decide to take. In addition, the model is flexible in terms of step order: it models any step order that observes the logical dependencies between the steps. As for the model's cognitive fidelity, as mentioned, the model partitions the knowledge based on the geometry theorem that is being applied. This decision is based largely on theoretical cognitive task analysis. It is a very reasonable decision. In order how to do even better, one would need to analyze student log data as described in the next section. In one place, the model has less than full cognitive fidelity, and this choice reflects engineering concerns (and resource limitations). In particular, the tutor does not *reify* detailed reasoning about how a particular geometry theorem applies to the diagram. It abstracts from such reasoning. For example, when applying *Alternate Interior Angles,* students do not communicate to the tutor which two lines are parallel, which line or segment is a *transversal* that intersects the two lines, and which angles formed form a pair of alternate interior angles. This kind of reasoning falls within the tutor's educational objectives, and presumably students engage in this kind of reasoning when deciding whether the theorem applies. The decision not to reify this reasoning followed primarily from a desire to keep down the cost of interface development. Not modeling this reasoning drastically simplified the model and also made it easier to author problems for the tutor, because the initial working memory configuration for each problem did not need to contain a representation of the diagram. Later research with the Geometry Cognitive Tutor however indicated that students *do* acquire a deeper understanding of geometry when reasoning about how the geometry theorems apply is reified in the tutor interface (Butcher and Aleven 2010; 2008 and 2007). This facility required very significant extensions of the tutor interface (i.e., an integrated, interactive diagram in which students could click to indicate how theorems apply). Therefore, it is fair to say that not reifying this reasoning in the original tutor was a reasonable trade-off.

As discussed, the geometry model illustrates a useful engineering technique that makes models easier to build. The model *pre-stores* certain information that a less omniscient being needs to compute while solving a geometry problem. Specifically, it pre-stores the quantitative relations for each problem and the values of the key quantities. Pre-storing solution aspects need not be detrimental to a model's flexibility. That is, it need not affect a model's ability to generate an appropriate range of solution paths with an appropriately flexible ordering. For example, the fact that the geometry model pre-stores quantitative relations does not negatively affect the model's ability to find the appropriate quantitative relations for a given problem step, even if (due to pre-storing) the diagrammatic reasoning that students go through in order to generate these quantitative relations is not modeled. In principle, pre-storing solution aspects should also not be detrimental to a model's cognitive fidelity. There is no principled reason that the rules that retrieve the pre-stored solution aspects could not be equally fine-grained as rules that would generate this information. Nonetheless, given that the typical purpose of pre-storing is to make model development easier, there is a risk that rules that retrieve pre-stored information do not receive sufficient careful attention during model development and are not sufficiently informed by cognitive task analysis. As a result, they may end up being simpler than they should be, failing to capture important distinctions within the student's psychological reality. With that caveat, pre-storing can be a useful engineering technique. As mentioned, within the geometry model, pre-storing quantitative relations leads to a very significant simplification of the model, since the diagram interpretation processes by which these quantitative relations are derived not need to be modeled. Without further empirical analysis, it is somewhat difficult to know with certainty that the model's cognitive fidelity is optimal. If the proof of the pudding is in the eating, however, the tutor's effectiveness (Koedinger et al. 2000) is an indication that the model is adequate if not better than that. The next section discusses techniques for analyzing a model's cognitive fidelity.

3.5 Concluding Remarks

In this final section, we discuss ways of evaluating the cognitive fidelity of a cognitive model, and we briefly mention some research that is underway to develop automated or semi-automated techniques to help create models with high cognitive fidelity. Such techniques are important, because, as discussed, the cognitive fidelity of a model used in a model-tracing tutor may influence the efficiency or effectiveness of student learning with that tutor. As we have argued, greater cognitive fidelity may lead to a more accurate student model and to better task selection decisions, which in turn may lead to more effective and/or efficient learning, by avoiding both under-practice and over-practice of skills. The conjecture that a model with greater fidelity pays off in terms of more effective or efficient learning has not been tested yet in a rigorous controlled study (e.g., a study evaluating student learning with multiple versions of the same tutor that have cognitive models of varying degrees of cognitive fidelity, though it seems only a matter of time.[8]

Throughout the chapter, we have emphasized that a model used in a tutor must have flexibility and cognitive fidelity. A model's flexibility is perhaps the less problematic requirement, as it deals with observable phenomena. A model must be complete and flexible enough to accommodate the alternative solution paths and variations in step order that students produce, at least the ones that are pedagogically acceptable. To illustrate this flexibility requirement, we discussed how both the fraction addition model and the geometry model accommodate a full range of student strategies and step order. Lack of flexibility occurs when a model is not correct or complete; perhaps students use unanticipated strategies in the tutor interface that the model does not capture. Perhaps the model developer has not done sufficient cognitive task analysis to be aware of all variability in student behavior, or has inadvertently restricted the order in which steps must be carried out.

In addition to flexibility, a model must have cognitive fidelity. It must partition knowledge into components (e.g., production rules) that accurately represent the psychological reality, that is, correspond closely to the knowledge components that the students are actually learning. For example, in the fraction addition model, we conjecture that students learn the different strategies for finding the common denominator separately, instead of learning a single (hypothetical) overarching strategy of which these strategies may be different surface level manifestations. This conjecture implies that a student may know one strategy but not know the other. It implies also that practice with one strategy does not help a student get better in using the other strategies (at least not to the same degree). Accordingly, the model contains separate rules for these different strategies, even though from an engineering perspective, it would have been easier to capture all strategies with a single rule. In general, lack of cognitive fidelity means that a model's rules do not correspond to actual student knowledge components. Lack of fidelity may occur when students "see" distinctions that the model developer has overlooked. For example, perhaps unbeknownst to the model developer, novice students view an isosceles triangle in its usual orientation (i.e., a fir tree) as distinct from one that is "upside down" (i.e., as an ice cream cone). In algebraic equation solving, they may view the term x as distinct from $1x$. In fractions addition, students may develop a separate strategy for dealing with the case where one denominator is a multiple of the other, which the model developer may not have captured. Alternatively, a model may fail to capture generalizations that students make. For example, in Excel formula writing, absolute cell references (i.e., using the "$" sign in cell references) can be marked in the same way for rows and columns (Mathan and Koedinger 2000). It is conceivable that a modeler would decide to model

[8] A classroom study by Cen et al. with the Geometry Cognitive Tutor tested part of this "causal chain:" it demonstrated that a more accurate student model can lead to more efficient learning (Cen et al. 2007). In this study, the improved accuracy of the student model was due not to greater cognitive fidelity, but to tuning the (skill-specific) parameters used by Bayesian knowledge-tracing algorithm that updates the student model after student actions.

these skills with separate rules, whereas there is empirical evidence that students can learn them as one. In addition to a model being too fine-grained or too coarse-grained, it could (at least in principle) happen that students' knowledge components are different from a model's without strictly being either more specific or more general, but it is difficult to think of a good example.

How does a modeler detect ensure that a cognitive model is sufficiently flexible and that it has sufficient cognitive fidelity? Careful task analysis in the early stages of model development, helps greatly in achieving flexibility. One result of cognitive task analysis should be an accurate and comprehensive understanding of the variety of ways in which students solve problems, which can then be captured in the model. It is helpful also to carefully pilot a tutor before putting it in classrooms. Any lack of flexibility remaining after model development is likely to result in a tutor that sometimes (incorrectly) rejects valid student input, which may be caught during piloting. If not caught during model development and tutor piloting, lack of flexibility will surely result in complaints from students and teachers who are using the system, and one obviously prefers not to catch it that way! Finally, careful scrutiny of tutor log data may help as well. For example, analysis of student errors recorded in the log data may help in detecting instances where the tutor incorrectly rejects correct student input, although again, one prefers not to find out that way.

Cognitive fidelity can be harder to achieve and ascertain than flexibility, given that it deals with unobservable phenomena rather than observable. Cognitive task analysis (such as the think-aloud methodology) may help in identifying different strategies that students use. But what is needed in addition is a determination whether seemingly different (observable) strategies represent a separate psychological reality or are in fact unified in the student's mind. Although think-aloud data may hold clues with regard to the psychological reality of different strategies, in general it is very difficult to make a definitive determination based on think-aloud data alone. A somewhat more definitive determination of the psychological reality of hypothesized skills can be made using a cognitive task analysis technique called "Difficulty Factors Analysis" (Baker et al. 2007). It also helps to test a model's cognitive fidelity using tutor log data, either "by hand" or through automated methods, currently an active area of research within the field of educational data mining (Baker and Yacef 2009). In essence, all methods test whether the transfer predictions implied by a model are actually observed in the log data. When a cognitive model has high cognitive fidelity, one expects to see a gradual increase over time in the performance of students on problem steps that – according to the model – exercise one and the same knowledge component. Psychological theories in fact predict the shape of the curve (we will use the term "learning curve") that plots the accuracy (or speed) of execution of a given knowledge component on successive opportunities (Heathcote et al. 2000; Newell and Rosenbloom 1981). A time-honored method for analyzing cognitive models is simply to extract learning curves from tutor log data, plot them, and inspect them visually. If the curve looks smooth, then all is well. On the other hand, when learning curves visibly deviate from the curves that cognitive theories predict, this deviation is an indication that the model has limited cognitive fidelity. For example, when a knowledge component in the model is overly general, compared to psychological reality, the learning curve for this knowledge component will include measuring points in which the actual knowledge component (as it exists in the student's head) is not involved. Instead, these measuring points represent use of one or more other knowledge components (as they exists in the student's head). Since students' mastery of these other knowledge components is, in principle, unrelated to that of the plotted knowledge component, these points will generally not align with the rest of the curve. The curve may look "ragged" Conversely, when a modeled knowledge component is overly specific, compared to the actual knowledge component in the student's head, the learning curve will not include all opportunities for exercising this knowledge component, and will not be fully smooth. When deviations are evident, the modeler might take a closer look at the tutor log data and the tutor problems to develop hypotheses for why it might be lacking. Good tools exist that help with this analysis, such as the DataShop developed by the Pittsburgh Science of Learning Center (Koedinger et al.).

A second method for evaluating the cognitive fidelity of a model is by fitting the learning curves extracted from log data against the theoretically-predicted function. If the fit is good, the model has high cognitive fidelity. This method is particularly useful when comparing how well alternative models account for given log data. The DataShop tools mentioned above also support "what-if" analyses to easily compare the fit of alternative models. Thus they help not only in spotting problems with cognitive fidelity, but also in finding models with greater fidelity. Recent and on-going research aims to automate (parts of) this process of model refinement. This work has led to semi-automated methods for revising models based on log data, such as Learning Factors Analysis (Cen et al. 2006). Much work is underway in educational data mining that focuses on learning models entirely from data (Baker and Yacef 2009). We expect that automated methods will become a very valuable addition to traditional cognitive task analysis. We do not foresee that they will replace traditional cognitive task analysis methods entirely, because the kinds of information that these methods produce are complementary. Also, cognitive task analysis can be done with small numbers of students, whereas data mining typically requires larger data sets.

To conclude, tutors using rule-based cognitive models have a long and rich history, and there is much reason to think that there will be many interesting developments yet to come. In particular, research focused on increasing the cognitive fidelity of models will help make this kind of tutoring system both easier to develop and (even) more effective in supporting student learning.

Acknowledgments. Albert Corbett and Laurens Feenstra gave very helpful comments on an early draft of the chapter. The work is supported by Award Number SBE-0836012 from the National Science Foundation to the Pittsburgh Science of Learning Center. Their support is gratefully acknowledged. All opinions expressed in this chapter represent the opinion of the author and not the funding agency.

References

Aleven, V., Koedinger, K.R.: An effective meta-cognitive strategy: learning by doing and explaining with a computer-based Cognitive Tutor. Cognitive Science 26(2), 147–179 (2002)

Aleven, V., McLaren, B.M., Sewall, J.: Scaling up programming by demonstration for intelligent tutoring systems development: An open-access website for middle-school mathematics learning. IEEE Transactions on Learning Technologies 2(2), 64–78 (2009)

Aleven, V., McLaren, B.M., Sewall, J., Koedinger, K.R.: The Cognitive Tutor Authoring Tools (CTAT): Preliminary evaluation of efficiency gains. In: Ikeda, M., Ashley, K.D., Chan, T.W. (eds.) ITS 2006. LNCS, vol. 4053, pp. 61–70. Springer, Heidelberg (2006)

Aleven, V., McLaren, B.M., Sewall, J., Koedinger, K.R.: A new paradigm for intelligent tutoring systems: Example-tracing tutors. International Journal of Artificial Intelligence in Education 19(2), 105–154 (2009)

Anderson, J.R.: Rules of the mind. Lawrence Erlbaum Associates, Hillsdale (1993)

Anderson, J.R., Lebiere, C.: The atomic components of thought. Lawrence Erlbaum Associates, Mahwah (1998)

Anderson, J.R., Pelletier, R.: A development system for model-tracing tutors. In: Birnbaum, L. (ed.) Proceedings of the international conference of the learning sciences. Association for the Advancement of Computing in Education, pp. 1–8, Charlottesville, VA (1991)

Anderson, J.R., Boyle, C.F., Reiser, B.J.: Intelligent tutoring systems. Science 228(4698), 456–462 (1985)

Anderson, J.R., Boyle, C.F., Yost, G.: The geometry tutor. In: Joshi, A. (ed.) Proceedings of the 9th international joint conference on artificial intelligence, pp. 1–7. Morgan Kaufmann, San Francisco (1985)

Anderson, J.R., Corbett, A.T., Koedinger, K.R., Pelletier, R.: Cognitive tutors: Lessons learned. The Journal of the Learning Sciences 4(2), 167–207 (1995)

Baker, R.S.J.d., Yacef, K.: The state of educational data mining in 2009: A review and future visions. Journal of Educational Data Mining, 1(1): 3-17 (2009)

Baker, R.S.J.d., Corbett, A.T., Koedinger, K.R.: The difficulty factors approach to the design of lessons in intelligent tutor curricula. International Journal of Artificial Intelligence and Education, 17(4): 341-369 (2007)

Butcher, K., Aleven, V.: Integrating visual and verbal knowledge during classroom learning with computer tutors. In: McNamara, D.S., Trafton, J.G. (eds.) Proceedings of the 29th Annual Conference of the Cognitive Science Society. Cognitive Science Society, Austin (2007)

Butcher, K., Aleven, V.: Diagram interaction during intelligent tutoring in geometry: Support for knowledge retention and deep transfer. In: Schunn, C. (ed.) Proceedings of the 30th Annual Meeting of the Cognitive Science Society, CogSci 2008, pp. 894–899. Lawrence Erlbaum, New York (2008)

Butcher, K., Aleven, V.: Learning during intelligent tutoring: when do integrated visual-verbal representations improve student outcomes? In: The 32nd Annual Meeting of the Cognitive Science Society, CogSci 2010 (in press, 2010)

Cen, H., Koedinger, K.R., Junker, B.: Learning factors analysis - A general method for cognitive model evaluation and improvement. In: Ikeda, M., Ashley, K.D., Chan, T.W. (eds.) ITS 2006. LNCS, vol. 4053, pp. 164–175. Springer, Heidelberg (2006)

Cen, H., Koedinger, K.R., Junker, B.: Is over practice necessary? Improving learning efficiency with the cognitive tutor through educational data mining. In: Luckin, R., Koedinger, K.R., Greer, J. (eds.) Proceedings of the 13th international conference on artificial intelligence in education (AIED 2007), pp. 511–518. IOS Press, Amsterdam (2007)

Clark, R.E., Feldon, D., van Merriënboer, J., Yates, K., Early, S.: Cognitive task analysis. In: Spector, J.M., Merrill, M.D., van Merriënboer, J.J.G., Driscoll, M.P. (eds.) Handbook of research on educational communications and technology, 3rd edn., Lawrence Erlbaum Associates, Mahwah (2007)

Corbett, A.T., Anderson, J.R.: Knowledge tracing: Modeling the acquisition of procedural knowledge. User Modeling and User-Adapted Interaction 4(4), 253–278 (1995)

Corbett, A., Kauffman, L., MacLaren, B., Wagner, A., Jones, E.: A Cognitive Tutor for genetics problem solving: Learning gains and student modeling. Journal of Educational Computing Research 42(2), 219–239 (2010)

Corbett, A., McLaughlin, M., Scarpinatto, K.C.: Modeling student knowledge: Cognitive tutors in high school and college. User Modeling and User-Adapted Interaction 10(2-3), 81–108 (2000)

Crowley, R.S., Medvedeva, O.: An intelligent tutoring system for visual classification problem solving. Artificial Intelligence in Medicine 36(1), 85–117 (2006)

Friedman-Hill, E.: JESS in action. Manning, Greenwich, CT (2003)

Heathcote, A., Brown, S., Mewhort, D.J.K.: The power law repealed: The case for an exponential law of practice. Psychonomic Bulletin & Review 7(2), 185–207 (2000)

Koedinger, K.R., Aleven, V.: Exploring the assistance dilemma in experiments with Cognitive Tutors. Educational Psychology Review 19(3), 239–264 (2007)

Koedinger, K.R., Anderson, J.R.: Abstract planning and perceptual chunks: Elements of expertise in geometry. Cognitive Science 14, 511–550 (1990)

Koedinger, K.R., Anderson, J.R.: Reifying implicit planning in geometry: Guidelines for model-based intelligent tutoring system design. In: Lajoie, S.P., Derry, S.J. (eds.) Computers as cognitive tool. Lawrence Erlbaum, Hillsdale (1993)

Koedinger, K.R., Corbett, A.T.: Cognitive tutors: Technology bringing learning sciences to the classroom. In: Sawyer, R.K. (ed.) The Cambridge handbook of the learning sciences. Cambridge University Press, New York (2006)

Koedinger, K.R., Nathan, M.J.: The real story behind story problems: Effects of representations on quantitative reasoning. The Journal of the Learning Sciences 13(2), 129–164 (2004)

Koedinger, K.R., Anderson, J.R., Hadley, W.H., Mark, M.A.: Intelligent tutoring goes to school in the big city. International Journal of Artificial Intelligence in Education 8(1), 30–43 (1997)

Koedinger, K.R., Baker, R., Cunningham, K., Skogsholm, A., Leber, B., Stamper, J.: A Data Repository for the EDM community: The PSLC DataShop. In: Romero, C., Ventura, S., Pechenizkiy, M., Baker, R.S.J.d. (eds.) To appear in Handbook of Educational Data Mining. CRC Press, Boca Raton (in press)

Koedinger, K.R., Corbett, A.T., Ritter, S., Shapiro, L.: Carnegie learning's cognitive tutor: Summary research results. White paper, Carnegie Learning Inc., Pittsburgh, PA (2000) e-mail: info@carnegielearning.com, web: `http://www.carnegielearning.com`

Mathan, S.A., Koedinger, K.R.: Fostering the intelligent novice: Learning from errors with metacognitive tutoring. Educational Psychologist 40(4), 257–265 (2005)

Mitrovic, A., Mayo, M., Suraweera, P., Martin, B.: Constraint-based tutors: A success story. In: Monostori, L., Vancza, J., Ali, M. (eds.) Proceedings of the 14th International Conference on Industrial and Engineering Applications of Artificial Intelligence and Expert Systems IEA/AIE-2001. Springer, Berlin (2001)

Mitrovic, A., Martin, B., Suraweera, P., Zakharov, K., Milik, N., Holland, J.: ASPIRE: An authoring system and deployment environment for constraint-based tutors. International Journal of Artificial Intelligence in Education 19(2), 155–188 (2009)

Murray, T.: An overview of intelligent tutoring system authoring tools: Updated analysis of the state of the art. In: Murray, T., Blessing, S., Ainsworth, S. (eds.) Authoring tools for advanced learning environments. Kluwer Academic Publishers, Dordrecht (2003)

NCTM, Professional Standards for Teaching Mathematics. VA, National Council of Teachers of Mathematics, Reston (1991)

Newell, A., Rosenbloom, P.S.: Mechanisms of skill acquisition and the law of practice. In: Anderson, J.R. (ed.) Cognitive skills and their acquisition. Erlbaum, Hillsdale (1981)

Newell, A., Simon, H.A.: Human problem solving. Prentice Hall, Englewood Cliffs (1972)

Ritter, S., Anderson, J.R., Koedinger, K.R., Corbett, A.: Cognitive tutor: Applied research in mathematics education. Psychonomic Bulletin & Review 14(2), 249–255 (2007)

Roll, I., Koedinger, K., Aleven, V.: The Invention Lab: Using a hybrid of model tracing and constraint-based tutoring to offer intelligent support in inquiry environments. To appear in the Proceedings of the 10th International Conference on Intelligent Tutoring Systems, ITS 2010 (in press, 2010)

Sweller, J., van Merriënboer, J.J.G.: Paas FGWC, Cognitive architecture and instructional design. Educational Psychology Review 10(3), 151–296 (1998)

VanLehn, K.: The behavior of tutoring systems. International Journal of Artificial Intelligence in Education 16(3), 227–265 (2006)

VanLehn, K., Lynch, C., Schultz, K., Shapiro, J.A., Shelby, R.H., Taylor, L., Treacy, D., Weinstein, A., Wintersgill, M.: The Andes physics tutoring system: Lessons learned. International Journal of Artificial Intelligence in Education 15(3), 147–204 (2005)

Woolf, B.P., Cunningham, P.: Building a community memory for intelligent tutoring systems. In: Forbus, K., Shrobe, H. (eds.) Proceedings of the sixth national conference on artificial intelligence. AAAI Press, Menlo Park (1987)

Chapter 4
Modeling Domains and Students with Constraint-Based Modeling

Antonija Mitrovic

Intelligent Computer Tutoring Group, Department of Computer Science & Software Engineering, Privte Bag 4800, University of Canterbury, Christchurch, New Zealand
tanja.mitrovic@canterbury.ac.nz

Abstract. Stellan Ohlsson proposed CBM in 1992 as a way to overcome some problems in student modeling. Since then, the approach has been extended and used in numerous intelligent tutoring systems, which we refer to as constraint-based tutors. CBM is now an established methodology for modeling instructional domains, representing students' domain knowledge and also higher-level skills. Authoring support for constraint-based tutors is now available, as well as mature, well-tested deployment environments. We present CBM, its foundation and extensions, various types of instructional domains we applied it to, and conclude with avenues for future research.

4.1 Introduction

It has been fifteen years since the initial work on SQL-Tutor, the first ever constraint-based tutor, started. Since then, we have extended CBM from the original, theoretical proposal (Ohlsson 1992) to a thoroughly tested and widely used methodology and have accumulated significant experience developing constraint-based tutors in a variety of instructional domains (Mitrovic et al. 2007). CBM has attracted significant attention and debate, and more and more researchers are adopting it for developing their own Intelligent Tutoring Systems (ITSs) – see e.g. (Galvez et al. 2009; Oh et al. 2009; Le et al. 2009; Siddappa and Manjunath 2008, Menzel 2006; Petry and Rosatelli 2006; Riccucci et al. 2005; Rosatelli and Self 2004).

In this chapter, we start from the basic idea and the psychological foundation of CBM, and present various extensions we have made over the years. We focus on how domain models can be represented in terms of constraints, and then present various kinds of student models based on them. We then turn to the implications of CBM for pedagogical decision making, and discuss various approaches to providing feedback, selecting problems, supporting higher-order skills and affect. CBM is a flexible approach which can be used in a wide variety of instructional domains, to support many teaching strategies.

R. Nkambou et al. (Eds.): Advances in Intelligent Tutoring Systems, SCI 308, pp. 63–80.
springerlink.com

4.2 CBM: The Basic Idea

At the time when CBM was proposed (Ohlsson 1992), model/knowledge tracing approach championed by the CMU researchers (Anderson et al. 1990; Anderson et al. 1995, Koedinger et al. 1997) was the clear winner on the ITS scene; in fact, it is still the most widely used approach for developing ITSs. Ohlsson proposed CBM as away to avoid some limitations of model-tracing, such as having runnable models of the expert and the student. There are several problems with developing such runnable models, expressed as sets of production rules. He noted the complexity of developing the production set, which is necessary to generate the solution to be compared to the student's actions. For some instructional tasks it might even be impossible to come up with the production set as the domain or the task itself may be ill-defined. Furthermore, if the system is to respond to errors intelligently, buggy rules are necessary. A buggy rule generates an incorrect action, and when it matches the student's action, the tutor can provide remedial feedback.

Enumerating mistakes students can (and do) make is time-consuming and intractable, as the space of incorrect knowledge is vast. Instead of capturing mistakes, CBM focuses on domain principles that *every* correct solution must follow. The fundamental observation CBM is based on is that all correct solutions (to any problems) share the same feature – they do not violate any domain principles. Therefore, instead of representing both correct and incorrect space, it is enough to represent the correct space by capturing domain principles in order to identify mistakes. Any solution (or action) that violates one or more domain principles is incorrect, and the tutoring system can react by advising the student on the mistake even without being able to replicate it.

CBM represents the solution space in terms of abstractions. All solutions states that require the same reaction from the tutor (such as feedback) are grouped in an equivalence class, which corresponds to one constraint. Therefore, an equivalence class represents all solutions that warrant the same instructional action. The advantage of this approach is in its modularity; rather than looking for a specific way of solving the problem (correct or incorrect), each constraint focus on one small part of the domain, which needs to be satisfied by the student's solution in order to be correct. An important assumption is that the actual sequence of actions the student took is not crucial for being able to diagnose mistakes: it is enough to observe the current state of the solution. The student model does not represent the student's action, but the effects of his/her actions instead.

A constraint is an ordered pair an ordered pair (C_r, C_s), where C_r is the relevance condition and C_s is the satisfaction condition. The relevance condition specifies a (simple or compound) test specifying the features of the student solution for which the constraint is of importance. For example, the constraint might be applicable to situations when the student has added two fractions which have the common denominator. The satisfaction condition specifies additional test(s) that must be met by the student solution in order for it to be correct. For the same example, the student's solution is correct if the denominator of the resulting fraction is equal to the denominators of the two given fractions, and the numerator is equal to the sum of the numerators of the given fractions. If the relevance condition is met, but the satisfaction condition is not, then the student's solution is incorrect. Therefore, the general form of a constraint is:

If <relevance condition> is true,
Then <satisfaction condition> had better also be true.

The origin of a mistake is not crucial for generating pedagogical interventions, as the tutoring system can give feedback to the student about the domain principle that was violated, without knowing exactly what kind of incorrect knowledge generated the mistake. Constraint-based tutors normally attach feedback messages directly to constraints.

It is important to point out the difference between constraints and production rules. Although the (English) statement above seems similar to an IF-THEN production rules, it is of a very different nature. A production rule proposes an action to be taken if the goal and the situation specified in the IF part are met. On the contrary, a constraint specifies conditions for a solution state to be correct. The two conditions in a constraint are not linked with logical implication, but with the "ought to" connective, as in if C_r is true, C_s ought to be true as well (Ohlsson and Mitrovic 2007). Production rules are of generative nature, while constraints are evaluative, and can be used for making judgment.

A set of constraints represent features of correct solutions explicitly. Any solution violating one or more constraints is incorrect; therefore the constraint set models errors indirectly, without enumerating them. As the consequence, CBM allows the student to explore the solution space freely; any correct approach to solving a problem will be supported, as there are no constraint violations. CBM is not sensitive to the radical strategy variability (Ohlsson and Bee 1991), which is the observation that students often use different strategies for solving the same problems. CBM allows for creativity: even those solutions that have not been considered by the system designers will be accepted by constraints, as they do not violate any domain constraints. Any problem-solving strategy resulting in a correct state will be recognized as such by CBM.

4.3 The Theoretical Foundations of CBM

CBM is based on Ohlsson's theory of learning from performance errors (Ohlsson 1996). This theory assumes the existence of procedural and declarative knowledge, as common to many theories of learning. Learning starts with accumulating declarative knowledge, which is later converted to procedural knowledge through practice. Procedural knowledge is necessary for generating actions, while declarative knowledge has an important function in evaluating consequences of actions.

The theory states that people make errors because their procedural knowledge is missing or is faulty. Faulty knowledge might be too general or too specific. The theory focuses on learning from errors, which consists of two phases: error detection and error correction. A person may be aware of the error he/she made because the actual outcomes of the action do not match the expected ones; this is the situation when the declarative knowledge (the expectancy of the results of the performed action) allows the person to identify the error themselves. In other situations, if declarative knowledge is missing, the person cannot identify the mistake on his/her own, and would need help, provided in terms of feedback. This feedback might come from the environment itself, or from the teacher, and enables the student to correct the procedural knowledge. The feedback from a tutor, be it a human or an artificial one, consists of identifying the part of the action/solution which is incorrect (blame assignment) and the domain principle that is violated by it. Therefore, the theory states that declarative knowledge is represented in the form of constraints on correct solutions. Such knowledge can be used to correct faulty knowledge, by making it more general or more specific. Computer simulations have shown that using constraints for correcting procedural knowledge is a plausible mechanism. For a detailed account of how constraints allow the corrections of procedural knowledge, please see (Ohlsson 1996).

4.4 Domain Modeling

Domain modeling s widely recognized as being time-consuming and critical for the success of an ITS. In this section, we discuss various types of constraints necessary for a constraint-based system, as well as some authoring considerations.

4.4.1 Syntax vs. Semantic Constraints

The original idea for CBM, as presented in (Ohlsson 1992), viewed constraints as only representing syntax knowledge, i.e. problem-independent domain principles. An example constraint presented in the 1992 paper is from the area of fraction addition:

> *If the current problem is a/b + c/d, and the student's solution is (a+c)/n, then it had better be the case that n=b=d*

The constraint is relevant for the situation when the student added two fractions by adding the numerators; this is only allowed when the denominators of the two fractions and the denominator of the resulting fraction are equal. We refer to such constraints as *syntax constraints*. These constraints allow an ITS to identify errors in student solutions which violate the general domain principles. Another simple example is *If you are driving in New Zealand, you better be on the left side of the road.*

Our research on CBM started with SQL-Tutor, a constraint-based tutor which teaches university-level students to specify queries in the SQL language. This task is a design task: the student is given a problem in the natural language, and the student needs to convert this into an SQL Select statement. There is no algorithm for performing the task. Furthermore, the natural language text could be ambiguous. Additional complexity inherent in the task is that the student needs to be familiar with the relational data model and the specific relational database the problem is based on. Students typically find the task very difficult (Mitrovic 1998b).

Some examples[1] of syntax constraints from SQL-Tutor are given in Figure 4.1. Constraint 2 is the simplest constraint in this system: the relevance condition is always true, and therefore the constraint is relevant to all solutions. Its satisfaction condition requires that the SELECT clause is specified. In other words, every query must list some expression(s) to be returned from the database.

Constraint 110 focuses on the FROM clause: the constraint is relevant when this clause contains the JOIN keyword. This keyword is used to specify a join condition, and the syntax requires the ON keyword to be specified in the same clause. Notice that this constraint does not check for other elements of the join condition – it simply specifies that those two keywords must appear in the same clause. There are other constraints in SQL-Tutor that check for other elements of the join condition. Such low-level of granularity allows for feedback messages to be very specific. In the case of constraint 110, the feedback message will remind the student that the JOIN and ON keywords need to be used at the same time.

This point is further illustrated by other constraints from Figure 4.1. Constraint 358 builds upon constraint 110: both conditions from 110 are in the relevance condition of constraint 358, and therefore its relevance condition is more restrictive than the one of constraint 110. Its satisfaction condition matches the FROM clause to the given pattern, which specified the general form of a join condition. It allows any number of elements at the beginning and at the end of the FROM clause (wildcards ?*d1 and ?*d2), but somewhere in the clause there must be a value that will be allocated to variable t1, optionally followed by another value (s1, which corresponds to an alias assigned to a table), which is followed by the JOIN keyword, another value (which will be assigned to t2), another optional value (which, if exists, will be assigned to s2), the ON keyword, one value (assigned to a1), the equal sign, and another value (a2). This constraint does not check the values assigned to the variables, but simply checks that the clause is of the specified form. The feedback message attached to this constraint can be more specific about how join conditions are specified.

The following constraint (399) is even more specific: now the relevance condition requires the student's FROM clause to match general form of the join condition, and the satisfaction condition checks that both t1 and t2 are valid table names from the current database. There are other constraints in SQL-Tutor that further check that a1 and a2 are valid attributes, and that the types of those two attributes match each other.

Constraint 11 has all the previously discussed tests in its relevance condition, and its satisfaction condition checks whether the types of attributes used in the join condition are the same. All of those checks can be done on the basis of syntax knowledge, independently of the problem requirements.

However, students do not make only syntax errors. Often their solutions are syntactically correct but the answer is not correct for the problem at hand, and the student should be alerted about that too. Therefore, it is necessary to check the semantic correctness of the solution, and for that reason we introduced semantic

[1] We present the constraints in their English form. The constraints as specified in the constraint language used in SQL-Tutor are given in (Mitrovic 1993).

Constraint 2:
C_r: t
C_s: the SELECT clause must be specified

Constraint 110:
C_r: the student's solution contains the JOIN keyword in the FROM clause
C_s: the ON keyword must also appear in the same clause.

Constraint 358:
C_r: the student's solution contains the JOIN and ON keywords in FROM
C_s: the FROM clause must match the following pattern
(?*d1 ?t1 ??s1 "JOIN" ?t2 ??s2 "ON" ?a1 "=" ?a2 ?*d2)

Constraint 399:
C_r: the student's solution contains the JOIN and ON keywords in FROM, and matches the following pattern:
'(?*d1 ?t1 ??s1 "JOIN" ?t2 ??s2 "ON" ?a1 "=" ?a2 ?*d2)
C_s: ?t1 and ?t2 must be valid tables from the current database.

Constraint 11:
C_r: the student's solution contains the JOIN and ON keywords in FROM, the clause matches the pattern and ?t1 and ?t2 are valid tables from the current database, and ?a1 is an attribute of table t1,
C_s: the types of attributes a1 and a2 must be the same.

Fig. 4.1 Syntax constraints from SQL-Tutor

constraints (Mitrovic 1998a, 1998b, 1998c; Mitrovic and Ohlsson 1999). A semantic constraint compares the student solution and the correct solution to the same problem, again focusing on a single domain principle. Semantic constraints check whether the student's solution is the correct solution for the problem at hand. The semantics of the particular problem is captured in the ideal solution; in SQL-Tutor, the teacher specifies an ideal solution for each problem. This ideal solution is carefully selected to illustrate some important features of the language.

However, in SQL there are often several correct solutions for the same problem, as the language is redundant. Therefore, semantic constraints cannot simply check whether the student's solution is identical to the ideal one; they need to check for equivalent ways of solving the problem.

Figure 4.2 illustrates some semantic constraints from SQL-Tutor. Constraint 207, for example, is relevant when the ideal solution has a join condition specified in the FROM clause, and the student's solution has an empty WHERE clause and more than one table in FROM. In general, if a query uses more than one table in FROM, a join condition is necessary. Join conditions in can be specified either in FROM or in WHERE; this constraints focuses on the FROM clause only, and requires (in the satisfaction condition) that the student specified the join condition in

Constraint 207:
C_r: the WHERE clause is empty in both the student's and ideal solutions,
there is more than one table in the student's FROM clause,
and the FROM clause of the ideal solution contains the JOIN keyword,
C_s: the JOIN keyword must appear in the student's FROM clause.
Constraint 387:
C_r: The student specified a join condition in FROM using valid tables t1 & t2,
The join condition is of form a1 = a2,
attribute a2 comes from table t1,
the ideal solution lists t1 and t2 in the FROM clause,
the join condition is not specified in FROM in the ideal solution,
its WHERE clause contains an attribute n1 from table t1,
and this attribute is compared to an attribute n2 from table t2,
C_s: attribute a1 should be equal to n2, and attribute a2 should be equal to n1.

Fig. 4.2 Semantic constraints from SQL-Tutor

FROM. Note that the satisfaction condition does not check for the complete join condition: it only requires the JOIN keyword to be used in FROM. If the student made a syntax error when specify the join condition in FROM, it would be caught by the syntax constraints we discussed previously

Constraint 387 is relevant when the student specified a join condition in the FROM clause, but the ideal solution contains a join condition in the WHERE clause. The relevance condition of this constraint establishes correspondences between the table and attribute names used in the ideal and the student solution, and the satisfaction condition checks that the corresponding attributes are equal. Note that there are other constraints that will be checking for different combination of attributes.

The constraints need to be specified on a low-level of granularity in order for feedback to be specific. As the consequence, a set of constraints is required to fully specify one domain principles. SQL-Tutor contains a large number of constraints (currently over 700) and it still does not cover all of SQL; the completeness of a constraint set is not necessary for the ITS to be useful. It is important that the constraint set allows for the diagnosis of solutions for a given set of problems. The addition of new problem types would require the constraint set to be extended. But it is easy to add problems of similar types: at the moment SQL-Tutor supports 13 databases, and about 300 problems defined for them To add another database and the corresponding set of problems, all that is necessary is to add problem text for each problem, its solution, and the information about the database.

4.4.2 Applying CBM to Ill-Defined and Well-Defined Tasks

As discussed previously, the task of specifying queries in SQL-Tutor is a design task, and such tasks are ill-defined. We developed other constraint-based tutors for design tasks. EER-Tutor (Suraweera and Mitrovic 2002, 2004; Mitrovic et al. 2004; Zakharov et al. 2005) teaches conceptual database design, another ill-defined task. The student needs to design an EER diagram for a specific situation starting from the requirements specified in the form of English text. Although the EER data model is well-designed and relatively simple, the actual task of designing a conceptual schema for a database is ill-defined (Suraweera and Mitrovic 2004). The requirements are often incomplete, and the student needs to use their general knowledge in order to design the EER diagram. There is no algorithm to use to identify the necessary components of the solution. Furthermore, the goal state (i.e. the solution) is defined in abstract terms, as an EER diagram that satisfies the requirements. An example syntax constraint from EER-Tutor checks that every regular entity type has at least one key attribute. Semantic constraints make sure that the student has identified all the necessary entities, relationships and attributes, at the same time allowing for alternatives. For example, an attribute of a 1:N relationship may be alternatively represented as an attribute of the entity on the N side of the relationship. Semantic constraints check for such equivalences between the student and the ideal solution.

Another constraint-based tutor for a design task is COLLECT-UML, a tutor that teaches object-oriented software design, by requiring students to design UML class diagrams from textual descriptions (Baghaei et al. 2006, 2007). In all of those instructional tasks, the domain itself is well defined (i.e. the domain theory is well-specified in the terms of the underlying model), but the instructional task is ill-defined.

However, CBM is not only capable of capturing domain knowledge for design tasks – it can also be used for procedural tasks, for which problem-solving algorithms are known. We have developed several tutors of this type. NORMIT (Mitrovic 2005) is a constraint-based tutor that teaches data normalization in relational databases. The domain is well-defined and so is the task: there is an algorithm that students need to learn and apply correctly. NORMIT breaks the task into a series of steps, and requires the student to apply a step correctly before moving on to the next step. This was a deliberate decision, as we wanted to stress the importance of the correct order in which the steps are applied. However, CBM could also be used in the same task in a less restrictive way, by specifying constraints in a slightly different way. We developed another version of NORMIT, in which the student can apply the steps in any order, but constraints check that the student has calculated all the necessary parts of the solution before attempting ones which depend on the previous steps. We refer to such constraints as the *path constraints*.

ERM-Tutor (Milik et al. 2006) is another example of a constraint-based tutor that teaches a procedural task – this time, the student needs to transform an EER diagram into a relational schema. We also developed a lot of tutors for procedural tasks within ASPIRE, our authoring system discussed later in this chapter. An example of such tutors is CIT, a constraint-based tutor that teaches students how to make decision on capital investments (Mitrovic et al. 2008).

In a recent paper (Mitrovic and Weerasinghe 2009), we presented a classification of ITSs in terms of two dimensions, instructional domains and instructional tasks. Both of these can be ill- or well-defined. CBM has previously been applied to well-defined domains, with both ill-defined tasks (SQL-Tutor, EER-Tutor, COLLECT-UML) and well-defined tasks (NORMIT, ERM-Tutor). In the case of ill-defined domains, the tasks can still be ill- or well-defined. CBM can be applied to well-defined tasks no matter what kind of domain is at hand; an example of ill-defined domain and a well-defined task is psychological assessment. In such cases, CBM would be applicable, but so would model-tracing. The latter approach however cannot be applied in the case of an ill-defined task and an ill-defined domain, such as essay writing or artistic tasks. CBM can still be used in such situations.

Let us consider architectural design. If the task is to design a house with three bedrooms for a given piece of land which needs to be eco-friendly and energy efficient, there can be a set of designs which satisfy the minimal requirements. Constraints that need to be satisfied involve the problem specification and the norms for energy consumption and ecological consequences – but the designs will differ in terms of aesthetics and personal preferences of the designer. The constraint set will capture the minimal requirements, and still allow for a variety of solutions. Therefore, in ill-defined domains the student has the freedom to include solution components to make the solution aesthetically pleasing or more to their preferences, and the ITS will still accept it as a good solution for the problem. It is also possible to have weights attached to constraints, with highest weights being assigned to mandatory constraints, and lower weights assigned to constraints that need not necessarily be satisfied as they correspond to optional elements.

In the case of ill-defined domains and ill-defined tasks, more attention needs to be devoted to the feedback provided to the student. In well-defined domains, feedback generation is straightforward: the student violates some constraints, and feedback on violated domain principles is provided. In model-tracing tutors, buggy production rules provide feedback on errors, and hints can be generated on the next step the student is to take. However, in ill-defined domains, the declarative knowledge is incomplete: the constraint set consists of a set of mandatory principles and some heuristics. Therefore, the feedback mechanism needs to be sophisticated, so that feedback does not confuse the student. If the solution is not complete, feedback becomes even more crucial, as the ITS should discuss only the issues the student has worked on so far.

Ill-defined domains and tasks are very complex, and therefore, ITSs need to scaffold learning, by providing as much information as possible without making it trivial. The common ITS techniques can also be used in ill-defined domains (e.g. visualizing goal structure and reducing the working memory load, providing declarative knowledge in the form of dictionaries or on-demand help etc). Furthermore, the ITS can simplify the process by performing one part of the task for the student automatically or by restricting the actions students can take. Furthermore, solution evaluation can be replaced with presenting consequences of student actions or supporting a related, but simpler task, e.g. peer review.

4.4.3 Authoring Domain Models: Constraint Granularity

Authoring domain models for constraint-based tutors consists of specifying syntax and semantic constraints. As discussed previously, constraint sets are different from production models, and generally easier to develop (Mitrovic et al. 2003). However, the process still requires a careful analysis of the target task.

Syntax constraints are generally easier to develop than semantic constraints, but they still require attention. The granularity of constraints is crucial for the effectiveness of the system. If the constraints are on a too coarse level, the feedback would be too general and not useful for the student. For example, we could have had only one constraint instead of a set of constraints such as those presented in Figure 4.1, which focus on the join conditions in the FROM clause. In that case, the only possible feedback message would be that there is something wrong with the FROM clause. Such feedback, of course, is not very useful.

Therefore, for constraints to be pedagogically effective, they need to focus on a very small aspect of a domain principle. There are potentially many constraints necessary to specify all the student should know about a single domain principle. The most important criterion in this process is the pedagogical importance: how specific is the feedback attached to a constraint?

4.4.4 Authoring Support for Constraint-Based Tutors

In addition to developing many constraint-based systems, we also devoted a lot of effort to providing authoring support (Mitrovic et al. 2007). WETAS (Martin and Mitrovic 2003) is the result of early research in this direction: it is an ITS shell that provides all the functionality necessary for a Web-based constraint tutor. In order to develop an ITS in WETAS, the author needs to provide the constraint set and the problems and their solutions. However, the development of a constraint set is still a demanding task. We developed ASPIRE (Mitrovic et al. 2006; Mitrovic et al. 2009), an authoring and deployment environment for constraint-based tutors, which supports the process of developing the constraint set. The author develops a domain ontology, and provides examples of problems with their solutions, from which ASPIRE generates constraints. ASPIRE is not capable of generating all constraints for any instructional domain, but the initial evaluation we performed shows that it is capable of generating the majority of necessary constraints (of the order of 90%). The quality and coverage of the developed constraint set, of course, depends critically on the quality of the author provided information (the ontology and problems/solutions). VIPER (Martin et al. 2009) further simplifies the authoring process by focusing on instructional domains with specific features, thus making the author's task easier.

4.5 Student Modeling

In the first paper on CBM, Ohlsson (1992) focused on short-term student modeling, or the diagnosis of the student's current action. The process starts by matching the relevance conditions of all constraints to the student solution. Then, for relevant constraints, the satisfaction conditions are matched as well. The same process can be applied incrementally, to isolated actions as the student is performing them, or to the whole solution. Therefore the short-term student model consists of the list of satisfied constraints and (potentially) the list of violated constraints. Violated constraints allow the constraint-based tutor to provide feedback to the student. The feedback states that the action/solution is wrong, points out the part of the solution which is wrong, and then specifies the domain principle that is violated. The error correction is left to the student to perform.

Feedback generation is only one of the pedagogical actions ITSs provide. Most often, feedback is generated on the basis of the last action the student performed, although previous performance can also be taken into account. However, ITSs also require long-term student model in order to generate other adaptive actions, such as selecting problems or topics to be covered in an instructional session. We therefore extended CBM by proposing several different ways of representing the long-term model of the student's knowledge.

In our early work (Mitrovic, 1998a, 1998b, 1998c), the long-term model of the student's knowledge was represented in terms of the overlay model. This is the logical extension of Ohlsson's initial proposal of CBM. A set of constraints represents what is true in the domain. Therefore the model of an expert would be equivalent to the whole constraint set, while a novice will only know some constraints. For each constraint the student has used, our tutors store the history of its usage, which allows us to track the student's progress on that constraint. Of course, over time the student's knowledge improves, and therefore the system cannot use the complete history always. A simple way to use such histories is to select a window of a particular size – say last five attempts on a constraint – and calculate the frequency of correct usage. This can be done for each constraint in the student model, and an estimate of the student's knowledge can be based on that. We have used such simple long-term models in most of our constraint-based tutors.

A more sophisticated approach is to develop a probabilistic, Bayesian model of the student's knowledge, as we have done for SQL-Tutor (Mayo and Mitrovic 2000) and CAPIT, a system that teaches elementary school children about punctuation and capitalization rules in English (Mayo and Mitrovic 2001). We also experimented with training an artificial neural network and using it for problem selection (Wang and Mitrovic 2002).

4.6 Pedagogy: What Can CBM Support?

CBM is neutral with respect to pedagogy. Ohlsson (1992) pointed out that CBM can be used offline, for diagnosing students' solution after each session, or online, to generate pedagogical actions. We have used CBM in our tutors with a variety of teaching strategies, to provide feedback to students, to select problems, support students' meta-cognitive skills and collaborative learning.

4.6.1 Feedback Generation in Constraint-Based Tutors

Constraint-based tutors use constraints to augment the student's declarative knowledge. If the student cannot detect errors by him/herself, the system will alert them to the domain principles which are violated. In this way, CBM can be used to provide feedback to students. Such feedback can be provided immediately, after each action the student performs, or in a delayed fashion, after the student is done with a problem. The choice of the timing of feedback is therefore flexible.

As stated previously, feedback is attached to constraints; when the student violates a constraint, the attached feedback message can be given to student. However, there are many additional considerations taken into account when presenting feedback, such as timing of feedback, content, type, presentation and adaptation.

Timing of feedback: In our tutors that teach design tasks, the student can decide when they want to get feedback. The solution is analyzed on student's request, and the whole solution is analyzed at once. The tutor does not diagnose individual student actions. Such an approach puts the student in control of their learning, while still providing necessary guidance when the student requests it. For procedural tasks, we typically break them into steps, and analyze a group of activities within a step. The student is required to complete one step correctly before going on to the next step. CBM can also be used to provide immediate feedback, by analyzing each action the student performs. The decision on the timing of feedback depends on the nature of the task performed: for design tasks it is more natural to diagnose the whole solution at once.

Amount of feedback: Another important pedagogical decision is related to feedback: how much information should be given to the student? Our tutors typically provide several levels of feedback. For example, SQL-Tutor offers the following feedback levels: *correct/incorrect, error flag, hint, all errors, partial solution,* and *complete solution*. On the first submission, the feedback only informs the student whether the solution is correct or not. The following submission points out the part of the solution which is incorrect; for example, the system might identify the FROM clause as being wrong. Such feedback is useful for the student to correct slips. If there is a deeper misunderstanding, the hint-level feedback will present the message attached to the violated constraint. If there are several violated constraints, the student can see the hint messages attached to them at the *All errors* level. The partial solution provides the correct version of the clause that was not right in the student's solution, while the full solution is available on the highest level of feedback. The system will automatically increase the feedback level on each submission until it reaches the *Hint* level; it will then stay on that level. The student, however, can ask for higher-level feedback whenever he/she desires.

Feedback content: The feedback messages we defined for early versions of our tutors were based on our intuition – we were thinking of what a good human tutor would say to the student violating a particular constraint. Such intuitive feedback does not have a theoretical foundation, and can result in feedback of variable quality. We therefore turned to the psychological theory of learning CBM was derived for. The theory says that effective feedback should tell the student here the error is, what constitutes the error and re-iterate the domain principle violated by the student. For example, the feedback message might say that the error is in the sum that the student computed, and point out that the sum is 93, but since the numbers are percentages, they should add up to 100.

We re-engineered feedback for EER-Tutor with these theoretical guidelines in mind (Zakharov et al. 2005). As an example, the original feedback message attached to constraint 23 was: "*Check whether each identifying relationship has an owner entity, which must be a regular entity type.*" The new, theory-based feedback for the same constraint is: "*An identifying relationship type must be connected to a regular entity type, which is the owner of the weak entity type. The highlighted identifying relationship is not connected to a regular entity type.*" When this feedback is given to the student, the incorrect construct (i.e. the identifying relationship) is highlighted in the diagram. The results of evaluation show that theoretical feedback is more effective in supporting learning than intuitive one, by increasing the learning rate.

Types of feedback: Feedback messages attached to constraints are feedback on errors – we refer to such feedback as *negative feedback*. Most feedback provided by ITSs is of this type. However, human teachers very often provide positive feedback, i.e. feedback on correct actions. Such feedback is useful as it confirms tentative actions and supports students in strengthening their knowledge. Positive feedback also helps the student to integrate newly acquired with existing knowledge. We developed a version of SQL-Tutor which provided positive feedback in addition to negative (Barrow et al. 2008). The content of positive feedback acknowledges the correct action, and re-iterates the domain principle that was satisfied by it. However, positive feedback is not given on each submission, as that would be overwhelming and repetitive. On the contrary, the system decides when to provide positive feedback, looking for evidence that the student was uncertain, but still managed to solve the problem (or a problem step). Other situations when positive feedback is provided is when he student learns a difficult constraint, uses the constraint correctly for the first time, or solves a difficult problem. The study has shown that students who received positive feedback solved the same number of problems and learnt the same amount of knowledge as the students in the control group but in half the time of the control group

Feedback presentation: if the student violated several constraints, the system needs to decide which constraints to target, and in which order. The simplest solution is to order the constraints within the domain model and use this order to present feedback, as we have done in early versions of SQL-Tutor. However, the ordering is difficult to implement. It is also possible to select constraints adaptively, on the basis of the student model, as we have done in several systems. Other researchers have suggested similar approaches, for example adding weights to constraints and selecting constraints with the highest weights (Le et al. 2009). Furthermore, constraints can be organized in terms of domain concepts, as done in ASPIRE (Mitrovic et al. 2009) and also in the context of EER-Tutor, when deciding on the tutorial dialogue to engage the student in (Weerasinghe et al. 2009).

Adapting feedback: Feedback provided on violated constraints corresponds to the current solution; however, if two students submit exactly the same solution, they would get the same feedback. For that reason, we enhanced SQL-Tutor to adapt the feedback to the particular student by changing the generality of feedback in relation to the student model (Martin and Mitrovic 2006).

4.6.2 CBM and Problem Selection

CBM also supports problem selection; as stated in the previous section, the long-term model stores the summary of the student's progress on a skill. Numerous problem-selection strategies can be formulated on the basis of such student models. In our early work, we started with a simple problem-selection strategy which focused on a single constraint that the student has most difficulty learning. Such a constraint can be easily identified from the long-term student model, and can guide the selection of the problem. For example, in early versions of SQL-Tutor (Mitrovic 1998a, 1998b, 1998c) the system identified the most often violated constraint, and then selected a problem which was new to the student, at the appropriate level of difficulty[2] (based on the student model) which exercised the chosen constraints. We later used decision theory and the probabilistic student model to select the best problem for the student (Mayo and Mitrovic 2000, 2001). In another version of SQL-Tutor, we trained an artificial neural network to predict the problem which will be at the right level of complexity for the student (Wang and Mitrovic 2002). Later on, we experimented with computing the problem difficulty dynamically, in relation to the student model; the corresponding problem-selection strategy proved to be superior to the one based on the static problem complexities (Mitrovic and Martin 2004). Finally, we also introduced problem templates and presented them to students, during problem selection (Mathews and Mitrovic 2007).

4.6.3 Supporting Higher-Level Skills and Affect with CBM

Constraints represent the domain knowledge, and are used as the basis for representing students' knowledge. We also explored the effect of opening the student model to the student on their higher-level skills, such as self-assessment. In a series of studies done in the context of SQL-Tutor (Mitrovic and Martin 2007), we have shown that students improve their self-assessment skills when having access to relatively simple open student models, in the form of skill meters. Although the student model was presented in a highly abstracted way, it helped student reflect on their knowledge and select more appropriate problems for further work.

Another important higher-level skill is self-explanation, which is known to promote deep learning (Chi 2000). In the studies with our database design tutor (Weerasinghe and Mitrovic 2006, 2008), the students participated in tutorial dialogues aimed at elicitating explanations from students. Such explanations enable students to relate their problem-solving skills to declarative knowledge (Mitrovic 2005).

Constraints can also be used to represent collaborative skills. COLECT-UML supports teams of students developing an UML diagram collaboratively (Baghaei et al. 2007). The system provides domain-level feedback based on the analysis of individual and group solutions. In addition, it also provides feedback on collaboration by analyzing the student's participation in group activities. The model of ideal

---

[2] Each problem in SQL-Tutor has a static problem complexity level specified by the teacher (ranges from 1 to 9).

collaboration was presented in the form of a set of meta-constraints. The system was successful in supporting both students' learning of domain knowledge and collaborative skills.

A lot of research has been done during the last decade on recognizing the student's affective state and responding to it actively. We developed an animated pedagogical agent which is capable of identifying the trend in which the student's affective state is changing. The system analyzes student's facial features and identifies changes from a positive to negative state or vice versa. This information is then combined with the information about the cognitive state, and used to modify the feedback provided to the student and also the agent's behaviour (Zakharov et al. 2008). Although there is a lot of research questions still to be answered, our initial experiences have been very positive. The students appreciated getting feedback from an affect-sensitive agent.

4.7 Conclusions

In the last fifteen years, CBM has grown from a theoretical proposal to a fully developed, mature methodology for building ITSs. We have used CBM successfully in many instructional domains, with students of different ages and backgrounds, in real classrooms at universities and in schools. Additionally, three of our database tutors have been available to students worldwide via the Addison-Wesley's DatabasePlace[3] Web portal since 2003, and have been used by more than 10,000 students worldwide. We have performed more than 30 evaluation studies, which proved the effectiveness of this modeling approach.

Constraint-based tutors, as discussed above, do not require runnable expert models in order to diagnose student solutions; this feature enables CBM to be applicable in ill-defined tasks/domains, where model-tracing tutors cannot be applied. Constraints can capture whatever is known about the ill-defined domain and the problem specification, thus begin able to evaluate the mandatory parts of the solution. Such a tutor can provide feedback to student, while still allowing for multiple solutions differing in non-essential elements, such as aesthetical and personal preferences.

CBM also does not require bug libraries and consequently constraint-based tutors require less effort than model-tracing ones CBM has a strong theoretical foundation, which provides guidelines for generating feedback content.

CBM is a flexible approach, but it is not the only possible way of developing ITSs. Many questions have been answered, but many more are still open. Our current work focuses on improving authoring support, modeling affect and student engagement, as well as meta-cognitive skills. We do not believe that only a single representation (constraint or production rules) is superior in all respects and situations. Hybrid systems, using combinations of different representations have a much higher probability of supporting different kinds of instructions and different learners.

[3] www.databaseplace.com

Acknowledgments. The work reported here could not have been done without the wonderful bunch of students and colleagues at the Intelligent Computer Tutoring Group (ICTG). Thank you all for all the discussions and friendship over the years – I have been privileged to work with you.

References

Anderson, J.R., Boyle, C.F., Corbett, A.T., Lewis, M.W.: Cognitive Modeling and Intelligent Tutoring. Artificial Intelligence 42, 7–49 (1990)

Anderson, J.R., Corbett, A.T., Koedinger, K.R., Pelletier, R.: Cognitive Tutors: Lessons Learned. Learning Sciences 4(2), 167–207 (1995)

Baghaei, N., Mitrovic, A., Irvin, W.: Problem-Solving Support in a Constraint-based Intelligent Tutoring System for UML. Technology, Instruction, Cognition and Learning 4(2), 113–137 (2006)

Baghaei, N., Mitrovic, A., Irwin, W.: Supporting collaborative learning and problem-solving in a constraint-based CSCL environment for UML class diagrams. Computer-Supported Collaborative Learning 2(2-3), 159–190 (2007)

Barrow, D., Mitrovic, A., Ohlsson, S., Grimley, M.: Assessing the Impact of Positive Feedback in Constraint-based ITSs. In: Woolf, B., et al. (eds.) ITS 2008. LNCS, vol. 5091, pp. 250–259. Springer, Heidelberg (2008)

Chi, M.T.H.: Self-explaining Expository Texts: The dual processes of generating inferences and repairing mental models. Advances in Instructional Psychology, 161–238 (2000)

Galvez, J., Guzman, E., Conejo, R., Millan, E.: Student Knowledge Diagnosis using Item Response Theory and Constraint-based Modeling. In: Dimitrova, V., Mizoguchi, R., du Boulay, B., Graesser, A. (eds.) Proc. 14th Int. Conf. Artificial Intelligence in Education, pp. 291–298 (2009)

Koedinger, K.R., Anderson, J.R., Hadley, W.H., Mark, M.A.: Intelligent Tutoring Goes to the Big City. Artificial Intelligence in Education 8, 30–43 (1997)

Le, N.-T., Menzel, W., Pinkwart, N.: Evaluation of a Constraint-Based Homework Assistance System for Logic Programming. In: Kong, S.C., Ogata, H., Arnseth, H.C., Chan, C.K.K., Hirashima, T., Klett, F., Lee, J.H.M., Liu, C.C., Looi, C.K., Milrad, M., Mitrovic, A., Nakabayashi, K., Wong, S.L., Yang, S.J.H. (eds.) Proc. 17th Int. Conf. Computers in Education, APSCE (2009)

Martin, B., Mitrovic, A.: Domain Modeling: Art or Science? In: Hoppe, U., Verdejo, F., Kay, J. (eds.) Proc. 11th Int. Conference on Artificial Intelligence in Education AIED. IOS Press, Amsterdam (2003)

Martin, B., Mitrović, A.: The Effect of Adapting Feedback Generality in ITSs. In: Wade, V., Ashman, H., Smyth, B. (eds.) AH 2006. LNCS, vol. 4018, pp. 192–202. Springer, Heidelberg (2006)

Martin, B., Kirkbride, T., Mitrovic, A., Holland, J., Zakharov, K.: An Intelligent Tutoring System for Medical Imaging. In: Bastiaens, T., Dron, J., Xin, C. (eds.) Proc. World Conf. E-Learning in Corporate, Government, Healthcare, and Higher Education. AACE, Vancouver, CA (2009)

Mathews, M., Mitrovic, A.: Investigating the Effectiveness of Problem Templates on Learning in ITSs. In: Luckin, R., Koedinger, K., Greer, J. (eds.) Proc. Artificial Intelligence in Education, pp. 611–613 (2007)

Mayo, M., Mitrovic, A.: Using a Probabilistic Student Model to Control Problem Difficulty. In: Gauthier, G., VanLehn, K., Frasson, C. (eds.) ITS 2000. LNCS, vol. 1839, p. 524. Springer, Heidelberg (2000)

Mayo, M., Mitrovic, A.: Optimising ITS Behaviour with Bayesian Networks and Decision Theory. Artificial Intelligence in Education 12(2), 124–153 (2001)

Menzel, W.: Constraint-based Modeling and Ambiguity. Artificial Intelligence in Education 16(1), 29–63 (2006)

Milik, N., Marshall, M., Mitrović, A.: Teaching Logical Database Design in ERM-Tutor M. In: Ikeda, M., Ashley, K., Chan, T.-W. (eds.) ITS 2006. LNCS, vol. 4053, pp. 707–709. Springer, Heidelberg (2006)

Mitrovic, A.: Learning SQL with a Computerized Tutor. In: 29th ACM SIGCSE Technical Symposium, pp. 307–311 (1998a)

Mitrovic, A.: A Knowledge-Based Teaching System for SQL. In: Ottmann, T., Tomek, I. (eds.) Proc. ED-MEDIA 1998, AACE, pp. 1027–1032 (1998b)

Mitrovic, A.: Experiences in Implementing Constraint-Based Modeling in SQL-Tutor. In: Goettl, B., Halff, H., Redfield, C., Shute, V. (eds.) ITS 1998. LNCS, vol. 1452, pp. 414–423. Springer, Heidelberg (1998c)

Mitrovic, A.: An Intelligent SQL Tutor on the Web. Artificial Intelligence in Education 13(2), 173–197 (2003)

Mitrovic, A.: The Effect of Explaining on Learning: a Case Study with a Data Normalization Tutor. In: Looi, C.-K., McCalla, G., Bredeweg, B., Breuker, J. (eds.) Proc. Conf. Artificial Intelligence in Education, pp. 499–506 (2005)

Mitrović, A., Martin, B.: Evaluating Adaptive Problem Selection. In: De Bra, P., Nejdl, W. (eds.) AH 2004. LNCS, vol. 3137, pp. 185–194. Springer, Heidelberg (2004)

Mitrovic, A., Martin, B.: Evaluating the Effect of Open Student Models on Self-Assessment. Artificial Intelligence in Education 17(2), 121–144 (2007)

Mitrovic, A., Koedinger, K., Martin, B.: A Comparative Analysis of Cognitive Tutoring and Constraint-based Modeling. In: Brusilovsky, P. Corbett, A., de Rosis, F. (eds.) UM 2003. LNCS (LNAI), vol. 2702, pp. 313–322. Springer, Heidelberg (2003)

Mitrovic, A., Martin, B., Suraweera, P.: Intelligent Tutors for all: Constraint-based Modeling Methodology, Systems and Authoring. IEEE Intelligent Systems 22(4), 38–45 (2007)

Mitrovic, A., Martin, B., Suraweera, P., Zakharov, K., Milik, N., Holland, J., McGuigan, N.: ASPIRE: an Authoring System and Deployment Environment for Constraint-based Tutors. Artificial Intelligence in Education 19(2), 155–188 (2009)

Mitrovic, A., McGuigan, N., Martin, B., Suraweera, P., Milik, N., Holland, J.: Authoring Constraint-based Systems in ASPIRE: A Case Study of a Capital Investment Tutor. In: Proc. ED-MEDIA 2008, pp. 4607–4616 (2008)

Mitrovic, A., Ohlsson, S.: Evaluation of a Constraint-Based Tutor for a Database Language. Artificial Intelligence in Education 10(3-4), 238–256 (1999)

Mitrovic, A., Suraweera, P., Martin, B., Weerasinghe, A.: DB-suite: Experiences with Three Intelligent, Web-based Database Tutors. Journal of Interactive Learning Research 15(4), 409–432 (2004)

Mitrovic, A., Suraweera, P., Martin, B., Zakharov, K., Milik, N., Holland, J.: Authoring Constraint-based Tutors in ASPIRE. In: Ikeda, M., Ashley, K., Chan, T.-W. (eds.) ITS 2006. LNCS, vol. 4053, pp. 41–50. Springer, Heidelberg (2006)

Mitrovic, A., Weerasinghe, A.: Revisiting the Ill-Definedness and Consequences for ITSs. In: Dimitrova, V., Mizoguchi, R., du Boulay, B., Graesser, A. (eds.) Proc. 14th Int. Conf. Artificial Intelligence in Education (2009)

Oh, Y., Gross, M.D., Ishizaki, S., Do, Y.-L.: Constraint-based Design Critic for Flat-pack Furniture Design. In: Kong, S.C., Ogata, H., Arnseth, H.C., Chan, C.K.K., Hirashima, T., Klett, F., Lee, J.H.M., Liu, C.C., Looi, C.K., Milrad, M., Mitrovic, A., Nakabayashi, K., Wong, S.L., Yang, S.J.H. (eds.) Proc. 17th Int. Conf. Computers in Education, APSCE (2009)

Ohlsson, S.: Constraint-based Student Modeling. Artificial Intelligence in Education 3(4), 429–447 (1992)

Ohlsson, S.: Learning from performance errors. Psychological Review 103, 241–262 (1996)

Ohlsson, S., Bee, N.: Strategy Variability: A Challenge to Models of Procedural Learning. In: Birnbaum, L. (ed.) Proc. Int. Conf. of the Learning Sciences, AACE, pp. 351–356 (1991)

Ohlsson, S., Mitrovic, A.: Fidelity and Efficiency of Knowledge representations for intelligent tutoring systems. Technology, Instruction, Cognition and Learning 5(2), 101–132 (2007)

Petry, P.G., Rosatelli, M.: AlgoLC: A Learning Companion System for Teaching and Learning Algorithms. In: Ikeda, M., Ashley, K.D., Chan, T.-W. (eds.) ITS 2006. LNCS, vol. 4053, pp. 775–777. Springer, Heidelberg (2006)

Riccucci, S., Carbonaro, A., Casadei, G.: An Architecture for Knowledge Management in Intelligent Tutoring Systems. In: Proc. IADIS Int. Cong. Cognition and Exploratory Learning in Digital Age (2005)

Rosatelli, M., Self, J.: A Collaborative Case Study System for Distance Learning. Artificial Intelligence in Education 14(1), 1–29 (2004)

Siddappa, M., Manjunath, A.S.: Intelligent Tutor Generator for Intelligent Tutoring Systems. In: Proc. World Congress on Engineering and Computer Science, pp. 578–583 (2008)

Suraweera, P., Mitrovic, A.: KERMIT: A Constraint-Based Tutor for Database Modelling. In: Cerri, S.A., Gouardéres, G., Paraguaçu, F. (eds.) ITS 2002. LNCS, vol. 2363, pp. 376–387. Springer, Heidelberg (2002)

Suraweera, P., Mitrovic, A.: An Intelligent Tutoring System for Entity Relationship Modelling. Artificial Intelligence in Education 14(3-4), 375–417 (2004)

Wang, T., Mitrovic, A.: Using neural networks to predict student's behaviour. In: Kinshuk, R., Lewis, K., Akahori, R., Kemp, T., Okamoto, L., Henderson, C.-H. (eds.) Proc. Int. Conf. Computers in Education, pp. 969–973 (2002)

Weerasinghe, A., Mitrovic, A.: Facilitating Deep Learning through Self-Explanation in an Open-ended Domain. Int. J. of Knowledge-based and Intelligent Engineering Systems 10(1), 3–19 (2006)

Weerasinghe, A., Mitrovic, A.: A Preliminary Study of a General Model for Supporting Tutorial Dialogues. In: Proc. Int. Conf. Computers in Education, pp. 125–132 (2008)

Weerasinghe, A., Mitrovic, A., Martin, B.: Towards individualized dialogue support for ill-defined domains. Artificial Intelligence in Education 14 (2009) (in print)

Zakharov, K., Mitrovic, A., Johnston, L.: Towards Emotionally-Intelligent Pedagogical Agents. In: Woolf, B.P., Aïmeur, E., Nkambou, R., Lajoie, S. (eds.) ITS 2008. LNCS, vol. 5091, pp. 19–28. Springer, Heidelberg (2008)

Zakharov, K., Mitrovic, A., Ohlsson, S.: Feedback Micro-Engineering in EER-Tutor. In: Looi, C.-K., McCalla, G., Bredeweg, B., Breuker, J. (eds.) Proc. 12th Int. Conf. Artificial Intelligence in Education, pp. 718–725. IOS Press, Amsterdam (2005)

Chapter 5
Building Intelligent Tutoring Systems for Ill-Defined Domains

Philippe Fournier-Viger[1], Roger Nkambou[1], and Engelbert Mephu Nguifo[2]

[1] Department of Computer Sciences, University of Quebec at Montreal,
201President-Kennedy Avenue, Montreal, Canada, H2X 3Y7
fournier_viger.philippe@courrier.uqam.ca,
nkambou.roger@uqam.ca

[2] Department of Mathematics and Computer Sciences,
Université Blaise-Pascal Clermont 2, BP 125, 63173 Aubière, cedex, France
mephu@isima.fr

Abstract. Domains in which traditional approaches for building tutoring systems are not applicable or do not work well have been termed "ill-defined domains." This chapter provides an updated overview of the problems and solutions for building intelligent tutoring systems for these domains. It adopts a presentation based on the following three complementary and important perspectives: the characteristics of ill-defined domains, the approaches to represent and reason with domain knowledge in these domains, and suitable teaching models. Numerous examples are given throughout the chapter to illustrate the discussion.

5.1 Introduction

In recent years more and more research has focused on building Intelligent Tutoring Systems (ITS) for domains that pose new challenges, i.e., where traditional approaches for building tutoring systems are not applicable or do not work well. Such domains have been termed "ill-defined domains" by the Artificial Intelligence in Education (AIED) community (Aleven 2003; Aleven et al. 2006; Aleven et al. 2007). Research on ill-defined domains has given rise to several solutions. Some are domain-specific, while others tend to be more generic. Despite the numerous papers published on this topic and the three workshops presented at recent conferences (Aleven 2003; Aleven et al. 2006; Aleven et al. 2007). the definition of an "ill-defined domain" is still under debate and none of the proposed solutions are appropriate for all domains (Aleven 2003; Aleven et al. 2006; Aleven et al. 2007). This chapter aims to clarify the definition of an "ill-defined domain," to offer various solutions for building tutoring systems for these domains, and to provide insight into current research. It provides an updated overview of the research on ill-defined domains, the previous synopsis having been published in 2006 (Lynch et al. 2006).

R. Nkambou et al. (Eds.): Advances in Intelligent Tutoring Systems, SCI 308, pp. 81–101.
springerlink.com

The chapter is organized into four main sections. Sections 2, 3 and 4 correspond to three complementary and important perspectives that need to be considered to build ITSs for ill-defined domains. Section 2 addresses the characteristics of ill-defined domains, which is important to identify in order to categorize solutions for domains with similar features. Sections 3 and 4 next discuss the approaches to represent and reason with domain knowledge in these domains and suitable teaching models. As will later be explained, some approaches and models are more appropriate for certain types of ill-defined domains, which ease considerably the task of building ITSs. Lastly, Section 5 provides a detailed case study of an ITS called CanadarmTutor. This case study illustrates the advantages and limitations of several approaches discussed in the chapter.

5.2 What Is an Ill-Defined Domain?

An "ill-defined domain" or "ill-structured domain" in the context of ITS research is a domain in which traditional approaches for building ITSs are not applicable or do not work well (Aleven 2003; Aleven et al. 2006; Aleven et al. 2007). A domain is ill-defined because its structure or content makes it less suitable for supporting tutoring services. It is important to note that a complex domain is not necessarily an ill-defined domain, although many ill-defined domains are complex. A domain may be complex because it contains numerous knowledge elements and/or relations, and is still well-defined. An example from the field of geography is the name of each country and its capital. This domain is complex because there are hundreds of capitals. However, it is a well-structured domain because the knowledge can be represented simply as a list of pairs.

It is also important to note that the notion of an ill-defined domain is vague. Indeed, there are no clear boundaries between ill-defined and well-defined domains. Rather, there is a continuum ranging from well-defined to ill-defined. To identify ill-defined domains, to compare domains on the basis of their ill-definedness, and to categorize the strategies and approaches for supporting tutoring services in these domains, we first need to identify what characterizes ill-defined domains. The following section addresses this issue.

5.2.1 Characteristics of Ill-Defined Domains

To provide a working definition of an ill-defined domain, Lynch et al. (Aleven et al. 2008) did an extensive review of the research in the field of ITS and of the literature on "ill-structured problems" in artificial intelligence and decision-making in uncertain conditions. Lynch et al. concluded that domains having one or more of the following characteristics are ill-defined (Lynch et al. 2006):

(1) Multiple and controversial solutions: Domains having problems with many controversial solutions and no clear procedures for evaluating a solution are ill-defined. An example of such a domain is the design of entity relationship diagrams from text descriptions. There is a potentially infinite number of diagrams with respects to a description, but the evaluation process of a diagram is partly subjective. A second example is law argumentation. Many legal arguments can be proposed for a legal case, yet there is not a single right solution even though some solutions may be preferable, depending on certain aspects, such as aesthetics and their success in a courtroom (Lynch et al. 2006). A third example is the domain of ethics. Ethical problems by definition have no right answer.

(2) No complete formal domain theory: Domains that do not have a clear or complete domain theory for determining a problem's outcome and testing its validity are ill-defined. For example, in the domains of music composition and architecture, there are only incomplete theories. In contrast, in the well-defined domain of geometry, there is a single theory applicable to all geometry problems.

(3) Ill-defined task structure: From the perspective of task structure, three domain types have been identified by Lynch et al. The first two are ill-defined. First, "design domains" contain tasks involving the design of new artefacts (e.g., writing a story and composing music). These domains are ill-defined as the goal is novelty, although some general rules or principles can provide guidance. The second main domain type is "analytical domains." Analytical tasks typically require performance analyses of incomplete and potentially incorrect information regarding a changing environment in order to make decisions. For this reason, these domains are also said to be ill-defined. Two examples of analytical tasks are stock trading and medical diagnoses. The third domain type is "problem-solving domains." These domains involve applying a formal theory to solve problems having a clear and complete definition in order to obtain a definite and verifiable answer. Such domains are considered well-defined. Most mathematical problems that do not involve the construction of new knowledge are examples of this domain type (e.g., calculating the volume of a sphere given its radius).

(4) Open-textured concepts: "Open-textured concepts" are abstract concepts that are partially undetermined or do not have absolute definitions. They are problematic when they need to be applied in concrete situations to carry out tasks. Domains, including open-textured concepts, are ill-defined. They include most domains that rely on natural language because words and sentences can have ambiguous meanings. Another example is the domain of law. In this domain, many domain concepts are abstract and can have several interpretations.

(5) Overlapping sub-problems: Domains having complex problems that cannot be divided into smaller independent sub-problems that are easier to solve are also said to be ill-defined. A general example is the problem of building a house. One must chose an appropriate site for construction and a plan for building the house. Since these two tasks are dependent, they cannot be planned separately without compromising the success of the whole task.

5.2.2 Ill-Definedness of Tasks and Domains

Recently, Mitrovic & Weerasinghe (2009) argued that ITS researchers need to consider two dimensions in ill-defined domains: "tasks" and "domains." They stated that both can be "ill-defined" or "well-defined" resulting in four different combinations of types of tasks/domains. A domain is viewed by Mitrovic & Weerasinghe as declarative domain knowledge or a domain theory that can be used in tasks (Ashley et al. 2002). For example, in the task of writing a story, the domain includes storytelling notions, whereas the task is to write a story.

However, Mitrovic & Weerasinghe did not provide the means to identify an example of an ill-defined domain containing well-defined tasks (Mitrovic and Weerasinghe 2009). Therefore, we argue that the domain dimension is debatable, especially from the point of view of ITS researchers, because the main goal of building an ITS is to provide support to the learner at the level of individual tasks (Woolf 2009). In fact, an ITS designer simply needs to consider the tasks and the subset of directly relevant domain knowledge in order to select appropriate techniques to provide tutoring services. Moreover, since a domain can contain both ill-defined and well-defined tasks, it does not seem relevant to discuss the ill-definedness of a domain as a whole. For example, this is the case for the domain of software engineering, in which an ITS could offer well-defined multiple-choice questions about the properties of UML diagrams, in addition to some ill-defined problems such as designing UML diagrams.

We believe that the confusion of "task" versus "domain" in the proposal of Mitrovic & Weerasinghe originates from the choice of the term "ill-defined domain" in the ITS context by Lynch et al. However, Lynch et al. (Lynch et al. 2006) are very clear to use the word "domain" instead of "problem" simply "to emphasize that the end goal of tutoring is typically general domain knowledge (...), not problem specific answers" and that the distinction between domains and problems "is immaterial." We therefore suggest considering only if a task is ill-defined and in what way it is ill-defined (including the domain knowledge directly used in the task) when choosing domain knowledge modelling and reasoning techniques for supporting tutoring services.

On the other hand, choosing appropriate tasks for teaching a domain of knowledge is, we believe, the responsibility of educational and domain experts. However, one should be aware that some ITS techniques are more appropriate for certain types of tasks and teaching models than others. For this reason, we suggest that educational/domain experts should ask ITS experts to evaluate the feasibility of supporting tutoring services for a given task and of a teaching model before deciding if it should be included in an ITS. Examples of teaching models that can be used in ill-defined domains are presented in Section 4.

5.2.3 Characteristics of Ill-Defined Tasks

To describe how tasks are ill-defined, we suggest using the definition of Simon (Simon 1978) as a complement to the definition of Lynch et al. The definition of Simon is based on the study of human problem-solving and on the research of building artificial problem-solvers. This definition is relevant to the ITS context because solving any exercise offered by an ITS can be viewed as a problem-solving task. Simon stated that a problem is ill-structured if it possesses one or more of the following features (Simon 1978):

(1) The indefinite starting point: The instructions or information necessary for solving the problem are incomplete or vague.
(2) The indefinite ending point: The criterion that determines if the goal has been attained is complex and imprecise. There could be multiple and controversial solutions (Aleven et al. 2008).
(3) Unclear strategies for finding solutions: There are no clear strategies for finding solutions at each step of the problem-solving process.
This definition is formulated in general terms and can thus cover a wide variety of domains. It is also consistent with the definition by Lynch et al. (whose definition was partly inspired by that of Simon) (Lynch et al. 2006). In the next section, we use both definitions to present the principal approaches to represent and reason with domain knowledge in ill-defined domains.

5.3 Representing and Reasoning on Domain Knowledge in Ill-Defined Domains

There are three traditional approaches for representing and reasoning on domain knowledge in problem-solving-based intelligent tutoring systems. This section reviews these approaches and highlights their limitations for ill-defined domains. It then presents two additional approaches for ill-defined domains.

5.3.1 Cognitive Approach and Model-Tracing

The first traditional approach for representing and reasoning on domain knowledge is the cognitive approach and model-tracing (MT). MT-tutors are generally built from cognitive task analysis. The process of cognitive task analysis consists of producing effective problem spaces or task models by observing expert and novice users (Koedinger et al. 1997) using different solving problems strategies.

A task model can be designed for a problem or a problem set. Task models are usually represented as sets of production rules (sometimes structured as a goal-decomposition tree (Woolf 2009)) or as state spaces in which each rule or transition corresponds to an action or an operation to perform a task. Some of the rules/transitions can be tagged as "buggy" or annotated with hints or other didactic information. The most well-known examples of model-tracing tutors are cognitive tutors (Koedinger et al. 1997) (see also chapter 3), which encode the operations for solving a problem as a set of production rules When a learner performs a task with a cognitive tutor, the latter follows the learner's reasoning by analyzing the rules being applied. This process is called model-tracing The MT paradigm is beneficial because the reasoning processes of the learner can be represented in great detail (in fact, authors of cognitive tutors claim to model the cognitive processes of the learner), and the models obtained can support a wide variety of tutoring services, such as: (1) suggesting to the learner the next steps to take, (2) giving demonstrations; (3) evaluating the knowledge that the learner possesses in terms of the skills that are applied; and (4) inferring learner goals.

Model-tracing tutors are recommended for tasks in which the goal is to evaluate the reasoning process rather than simply determining if the learner attained the correct solution. MT-Tutors generally assume that the starting and ending points of a problem are definite. The main limitation of model-tracing with respect to ill-defined domains is that, for some domains, there are no clear strategies for finding solutions, and it can therefore be difficult to define an explicit task model. Moreover, for complex domains, one would need to determine a large number of rules and solution paths, and designing a set of rules or a state space for a task would be very time-consuming.

A strategy for using the MT paradigm in ill-defined domains is to describe only the well-defined part of a task with a task model (Lynch et al. 2006). This strategy is used, for example, in CanadarmTutor, an ITS used to teach astronauts how to operate a robotic arm [11]. Another strategy for enabling the MT to operate ITSs in ill-defined domains is to use an iterative approach to design task models (Ogan et al. 2006). Initially, a rough model is created by domain experts. The model is then evaluated empirically several times to see if it correctly predicts user-behavior. It is subsequently adjusted until a satisfactory model is attained. This strategy was used to build an ITS which enables the learner to distinguish verb tenses in the French language (Ogan et al. 2006).

5.3.2 Constraint-Based Modeling Approach

The second approach for representing and reasoning on domain knowledge is constraint-based modeling (CBM) (Mitrovic et al. 2007; Mitrovic and weerasinghe 2009). This approach is thoroughly described in chapter 4. It consists of specifying sets of constraints on what is a correct behavior or solution rather than to provide an explicit task model. When the learner violates a constraint during a task, the CBM Tutor diagnoses that an error has been made and provides help to the learner regarding the violated constraint.

Unlike MT-Tutors, CBM-Tutors do not support tutoring services such as to present demonstrations or suggest the next steps to perform to the learner. This is one of the principal limitations of the CBM approach. CBM is recommended for domains in which the goal is to validate states/solutions regardless of the strategies a learner uses to provide solutions (it can be applied to domains in which there are no clear strategies). However, in order for CBM to be applicable, one needs to define relevance and satisfaction conditions that characterize optimal states/solutions for a task. But, designing and selecting a set of constraints is not always easy. Moreover, for some tasks, states/solutions sometimes do not provide enough information to permit the specification of a set of relevant constraints, and, in some cases, a large number of constraints is required because there are too many diverse solutions (Kodaganallur et al. 2006). Another limitation of CBM is that CBM-Tutors do not take into account the solution path leading to a constraint violation. This limitation can have two negative implications. First, the help generated may be inadequate, especially when a solution path differs significantly from the ideal solutions (Woolf 2009). Second, if the reasoning that led to the solution is not evaluated, the CBM-Tutor may be unable to distinguish that a learner may have intentionally or unintentionally answered a question correctly.

Despite these limitations, CBM has been applied successfully to some ill-defined domains, in particular, to problem-solving and design tasks (see (Mitrovic et al. 2007) for an overview of CBM applications). An example of a CBM-Tutor is KERMIT/EER-Tutor, an ITS used to teach design entity relationship diagrams (Mitrovic et al. 2007). Designing such diagrams is an ill-defined task because it involves creativity (it is a design task), there is no clear strategy for performing it, problems statements are generally ambiguous and for one problem there can be many acceptable and controversial solutions. When applying CBM in KERMIT, approximately 100 constraints were defined and KERMIT's interface was designed to restrict the learner to select terms from problem descriptions to name diagram elements. The latter restriction greatly reduces the scope of possible solutions. It also illustrates a popular design strategy for building ITSs in ill-defined domains, in general, i.e., forcing a structure into a domain so as to transform it into a better defined form. However, a drawback of this strategy, from a pedagogical perspective, is that the learner eventually has to learn to perform tasks in less-structured problem-solving environments (Moritz and Blank 2008).

5.3.3 The Expert System Approach

The third approach for representing and reasoning on domain knowledge consists of integrating an expert system in an ITS (Clancey 1984; Graesser et al. 2000; Kabanza et al. 2005; Moritz and Blank 2008). This approach is very broad because many forms of expert systems can be used, such as rule-based, neural networks, decision trees and case-based reasoning systems. The advantage of this approach is that tailored expert systems are particularly well-suited for some domains, unlike CBM and MT that are general approaches. There are two principal and complimentary means of using an expert system in an ITS.

First, an expert system can be used to generate expert solutions. The ITS can then compare these solutions with learner solutions. It can subsequently use them as demonstrations or to suggest to the learner the problem-solving steps he/she should take. For example, this approach was used in GUIDON (Clancey 1984), an ITS used to teach the learner how to diagnose infectious diseases based on a patient's history of clinical tests. GUIDON relies on MYCIN, a rule-based expert system containing approximately 500 rules to generate expert solutions. These solutions are then compared with the learner's solutions in order to diagnose mistakes, generate demonstrations, and produce mixed initiative dialogues. MYCIN is particularly well-suited for use in an ITS, as the rules in its domain are meaningful to the learner. GUIDON uses this property extensively to present rules to the learner that support or refute his/her reasoning.

The second principal means for using an expert system in an ITS is to compare ideal solutions with learner solutions (Clancey 1984, Graesser et al. 2000). The first example is AutoTutor, (Graesser et al. 2000) an ITS which was applied to several domains, including Newtonian physics and computer literacy. AutoTutor teaches by conducting conversations with learners in natural language. To evaluate the student's natural language answers, AutoTutor uses Latent Semantic Analysis (LSA) to compute the semantic distance with the expected answers. LSA is a black-box technique in the sense that it cannot explain its "reasoning," in contrast to expert systems such as MYCIN. Despite this limitation, it is very effective in assessing natural language answers.

The second example of an ITS that relies on an expert system to compare learner solutions with ideal solutions is DesignFirst-ITS (Moritz and Blank 2008), an ITS which assists the learner in designing UML diagrams from text descriptions. The design of UML diagrams is an ill-defined task because it is a design task, the solution space is very large, there is no right solution, problem statements are ambiguous, and it uses natural language (Moritz and Blank 2008). The uniqueness of DesignFirst-ITS is that it offers an almost completely unconstrained problem-solving environment with the learner, unlike other ITS for UML, such as Collect-UML, which relies on CBM (Moritz and Blank 2008). DesignFirst-ITS evaluates the step by step construction of UML diagrams by comparing them with templates of acceptable solutions. The matching process is not insignificant as it searches for synonyms, spelling errors, adjective use, etc.

In all of these cases, the expert systems approach provides advanced tutoring services that would be hard to offer for the same domains with the MT or CBM approaches. However, the limitations of the expert system approach are the following: (1) developing or adapting an expert system can be costly and difficult, especially for ill-defined domains; and (2) some expert systems cannot justify their inferences, or provide explanations that are appropriate for learning.

5.3.4 The Partial Task Modeling Approach

The three aforementioned classical approaches for representing and reasoning on domain knowledge can be very time consuming and difficult to apply in some ill-defined domains. As an alternative to these approaches, a promising approach for ill-defined domains is to apply data-mining or machine-learning techniques to automatically learning partial task models from user solutions. The rationale for this approach is that a partial task model could be a satisfactory alternative to an exhaustive task model in domains in which a task model is difficult to define.

This approach was demonstrated in CanadarmTutor (Fournier-Vigier et al. 2009; Kabanza et al. 2005) (mentioned in section 3.1), an ITS for learning to operate Canadarm2, a robotic arm deployed on the international space station which has seven degrees of freedom. In CanadarmTutor, the main learning activity is to move the arm from an initial configuration to a goal configuration in a 3D-simulated environment. To move the arm, the operator must select every minute the three best cameras to view the operation scene. He must next select and perform appropriate joint rotations to move the arm, while avoiding collisions and dangerous configurations. Moving Canadarm2 is an ill-defined task because there is an almost infinite number of solution paths and no simple "legal move generator" to find the best operations to perform at each step.

To allow the system to learn partial task models automatically, CanadarmTutor was modified to record each attempt to solve a problem in a database as a sequence of actions annotated with contextual information, such as the success or failure of the attempt (Fournier-Vigier et al. 2009) Data-mining algorithms were then applied to extract partial solution paths that regularly occur for each problem. The idea is that even if there are many solution paths for a problem, some parts occur frequently, and, thus, could be used to support tutoring services. In CanadarmTutor, exploiting these frequent paths allows for supporting tutoring services that are comparable to what a MT-tutor can provide for a well-defined task (Fournier-Vigier et al. 2009).

The approach of learning partial task models is appropriate for problem-solving tasks in which: 1) the initial state and the goal state are clear; 2) there is a large number of possibilities; 3) there is no clear strategy for finding the best solution; and 4) solution paths can be expressed as sequences of actions. The advantages of this approach are that it does not require any specific background knowledge by the domain expert, and the system can enrich its knowledge base with each new solution. Also, unlike the expert system approach, this approach is based on human solutions. On the other hand, no help is provided to the learner if part of a solution path was previously unexplored. One way to address this limitation is to combine this approach with other approaches, such as CBM or MT (see next subsection). A second limitation of the approach is that it needs to be applied to each problem. However, this may be a compromise worth accepting if a collection of exercises can be set up and administered to many students.

5.3.5 The Hybrid Approach

The last approach for representing and reasoning on domain knowledge is to combine two or more of the aforementioned approaches. The goal of this approach is to combine the advantages of different approaches in order to overcome their limitations for ill-defined tasks. The motivation behind having the hybrid approach is that different approaches can be better suited for different parts of the same ill-defined task so as to offer common or complementary tutoring services. For example, in CanadarmTutor, an expert system (Kabanza et al. 2005), a cognitive model (Fournier-Vigier et al. 2008) and the partial task model approach (Fournier-Vigier et al. 2009) are integrated to provide assistance for different parts of the arm manipulation task. The result is tutoring services which greatly exceed what was possible to offer with each individual approach, for this domain (Fournier-Vigier et al. 2009). According to Lynch et al. (Lynch et al. 2006), the hybrid approach is one of the most promising ways of supporting tutoring services in ill-defined domains.

In conclusion, Section 4.3 describes five approaches for representing and reasoning on domain knowledge in ill-defined domains for ITSs. The description of these five approaches provides a general overview of the techniques available.

5.4 Teaching Models for Building Intelligent Tutoring Systems in Ill-Defined Domains

Section 4 presents the challenge of building ITS in ill-defined domains from a complementary perspective, that is to say, suitable teaching models. Choosing a teaching model for a domain is an important decision because it can considerably facilitate or impede the conception of an ITS.

5.4.1 Structuring Learning around Case Studies

The first effective teaching model for ill-defined domains, which has partial domain theories and in which practitioners may have to draw analogies with past cases (e.g., medical diagnoses and law argumentation), is a model in which the learning is structured around the study of cases.

An example of an ITS which implements this strategy is CATO (Aleven 2003). The goal of CATO is to teach beginning law students the basic skills of developing arguments with cases (Ashley et al. 2002). In particular, it is designed to teach the skill of distinguishing legal cases, which consists in demonstrating that a case is significantly different from another, to suggest that it needs to be decided differently. Identifying positive distinctions among cases for the person invoking the argument is a difficult and ill-defined task as natural language is used and since verifiable and accurate solutions are rarely determined (Ashley et al. 2002). CATO presents good and bad examples with interactive explanations to teach the learner the skills to distinguish among cases, Lynch et al. 2006; Simon 1978. To support these tutoring services, CATO uses an expert system which incorporates a set of cases indexed with legal factors (Aleven 2003). The latter is also used in CATO-Dial (Ashley et al. 2002), a variation of CATO that engages the student in simulated courtroom arguments.

CATO and CATO-Dial implement elaborate tutoring services to support learning with cases and to teach learners how to compare cases. However, this is not the only way to support learning with cases. A simpler strategy exists to enable the learner to analyse or practice one case at a time, such as is the case in GUIDON (Clancey 1984).

5.4.2 Supporting Metacognition

Another teaching model that has successfully been used in ITSs for ill-defined domains is a model which provides metacognitive support to the learner so as to improve his/her ability to acquire knowledge. This model also gives him/her some support to learn domain knowledge. The advantage of this teaching model is that it does not require a detailed model of domain expertise.

An example of an ITS that supports metacognition in an ill-defined domain is the experimental tutoring system designed by Walker et al. (2008) for teaching intercultural competence skills. This domain is ill-defined because explaining cultural behavior is based on the interpretation of events and language, and, although some rules exist, there is not a complete formal theory (Walker et al. 2008). A student who uses the system by Walker et al. is asked to watch clips of a French movie that relate to immigration. He/she is then required to answer questions about the clips. Lastly, he/she must contribute to a forum discussion. Each time the learner contributes to the forum, the system uses keyword analysis algorithms to evaluate the writing. The evaluation is based on five quality measures, for example, if the writing is on topic and if it shows awareness of different points of views. The ITS then generates advice for improvement. The learner is subsequently required to make at least one revision to the post before publishing it. Although the system has no domain model of what constitutes intercultural competence skills, it fosters learning by promoting good learning behavior (writing posts which satisfy quality measures) (Walker et al. 2008). The system of Walker et al. uses an expert system approach to support metacognition. However, MT and CBM (Mitrovic et al. 2007) can also be used to support metacognition.

5.4.3 Supporting Inquiry Learning

A third teaching model used in ITSs for ill-defined domains is a model which supports inquiry-learning i.e., a constructivist approach to learning (Woolf 2009). Inquiry learning consists of constructing knowledge by discovery, gathering data and testing hypotheses. The role of an ITS for inquiry learning is to support the inquiry process rather than to provide knowledge to the learner. Thus, the ITS does not require a sophisticated domain model. For this reason, inquiry learning is an appropriate teaching model for many ill-defined domains in which there is no clear or complete formal theory. ITSs for inquiry learning can provide different types of support, i.e., controlling the inquiry process, providing help at a metacognitive level concerning what is a good inquiry behavior and evaluating the reasoning of the learner to give tailored advice (Woolf 2009).

An example of an ITS for inquiry learning is Rashi (Dragon et al. 2006), a generic ITS which was applied to several domains, including forestry, human biology, history and geology. Rashi presents cases to the learner and enables him/her to gather data, formulate hypotheses, and develop arguments in order to verify his/her hypotheses. The learner can collect information about a case with data-collecting tools, such as an interview tool or an image explorer tool. He can next construct his/her arguments using the argument-building tool. Rashi supports learning at a metacognitive level, for example, by promoting the consideration of multiple hypotheses, or by encouraging top-down or bottom-up argument construction. Rashi also uses a limited domain model to detect faulty relationships in arguments and to suggest missing relationships.

An interesting feature of Rashi for ill-defined domains is that it allows the learner to enter additional evidence that is not predefined in the domain model to support arguments. The importance of building ITSs that take into account the impact of the learner's background knowledge on his/her reasoning has also been noted by Easterday et al. (2007), in the domain of policy problems. A policy problem consists of judging the likelihood that a policy will lead to a desired outcome

(Easterday et al. 2007). Policy problems are ill-defined because there are no objectively correct answers and no agreed upon strategies for representing problems (Easterday et al. 2007). Easterday et al. observed that novices and experts who solve policy problems sometimes: (1) disregard the overriding evidence because they rely on their background knowledge; and (2) speculate about the outcomes because they are influenced by their background knowledge. To handle these types of errors, Easterday et al. proposed several solutions: (1) to ask the student to reason with the evidence and with his/her background knowledge separately before reasoning with both; (2) to design user interfaces that allow the learner to add his/her background knowledge; and (3) to ask the learner to articulate his/her reasoning and to identify at what point his/her background knowledge intervenes.

Recently, Easterday et al. proposed PolicyWorld (Easterday), an ITS for teaching deliberation for policy problems. PolicyWorld supports deliberation for policy problems by supporting searching for evidences and comprehending them, building interpretation of policy problems as causal diagrams, relating these diagrams to evidences, synthesizing diagrams, and taking decisions based on them. PolicyWorld does not allow the learner to enter his/her background knowledge. However, the tutoring services provided can indirectly handle errors that are caused by the learner's background knowledge. For example, they can check if the learner's beliefs appear consistent with the evidence collected. PolicyWorld uses the hybrid approach to represent and reason on domain knowledge by combining MT, CBM and the expert system approach. It relies on an inquiry-based teaching model.

5.4.4 Using Interactive Narratives

A fourth teaching model that has been used in ITSs for ill-defined domains is a model which offers "interactive narratives." A system that provides interactive narratives puts the learner in stories and enables him/her to make decisions that directly affect the story's direction and/or outcome (Hodhod and Kudenko 2008). This approach is demonstrated in AEINS (Hodhod and Kudenko 2008), an ITS for learning ethics through moral dilemmas. AIENS takes part in narratives in which the learner is faced with moral dilemmas. Making decisions in moral dilemmas is an ill-defined task because there is no right answer. The learner makes the final judgement of what is right or wrong. AEINS gives freedom to the learner to act and make decisions which permits the learner to learn from his/her decisions. When there are dilemmas, AEINS uses the Socratic Method to conduct a dialogue so as to encourage the learner reflect on the implications of the decisions made. AIENS has the capability to generate and adapt narratives spontaneously in order to provide situations that are tailored to each learner. AEINS relies on the MT approach to update a student model.

Learning with interactive narratives is similar to inquiry learning as it is also a form of learning by discovery. However, it differs from inquiry learning in that the learner does not have to gather data, nor to build and test hypotheses in a structured fashion.

5.4.5 Structuring Learning around Collaboration

A fifth teaching model for ill-defined domains is a model which structures learning around collaboration. The goal of this model is to make the student learn by working and exchanging information and ideas with his/her peers. The benefit of this strategy for ill-defined domains is that supporting collaboration can replace the need to build a domain model.

An ITS which supports collaboration can offer various levels of control in interactions. It can guide collaboration, provide advice on how to improve interactions, and display aggregated data regarding interactions so that the participants can decide on appropriate remedial actions (Soller et al. 2005). One example of a collaborative system in ill-defined domains is that of Walker et al. (Walker et al. 2008). As previously mentioned, Walker et al.'s system fosters learning by helping learners make useful contributions to forum discussions.

One approach to collaboration that merits mention is that of creating virtual companions that interact with the learner (Chou et al. 2003). The learner can compete, collaborate and even learn by teaching his/her virtual companions. Although building a virtual companion may not require a detailed domain model, designing a virtual companion can be challenging.

5.4.6 Other Teaching Models

The list of teaching models presented in this section is not exhaustive and many variants of the presented teaching models are possible. The intent of this section is to present a set of less conventional teaching models. But it is also possible to adopt more traditional models. For instance, CanadarmTutor (Fournier-Vigier et al. 2008 and 2009; Kabanza et al. 2005) adopts a problem-solving teaching model in which the principal learning activity is to solve problems akin to most ITSs which teach problem-solving tasks.

5.5 A Case Study: CanadarmTutor

The two previous sections listed several examples of ITSs and discussed the strategies used for representing and reasoning on domain knowledge in ill-defined domains and suitable teaching models. The focus of Section 5 is to look at the case study of CanadarmTutor in more detail. It was deemed appropriate to analyse CanadarmTutor in greater depth due to the fact that there have been several studies on the different approaches for representing and reasoning on domain knowledge with this ITS. This case study, therefore, allows for the comparison of various approaches in the same domain.

5.5.1 CanadarmTutor

CanadarmTutor is an ITS that we have developed (Fournier-Vigier et al. 2008 and 2009; Kabanza et al. 2005) (depicted in Figure 5.1.a). It is a simulation-based

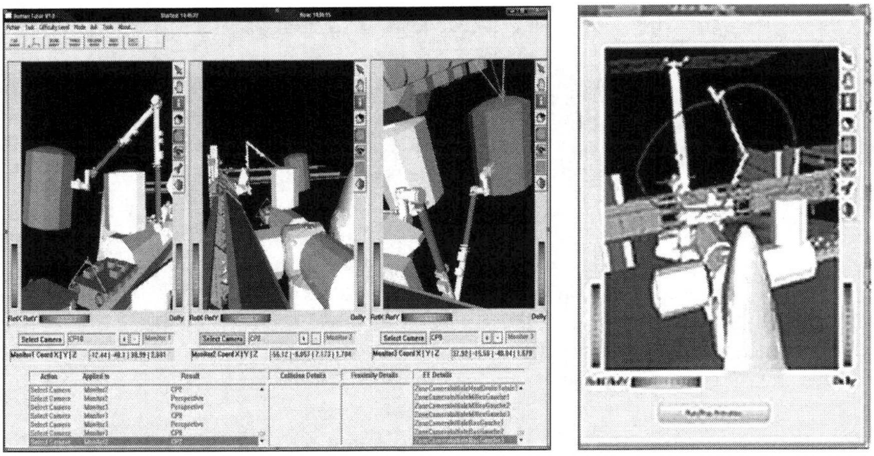

Fig. 5.1 (a) The CanadarmTutor user interface, (b) a path-planner demonstration

tutoring system to teach astronauts how to operate Canadarm2, a robotic arm deployed on the International Space Station (ISS) that provides 7 degrees of freedom. As previously explained, the main learning activity in CanadarmTutor is to move the arm from a given configuration to a predetermined configuration. Operating Canadarm2 is a difficult task since operators do not have a direct view of the scene of operation at the space station and must rely on cameras mounted on the manipulator and on strategic places in the environment where it operates. To move the arm, the operator must select at every moment the best cameras for viewing the scene of operation. Moreover, an operator has to select and perform appropriate joint rotations to move the arm, while avoiding collisions and dangerous configurations. To provide domain expertise to CanadarmTutor, the three following approaches were applied (Fournier-Vigier et al. 2008 and 2009; Kabanza et al. 2005).

5.5.2 The Expert System Approach

Initially, a special path-planner was integrated into CanadarmTutor. The path-planner is based on a probabilistic roadmap approach (Kabanza et al. 2005) and can produce examples of correct and incorrect motions in training. It acts as a domain expert and can calculate the arm's moves avoiding obstacles, and consistent with the best available camera views to achieve a given goal (Kabanza et al. 2005). The path-planner makes it possible to track the learner's solution step- by-step, and to generate demonstrations when necessary (as depicted in Figure 5.1.b). However, the generated paths are not always realistic or easy to follow since they are not based on human experience and cannot provide help on important aspects of the manipulation task, such as selecting cameras and adjusting their parameters. Also, the path-planner cannot support important tutoring services, such as estimating the learner's knowledge gaps, as there is no knowledge or skills representation.

5.5.3 The Model-Tracing Approach

To sustain more effective learning, a cognitive task analysis was performed, and an effective problem space that captures real user-knowledge was defined (Fournier-Vigier et al. 2008). To model the manipulation of Canadarm2, we were inspired by the research on spatial cognition, which states that spatial representations used for complex spatial reasoning are encoded as semantic knowledge (Fournier-Vigier et al. 2008). We used a custom cognitive model including a semantic retrieval mechanism to model the recall of spatial knowledge during the task (Fournier-Vigier et al. 2008). To model the spatial knowledge, we discretized the 3D space into 3D sub spaces called elementary spaces (ES). Spatial knowledge was then encoded as relationships, such as: (1) a camera can see an ES or an ISS module; (2) an ES contains an ISS module; (3) an ES is next to another ES; and (4) a camera is attached to an ISS module. The procedural knowledge of how to move the arm to a goal position was modeled as a loop in which the learner must recall a set of cameras to view the ESs containing the arm, select the cameras, adjust their parameters, retrieve a sequence of ESs to go from the current ES to the goal, and move to the next ES. CanadarmTutor detects all the atomic actions, such as camera changes and entering/leaving an ES. Performing model tracing with the cognitive model allows CanadarmTutor to offer the following six important tutoring services (Fournier-Vigier et al. 2008).

First, the learner can freely explore the task model to learn how to operate the arm. The learner can also consult the declarative knowledge associated with the task model to learn about the properties of the space station, the cameras and Canadarm2. Second, model-tracing allows CanadarmTutor to evaluate the knowledge acquired by the learner during the arm manipulation exercises. After a few exercises, CanadarmTutor builds a detailed learner profile that shows the strengths and weaknesses of the learner in terms of mastered, missing and buggy knowledge. The third tutoring service evaluates the declarative knowledge linked to the task model by asking direct questions, such as "Which camera can be used to view the Node02 ISS module?" The fourth tutoring service provides hints and demonstrations on request during arm manipulation exercises. The next step is suggested by model-tracing. CanadarmTutor can give messages, such as: "If you have finished adjusting the third monitor, then you should start moving the arm." Demonstrations are generated dynamically due to model-tracing. The fifth tutoring service generates personalized exercises based on the student model. During a training session, CanadarmTutor relies on the student model to generate exercises that progressively involve new knowledge and knowledge judged to be not yet mastered by the learner. For instance, CanadarmTutor can suggest an exercise involving a camera that has been rarely used by the learner. The sixth and last tutoring service offers proactive help to the learner. When the proactive help is activated, CanadarmTutor, for example, warns the learner that another camera should be selected if the arm has moved to an area that is not visible by the currently selected cameras. This type of help is particularly appreciated by beginners.

5.5.4 The Automatic Acquisition of Partial Task Models

Although the task model specified by the hand provides a precise cognitive assessment of the learner's knowledge for the main steps of the manipulation task, it does not go into finer details such as how to select joint rotations for moving Canadarm2. Being that Canadarm2 is an arm with seven joints, there are multiple possibilities on how to move the robot to a goal configuration for a given robot manipulation problem. Since one must also consider the safety and the facility of each manoeuvre, it is very difficult to define a "legal move generator" to generate the joint rotations that a person could execute. In fact, some joint manipulations are preferable to others, based on several criteria, which are difficult to determine, such as the view of the arm given by the cameras chosen at a given moment, the relative position of obstacles to the arm, the arm configuration (e.g. avoiding singularities) and the familiarity of the user with certain joints manipulations. It is thus not possible to define a complete and explicit task model for this task (this task is ill-defined according to the definition of Simon (1978)). On the other hand, the path-planner sometimes provides paths that are too complex and difficult to be executed by the user, as the paths are not based on human solutions.

Due to these difficulties, we have applied the fourth approach, the automatic acquisition of partial task models (Fournier-Vigier et al. 2009). This approach is carried out in three phases in CanadarmTutor.

The first phase records the user's solutions when he/she attempts an exercise. In CanadarmTutor, one exercise is to move the arm from an initial configuration to a goal configuration. For each attempt, a *sequence of events* is created in a database. We define an event as a set of actions carried out by the learner that are considered to be simultaneous. In CanadarmTutor, we defined 112 actions that can be recorded, including: (1) selecting a camera; (2) performing an increase or decrease of the pan/tilt/zoom of a camera; and (3) applying a rotation value to one of the seven arm joints. An example of a partial action sequence recorded for a user in CanadarmTutor is <(0, *rotateSP*{2}), (1, *selectCP3*), (2, *panCP2*{4}), (3, *zoomCP2*{2})>, which represents decreasing the rotation value of joint SP by two units, selecting camera CP3, increasing the pan of camera CP2 by four units and then its zoom by two units. Furthermore, each sequence can have annotations called "dimensions" (Fournier-Vigier et al. 2009). Table 5.1 shows an example of a toy database containing six learner plans annotated with five dimensions. In this table, the single letters *a*, *b*, *c*, and *d* denote actions. The first dimension "*Solution state*" indicates if the learner plan is a successful or a buggy solution. In the case of CanadarmTutor, values for this dimension are produced by the tutoring system. The four other dimensions are examples of dimensions that can be added manually. While the dimension "Expertise" denotes the expertise level of the learner who performed a sequence, "Skill_1", "Skill_2" and "Skill_3" indicate if any of the three specific skills were demonstrated by the learner when solving the problem. This example illustrates a five dimensional database. However, any kind of learner information or contextual information can be encoded as a dimension. In CanadarmTutor, we used 10 skills, and the dimensions of "solution state" and "expertise level" to annotate sequences (Fournier-Vigier et al. 2009).

Table 5.1 An example toy database containing 6 user solutions

ID	Dimensions					Sequence of actions
	Solution state	Expertise	Skill_1	Skill_2	Skill_3	
S1	successful	expert	yes	yes	yes	<(0,a),(1,bc)>
S2	successful	novice	no	yes	no	<(0,d) >
S3	buggy	expert	yes	yes	yes	<(0,a),(1,bc)>
S4	buggy	intermediate	no	yes	yes	<(0,a),(1,c), (2,d)>
S5	successful	expert	no	no	yes	<(0,d), (1,c)>
S6	successful	novice	no	no	yes	<(0,c), (1,d)

Table 5.2 Some frequent patterns extracted from the dataset of Table 1 with a minsup of 33 %

ID	Dimensions					Sequence of actions	Support
	Solution State	Expertise	Skill_1	Skill_2	Skill_3		
P1	*	expert	yes	yes	yes	<(0,a)>	33 %
P2	*	*	*	yes	yes	<(0,a)>	50 %
P3	*	expert	yes	yes	yes	<(0,a), (1,b)>	33 %
P4	successful	*	no	*	*	<(0,d)>	50 %
P5	successful	expert	*	*	yes	<(0,c)>	33 %
P6	successful	novice	no	*	no	<(0,d)>	33 %

The second phase extracts partial task models from user plans. In order to accomplish this, CanadarmTutor applies a custom sequential pattern mining algorithm that we developed (see (Fournier-Vigier et al. 2009) for more details), which takes a database of user solutions as input and a user-defined threshold called *minsup*. The output is the set of all subsequences that appears in at least *minsup* sequences of the database. When applying the algorithm, it is possible to specify time constraints on subsequences to be discovered (see (Fournier-Vigier et al. 2009) for detailed information). For example, Table 5.2 shows some subsequences (also called patterns) found from the database shown in Table 5.1 with *minsup* = 33 %. When taking pattern P3 into consideration, pattern P3 represents doing action *b* one time unit immediately after action *a*. The pattern P3 appears in sequences S1 and S3 of Table 1. It has thus a support of 33 % or two sequences. Moreover, the annotations for P3 tell us that this pattern was performed by expert users who possess skills "Skill_1", "Skill_2" and "Skill_3" and that P3 was found in plans that failed, as well as plans that succeeded.

The third phase uses the partial task model for supporting tutoring services. We implemented three tutoring services (Fournier-Vigier et al. 2009). The two first tutoring services rely on a plan recognition algorithm, which after each learner action during an exercise, identifies the patterns that best match the last actions performed (see (Fournier-Vigier et al. 2009) for details on the algorithm).

The first tutoring service is to assess the profile of the learner (expertise level, skills, etc.) by looking at the patterns applied. If, for example, 80 % of the time a learner applies patterns with a value "intermediate" for the dimension "expertise," then CanadarmTutor can assert with confidence that the learner's expertise level is "intermediate." In the same way, CanadarmTutor can diagnose mastered and missing/buggy skills for the user who demonstrated a pattern by looking at the "skills" dimensions of patterns applied (e.g., "Skill_1" in Table 2).

The second tutoring service is to guide the learner. This tutoring service consists of determining the possible actions from the set of patterns and proposing one or more actions to the learner. In CanadarmTutor, this functionality is triggered when the student selects "What should I do next?" in the interface menu. CanadarmTutor then identifies the set of next actions possible according to the matching patterns found by the plan recognition algorithm.

Finally, a tutoring service was implemented in CanadarmTutor to enable the learner to explore patterns so as to discover possible methods to solve problems. Currently, the learner can explore a pattern with an interface that lists the patterns and their annotations and provides sorting and filtering functions (for example to display only patterns leading to success).

5.5.5 The Hybrid Approach

Although learning partial task models from user solutions in CanadarmTutor is useful in helping the learner to manipulate the joints –a task which was impossible with the cognitive model or the path-planner (Fournier-Vigier et al. 2009), no assistance is provided to the learner if part of a solution path has not been explored by other users. Since the three approaches that we applied to CanadarmTutor have advantages and disadvantages, we decided to combine them to create a hybrid model. This hybrid model works as follows.

When the learner performs an exercise with CanadarmTutor, the model-tracer uses the cognitive model to update the student model (as previously explained). This latter contains probabilities that knowledge units defined in the cognitive model are mastered by the learner. When the learner answers the questions asked by CanadarmTutor, the student model is also updated.

When a user completes, either successfully or unsuccessfully, an arm manipulation exercise), the solution is added to the sequence database of user solutions for that exercise. The knowledge mastery levels from the student model are used to annotate the sequence. Other annotations can be added, for example, information regarding the success or failure of the student and manual annotations. When a minimum of a few sequences have been recorded, the data mining algorithm is applied to extract a partial task model for the exercise. The patterns extracted include skills from the cognitive model as annotations.

When CanadarmTutor detects that that learner is following a pattern from the partial task model during an exercise, the annotations of the pattern are also used to update the student model. For example, if a learner applies a pattern common to the learners possessing "skill_1," the mastery level of "skill_1" in the student model will be increased. As a result, the partial task model is also used to update the student model (the student model is now shared by the cognitive model and the partial task model).

During a learning session, CanadarmTutor uses the student model to generate exercises that progressively integrate knowledge that is judged to be not yet mastered by the learner. Generated exercises are questions about the cognitive model and its declarative knowledge, or manipulation exercises.

When the learner asks for help regarding the next step of a manipulation exercise, three different types of system assistance is provided. First, the cognitive model gives a general procedure that should be followed to move the arm. For example, the cognitive model might say to the student, "If you have finished adjusting the third monitor, then you should start moving the arm." This assistance is generated by model-tracing with the cognitive model. In the same window, the patterns from the partial task model that match the current user solution are then displayed to the learner. These patterns provide information pertaining to the joint rotations that should be performed to move the arm. If no pattern matches the current learner solution, a demonstration generated by the path-planner is shown to the learner to illustrate a possible solution path. This information is complementary.

The learner can also explore patterns from the partial task models, as explained earlier, which illustrate possible alternatives for solving problems. The learner can also explore the cognitive model to learn the general procedure for moving the arm. In addition, at any time, the learner can ask for demonstrations from the path-planner or the cognitive model.

Lastly, as previously discussed, CanadarmTutor can use the cognitive model to generate proactive help to the learner, such as giving suggestions on selecting cameras.

We recently conducted a preliminary evaluation of the new version of CanadarmTutor with five users who had experience using the previous versions of CanadarmTutor. The study consisted of having each user try the new version for one hour. We then interviewed them regarding their experience using the new version of CanadarmTutor. All users agreed that the new version of CanadarmTutor is more comprehensive in all aspects of the manipulation task than are the previous versions they had used. They unanimously agreed that integrating the three approaches into CanadarmTutor resulted in better tutoring services. We are currently working on enhancing CanadarmTutor with more elaborated pedagogical strategies. We are also preparing an extensive empirical study to evaluate formally how the newest version of CanadarmTutor contributes to fostering learning.

5.6 Conclusion

Research on ill-defined domains presents many challenges to ITS researchers. However, it is an exciting area of research as many domains are unexplored and remain ill-defined. Research on ill-defined domains will undoubtedly result in the creation of ITSs in domains which, until recently, have been overlooked in ITS research.

In this chapter, we provided a synopsis of the problems and solutions for building ITSs for ill-defined domains. Three complementary and important perspectives were taken into consideration when creating ITSs for ill-defined domains: (1) the characteristics of ill-defined domains; (2) the approach for representing and reasoning on domain knowledge; and (3) the teaching models. Throughout the chapter, we presented several examples of ITSs in order to illustrate their different approaches and to highlight some of their advantages and disadvantages. We specifically presented the case-study of CanadarmTutor in order to compare four approaches for representing and reasoning on domain knowledge in the same ITS.

We believe it is important for future researchers, to further investigate domain-specific and general approaches to represent and reason on domain knowledge, since the former can be more effective in specific domains. We also believe that creating hybrid models for representing and reasoning on domain knowledge, such as that used in CanadarmTutor, is a promising approach which needs to be explored in greater depth.

Acknowledgments. Our thanks go to the NSERC (Natural Sciences and Engineering Research Council), and the FQRNT (Fond Québécois de Recherche sur la Nature et les Technologies) for their logistic and financial support.

References

Aleven, V.: Using background knowledge in case-based legal reasoning: a computational model and an intelligent learning environment. Artif. Intell. 150, 183–237 (2003)

Aleven, V., Ashley, K., Lynch, C., Pinkwart, N.: Proc. ITS for Ill-Defined Domains Workshop. ITS 2006 (2006)

Aleven, V., Ashley, K., Lynch, C., Pinkwart, N.: Proc. Workshop on AIED applications in ill-defined domains. AIED 2007, Los Angeles, USA (2007)

Aleven, V., Ashley, K., Lynch, C., Pinkwart, N.: Proc. ITS for Ill-Defined Domains Workshop. ITS 2008, Montreal, Canada (2008)

Ashley, K.D., Desai, R., Levine, J.M.: Teaching Case-Based Argumentation Concepts Using Dialectic Arguments vs. Didactic Explanations. In: Cerri, S.A., Gouardéres, G., Paraguaçu, F. (eds.) ITS 2002. LNCS, vol. 2363, pp. 585–595. Springer, Heidelberg (2002)

Chou, C.-Y., Chan, T.-W., Lin, C.-J.: Redefining the learning companion: The past, present, and future of educational agents. Comput. and Educ. 40, 255–269 (2003)

Clancey, W.: Use of MYCIN's rules for tutoring. In: Buchanan, B., Shortliffe, E.H. (eds.) Rule-Based Expert Systems. Addison-Wesley, Reading (1984)

Dragon, T., Woolf, B.P., Marshall, D., Murray, T.: Coaching Within a Domain Independent Inquiry Environment. In: Ikeda, M., Ashley, K.D., Chan, T.-W. (eds.) ITS 2006. LNCS, vol. 4053, pp. 144–153. Springer, Heidelberg (2006)

Easterday, M.W.: Policy World: A cognitive game for teaching deliberation. In: Pinkwart, N., McLaren, B. (eds.) Educational technologies for teaching argumentation skills. Bentham Science Publ., Oak Park (in press)

Easterday, M.W., Aleven, V., Scheines, R.: The logic of Babel: Causal reasoning from conflicting sources. In: Proc. ITS for Ill-Defined Domains Workshop. AED 2007, Los Angeles, USA (2007)

Fournier-Viger, P., Nkambou, R., Mayers, A.: Evaluating Spatial Representations and Skills in a Simulator-Based Tutoring System. IEEE Trans. Learn. Technol. 1(1), 63–74 (2008)

Fournier-Viger, P., Nkambou, R., Mephu Nguifo, E.: Exploiting Partial Problem Spaces Learned from Users' Interactions to Provide Key Tutoring Services in Procedural and Ill-Defined Domains. In: Proc. AIED 2009, pp. 383–390. IOS Press, Amsterdam (2009)

Graesser, A., Wiemer-Hastings, P., Wiemer-Hastings, K., Harter, D., Person, N.: Using Latent Semantic Analysis to evaluate the contributions of students in AutoTutor. Interact. Learn, Environ. 8, 149–169 (2000)

Hodhod, R., Kudenko, D.: Interactive Narrative and Intelligent Tutoring for Ethics Domain. In: Proc. ITS for Ill-Defined Domains Workshop. ITS 2008, Montreal, Canada (2008)

Kabanza, F., Nkambou, R., Belghith, K.: Path-Planning for Autonomous Training on Robot Manipulators in Space. In: Proc. 19th Intern. Joint Conf. Artif. Intell., pp. 1729–1731 (2005)

Kodaganallur, V., Weitz, R., Rosenthal, D.: An Assessment of Constraint-Based Tutors: A Response to Mitrovic and Ohlsson's Critique of "A Comparison of Model-Tracing and Constraint-Based Intelligent Tutoring Paradigms. Intern. J. Artif. Intell. Educ. 16, 291–321 (2006)

Koedinger, K.R., Anderson, J.R., Hadley, W.H., Mark, M.A.: Intelligent Tutoring goes to school in the big city. Intern. J. Artif. Intell. Educ. 8, 30–43 (1997)

Lynch, C., Ashley, K., Aleven, V., Pinkwart, N.: Defining Ill-Defined Domains; A literature survey. In: Proc. Intelligent Tutoring Systems for Ill-Defined Domains Workshop, ITS 2006, pp. 1–10 (2006)

Mitrovic, A., Weerasinghe, A.: Revisiting Ill-Definedness and the Consequences for ITSs. In: Proc. AIED 2009, pp. 375–382 (2009)

Mitrovic, A., Martin, B., Suraweera, P.: Intelligent Tutors for All: The Constraint-Based Approach. IEEE Intell. Syst. 22, 38–45 (2007)

Moritz, S., Blank, G.: Generating and Evaluating Object-Oriented Design for Instructors and Novice Students. In.: Proc. ITS for Ill-Defined Domains Workshop, ITS 2008, Montreal, Canada (2008)

Ogan, A., Wylie, R., Walker, E.: The challenges in adapting traditional techniques for modeling student behavior in ill-defined domains. In: Proc. Intelligent Tutoring Systems for Ill-Defined Domains Workshop (2006)

Simon, H.A.: Information-processing theory of human problem solving. In: Estes, W.K. (ed.) Handbook of learning and cognitive processes (5). Human information

Soller, A., Martinez-Monez, A., Jermann, P., Muehlenbrock, M.: From mirroring to guiding: A review of state of the art technology for supporting collaborative learning. Intern. J. Artif. Intell. Educ. 15(4), 261–290 (2005)

Walker, E., Ogan, A., Aleven, V., Jones, C.: Two Approaches for Providing Adaptive Support for Discussion in an Ill-Defined Domain. In: Proc. ITS for Ill-Defined Domains Workshop, ITS 2008, Montreal, Canada (2008)

Woolf, B.P.: Building Intelligent Interactive Tutors. Student Centered Strategies for revolutionizing e-learning. Morgan Kaufmann Publ., Mass. (2009)

Chapter 6
A Survey of Domain Ontology Engineering: Methods and Tools

Amal Zouaq[1] and Roger Nkambou[2]

[1] Simon Fraser University Surrey, 13450 102 Ave. Surrey, BC V3T 5X3 Canada
azouaq@sfu.caa
[2] University of Québec at Montréal, 201 Du Président-Kennedy Avenue, PK 4150, Montréal, QC, H2X 3Y7, Canada
Nkambou.roger@uqam.ca

Abstract. With the advent of the Semantic Web, the field of domain ontology engineering has gained more and more importance. This innovative field may have a big impact on computer-based education and will certainly contribute to its development. This chapter presents a survey on domain ontology engineering and especially domain ontology learning. The chapter focuses particularly on automatic methods for ontology learning. It summarizes the state of the art in natural language processing techniques and statistical and machine learning techniques for ontology extraction. It also explains how intelligent tutoring systems may benefit from this engineering and talks about the challenges that face the field.

6.1 Introduction

As described in chapter 2, the expert module is responsible for the learning content which indicates what can be taught by the ITS (the domain model). In this regard, some of the most important research issues that need to be addressed are: how the expert module can be effectively modeled, what kinds of knowledge representations are available and what kind of knowledge acquisition techniques are applicable. In fact, one of the main obstacles to ITSs development and wide dissemination is the cost of their knowledge base and particularly the cost of producing the domain model from scratch. Faced with these knowledge acquisition challenges, many attempts have been made to create automated methods for domain knowledge creation. However, these attempts have not been as successful as one would wish them to be. Moreover, these efforts have not led to reusable and standard methods and formalisms for knowledge base creation and update.

With the advent of the Semantic Web, new research avenues have been created, especially within the domain ontology engineering field. The research community now acknowledges the need to create domain ontologies in a (semi)automatic way. Representing knowledge using domain ontologies has two main advantages: first, their standard formalism makes it possible to share and reuse ontologies

R. Nkambou et al. (Eds.): Advances in Intelligent Tutoring Systems, SCI 308, pp. 103–119.
springerlink.com

between any ontology-friendly environments. Second, their formal structure makes it possible to work out how to obtain knowledge representations and figure out how to automatically extract ontological components in modular layers. This automatic ontological extraction is known as "**Ontology Learning**".

In general, the entire knowledge acquisition process is a tedious task, including such difficulties as building, reusing and propagating intelligent tutoring systems. In fact, we need to use standard representations to modularize the creation, evolution and maintenance of intelligent tutoring systems. It is also a necessary to provide explicit semantic relationships between the learning content concepts and to develop pedagogical activities that are built on this domain knowledge. With the advent of the Semantic Web, many questions have been raised about how the domain model could benefit from the Semantic Web languages and techniques. A successful integration of the Semantic Web and the ITS philosophy could ensure better reuse of ITS components and better sharing and engineering of domain knowledge. Because ontologies are the backbone of the Semantic Web, using them to represent domain and instructional knowledge can be an interesting avenue. These questions have been particularly high on the list with the rise of the *Educational Semantic Web* (Aroyo and Dicheva, 2004), which comes from the eLearning field and which has proposed the use of ontologies to index and structure the learning content. Intelligent tutoring systems have been slower in adopting the 'ontology' concept, especially for modeling domain knowledge but this is now an undeniable fact. Intelligent tutoring systems can benefit from ontology engineering because ontologies represent a standard way for modeling knowledge. They are expressed using formal and standard languages which facilitate sharing and reasoning. Moreover, there is a growing awareness within the ITS and eLearning communities of the importance of adopting common methods for domain knowledge acquisition and representation. As a result, ITS stands to benefit from the huge number of available eLearning resources. Similarly, eLearning systems will benefit from ITS domain modeling and reasoning. Finally, since ITSs are domain-dependent, it is important to develop easy and reusable knowledge acquisition tools and to integrate automatic methods for this acquisition and evolution. Ontology engineering can provide an answer to these needs and the following section introduces the reader to domain ontology engineering.

This chapter presents an overview of domain ontology engineering and focuses particularly on automatic methods for ontology learning, especially from texts. It is organized as follows. After the introduction, section 2 briefly explains the field of ontology engineering. Section provides an overview of the ontology learning process from text. Each task and component of this process is explained. We also present, for each task, the natural language processing (NLP) techniques and the statistical and machine learning techniques. Sections four and five, in addition to an ontology update task, briefly introduce the ontology learning process from other non text-based sources.. Section six highlights the more general challenges that face domain ontology engineering as well as more specific ITS-related ones. The entire chapter is summarized in the conclusion.

6.2 Ontology and Ontology Engineering

Before going into further detail, it is important to first define the notion of ontology. Very briefly, ontology is a formal specification of a conceptualization (in this case, a domain) and it includes the definition of classes, objects, properties, relationships and axioms. Ontologies are expressed using a formal language such as RDF or OWL and support automatic inference. Generally, ontologies involve a kind of consensus within a community, meaning that they formalize concepts that are generally accepted within this community. There are many kinds of ontologies such as upper-level ontologies, task ontologies and domain ontologies. We are especially interested here in domain ontologies.

As previously pointed out, the concept of a domain ontology as envisioned by the eLearning community is relatively new in the field of ITS. However, domain ontology engineering is a growing research area that has received much attention in other fields and it is the corner stone of the Semantic Web. Ontology engineering is a field that explores the methods and tools for handling the ontology lifecycle. It requires a general and domain-independent methodology that provides guidance for ontology building, refinement and evaluation (Guarino and Welty, 2002). The ontology life-cycle can be schematized in four main stages: the specification stage, the formalization stage, the maintenance stage, and the evaluation stage.

- *The specification stage* identifies the purpose and scope of the ontology. Generally, this relies a lot on domain experts and needs to define the competency questions that the ontology has to to answer. It is also dependent on the application that is going to be used by the ontology;
- *The formalization stage* produces a conceptual and formal model that meets the requirements of the specification stage;
- *The maintenance stage* keeps track of the ontology's updates and evolution, and checks its consistency;
- Finally, the *evaluation stage* analyzes the resulting ontology and checks if it meets the initial needs and has the desired features.

At this point, we are especially interested in the formalization stage and how it can benefit from automated methods for knowledge acquisition. In fact, the most common and successful techniques for domain engineering are generally manual and the best ITS authoring tools can help the expert formalize his knowledge but these tools are generally far from being part of an automated procedure (see chapter 18). It is therefore worthwhile to explicitly state the steps that can be automated to alleviate the task of human experts and the burden of knowledge acquisition. Ontology learning techniques have been adopted to achieve this goal (Aussenac-Gilles et al., 2000). These learning techniques can vary according to the degree of automation (semi-automatic, fully automatic), the ontological knowledge that has to be extracted (concepts, taxonomy, conceptual relationships, attributes, instances, axioms), the knowledge sources (texts, databases, xml documents, etc.) and finally the purpose (creating ontologies from scratch and/or updating existing ontologies).

6.3 Building Domain Ontologies from Texts

This section focuses on ontology learning from texts. At the specification stage, the knowledge engineer should prepare a corpus related to the domain of interest. Of course, this corpus has to be carefully chosen and should properly describe the domain. A number of sub-tasks have to be performed in order to learn a domain ontology including concepts, taxonomy, conceptual relationships, attributes, instances and axiom learning. Examples of systems that completethe ontology learning task include: Text-2-Onto (Cimiano and Volker, 2005a), TEXCOMON (Zouaq and Nkambou, 2009a; Zouaq and Nkambou, 2009b), OntoLearn (Velardi et al., 2005), and OntoGen (Fortuna et al., 2007). In the following sections, we highlight state-of-the-art knowledge extraction techniques used in each of the ontology learning sub-tasks. Each of these sub-tasks has an NLP-based technique and a statistical and machine learning technique.

6.3.1 Concept Extraction

The first task that has to be performed in ontology engineering is the identification of concepts. Concepts can be described as complex mental objects that are characterized by a number of features. Concept extraction refers to the identification of important domain classes.

Concepts are terms that are particularly important for the domain when using terminological approaches. These terms are generally extracted from the corpus as outlined by Buitelaar et al. (2005) who consider that a concept should have a linguistic realization. In this case, it is quite challenging to differentiate domain terms from non domain terms, especially when using statistical filtering. The identified terms—composed from single or several words—can then be either considered as specific concepts/classes or they can be classified according to broad classes already available in thesauri and vocabularies. Other approaches rely on clustering and machine learning as a way of learning semantic classes. In this case, a concept may have no corresponding term in the corpus. This is further explained in the following paragraphs.

6.3.1.1 NLP-Based Techniques

NLP-based techniques for concept learning consider terms as candidate concepts. These approaches rely on linguistic knowledge and use parsers and taggers to determine the syntactic roles of terms or to unveil linguistic patterns. Some works typically adopt a surface analysis by running a part-of-speech tagger over the corpus and identifying manually defined patterns (Sabou, 2005; Moldovan and Girju, 2001) while others use a deep-level analysis and use a NLP parser (Reinberger and Spyns, 2005; Zouaq and Nkambou, 2009a). In general, the syntactic analysis identifies the nominal phrases that may be important for the domain. For example, Zouaq and Nkambou (2009a) use dependency relationships indicating nominal phrases such *nominal subject*, *direct object* and *noun compound modifier* to detect these nominal phrases. Most of the time, there is also a list of manually defined seed words that triggers the ontology learning process. However, Zouaq and Nkambou (2009a) proposed the use of an automatic keyword extractor to help automate this task.

6.3.1.2 Statistical and Machine Learning Techniques

Usually, NLP-based approaches are not used alone and require statistical filtering. Statistical approaches consider all important terms in a domain as potential concepts and require quantitative metrics to measure the weight of a term. Such quantitative measurements include the popular TF*IDF (Salton and Buckley, 1988) and C-value/NC-value (Frantzi et al., 1998). The employed measurements can differ depending on the application.

Clustering techniques based on Harris' distributional hypothesis (Harris, 1954), can also be used to induce semantic classes (Almuraheb and Poesio, 2004; Lin and Pantel 2001). Here, a concept is considered as a cluster of related and similar terms. Harris' hypothesis, which is the basis of word space models, states that words that occur in similar contexts often share related meaning (Sahlgren 2006). Term similarity can be computed using collocations (Lin 1999), co-occurrences (Widdows and Dorow, 2002) and latent semantic analysis (Hearst and Schutze, 1993). For example, Lin and Pantel (2001) represent each word by a feature vector that corresponds to a context in which the word occurs. The features are specific dependency relationships coupled with their occurrence in the corpus. The obtained vectors are then used to calculate the similarity of different terms using measurements such as mutual information (Hindle, 1990; Lin, 1998) and to create clusters of similar terms. Comparable approaches include Formal Concept Analysis (such as the approach presented in (Cimiano, 2006)) and Latent Semantic Indexing algorithms (e.g. Fortuna et al., 2005). These approaches build attributes/values pairs that correspond to concepts.

Statistical approaches can also be used on top of NLP-based approaches to identify only relevant domain terms by comparing the distribution of terms between corpora (Navigli and Velardi, 2004). Another approach used by (Velardi et al., 2005) linguistically analyses WordNet glosses (textual description) in order to extract relevant information about a given concept and enrich its properties. This analysis can help detect synonyms and related words and can contribute to concept definition. In fact, concept learning requires not only that conceptual classes be identified but also makes it necessary to describe concepts through the identification of attributes, sub-classes and relationships. This is further explained in the following sections.

6.3.2 Attribute Extraction

Since concepts are characterized by a number of features, it is important to unveil the distinctive attributes or properties that define a concept. In his ontology, Guarino (1992) distinguishes between relational and non-relational attributes. Relational attributes include qualities and relational roles, while non-relational attributes include parts. Following Guarino (1992) and Pustejovsky (1995), Almuraheb and Poesio (2005) presented another scheme for classifying attributes into qualities, parts, related-objects, activities and related-agents.

In this chapter, attributes designate a data type property such as *id, name*, etc., in contrast with object properties which are considered as conceptual relationships—these are addressed in the Conceptual Relationships Extraction section.

6.3.2.1 NLP-Based Techniques

According to Poesio and Almuhareb (2005), the right meaning of attributes can be found by looking at Wood's linguistic interpretation (Wood, 1975): *Y is a value of the attribute A of X if it is possible to say that Y is an A of X (or the A of X).* If it is not possible to find a Y then A cannot be an attribute. In order to comply with this linguistic interpretation, linguistic patterns are also proposed for the detection of attributes. Following Woods (1975), Almuhareb and Poesio (2005) suggested the use of the following patterns in order to search for attributes of a concept C:

- "(a|an|the) * C (is|was)" (e.g.: *a red car is…*).
- "The * of the C (is|was)" (e.g.: *the color of the car is…*)
- "The C's * R" (e.g.: *The car's price is…*) where R is a restrictor such as "is" and the wildcard denotes an attribute.

Cimiano (2006) proposed another set of patterns for attribute extraction based on adjective modifiers and WordNet and presented a number of interesting patterns describing attributes and their range according to syntax (parts-of-speech).

6.3.2.2 Statistical and Machine Learning Techniques

As indicated earlier, natural-language processing techniques are generally coupled with statistical filtering and machine learning. Poesio and Almuraheb, (2005) proposed a supervised classifier for learning attributes based on morphological information, an attribute model, a question model, and an attributive-usage model. These models are used to differentiate types of(different kind of) attributes based on a specific classification scheme. In Poesio and Almuhareb (2008), the Web is used to extract concept descriptions. Another approach, proposed by Ravi and Pasca (2008), describes a weakly supervised classifier for learning attributes and values, based on a small set of examples.

6.3.3 Taxonomy Extraction

One of the most important tasks in knowledge engineering is the organization of knowledge into taxonomies which indicate generalization/specialization relationships between classes. These relationships enable inheritance between concepts and automated reasoning (Corcho & Gomez-Perez, 2000).

6.3.3.1 NLP-Based Techniques

The most common way of extracting taxonomical links is the use of specific lexico-syntactic patterns as proposed by Hearst (Hearst, 1992). In Pattern-based techniques, the text is scanned for instances of distinct lexico-syntactic patterns that indicate a taxonomical link. Patterns are usually expressed as regular expressions (Cimiano and Volker, 2005b) but they can also be represented by dependency relationships (Zouaq and Nkambou, 2009a; Lin and Pantel, 2001).

Since a domain corpus is sparse and because hierarchical patterns are rare in domain-specific corpora, many approaches extend the corpus by a search of taxonomical links in dedicated resources such as WordNet (Snow et al., 2004) or on the Web (Cimiano et al, 2004; Maedche and Staab, 2001) so as to increase their recall (Etzioni et al., 2004). A remedy for the burden of manually defining patterns is proposed by Snow et al. (2004) using a classifier for automatically learning hyponym (is-a) relations from text based on dependency paths and using WordNet.

Other linguistic approaches use the internal structure of multiple-word terms (nouns phrases) in order to deduce taxonomical links. For example, there is a taxonomical link between a term and the same term modified by an adjective (e.g.: an intelligent man is-a man). This approach is quite popular (Buitelaar et al., 2003) (Velardi et al., 2005; Zouaq and Nkambou, 2009a).

6.3.3.2 Statistical and Machine Learning Techniques

Statistical and machine learning approaches for taxonomy learning rely on Harris' distributional hypothesis, just as those used in concept learning. Hierarchical clustering algorithms are used to extract taxonomies from text and produce hierarchies of clusters. Maedche et al. (2002) describe the two main approaches that can be used to implement hierarchical clustering: the bottom-up approach which starts with individual objects and groups the most similar ones, and the top-down approach, where all the objects are divided into groups. This approach has been used in many works such as Bisson et al. (2000), Carabello (1999), and Faure and Nedellec (1998). Typically, as highlighted by Cimiano et al. (2004), a term t is a subclass of t2 if all the syntactic contexts in which t appears are also shared by t2. The syntactic contexts are used as feature vectors and a similarity measure is applied. For example, in order to compute the relation *is_a (t, t2)*, Cimiano et al. (2004) applied a directed Jaccard coefficient computing the number of common features divided by the number of features of term t.

Cimiano et al. (2004) propose also the use of multiple sources of evidence and techniques in order to learn hierarchical relationships. Similarly, Widdows (2003) proposes the use of unsupervised methods combining statistical and syntactic information to update an existing taxonomy with new terms.

6.3.4 Conceptual Relationships Extraction

Conceptual relations refer to any relationship between concepts aside from taxonomic relations. Specific conceptual relationships may include synonymy, part-of, possession, attribute-of, causality, as well as more general relationships referring to any labeled link between a source concept (the domain of the relation) and a destination concept (the range of the relation). In the following sections, we identify the different techniques used to describe specific relationships and generic relationships.

6.3.4.1 NLP-Based Techniques

In the information extraction community, conceptual relation extraction is known as template filling, frame filling, semantic role labeling or event extraction. In this case, it relies on lexico-semantic lexicons such as FrameNet (Baker et al., 1998) and VerbNet (Kipper et al., 2000) to extract particular relationships and to assign roles (such as Agent, Theme, etc.) to the arguments of the relation. Approaches based on frames include ASIUM (Faure and Nédellec, 1998) which enables an acquisition of relations between concepts based on triggering words. Another work related to roles is the identification of Qualia structures by Pustejovsky (1995). These qualia structures can help identify particular relationships as shown by Cimiano and Wenderoth (2005) who proposed a number of linguistic patterns indicating the different roles defined by Pustejovsky.

There is quite a lot of work on the use of linguistic patterns to unveil ontological relations from text. Following Hearst's work (Hearst, 1992) on taxonomic relations, different researchers created patterns for non-hierarchical relationships (Iwanska et al., 2000; Zouaq and Nkambou, 2009a), for part-of relations (Charniak and Berland, 1999; Van Hage et al., 2006) or causal relations (Girju et al., 2003). In fact, many works consider that ontological relationships are mostly represented by verbs and their arguments. In the same line of research, Navigli and Velardi (2004) use patterns expressed as regular expressions and restricted by syntactic and semantic constraints. Finally, WordNet can be used to extract synonyms, antonyms and other kinds of relationships. This also involves the detection of the right meaning of the term and thus the use of word meaning disambiguation algorithms.

6.3.4.2 Statistical and Machine Learning Techniques

Most of the work on relation extraction combines statistical analysis with more or less complex levels of linguistic analysis. For example, Zouaq and Nkambou (2009b) use typed dependencies to learn relationships and statistical measurements which in turn are used to determine whether or not the relationships should be included in the ontology.

Other machine learning techniques for learning qualia structures include the work of Claveau (2003) using inductive logic programming or Yamada and Baldwin (2004) whose work relies not only on lexico-syntactic patterns but also on a maximum entropy model classifier. Cimiano and Wenderoth (2007) developed an algorithm for generating a set of clues for each qualia role: download the snippets of the first 10 Google hits matching the generated clues, part-of-speech-tagging of the downloaded snippets, matching regular expressions conveying the qualia role of interest and finally weighting the returned qualia elements according to some measure.

An interesting technique for learning non labeled relationships is the use of association rule learning, where association rules are created from the co-occurrence of elements in the corpus. This technique has been adopted by the Text-to-Onto system (Maedche and Staab, 2001). However, these relationships should be manually labeled later and this task is not always easy for the ontology engineer.

6.3.5 *Instance Extraction*

Instance extraction, also known as Ontology Population (OP), is a classification task which aims at finding instances of concepts defined in an ontology. It is similar to Named Entity Recognition (e.g. Person, Location, Organization, etc.), which is often used in information extraction. Examples of systems especially devoted to instance extraction include WEB→KB (Craven et al., 2000) and Know-it-All (Etzioni et al., 2004).

6.3.5.1 NLP-Based Techniques

There are a number of approaches that use NLP-based techniques for ontology population. A pattern-based approach similar to the one presented in the taxonomy extraction section relies on Hearst patterns (Hearst, 92; Schlobach et al., 2004; Zouaq and Nkambou, 2009b; Etzioni et al., 2004) or on the structure of words (Velardi et al., 2005). These approaches try to find explicitly stated "is-a" relationships. Other linguistic approaches are based on the definition or the acquisition of rules. For example, the work of (Amardeilh et al., 2005) proposes the definition of acquisition rules that are activated once defined linguistic tags are found. These tags are mapped onto concepts, attributes and relationships from the ontology and help find instances of these elements.

6.3.5.2 Statistical and Machine Learning Techniques

There are supervised and weakly supervised techniques for ontology population (Tanev and Magnini, 2006). Among the weakly supervised techniques, Cimiano and Volker (2005b) used vector-feature similarity between each concept c and a term to be categorized t. Cimiano and Volker evaluated different context features (word windows, dependencies) and showed that syntactic features work better. Their algorithm assigned a concept to a given instance by computing the similarity of this instance feature vector and the concept feature vector. Tanev and Magnini (2006) used syntactic features extracted from dependency parse trees. Their algorithm required only a list of terms for each class under consideration as training data.

Supervised techniques for ontology population ensure higher accuracy. However, they require the manual construction of a training set, which is not scalable (Tanev and Magnini, 2006). An example of a supervised approach is the work of Fleischman (2001); Fleischman and Hovy (2002) which involved designing a machine learning algorithm for fine-grained Named Entity categorization. Web->KB (Craven et al., 2000) also relies on a set of training data, which consists of annotated regions of hypertext that represent instances of classes and relations, used to extract named entities. Based on the ontology and the training data, the system learns how to classify arbitrary Web pages and hyperlink paths.

6.3.6 Axioms Extraction

Axiom extraction represents one of the most difficult tasks of ontology learning. Axioms express necessary and sufficient conditions that are used to constrain the information contained in the ontology and to deduce new information (Shamsfard and Barforoush, 2003). Few systems have tackled the problem of axiom extraction. HASTI is a system that translates explicit axioms in conditional and quantified natural language sentences to logically formatted axioms in KIF (Shamsfard and Barforoush, 2002). LExO2 (Volker et al, 2008) is another initiative for transforming natural language sentences (definitions) into description logic axioms.

6.3.6.1 NLP-Based Techniques

Natural language techniques for axiom extraction rely on the syntactic transformation of natural language (definitions) into description logic axioms (Volker et al, 2008). This supposes the availability of such definitions. Volker et al (2008) also focus on learning a particular axiom which is disjointedness by a lexico-syntactic pattern used to detect enumerations. Their underlying assumption is that terms which are listed separately in an enumeration mostly denote disjointed classes. Zouaq and Nkambou (2009b) describe a pattern for defining equivalent classes. This pattern is based on the appositive grammatical relationship between two terms to indicate that these terms are similar and denote the same concept. Another interesting work is the Lin and Pantel (2001) approach (which uses of paths in dependency trees to learn similar relationships. This makes it possible to create inverse properties for these relationships, such as *X solves Y* and *Y is solved by X*.

6.3.6.2 Statistical and Machine Learning Techniques

To the best of our knowledge, there are very few machine learning approaches for learning axioms. A machine learning classification approach has also been used by Volker et al. (2008) to determine disjointedness of any two classes. They automatically extract lexical and logical features that provide a basis for learning disjointedness by taking into account the structure of the ontology, associated textual resources, and other types of data. The features are then used to build an overall classification model.

6.4 Ontology Learning from Non-text Sources

Ontology learning from non text sources involves the use of structured or semi-structured data as input to the learning process. Structured data refer to already defined knowledge models including database schemas or existing ontologies. Semi-structured data designates the use of some mixed structured data with free text such as Web pages, Wikipedia, dictionaries and XML documents. The existence of a structure helps direct the ontology learning process towards relevant parts of data.

6.4.1 NLP-Based Techniques

Most of the approaches for ontology learning from non text sources rely on linguistic techniques or on the underlying schema already available in the structure. As previously indicated, some researchers such as Cimiano and Staab (2004) have used the Web to deal with the data sparseness problem within domain corpora. In this case, the Web is used to retrieve and learn patterns. Other works rely on dictionaries such as WordNet and try to parse natural language definitions and WordNet Synsets. Examples of such works include OntoLearn (Navigli et al., 2004; Velardi et al., 2005; and Rigau et al., 1998). Others rely on thesauri as the knowledge source (Van Assem et al., 2004).

The approach of Volz et al. (2003) converts an xml schema into a domain ontology by translating non-terminal and terminal symbols into concepts and roles using a set of rules. Similarly, the work of Stojanovic et al. (2002) uses a rule mapping scheme to convert an xml schema or a relational database schema into a domain ontology. Finally, we can cite the work of Delteil et al. (2001) who created an ontology learning procedure from RDF annotations and Nyulas et al. (2007) who created a plug-in (for Protégé) for importing relational databases into an ontology editing environment.

6.4.2 Statistical and Machine Learning Techniques

The approach of Suryanto and Compton (2001) aims at learning an ontology from a rule knowledge base in the medical domain. Their algorithm creates a set of classes and uses statistical measures to determine the relationships between the classes (subsumption, similarity, mutual-exclusivity). The work of Jannink and Wiederhold (1999) extracts a graph structure from dictionaries and uses statistical filtering and the PageRank algorithm to determine important relationships and concepts. Another example is the work of Papatheodorou et al. (2002), who build taxonomies using cluster mining from xml or RDF domain repositories.

6.5 Ontology Update and Evolution

Despite the important number of initiatives for ontology learning, the results are not still completely satisfactory and the field has to gain more maturity. Moreover, the evolution of ontologies seems to be even less supported in the research community. In fact, enabling this evolution (semi)automatically is a key factor for the Semantic Web and this involves the ability to update an ontology with new concepts, relationships, properties and axioms, the ability to appropriately place a concept in the taxonomy and the ability to perform mapping and alignment between existing ontologies. Here again, we have only provided a brief and incomplete overview of the NLP-based and statistical and machine learning techniques. We also want to point out that we have notdealt with change, versioning and consistency management during the evolution process, as we prefer to refer the reader to Haase and Sure (2004) and Flouris et al. (2006) to gain more insight into this question. Other interesting questions not dealt with here include: ontology matching and alignment (Shvaiko and Euzenat, 2008).

Ontology evolution can target each component of the ontology learning process. From the NLP side, enriching an existing concept with new attributes and relationships has been done in the work of Velardi et al. (2005) by searching the concept in WordNet and reusing its Synsets to enrich the ontology. This involves word meaning disambiguation. For more on instance extraction, we refer the reader to section 3.6.

From the statistical and machine learning side, there has been attempts to add new concepts to the ontology taxonomy. Updating the ontology with a new concept involves placing it correctly in the hierarchy and retrieving appropriate parents. A number of categorization techniques have been used to augment an ontology with a new concept: the *k-nearest neighbor method (kNN),* the *category-based method* and the *centroid-based method* (Maedche et al., 2002). These methods use vector-based features for representing concepts based on co-occurrence and word windows. The new concept can then be placed in the hierarchy according to similarity metrics with existing concepts in the ontology. Maedche et al. (2002) provide a good review about these methods.

6.6 Current Challenges

There are many challenges that face the ontology engineering as well as computer-based educational communities which consider using ontologies for domain knowledge representation. These can be divided into general and ITS-specific challenges.

Despite the large number of available systems, there is still room for more development in ontology engineering. More importantly, a reusable framework has to be set up to make combining and comparing different extraction methods easier. In fact, there is a lack of reusable services for ontology learning, updating and evaluation. There is also a lack of a comprehensive framework that highlights the available methods for each subtask of the ontology learning process based on various criteria (corpus, task, etc.). Such a framework would provide informed choices for ontology learning. From my point of view, a service-oriented architecture is essential for a broader development and reuse of automatic methods for ontology learning.

One other issue relating to automatic methods for ontology learning is that these methods can produce inconsistent or duplicate entries and dealing with such inconsistencies is particularly challenging (Volker et al., 2008). Inconsistencies can arise not only from the methods used, but also from the input data, which may be too sparse or may contain contradictions. Volker et al. (2008) propose three alternatives: using a reasoning-supported process to guarantee that the learned ontologies are kept consistent over time; repairing consistencies following the production of an ontology; or setting up reasoning mechanisms that can deal with these inconsistencies.

Adding to the list of challenges is the limited support regarding several key aspects of ontology engineering, especially ontology evolution, reuse, merging, alignment and matching. These different areas still need to mature. It is especially important to make available an environment entirely dedicated to ontology engineering involving all the different aspects of the ontology lifecycle.

The general challenges of ontology engineering impact the specific ITS-related challenge of building a domain model. In fact, successful attempts to build an ITS domain model automatically have been limited (Suraweera et al., 2004). Ontology engineering can help satisfy this need and contribute to the wider adoption of Intelligent Tutoring Systems. Moreover, ontology engineering can contribute to building a bridge with the eLearning community by making eLearning resources the main material for building the ITS domain model (Zouaq and Nkambou, 2009a). Similarly, eLearning can benefit from this domain model for indexing learning resources and developing more "intelligent" techniques for training learners.

6.7 Conclusion

We have described the field of domain ontology engineering, focusing on ontology learning techniques and highlighting how intelligent tutoring systems may benefit from this ontology engineering. One of the main advantages of this engineering is that it can provide a solution to two issues: first, the difficulty of building an ITS domain model from scratch for each domain and second, the difficulty of sharing and reusing the available representations. As standard knowledge representations, ontologies can support the ITS community in producing ITS components more easily and at lower costs. However, this involves the availability of a unified framework for the entire ontology lifecycle, including ontology learning, evolution, alignment, matching and evaluation.

References

Almuhareb, A., Poesio, M.: Attribute-based and value-based clustering: an evaluation. In: Proc. of EMNLP, Barcelona (July 2004)

Almuhareb, A., Poesio, M.: Finding Concept Attributes in the Web. In: Proc. of the Corpus Linguistics Conference, Birmingham (July 2005)

Amardeilh, F., Laublet, P., Minel, J.-L.: Document annotation and ontology population from linguistic extractions. In: Proc. of the 3rd international conference on Knowledge capture, Banff, Alberta, Canada, pp. 161–168 (2005)

Aroyo, L., Dicheva, D.: The New Challenges for E-Learning: The Educational Semantic Web. Educational Technology & Society 7(4), 59–69 (2004)

Aussenac-Gilles, N., Biebow, B., Szulman, S.: Revisiting Ontology Design: A Methodology Based on Corpus Analysis. In: Dieng, R., Corby, O. (eds.) EKAW 2000. LNCS (LNAI), vol. 1937, pp. 172–188. Springer, Heidelberg (2000)

Baker, C.F., Fillmore, C.J., Lowe, J.B.: The Berkeley FrameNet project. In: Proc. of the COLING-ACL, Montreal, Quebec, Canada (1998)

Bisson, G., Nedellec, C., Canamero, L.: Designing clustering methods for ontology building – The Mo'K workbench. In: Proc. of the ECAI Ontology Learning Workshop, pp. 13–19 (2000)

Buitelaar, P., Olejnik, D., Sintek, M.: A protege plug-in for ontology extraction from text based on linguistic analysis. In: Fensel, D., Sycara, K., Mylopoulos, J. (eds.) ISWC 2003. LNCS, vol. 2870. Springer, Heidelberg (2003)

Buitelaar, P., Cimiano, P., Magnini, B.: Ontology Learning from Text: An Overview. In: Buitelaar, P., Cimiano, P., Magnini, B. (eds.) Ontology Learning from Text: Methods, Evaluation and Applications. Frontiers in Artificial Intelligence and Applications Series, vol. 123. IOS Press, Amsterdam (July 2005)

Caraballo, S.A.: Automatic construction of a hypernym-labeled noun hierarchy from text. In: Proc. of the 37th Annual Meeting of the Association for Computational Linguistics, pp. 120–126 (1999)

Charniak, E., Berland, M.: Finding parts in very large corpora. In: Proc. of the 37th Annual Meeting of the ACL, pp. 57–64 (1999)

Cimiano, P.: Ontology Learning Attributes and Relations. In: Ontology Learning and Population from Text, pp. 185–231. Springer, Heidelberg (2006)

Cimiano, P., Wenderoth, J.: Automatic Acquisition of Ranked Qualia Structures from the Web. In: Proc. of the 45th Annual Meeting of the Association for Computational Linguistics (ACL), Prague (2007)

Cimiano, P., Wenderoth, J.: Automatically Learning Qualia Structures from the Web. In: Proc. of the ACL Workshop on Deep Lexical Acquisition, pp. 28–37 (2005)

Cimiano, P., Völker, J.: Text2Onto–A Framework for Ontology Learning and Data-driven Change Discovery. In: Montoyo, A., Muńoz, R., Métais, E. (eds.) NLDB 2005. LNCS, vol. 3513, pp. 227–238. Springer, Heidelberg (2005a)

Cimiano, P., Volker, J.: Towards large-scale, open-domain and ontology-based named entity classification. In: Proc. of RANLP 2005, Borovets, Bulgaria, pp. 166–172 (2005b)

Cimiano, P., Pivk, A., Schmidt-Thieme, L., Staab, S.: Learning Taxonomic Relations from Heterogeneous Sources of Evidence. In: Ontology Learning from Text: Methods, Evaluation and Applications. IOS Press, Amsterdam (2004)

Cimiano, P., Staab, S.: Learning by googling. ACM SIGKDD Explorations 6(2), 24–33 (2004)

Claveau, V.: Acquisition automatique de lexiques sémantiques pour la recherche d'information. Thèse de doctorat, Université de Rennes-1, Rennes (2003)

Corcho, O., Gómez-Pérez, A.: A Roadmap to Ontology Specification Languages. In: Dieng, R., Corby, O. (eds.) EKAW 2000. LNCS (LNAI), vol. 1937, pp. 80–96. Springer, Heidelberg (2000)

Craven, M., Di Pasquo, D., Freitag, D., McCallum, A., Mitchell, T.M., Nigam, K., Slattery, S.: Learning to construct knowledge bases from the World Wide Web. Artificial Intelligence 1-2(118), 69–113 (2000)

Deiltel, A., Faron-Zucker, C., Dieng, R.: Learning ontologies from rdf annotations. In: Proc. of the IJCAI Workshop in Ontology Learning (2001)

Etzioni, O., Cafarella, M., Downey, D., Kok, S., Popescu, A.-M., Shaked, T., Soderland, S., Weld, D.S., Yates, A.: Web-scale information extraction in KnowItAll (preliminary results). In: Proc. of the 13th World Wide Web Conference, pp. 100–109 (2004)

Faure, D., Nedellec, C.: A corpus-based conceptual clustering method for verb frames and ontology. In: Velardi, P. (ed.) Proc. of the LREC Workshop on Adapting lexical and corpus resources to sublanguages and applications, pp. 5–12 (1998)

Frantzi, K., Ananiadou, S., Tsuji, J.: The c-value/nc-value method of automatic recognition for multi-word terms. In: Nikolaou, C., Stephanidis, C. (eds.) ECDL 1998. LNCS, vol. 1513, pp. 585–604. Springer, Heidelberg (1998)

Fleischman, M.: Automated Subcategorization of Named Entities. In: 39th Annual Meeting of the ACL. In: Student Research Workshop, Toulouse, France (July 2001)

Fleischman, M., Hovy, E.H.: Fine Grained Classification of Named Entities. In: COLING 2002 (2002)

Flouris, G., Plexousakis, D., Antoniou, G.: Evolving Ontology Evolution. In: Proc. of the 32nd International Conference on Current Trends in Theory and Practice of Computer Science (SOFSEM-06), pp. 14–29 (2006) (invited talk)

Fortuna, B., Grobelnik, M., Mladenic, D.: OntoGen: Semi-automatic Ontology Editor. In: HCI International 2007, Beijing (2007)

Fortuna, B., Mladevic, D., Grobelnik, M.: Visualization of Text Document Corpus. In: ACAI 2005 Summer School (2005)

Automatic Discovery of Part–Whole Relations. In: Proc. of the, Conference of the North American Chapter of the Association for Computational Linguistics on Human Language, pp. 1–8. Association for Computational Linguistics, Morristown (2003)

Guarino, N.: Concepts, attributes and arbitrary relations: some linguistic and ontological criteria for structuring knowledge base. Data and Knowledge Engineering 8, 249–261 (1992)

Guarino, N., Welty, C.: Evaluating Ontological Decisions with OntoClean. Communications of the ACM 45(2), 61–65 (2002)

Haase, P., Sure, Y.: State-of-the-Art on Ontology Evolution, SEKT deliverable 3.1.1.b (2004), http://www.aifb.uni-karlsruhe.de/WBS/ysu/publications/SEKT-D3.1.1.b.pdf

Harris, Z.: Distributional structure. Word 10(23), 146–162 (1954)

Hearst, M.: Automatic Acquisition of Hyponyms from LargeText Corpora. In: Proc. of the Fourteenth International Conference on Computational Linguistics, Nantes, pp. 539–545 (1992)

Hearst, M., Schutze, H.: Customizing a lexicon to better suit a computational task. In: ACL SIGLEX Workshop, Columbus, Ohio (1993)

Hindle, D.: Noun classification from predicate-argument structures. In: Proc. of ACL 1990, Pittsburg, Pennsylvania, pp. 268–275 (1990)

Iwanska, L.M., Mata, N., Kruger, K.: Fully automatic acquisition of taxonomic knowledge from large corpora of texts. In: Natural Language Processing and Knowledge Processing, pp. 335–345. MIT/AAAI Press (2000)

Jannink, J., Wiederhold, G.: Ontology maintenance with an algebraic methodology: A case study. In: Proc. of AAAI workshop on Ontology Management (1999)

Kipper, K., Dang, H.D., Palmer, M.: Class-Based Construction of a Verb Lexicon. In: Proc. of AAAI-2000 Seventeenth National Conference on Artificial Intelligence, pp. 691–696 (2000)

Lin, D., Pantel, P.: Induction of semantic classes from natural language text. In: Proc. of SIGKDD 2001, San Francisco, CA, pp. 317–322 (2001)

Lin, D.: Automatic identification of non-compositional phrases. In: Proc. of ACL 1999, pp. 317–324 (1999)

Lin, D.: Automatic Retrieval and Clustering of Similar Words. In: Proc. of COLING-ACL 1998, Montreal, Canada, pp. 768–774 (1998)

Maedche, A., Pekar, V., Staab, S.: Ontology Learning Part One—On Discovering Taxonomic Relations from the Web. Web Intelligence, pp. 301–322. Springer, Heidelberg (2002)

Maedche, A., Staab, S.: Ontology Learning for the Semantic Web. IEEE Intelligent Systems 16(2), 72–79 (2001)

Moldovan, D.I., Girju, R.C.: An interactive tool for the rapid development of knowledge bases. International Journal on Artificial Intelligence Tools (IJAIT) 10(1-2) (2001)

Navigli, R., Velardi, P., Cucchiarelli, A., Neri, F.: Quantitative and Qualitative Evaluation of the OntoLearn Ontology Learning System. In: Proc. of the 20th international conference on Computational Linguistics, Switzerland (2004)

Navigli, R., Velardi, P.: Learning Domain Ontologies from Document Warehouses and Dedicated Web Sites. Computational Linguistics 30(2), 151–179 (2004)

Nyulas, C., O'Connor, M., Tu, S.: Datamaster–a plug-in for importing schemas and data from relational databases into protégé. In: Proc. of 10th Intl. Protégé Conference, Budapest (2007)

Papatheodorou, C., Vassiliou, A., Simon, B.: Discovery of Ontologies for Learning Resources Using Word-based Clustering. In: Proc. of ED-MEDIA 2002, AACE, Denver, USA (2002)

Poesio, M., Almuhareb, A.: Identifying Concept Attributes Using A Classifier. In: Proc. of the ACL Workshop on Deep Lexical Acquisition, pp. 18–27. Association for Computational Linguistics, Ann Arbor (2005)

Poesio, M., Almuhareb, A.: Extracting concept descriptions from the Web: The importance of attributes and values. In: Buitelaar, P., Cimiano, P. (eds.) Bridging the Gap between Text and Knowledge, pp. 29–44. IOS Press, Amsterdam (2008)

Polson, M.C., Richardson, J.J. (eds.): Foundations of Intelligent Tutoring Systems. L. Erlbaum Associates Inc., Mahwah (1988)

Pustejovsky, J.: The generative lexicon. MIT Press, Cambridge (1995)

Ravi, S., Pasca, M.: Using Structured Text for Large-Scale Attribute Extraction. In: Proc. of the 17th ACM Conference on Information and Knowledge Management, CIKM-2008 (2008)

Reinberger, M.-L., Spyns, P.: Unsupervised Text Mining for the learning of DOGMA-inspired Ontologies. In: Buitelaar, P., Cimiano, P., Magnini, B. (eds.) Ontology Learning from Text: Methods, Applications and Evaluation. Advances in Artificial Intelligence, pp. 29–43. IOS Press, Amsterdam (2005)

Rigau, G., Rodríguez, H., Agirre, E.: Building Accurate Semantic Taxonomies from Monolingual MRDs. In: Proc. of the 17th International Conference on Computational Linguistics COLING-ACL 1998, Montreal, Quebec, Canada (1998)

Sabou, M.: Learning Web Service Ontologies: an Automatic Extraction Method and its Evaluation. In: Buitelaar, P., Cimiano, P., Magnini, B. (eds.) Ontology Learning from Text: Methods, Evaluation and Applications. IOS Press, Amsterdam (2005)

Sahlgren, M.: The Word-Space Model: Using distributional analysis to represent syntagmatic and paradigmatic relations between words in high-dimensional vector spaces. Ph.D. dissertation, Department of Linguistics, Stockholm University (2006)

Salton, G., Buckley, C.: Term-weighting approaches in automatic text retrieval. Information Processing & Management 24(5), 515–523 (1988)

Schlobach, S., Olsthoorn, M., de Rijke, M.: Type Checking in Open-Domain Question Answering. In: Proc. of European Conference on Artificial Intelligence, pp. 398–402. IOS Press, Amsterdam (2004)

Shamsfard, M., Barforoush, A.A.: The State of the Art in Ontology Learning: A Framework for Comparison. The Knowledge Engineering Review 18(4), 293–316 (2003)

Shamsfard, M., Barforoush, A.A.: An Introduction to HASTI: An Ontology Learning System. In: Proc. of 6th Conference on Artificial Intelligence and Soft Computing (ASC 2002), Banff, Alberta, Canada (2002)

Shvaiko, P., Euzenat, J.: Ten Challenges for Ontology Matching. In: Meersman, R., Tari, Z. (eds.) OTM 2008, Part I. LNCS, vol. 5331, pp. 1164–1182. Springer, Heidelberg (2008)

Snow, R., Jurafsky, D., Ng, A.Y.: Learning syntactic patterns for automatic hypernym discovery. In: Advances in Neural Information Processing Systems (NIPS 2004), Vancouver, British Columbia (2004)

Stojanovic, L., Stojanovic, N., Volz, R.: Migrating data-intensive Web Sites into the Semantic Web. In: Proc. of the 17th ACM symposium on applied computing (SAC), pp. 1100–1107. ACM Press, New York (2002)

Suraweera, P., Mitrovic, A., Martin, B.: The role of domain ontology in knowledge acquisition for ITSs. In: Lester, J.C., Vicari, R.M., Paraguaçu, F. (eds.) ITS 2004. LNCS, vol. 3220, pp. 207–216. Springer, Heidelberg (2004)

Suryanto, H., Compton, P.: Discovery of Ontologies from Knowledge Bases. In: Gil, Y., Musen, M., Shavlik, J. (eds.) Proc. of the First International Conference on Knowledge Capture, Victoria, British Columbia Canada, pp. 171–178. The Association for Computing Machinery, New York (2001)

Tanev, H., Magnini, B.: Weakly Supervised Approaches for Ontology Population. In: Proc. of EACL 2006, Trento, Italy, pp. 3–7 (2006)

Van Assem, M., Menken, M.R., Schreiber, G., Wielemaker, J., Wielinga, B.J.: A Method for Converting Thesauri to RDF/OWL. In: International Semantic Web Conference, pp. 17–31 (2004)

Van Hage, W.R., Kolb, H., Schreiber, G.: A Method for Learning-Part-Whole Relations. In: Cruz, I., Decker, S., Allemang, D., Preist, C., Schwabe, D., Mika, P., Uschold, M., Aroyo, L.M. (eds.) ISWC 2006. LNCS, vol. 4273 pp. 723–735. Springer, Heidelberg (2006)

Velardi, P., Navigli, R., Cuchiarelli, A., Neri, F.: Evaluation of ontolearn, a methodology for automatic population of domain ontologies. In: Buitelaar, P., Cimiano, P., Magnini, B. (eds.) Ontology Learning from Text: Methods, Applications and Evaluation. IOS Press, Amsterdam (2005)

Volker, J., Haase, P., Hitzler, P.: Learning Expressive Ontologies. In: Buitelaar, P., Cimiano, P. (eds.) Ontology Learning and Population: Bridging the Gap between Text and Knowledge. Frontiers in Artificial Intelligence and Applications, vol. 167, pp. 45–69. IOS Press, Amsterdam (2008)

Volz, R., Oberle, D., Staab, S., Studer, R.: OntoLiFT Prototype. IST Project 2001-33052 WonderWeb Deliverable 11 (2003)

Widdows, D.: Unsupervised methods for developing taxonomies by combining syntactic and statistical information. In: Proc. of HLT-NAACL, pp. 197–204 (2003)

Widdows, D., Dorow, B.: A graph model for unsupervised lexical acquisition. In: 19th International Conference on Computational Linguistics, Taipei, Taiwan, pp. 1093–1099 (2002)

Woods, W.A.: What's in a link: Foundations for semantic networks. In: Bobrow, D.G., Collins, A.M. (eds.) Representation and Understanding: Studies in Cognitive Science, pp. 35–82. Academic Press, New York (1975)

Yamada, I., Baldwin, T.: Automatic discovery of telic and agentive roles from corpus data. In: Proc. of the 18th Pacific Asia Conference on Language, Information and Computation (PACLIC), Tokyo, pp. 115–126 (2004)

Zouaq, A., Nkambou, R.: Enhancing Learning Objects with an Ontology-Based Memory. IEEE Transactions on Knowledge and Data Engineering 21(6), 881–893 (2009a)

Zouaq, A., Nkambou, R.: Evaluating the Generation of Domain Ontologies in the Knowledge Puzzle Project. IEEE Transactions on Knowledge and Data Engineering 21(11), 1559–1572 (2009b)

Part II
Tutor Modeling and Learning Environments

Chapter 7
Modeling Tutoring Knowledge

Jacqueline Bourdeau[1] and Monique Grandbastien[2]

[1] Télé-université, University of Quebec in Montreal, 100, Sherbrooke W., Montreal (Quebec), H2X 3P2, Canada
bourdeau@teluq.uqam.ca
[2] LORIA, Université Henri Poincaré Nancy1, Campus scientifique, BP 239, 54506 Vandoeuvre Cedex, France
monique.grandbastien@loria.fr

Abstract. This chapter introduces Part II on modeling tutoring knowledge in ITS research. Starting with its origin and with a characterization of tutoring, it proposes a general definition of tutoring, and a description of tutoring functions, variables, and interactions. The Interaction Hypothesis is presented and discussed, followed by the development of the tutorial component of ITSs, and their evaluation. New challenges are described, such as integrating the emotional states of the learner. Perspectives of opening the Tutoring Model and of equipping it with social intelligence are also presented.

7.1 Introduction

In the field of Artificial Intelligence in Education (AIED), the research on ITSs has attempted to develop, implement, and test systems that contain AIED principles and techniques. The "I" in ITS represents "intelligence" (AI techniques), the "T" is an abbreviation for "tutor," and the "S" signifies the application "software." In this chapter, the challenges of modeling tutoring knowledge are addressed, along with the problems that have arisen and have been dealt with, the solutions that have been tested, and the open questions that have been raised. This chapter begins by defining the term "tutoring", first, to distinguish it from education, instruction, and teaching, and, second, to place it in the context of ITS research. This chapter also provides insight into the following three main processes involved in building ITSs: modeling, developing, and evaluating the aspects and behavior of the tutoring system. This chapter concludes with a series of open questions.

7.2 What Is Tutoring?

Since tutoring is a concept from the field of education, we start by looking at the origins and meaning of tutoring. We then analyze tutoring in the context of ITSs,

R. Nkambou et al. (Eds.): Advances in Intelligent Tutoring Systems, SCI 308, pp. 123–143.
springerlink.com

as well as its functions and variables. We lastly attempt to understand better why tutoring is the foundation of ITS research.

7.2.1 Tutoring in Education

Tutoring has always been a form of education and a means of instruction. It has two main properties: 1) the tutor/student ratio is 1/1-3 (in most cases 1/1), so tutoring is often interpreted as individualized instruction since the attention of the tutor is totally focused on one student; 2) guidance or tutor control occurs, although this control may be shared with the student by means of guided discovery or cognitive apprenticeship (Collins 2006). To understand tutoring better, two aspects of education need to be clarified. Whereas instruction focuses on the role of the instructor, learning focuses on the processes conducted by the learner, which may occur with or without formal instruction. Tutoring focuses on the personalization of the instruction-learning process through various types of interactions. Although instruction and learning are often considered two different processes, they can also be considered one entity defined by the interactions between the two: This is called "Instructional-Learning Event" in the OMNIBUS ontology (see Chapter 11). Tutoring is also seen as a locus of control, fluctuating between single or mixed initiatives. From this perspective, the Socratic dialogue is led by the tutor, who keeps questioning the learner until she becomes knowledgeable and aware of this process, i.e. learning. In terms of adaptation, tutoring plays a contrary role to teaching: A teacher requires the students to adapt to the class, whereas a tutor adapts to the student. Both tutoring and teaching are roles, which may overlap in real settings and be played by the same person.

7.2.2 Tutoring in ITSs

After extensive research, the ITS community has come to realize that tutoring with ITSs involves, in fact, not merely guidance from the tutor, but also interaction between the tutor and the learner, who work together as a duo (Graesser et al. 2001). This interaction may be a single or mixed initiative. Whereas teaching can happen without explicit interaction (and ultimately with little learning), such as is the case when lecturing, tutoring cannot. Interaction between the tutor and student, as well as ongoing and active adaptation on the part of the tutor, is a fundamental feature of tutoring. Consequently, the main challenge for the tutoring component of an ITS is to design interactions so as to obtain a precise adaptation and to shape the tutoring behavior by reasoning about rapidly real-time data from the regular interactions between the learner and the system, for example, with machine learning techniques. While the work on student and agent models is continuing to evolve in ITS research design, the tutoring function is constantly being revisited and occupies a central position in ITSs and ITS research (Graesser et al. 2001; Heffernan 2001; Lajoie et al 2001; Virvou 2001). The research on the Interaction Hypothesis is continuing to grow: For instance, VanLehn (2008) interpreted the Interaction Hypothesis as predicting that the effectiveness of tutoring increases monotonically with the degree of interactivity. That is, the more interactive the tutoring, the larger the learning gains produced. Thus, the most interactive form of tutoring (face-to-face human tutoring) should be the most effective. However, VanLehn's review suggested that there may be diminishing returns, and that after interactivity reaches the level of step-based tutors (VanLehn 2006), learning gains no longer increases with interactivity. That is, there is an Interaction Plateau. This controversy indicates how active this area research has become.

Nonetheless, the original challenge of matching the success of good human tutors remains constant in ITS research. The research agenda is vast: The objectives are, first, to define and, as much as possible, to reproduce the tutoring functions and the tutoring variables; second, to find out what "causes" success; and, third, to determine how to measure success within and while interacting with an ITS.

7.2.3 Tutoring Functions

The two main functions of instruction and, therefore, of tutoring are to foster and assess learning, both of which ITS research has mainly addressed separately and sometimes together, as in (VanLehn 2007) and in Heffernan and Koedinger's ASSISTment system (see Chapter 20). The term "ASSISTment" was coined by Koedinger to combine "assistance" and "assessment." Although "assistance" implies that the control is with the learner, in the case of "ASSISTment," the definition is broadened to mean that the learner is helped through guided assistance or guided discovery. Tutorial dialogue design and management have been a focus for the team at the University of Memphis for a decade and are reported in Chapter 8.

When and how to assist or tutor learners is the fundamental dilemma of tutoring, as discussed in the papers entitled *To Tutor or Not to Tutor: That is the Question* (Razzaq & Heffernan, 2009b), and *Does Help Help*? (Beck & al., 2008).

Helping and tutoring the learner can be divided into two sub-functions: cognitive diagnosis, defined as the detection of the sources of errors, and the selection of tutoring or remediation strategies.

By considering the emerging capacity of ITSs to compute the learner's affective states (see Chapters 10 & 17), new challenges are being addressed by researchers, such as how to balance cognitive and motivational dimensions in tutorial strategies (Boyer et al. 2008). This latest development has added to the complexity of computing the learner's states (cognitive, meta-cognitive, and affective) and reasoning abilities in order to make optimal tutoring decisions.

7.2.4 Tutoring Variables

An ITS researcher is confronted with a multitude of variables when designing a tutoring system: Who, What, How and Where does the tutoring happen? Each decision presents its own questions and challenges. Managing the interdependencies of a tutoring system is a challenge in itself.

The first question that an ITS researcher should ask is: Who is the learner? The first variable is the learner's characteristics: age, previous knowledge and skills, motivation, goals, culture, learning difficulties, etc. Knowing the learner is a necessary condition for adapted and adaptive tutoring. ITS research has long focused on how to induce the cognitive states of the learner. The emotional state of the learner is a new variable that has only recently been taken into consideration. Detecting emotions in real time and taking them into account in adapting the tutor is a broad field for investigation.

The second question that an ITS researcher should ask is: What is being tutored? The second variable is defined as the contents, subject matter, curriculum, or the "what-to-learn" in the OMNIBUS ontology (see 11). Classifications of this variable include the characteristics of the domain (math, science, language, attitude learning), the topics (concept or rule learning), the skills (problem solving, reading, writing), and the competencies.

The third question that an ITS researcher should ask is: How is the content to be tutored? This variable corresponds to tutoring strategies, such as scaffolding/fading, single or mixed initiative in dialogue, eliciting knowledge, errors or plans, diagnosing, planning, feedback/remediation, steps/hints, intervening/refraining, stimulated and support learning from errors, providing multiple visualizations, orchestrated collaborative learning, manipulation of variables, and selecting an assessment strategy. Most of the fundamental questions concern not only the adaptability of the system, but also the effectiveness of the tutoring, such as: "When are tutorial dialogues more effective than reading?" (VanLehn et al. 2007).

The last question that an ITS researcher should ask is: In what context is the tutoring taking place? This last variable can be interpreted in several ways: the physical context (i.e., classroom, workplace, home environment, distance learning, mobile learning, and web-based learning), the cultural context (see Chapter 24), the cognitive context, or the affective context. Whereas the first three categories of variables have been extensively used in existing ITSs, the fourth one is more recent. Indeed, the context has been implicitly limited to one learner in front of one computer. The limited knowledge of the context appears to be a major drawback in ITS development at this point.

7.2.5 Tutoring as a Foundation for ITS

AIED was emerging as a field of study (Wenger 1987) when Bloom (1984) published the results of his 10-year studies, and Cohen et al. (1982) published their meta-analysis on tutoring. Bloom's studies revealed a 2-sigma effect of tutoring over group instruction. Consequently, Bloom called for effective group instruction since the cost of tutoring was too high. However, AIED researchers saw promising possibilities for building effective systems that could foster and assess learning effectively and efficiently with tutoring, since using costly human tutors for each student was not an option, except in exceptional situations such as private tutoring or distance learning[1]. Developing appropriate and effective tutoring systems, thus,

---

[1] In distance learning (e.g., Tele-University), the learning activities and contents of a course are designed by a professor to allow for independent study, and the dialogue with the student is conducted by a human tutor over the phone or by e-mail.

became the challenge of many AIED researchers. However, this focus soon became the subject of controversy due to its tutor-centered view. The ITS community has since been open to different perspectives (Vassileva and Wasson 1996; Lajoie 2000) and has evolved in such a way that tutoring has come to include many approaches to foster and assess learning. These approaches are presented in the following chapters.

7.3 Modeling Tutoring Knowledge

If the goal of an ITS is to be at least as effective as a good human tutor, and if being so requires that ITSs mimic human tutors, what should the tutoring model consist of in ITSs? Human tutors generally have a comprehensive knowledge of the curriculum, cognition and learning, and tutoring strategies. They have: a) facts about the student that they update regularly, b) perceptions of the student's personality and moods, and c) the ability to modify their tutoring strategies in order to optimize the learning experience with the student. The objective of this chapter is not to provide an in-depth analysis of past and present attempts to model tutoring in ITSs, but rather to highlight key issues.

Section 7.3 looks at four aspects of tutoring knowledge. First, it presents the different sources of tutoring knowledge and their role in modeling the tutor. Second, it characterizes tutoring. Third, it defines tutoring. Lastly, it briefly discusses the issues raised by the design of different tutorial interactions.

7.3.1 *Sources of Tutoring Knowledge*

In their paper entitled, *Modeling human teaching tactics and strategies for tutoring systems*, duBoulay and Lukin (2001) noticed "three principled methodologies for developing the teaching expertise in AIED systems, namely the observation of human teachers, the study of learning theories and the observation of real students interacting with online systems." These three procedures must still be considered today. As the challenge is to reach the success of good human teachers, it is worthwhile to observe their teaching behavior. However, it is important to note the differences between teaching in a classroom setting with a human teacher and tutoring with an ITS. First, tutoring 1/1-3 students is different from teaching a class of 30 or more students. Second, duBoulay and Luckin (2001) question "...whether tactics that are effectively applied by human teachers can be as effective when embodied in machine teachers." Third, Grandbastien (1999) mentions that: "Last, but not least, existing teaching expertise was mainly built without computers in the schools." Indeed, although computers have been used for learning purposes for several decades now, there is a lack a studies focusing on how tutoring guidance and decisions differ when they are to be implemented out of or within interactive environments. An attempt is being made to bridge this gap by keeping teachers in the design loop when building ITSs. The observation of novice tutors in schools was a primary source for designing AutoTutor, just as the observation of expert tutors is the main one for developing the Guru system (Chapter 9).

Several researchers in the field of ITSs have based their work directly on cognitive theories (see Chapter 8), or on collaborative and social learning theories (Greer et al. 1998). The main theories that have been used as sources of tutoring knowledge are the following: Bloom's Mastery Learning, Anderson's Cognitive Theory – ACT-R- (Koedinger and Corbett 2006), Vigotsky's Zone of Proximal Development (Luckin and duBoulay 1999), and Gagne's instructional design theory (Pirolli and Greeno 1988; Nkambou et al. 2003; Murray 2003). Sources of theoretical knowledge on learning and instruction were reviewed in Greeno et al. (1996) and, more recently, in the collection of theories edited by Sawyer (2006). Sources of practical teaching knowledge pose serious difficulties in terms of formalizing and modeling (Grandbastien 1999; duBoulay and Luckin 2001), since teaching practice is ill-defined and hard to systematize. Moreover, the idea of merging theoretical knowledge with practical knowledge into a united knowledge model remains an open problem. However, studies continue to be conducted in the areas of the observation and systemization of the behavior of tutors and teachers in ITS research. The structuring of a simple tutor dialogue was the basis for AutoTutor (Graesser et al. 2001), and the concept of map modeling has been proposed for building the Guru system (Chapter 9).

Another source of knowledge that may improve ITS behavior originates from the data collected from the learner's interactions with the system. The challenges for data mining and machine learning techniques in ITS research are well-documented in Part III and in Woolf's recent book (2009). Educational Data Mining (EDM) has recently evolved as an autonomous field of research. This field gives its own conferences, even if most researchers contribute to both ITS and EDM.

Making knowledge explicit from all sources contributes to the domain of AIED, just as it would for any domain, and contributes to the advancement of the learning sciences; it is also instrumental in building tutoring systems. Declarative knowledge, as in ontology, provides a better opportunity for criticism than procedural knowledge, and promotes discussions which can be consensus-building.

Given the heterogeneity of the various sources of knowledge, it is difficult to characterize and define tutoring.

7.3.2 Characterizing Tutoring

In contrast to the tradition in ITS research, which is to define and describe ITSs through their structure and components, VanLehn (2006) characterized ITSs through their behaviors. He considered two loops in a tutoring system: an outer loop at the task level and an inner loop at the step level. His claim was that systems that have inner loops should be called ITSs, since they demonstrate the adaptive capability of the systems, while systems that have only an outer loop should be called CAI (Computer Aided Instruction), CBT (Computer Based Training), CAL (Computer Aided Learning), etc. This claim was criticized by both duBoulay (2006) and Lester (2006). duBoulay stated that this characterization "…applies only to a certain subclass of ITSs (…) whose intelligence is devoted to maximizing the chances [the student has of learning] how to solve a certain kind of multi-step problem in technical domains. (…) [T]he two loop structure can still be imposed on the behavior of the system, but the nature of what counts as a step and a task may differ." Lester supported duBoulay's statement by describing what he called "the KVL Tutoring Framework" He stated that VanLehn has "distilled more than three decades of research into a generic procedure defined by two nested loops that iterate over tasks and steps," and this framework "marks the arrival of AI in education as a mature field." However, he imagined that, in the future, one could question "the lack of ill-defined task domains and the dearth of fielded systems." He concluded by saying that this framework implicitly provides a research agenda and could serve as a roadmap for AIED researchers.

In an earlier article, Graesser, VanLehn et al. (2001) reflected on the complex nature of tutoring and proposed the Interaction Hypothesis. They conducted a detailed analysis of interactions in several developed ITSs. They described how these interactions were structured and expressed and how the student's expressions were predicted, interpreted and assessed. The tutoring strategy in Graesser's AutoTUTOR is based on structured dialogues in natural language with an "understanding" of the student's expression. The dialogue is guided by an agent that gives encouragement, approval, or other feedback. The interactive strategy in VanLehn's ANDES for problem solving in physics relies upon model tracing. It tracks the student's actions and adapts the dialogue accordingly. It should be noted that Graesser et al.'s (2001) conclusions drew upon dialogues in natural language and may not apply to systems that use speech recognition, such as LISTEN (Mostow et al. 2001), or sketch recognition (see Chapter 12).

Characterization of ITSs remains a constant concern, as illustrated above by VanLehn and Graesser's efforts. Its definition, therefore, continues to evolve, as presented in the next section.

7.3.3 Towards a Definition of Tutoring in ITSs

In 1999, John Self distinguished ITSs and ITS research; he characterized ITSs as systems that support learners:

> ITSs are computer-based learning systems which attempt to adapt to the needs of learners and are therefore the only such systems which attempt to 'care' about learners in that sense. Also, ITS research is the only part of the general IT and education field which has as its scientific goal to make computationally precise and explicit forms of educational, psychological and social knowledge which are often left implicit.

The first part of the definition of tutoring highlights the adaptive nature of ITSs as being central. The last part of the definition stresses the need for ITS researchers to contribute to the learning sciences. If the notion of "caring" can be interpreted as an attentive and sensitive adaptation to the cognitive and emotive states of the learner, it is somehow equivalent to good "tutoring."

Both this perspective and that of the Interaction Hypothesis (Graesser et al. 2001) permit tutoring with ITSs to be defined as <u>fostering and assessing learning through adaptive interaction between the student and the system</u>. Other people (e.g., teachers, students, workers) can be in the loop, but they are external and unrecognized by the system, except in Computer Supported Collaborative Learning (CSCL) and in Social Learning.

7.3.4 The Design of the Tutorial Interaction

As a consequence of this hypothesis, the key to the success (performance) of tutoring in ITSs is the design of the tutorial interaction. The design of ITSs must enable the designers to answer the question: Will the ITS enable dialogue, or not? Since dialogue is the most natural form of interaction, and, since it enables fine-tuned adaptation, dialogue in natural language has become central to many ITSs, despite the many challenges associated with it (Graesser et al. 2001).

An alternative to tutoring through dialogue is tutoring through actions by the use of a rich interface, such as in Andes (see Chapter 21). In this case, the following question arises: At what level is dialogue effective (VanLehn et al. 2006)? Putting agents or animated agents in charge of the tutoring is another alternative, as demonstrated by Johnson et al. (2000) and Chen and Wasson (2002).

In Discovery Learning Environments (DLE), tutoring is embedded in: 1) the simulation itself, which is a representation of a world or a phenomenon; 2) the allowed manipulation of variables by the learner; and 3) the feedback provided by the system. In a DLE, very little, if any, of the interaction between the tutor and the student occurs by means of a dialogue. Rather, the student and tutor communicate to one another via the student's actions (the manipulation of variables) and the feedback of the system (behavior or data). Most DLEs combine both dialogue and simulation.

7.3.5 Tutoring and Collaborative Learning

In 1988, Chan proposed tutoring one student first with a computer agent called "The computer as a learning companion," then with "Reciprocal Tutoring" and then with "Distributed learning companion systems" (Chan and Buskin 1988; Chan 1992; Chan et al. 1996). This line of research has further developed within ITS research under CSCL, which was presented in the "Special Issue on Computer Supported Collaborative Learning" of *IJAIED Journal* (1998), and has since been presented in all *ITS Proceedings*. It has evolved into an independent field, with a number of researchers working in both ITS and CSCL, for example, on ontologies for structuring group formation (Isotani and Mizoguchi 2008) and intelligent agents to support collaborative learning (Wasson 1998; Chen and Wasson 2003). In CSCL environments, learning activities are often not seen as "tutored." However, several teams have worked on activities in scenarios or scripts so as to create a way of integrating guidance. In CSCL research, an attempt to build adaptive scripts and to use learners' data for this adaptation is seen as equivalent as seeking adaptability in ITS research (see Chapter 22).

7.3.6 Tutoring and Adaptive Web-Based Learning

Brusilovsky started the area of research on Adaptive Hypermedia systems (Brusilovsky 1995) and edited the 2003 special issue of the *IJAIED Journal* on "Adaptive and Intelligent Web-Based Systems" (Brusilovsky and Peylo 2003). These systems attempt to be more adaptive than other systems as they are able to build a model of the goals, preferences and knowledge of each individual student and use this model throughout the interaction with the student in order to adapt to his/her needs. They integrate individual tutoring by incorporating and performing some activities traditionally executed by a human teacher - such as coaching students or diagnosing their misconceptions. The tutoring decisions result in few dialogues between the tutor and the learner and the decisions are mainly based on changing the way the system displays the subject matter to the learner (presentation adaptation) and on the availability/advice of the links to be followed from the page presented (navigation adaptation). The model used in Adaptive Hypermedia systems provides most of the control to the learner. As did EDM and CSCL, Adaptive Hypermedia has developed as an autonomous field and holds biannual conferences. In 2007, Israel & Aiken proposed an integration of CSCL and Intelligent Web-based Systems, which seems promising.

7.4 Developing a Model of Tutoring Knowledge

Although the term "modeling" can be interpreted in different ways (Baker 2001), this section defines "tutor modeling" as any form by which researchers try to conceptualize and to operationalize tutoring functions and variables. The first point considered is the location of the tutoring model in the conceptual architecture of an ITS. Next, a brief review of the current work and perspectives on the modeling of tutoring knowledge is examined. This section concludes by providing a perspective on the opening of a tutoring model.

7.4.1 A Tutoring Model in an ITS Architecture

A key issue in tutor modeling depends on various perspectives and, as such, the following questions need to be considered: Where in the architecture is the tutoring model? Is it at the core or at the periphery of the system? Does the tutoring model have one distinct component? Are its tutoring functions distributed throughout in the system? Do the tutoring functions emerge strictly from the student's data? Are the tutoring functions integrated with the student model? Should the student model be one component of the tutoring model, as is the case in human tutoring? Does the system have a human teacher in the loop? If so, the tutoring functions are shared between the human teacher and the system. Thus, these two actors may or may not have a communication channel. Woolf's (2009) illustration of architectures that can combine classic components and emerging knowledge, while having two humans in the loop, highlights the complexity of the architectural design of ITSs (Fig. 7.1).

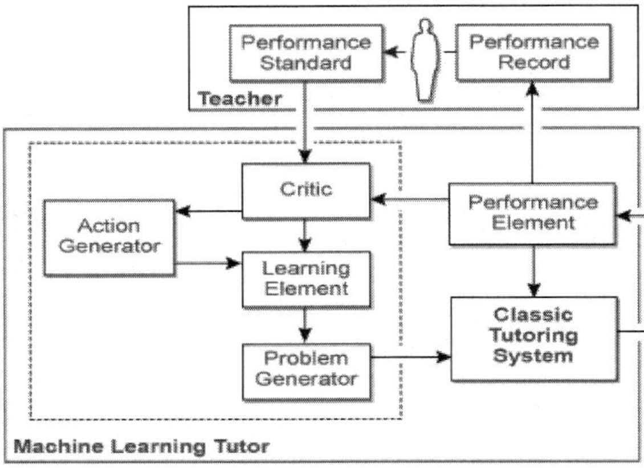

Fig. 7.1 Components of machine learning techniques integrated into an intelligent tutor (Woolf, 2009, reprinted with permission)

The issue regarding architecture arises from the following questions and relates is to the paradigms used to create the IT'S: 1) Which cognitive architecture would be the most beneficial in the view of the designer? 2) What does she think are the appropriate sources for tutoring knowledge? 3) Should tutoring exist autonomously, or should it be incorporated in the student model or in the curriculum model? These fundamental choices govern the architectural design, the design of the tutoring functions and the consideration given to the tutoring variables. Tutoring is defined as having two main functions: fostering and assessing. It is modeled according to the cognitivist, contructivist paradigms used, the sub-functions (whom, what, where, when and how), and the design of tutoring interactions or dialogues. The instructional design decisions are modeled depending on one's theory, be it ACT-R, piagetian, vitgoskyan, or any other cognitive, learning or assessment theory. The challenges regarding the coordination of the interdependence between the components of emerging and specific knowledge also depend on one's fundamental choices, i.e., paradigms. In other words, modeling tutoring can be defined as a series of decisions made on how to foster and assess the learning of knowledge, skills and competencies. These choices emphasize either the cognitive, meta-cognitive, or emotive dimensions; the individual consciousness, the context, or the collaboration; or very complex issues, such as cognitive diagnosis.

7.4.2 Reclaiming Tutoring Knowledge

Most research teams in the field of ITSs often incorporate tutoring knowledge in the domain model or the student model, depending on where they believe it would be the most appropriate. Heffernan's dissertation title explicitly expresses the concern that some researchers have regarding the weakening role of the tutor in ITSs: "ITS have Forgotten the Tutor" (2001). As a result, researchers are once again acknowledging the key role that tutoring knowledge plays in making tutoring decisions. Grandbastien (1999) reclaimed practical teaching knowledge; Heffernan (2001) as well as Virvou (2001) modeled human instructors; Pirolli & Greeno (1988) and Murray (1999, 2003) had instructional design knowledge as being central to ITSs; Hayashi et al. (2009) claimed that theoretical tutoring knowledge is a foundation to decision-making regarding tutorial strategies and instructional design tasks. This debate has been a productive one and has allowed for diversity. However, it has also limited the sharing and reusing when authoring ITSs, as this debate has fostered 'idiosyncratic' implementations.

7.4.3 Authoring the Tutor

This section provides an overview of current work and perspectives in the authoring of tutoring functions in ITSs (for an overview on authoring systems, see Chapter 18). As the previous section discussed, the tutoring functions are neither uniformly represented, nor consistently distinct from other functions or components. Some or all tutoring functions may be pervasive or prevalent in their components or agents, such as the diagnosis in the student model or the dialogue in the communication component. One possible exception is Heffernan's tutoring function (2001), in which he intentionally built and demonstrated a distinct tutoring component with reasoning capabilities. In addition to the paradigmatic issue, two current orientations, providing authoring tools rather than complete systems and increasing the accessibility and affordability of authoring ITSs, are discussed in Chapter 18.

The issue of strong dependence of tutoring functions on the ITS architecture is also true of authoring systems or tools, as illustrated by Woolf (2009). The availability of authoring processes and tools is determined by whether or not a tutoring function is distinct, autonomous, or pervasive.

Where is research in this field heading? A decade ago, Murray (2003) suggested that ontologies would be instrumental in authoring ITSs. In 2000, Mizoguchi and Bourdeau (2000) anticipated that ontological engineering would serve this purpose and expressed their desire for an ontology that would respect the diversity of ITSs, while serving the field with tutoring strategies explicitly anchored in either theoretical or empirical knowledge. Will this be a reality in the next 10 years? Hayashi et al. (2009) paved the way in this direction, suggesting that the ITS community commits to ontologies that would allow for sharing and reuse, as well as making tutoring strategies explicit and capable of being examined and edited.

7.4.4 Opening the Tutoring Model

Although some systems have been widely distributed, there are few products available for teachers and little enthusiasm from teachers to use them. One of the many reasons is that adaptation to the learning context is still weak compared to what a human actor can actually do. Grandbastien (1999) observed that teachers are willing to use products that enable them to be creative: Teachers need professional tools in the same way that craftsmen need theirs. In order to give human tutors these tools, the tutor model should be opened to them. However, the following questions would need to be addressed:

In what context could the tutor model be opened?

The opening could involve the preparation of the learning sequence (e.g., adding or replacing exercises), and of the behavior of the system during the learning sequence itself (e.g., weighing strategies in accordance with personal preferences), and the exploitation of post-session feedback.

For whom and for what purpose could the tutor model be opened?

Attempts to give more initiative to teachers to tailor ITSs to their needs are not new. For instance, in their paper entitled Teachers implementing pedagogy through REDEEM, Ainsworth et al. (1999) explained that REDEEM could be used as an authoring tool by teachers who had no previous experience, but it was still authoring. Current initiatives aim at providing flexible environments to meet the various needs expressed by educators: the PEPITE system (Delozanne et al. 2008), which includes a multi-criteria automatic assessor for high school algebra, and the ASSISTment Builder (Razzaq et al. 2009a; see Chapter 20). The PEPITE system provides the teacher a with a range of possibilities starting from building a complete set of assessment exercises to retrieving an exercise from an existing bank and using it as provided, or adapting the statement of the exercise by filling out forms. In addition to the raw answers, the teacher obtains an overall view of each student's competency in algebra, as well as the competencies of groups of students with similar student profiles. Once tailored to the user's needs, the environment may also be used independently by the learner to check her competencies and to access exercises adapted to her profile.

In the aforementioned specialized authoring tools, the tutor modeling aims to give more initiative to teachers. Moreover, opening the tutor model emphasizes the significance of opening the learner model, which has been investigated in much more depth (see Chapter 15). Consequently, further investigation is required to determine in which contexts and for which categories of learners the opening of the tutor model could improve learning.

7.5 Evaluating the Tutoring Model

The methodologies and tools used when studying and evaluating tutoring are of the utmost importance in ITS research. In her recent book, Woolf (2009) provides an extensive review of evaluation methods for ITSs. In addition, the *ITS '08 Best Paper Award* went to Beck. & al. (2008) for their paper comparing three methods of evaluating the efficacy of an ITS. When evaluation specializes in tutoring functionalities, the goal is to test the accuracy of the strategy, the effectiveness of the tools (error diagnosis, etc.), and/or the adaptability to individual situations. Making comparisons does not necessarily lead to the most meaningful results, and large *in situ* studies are often out of reach for small research teams. However, when available, the data gathered from large studies are of great significance in the field of ITSs and provide justification for their use in classrooms and other settings (Koedinger and Corbett 2006).

Causality underlies many projects which aim at determining what might "cause" learning (VanLehn et al. 2003; VanLehn et al. 2007). However, it is difficult to prove or demonstrate a causal relationship; thus, hypotheses are presented as associations or correlations between events and situations. Current pitfalls in evaluating the tutoring function of ITSs are ill-defined hypotheses, biased and confounded variables, and weaknesses in the evaluation design and/or the interpretation of results (Woolf 2009). In addition to the goal of testing the tutoring function, the goals of testing new techniques of visualization and detecting emotion need to be studied. As researchers are now able to compare methods of evaluating the same tutorial behaviors, exploration in this field of study is expanding (Zhang et al. 2008).

Advanced research teams (Cognitive Tutors, Constraint-based Tutors, DLE systems, AutoTUTOR systems) have constituted large corpora of data (from log files and other sources) together with analytical tools specific to their line of research. For example, the Pittsburgh Science of Learning Center has developed a data storage and analysis facility called Data shop, which has a series of analytical tools, such as the Error Report and the Learning Curve Generator (VanLehn 2007). As a result, the potential for analysis has greatly improved and has developed the study of representations in ITSs. In addition, the concept of "steps" (the smallest possible correct entry that a student can make) has been revisited and become a key concept for cognitive tutors and their evolution.

Although Design-Based Research (DBR) is only in its early stages, it has much to offer to ITS research. This research paradigm has grown out the need to bring together design and *in situ* studies. It was built upon the general "design loop" with a series of sequences added to improve the design. The result is less control, but better R-D for educational innovations, with or without the use of technology (Barab 2006). This approach might be more in line with system science and design science than methodologies originating from psychology or sociology.

In conclusion, even if the methodologies and tools for evaluating the tutoring functions in and with ITSs have developed extensively and even permitted progress in ITS research, such as the representation of tutor modeling, many questions remain unanswered. The following section describes some open questions that need further investigation.

7.6 Open Questions

Discoveries and development have been transferred from research labs to real educational settings during the last few decades. However, as ITSs are being implemented in schools, the interest, as was expressed at the workshop presented at the AIED conference in 2009, in scalability issues is growing among the community. Some of the scaling dimensions identified include: addressing curricula over long periods (e.g., one semester), learning product lines that interoperate with each other and span over a variety of learning platforms, and supporting large numbers of stakeholders (learners, teachers, institutions). Moreover new expectations continue to arise, and many challenging research questions require further investigation.

In this section, we consider the following viewpoints: technology; affective, cognitive and contextual dimensions; and digital games.

7.6.1 Technology

The technological dimension of ITS tutoring continues to benefit from the progress made in Natural Language Processing (NLP software libraries), computer graphics (new interfaces), virtual agents and human-like avatars. Significantly improvement in natural language dialogue facilities has been and remains a crucial issue since dialogue is an essential component of human tutoring (Graesser et al., 2001). In addition to natural language, sketching is a common means for people to interact with each other in certain domains, and sketch modeling has become a new model for tutoring dialogues (see Chapter 12). Systems, such as Cog Sketch by Forbus (2009), also offer software that provides human-like visual and spatial representations.

New interfaces (e.g., complete immersion, haptic systems with force feedback) allow for the creation of new professional training environments which permit the authentic practice for real world skills that were not previously within the scope of computer-supported learning. Some interfaces include tutoring components, such as described in *VR for Risk Management* (Amokrane et al. 2009a, 2009b).

Johnson, Rickel and Lester (2000) were the first to propose animated pedagogical agents. Recent techniques allow for the use of animated agents which have sophisticated and realistic facial expressions. In conjunction with virtual reality and virtual agents which act in the cultural environment of the learner, Johnson has developed Alelo's Tactical Language and Cultural Training Systems (TLCTS). Some of these systems, such as Tactical Iraqi ™ and Tactical French ™, are widely being used (Johnson and Valente 2009). As a partial summary of their experiences, Lane & Johnson (2008) summarize principles that provide guidance in immersive and virtual learning environments. These systems have gone far beyond the aforementioned learning applications. It is due to the development of the videogame industry that the use of virtual reality and avatars is so widespread. The potential use of digital games in education is further discussed in Section 6.3. More research in the field of the use of virtual agents and human-like avatars in ITSs is of increasing interest.

7.6.2 Affective, Cognitive and Contextual Dimensions

Affect has been shown to have significant effects on learning. Recent advances in the tracking of facial features and other affect recognition techniques now make it possible to devise interactive tools that recognize and respond according to a student's affective states. Arroyo et al. (2009) used sensors to record physiological activity and compare them to students' self reports of emotions. Other achievements in detection and response to a student's emotions are described in Chapters 10 and 17. In addition, when ITSs are on the Internet, it is possible to obtain data from many real users, to use data mining techniques, and to use the results to determine and to record what the student is doing. It is also possible to improve the given product. Indeed, an in-depth understanding of a student's interests, intentions and emotions obtained from observing him/her while interacting on a website can improve the effectiveness of tutoring. Research in this area is significant and promoting new research will undoubtedly result in a new generation of systems that recognize the affective dimension of human behavior. However, (Conati and Mitrovic 2009) point out that "we are still lacking from [sic] strong theories on how to use affect in pedagogical interactions and strong evidence that taking affect into account when interacting with student will actually improve learning." The situated cognition dimension is increasingly being implemented in virtual learning environments so as to promote learning activities in authentic social and physical contexts (Lane and Johnson 2008).

As tutoring decisions may lead to a wide variety of activities for the learner, developing environments in which the built-in tutor and the classroom teacher complement each other, so as to optimize the support offered to the learner, is another possible means of implementing tutoring decisions. Integrating ITSs and the teacher results in opening the systems to the human tutor, and keeping the human tutor in the ITS design loop and the learning phase. It also means providing the human tutor with an overview of the learner's previous activities and results, as performed in the PEPITE system (Delozanne et al. 2008) and the ASSISTment system (Razzaq et al. 2009a) .

Although tutoring is mostly seen as a 1/1-3 process, from a social perspective, more collaborative environments are being proposed, thus making it possible to bridge tutoring systems and social web systems (Greenhow et al., 2009). Tutoring groups of online learners need new tutoring skills that can be recorded and implemented. ITSs should be instrumental in creating what McCalla called a "learning ecology" in his 2000 vision paper (McCalla 2000).

With the semantic Web, tutoring knowledge is no longer limited to the tutoring module in ITSs. It is spreading. For instance, it can be embedded in the Learning Management System (LMS) and in Learning Objects Repositories (LOR) through tagging learning resources to fill metadata fields.

Lastly, investigating the empirical research in the neurosciences would be pertinent to researchers in the field of ITSs so they could better understand the overall learning process (cognitive, meta-cognitive, affective, social dimensions), e.g.: IRMf techniques. In fact, efforts to bridge education and neuroscience are underway (Varma et al. 2009) so as to bring a new light on the physical phenomenon of learning on the human brain by providing dynamic data and a systemic view of brain functions.

7.6.3 Digital Games

Another dimension to consider when modeling tutoring knowledge is how learning occurs in contexts that radically differ from schooling, e.g., video games and support systems in the workplace. Since there are many specialized terms, concepts and techniques in this field, we will begin by presenting some definitions. First, since educational games existed long before videogames, it is important to understand the principles and lessons to be gained by educational games. Educational digital games are based on the attraction and motivation observed when children are engaged in games and learning activities in digital environments. For example, in order for a child to open a door for one of his/her favorite avatars, he/she has to perform a mental calculation to arrive at the answer. The more quickly he/she can determine the answer, the better his/her results. Following the educational digital games, came what was initially called "serious games." The term "serious game" is now used to name a variety of systems, most of which are simulation-based micro-worlds often containing complemented scoring systems. Alvarez and Rampnoux, (2007) from the European Center for Children's Products (University of Poitiers, France), have attempted to classify serious games into 5 main categories: Advergaming, Edutainment, Edumarket game, Diverted games and Simulation games. A serious game can be defined as a mental contest, played with a computer in accordance with specific rules that uses entertainment to further develop government or corporate training, education, health, public policy, and strategic communication objectives. Serious games are built with the principles and techniques originating from the video game industry and are inspired from the motivation that adults or groups have when playing these games. The idea is simple: Videogames which are designed for entertainment are being transformed into serious games. The hypothesis is that the people playing these games will have fun and, consequently, will learn through playing. However, the reality of the situation is complex and presents many unanswered questions. The main feature of entertainment videogames is their high motivation value. Consequently, the main question for serious game designers is how to foster learning without decreasing the intrinsic motivation to participate in the games. Moreno-Ger & al. (2009) provide an interesting overview of the synergies between digital games and e-learning. Johnson et al. explored the question: "Can AIED technologies complement and enhance serious games design techniques or does good serious games design render AIED techniques superfluous?" (Johnson et al., 2007). In his article *Deep learning Properties of Good Digital Games: How Far Can They Go*? Gee 2009) argues that learning occurs mainly because the design of such games is based on good learning principles and not due to gaming. Lane et al. support this in their papers (2008, 2009). However, they question the effectiveness of digital games as learning tools. Consequently, fundamental questions on this topic continue to emerge. As there is a growing interest in serious games, it is crucial to understand in which contexts, for which kinds of learners, and for which subjects digital games may be beneficial. Other questions include: How should assessments be integrated into the games? How should guidance (if any) be incorporated into the games? What is the nature of motivation in games? How does playing the games translate into learning for the student?

Tutoring strategies should include additional data about the learner (*e.g.*, modeling and tracking his/her activity over a longer period of time), as noted by Johnson et al. (2007). Another question to take into consideration is: Would it be better to alternate gaming and other activities, or would it be better to introduce new kinds of tutoring rules within a digital game architecture as a solution?

Besides the specific dimensions discussed above, more general open questions continue to be posed in this field: What has been learned from new settings involved in human instruction and integrated learning that could improve tutoring systems? Can they develop the ability to tutor? How can tutor modeling help researchers to understand more about human instruction? Should the tutor demonstrate emotions and a sense of consciousness? Can the fields of neuroscience and brain science inspire researchers in their future studies of ITSs?

7.7 Conclusion

In 2010, ITS research is developing with respect to the design of tutoring functions in planning, dialogue, and tutorial interaction design. Recent progress has been obtained with the design of systems based on machine-learning techniques that enable reasoning on student data. Among the challenges that remain are: a) integration as a means to obtain adaptability and effectiveness in tutoring; b) using real-time and other elements of context; c) improving accessibility and affordability; and, c) evaluating methodologies and tools to assess the many dimensions of tutoring in and with ITSs.

After a generation of research in cognitive-oriented ITSs, the use of emotions in ITSs is now being studied so as to have a deeper understanding of the student's cognitive-emotive state so that the tutor can better adapt to her state. With the advances in brain research in relation to social intelligence, are we going to see more effort to include this dimension in ITS research? Although the social Web has allowed for multiple (both teaching and non-teaching) activities based on social interactions, we have not yet seen research that focuses on the social dimension of learning with the same depth as the cognitive and emotive dimensions of learning. Integrating the cognitive and emotive aspects in ITSs is already proving to be quite a challenge. Research in the field of IT'S could prove to be all the more challenging if social intelligence were added to the equation. If the basis of ITS research pertains to models (Baker 2001), and if "social intelligence is the new science of human relationships" (Goleman 1995), then the field of ITS needs computational models of social intelligence, either as part of the student model, the tutoring model, or the interaction model.

In conclusion, our goal is to achieve a tutoring intelligence that gives rise to and evaluates learning, provides complete, in-depth support and is able to provide a service that is at least as good as that of an excellent human tutor.

Acknowledgements. Figure 10.1 is reprinted from *Building Intelligent Interactive Tutors*, Beverly Park Woolf, Chapter *Machine Learning*, Page 229, Copyright 2010, with permission from the author and from Elsevier.

References

Alvarez, J., Rampnoux, O.: Serious Game: Just a question of posture? In: Proc. from Artificial & Ambient Intelligence AISB 2007, Newcastle - UK, vol. 5(1), pp. 420–423 (2007)

Amokrane, K., Lourdeaux, D.: Pedagogical system in virtual environments for high-risk sites. In: Proc. from ICAART 2009: the International Conference on Agents and Artificial Intelligence, pp. 371–376. INSTICC Press, Valencia (2010a)

Amokrane, K., Lourdeaux, D.: Virtual Reality Contribution to training and risk prevention. In: Proc. from ICAI 2009: the International Conference on Artificial Intelligence, pp. 466–472. CSREA Press, Las Vegas (2009b)

Arroyo, I., Cooper, D.G., Burleson, W., Woolf, B., Muldner, K., Christopherson, B.: Emotion Sensors Go To School. In: Dimitrova, V., Mizoguchi, R. (eds.) Proceedings from 14th International Conference on AIED. IOS Press, Amsterdam (2009)

Baker, M.: The roles of models in Artificial Intelligence and Education research: a prospective view. IJAIED 11(2), 122–143 (2000)

Barab, S.: Design-Based Research. In: Sawyer, K. (ed.) The Cambridge Handbook of the Learning Sciences. Cambridge University Press, Cambridge (2006)

Beck, J., Chang, K., Mostow, J., Corbett, A.: Does Help Help? Introducing the Bayesian and Assessment Methodology. In: Woolf, B.P., Aïmeur, E., Nkambou, R., Lajoie, S. (eds.) ITS 2008. LNCS, vol. 5091, pp. 383–394. Springer, Heidelberg (2008)

Bloom, B.: The 2 Sigma Problem: The Search for Methods of Group Instruction as Effective as One-to-One Tutoring. Educational Researcher 13(6), 3–16 (1984)

Boyer, K., Phillips, R., Vouk, M., Lester, J.: Balancing Cognitive and Motivational Scaffolding in Tutorial Dialogue. In: Woolf, B., et al. (eds.) Proc. of ITS 2009. Springer, New York (2008)

Brusilovsky, P.: Intelligent tutoring systems for World-Wide Web. In: Holzapfel, R. (ed.) Poster Proceedings of Third International WWW Conference, Darmstadt, pp. 42–45 (1995)

Brusilovsky, P., Peylo, C.: Adaptive and intelligent Web-based educational systems. In: Brusilovsky, P., Peylo, C. (eds.) International Journal of Artificial Intelligence in Education, Special Issue on Adaptive and Intelligent Web-based Educational Systems, vol. 13(2-4), pp. 159–172 (2003)

Chan, T.-W., Baskin, A.B.: Studying with the prince the computer as a learning companion. In: Proc. of ITS, pp. 194–200. IOS Press, Amsterdam (1988)

Chan, T.-W., Chung, Y.-L., Ho, R.-G., Hou, W.-J., Lin, G.-L.: Distributed learning companion systems - WEST revisited. In: Frasson, C., McCalla, G.I., Gauthier, G. (eds.) ITS 1992. LNCS, vol. 608, pp. 643–650. Springer, Heidelberg (1992)

Chan, T.-W.: Learning companion systems, social learning systems, and the global social learning club. IJAIED 7(2), 125–159 (1996)

Chen, W., Wasson, B.: Coordinating Collaborative Knowledge Building. International Journal of Computer and Applications 25(1), 1–10 (2003)

Chen, W., Wasson, B.: An Instructional Assistant Agent for Distributed Collaborative Learning. In: Lester, J.C., Vicari, R.M., Paraguacu, R.F. (eds.) ITS 2002. LNCS, vol. 2363, p. 609. Springer, Heidelberg (2004)

Cohen, P., Kulik, J., Kulik, C.: Educational outcomes of tutoring: A meta-analysis of findings. American Educational Research 19(2), 237–248 (1982)

Collins, A.: Cognitive Apprenticeship. In: Sawyer, K. (ed.) The Cambridge Handbook of the Learning Sciences. Cambridge University Press, Cambridge (2006)

Conati, C., Mitrovic, T.: Closing the Affective Loop in Intelligent Learning Environments (Workshop). In: Dimitrova, V., et al. (eds.) Proc. of the AIED 2009 Conference. IOS Press, Brighton (2009)

Delozanne, E., Prévit, D., Grugeon, B., Chenevotot, F.: Automatic Multi-criteria Assessment of Open-Ended Questions: A case study in School Algebra. In: Woolf, B.P., Aïmeur, E., Nkambou, R., Lajoie, S. (eds.) ITS 2008. LNCS, vol. 5091, pp. 101–110. Springer, Heidelberg (2008) du Boulay, B., Luckin, R.: Modeling human teaching tactics and strategies for tutoring systems. IJAIED, Special Issue on Modeling Teaching 12(3), 1–24 (2001)

du Boulay, B.: Commentary on Kurt VanLehn's "The Behavior of Tutoring Systems. IJAIED 16, 267–270 (2006)

Forbus, K.: Open-Domain Sketch Understanding for AI and Education. In: Dimitrova, V., et al. (eds.) Proc. of AIED 2009 Conference. IOS Press, Amsterdam (2009)

Gee, P.: Deep Learning Properties of Good Digital Games: How Far Can They Go? In: Ritterfeld, U., Cody, M., Vorderer, P. (eds.) Serious Games: Mechanisms and Effects. Taylor & Francis Group, Routledge (2009)

Goleman, D.: Emotional Intelligence: Why it ca matter more than Iq. Bantam Dell, New York (1996)

Grandbastien, M.: Teaching expertise is at the core of ITS research. International Journal of Artificial Intelligence in Education 10, 335–349 (1999)

Graesser, A., VanLehn, K., Rosé, C., Jordan, P., Harter, D.: Intelligent Tutoring Systems with Conversational Dialogue. AI Magazine 22(4), 39–51 (2001)

Greenhow, G., Robelia, B.: Learning, Teaching, and Scholarship in a Digital Age. Web 2.0 and Classroom, Educational Researcherv 38, 246–259 (2009)

Greer, J., McCalla, G., Collins, J., Kumar, V., Meagher, P., Vassileva, J.: Supporting Peer Help and Collaboration in Distributed Workplace Environments. IJAIED, Special Issue on Computer Supported Collaborative Learning 9(2), 159–177 (1998)

Greeno, J.G., Collins, A.M., Resnick, L.B.: Cognition and Learning. In: Berliner, D.C., Calfee, R.C. (eds.) Handbook of Educational Psychology. Macmillan, New York (1996)

Hayashi, Y., Bourdeau, J., Mizoguchi, R.: Using Ontological Engineering to Organize Learning/Instructional Theories and Build a Theory-Aware Authoring System. IJAIE 19(2), 211–252 (2009)

Heffernan, N.: Intelligent Tutoring Systems Have Forgotten the Tutor: Adding a Cognitive Model of Human Tutors (Doctoral Dissertation, School of Computer Science, Carnegie Mellon University) (2001)

Heraz, A., Frasson, C.: Predicting Learner ANSwers Correctness through Brainwaves Assessment and Emotional Dimensions. In: Dimitrova, V., et al. (eds.) Proc. of AIED 2009, pp. 49–56. IOS Press, Brighton (2009)

Isotani, S., Mizoguchi, R.: Theory-Driven Group Formation through Ontologies. In: Woolf, B., et al. (eds.) ITS 2008. LNCS, vol. 5091, pp. 646–655. Springer, Heidelberg (2008)

Israel, J., Aiken, R.: Supporting Collaborative Learning with an Intelligent Web-Based System. IJAIED 17, 1 (2007)

Johnson, W.L., Rickel, J.W., Lester, J.C.: Animated Pedagogical Agents: Face-to-Face Interaction in Interactive Learning Environments. IJSIE 11, 47–78 (2000)

Johnson, W.L., Vilhjalmsson, H., Marsella, S.: Serious games for language learning: How much game, how much AI? In: Proc. of the 12th International Conference on Artificial Intelligence in Education. IOS Press, Amsterdam (2005)

Johnson, W.L., Valente, A.: Tactical Language and Culture Training Systems: Using AI to Teach Foreign Languages and Cultures. AI Magazine 30(2), 60–72 (2009)

Koedinger, K., Corbett, A.: Cognitive Tutors. In: Sawyer, K. (ed.) The Cambridge Handbook of the Learning Sciences. Cambridge University Press, Cambridge (2006)

Lajoie, S.: Computers as Cognitive Tools: No More Walls. Lawrence Erlbaum Associates Inc., Mahwah (2000)

Lajoie, S., Faremo, S., Wiseman, J.: Tutoring strategies for effective instruction in internal medicine. IJAIED 12, 293–309 (2001)

Lane, H.C., Johnson, W.L.: Intelligent tutoring and pedagogical expertise manipulation. In: Schmorrow, D., Cohn, J., Nicholson, D. (eds.) The PSI Handbook of Virtual Environments for Training and Education: Developments for the Military and Beyond, Praeger Security International, Westport, CT (2008)

Lane, H.C., Ogan, A.E., Shute, V.: Intelligent Educational Games. In: Proc. of the AIED 2009. IOS Press, Brighton (2009)

Lester, J.: Reflections on the KVL Tutoring Framework: Past, Present, and Future. IJAIED 16, 271–276 (2006)

McCalla, G.: The Fragmentation of Culture, Learning, Teaching and Technology: Implications for the Artificial Intelligence in Education Research Agenda in 2010. IJAIED 11(2) (2000)

Mizoguchi, R., Bourdeau, J.: Using Ontological Engineering to Overcome Common AI-ED Problems. IJAIED, Special Issue on AIED 11(2), 107–121 (2000)

Moreno-Ger, P.D., Burgos, J., Torrente: Digital Games in eLearning Environments. Current Uses and Emerging Trends, Simulation & Gaming 40(5), 669–687 (2009)

Mostow, J., Aist, G.: Evaluating tutors that listen: An overview of Project LISTEN. In: Forbus, K., Feltovitch, P. (eds.) Smart Machines in Education. MIT/AAAI Press, Menlo Park (2001)

Nkambou, R., Frasson, C., Gauthier, G.: CREAM-Tools: An Authoring Environment for Knowledge Engineering in ITS. In: Murray, T., Blessing, S., Ainsworth, S. (eds.) Authoring Tools for Advances Technology Learning Environment. Kluwer Academic Publishers, Dordrecht (2003)

Murray, T.: Principles for Pedagogy-Oriented Knowledge-Based Tutor Authoring Systems: Lessons Learned and a Design Meta-Model. In: Murray, T., Blessing, S., Ainsworth, S. (eds.) Authoring Tools for Advances Technology Learning Environments. Kluwer Academic Publishers, Dordrecht (2003)

Person, N., Graesser, A.: Fourteen facts about human tutoring: Food for thought for ITS developers. In: AIED 2003 Workshop Proceedings on Tutorial Dialogue Systems, pp. 335–344 (2003)

Pirolli, P., Greeno, J.: The Problem Space of Instructional Design. In: Psotka, J., Massey, L., Mutter, S. (eds.) Intelligent tutoring systems: lessons learned, LEA, New Jersey (1988)

Razzaq, L., Patvarczki, J., Almeida, S., Vartak, M., Feng, M., Heffernan, N., Koedinger, K.: The ASSISTment Builder: Supporting the Life Cycle of Tutoring System Creation. IEEE Transaction on Learning Technologies 2(2), 157–166 (2009a)

Razzaq, L., Heffernan, N.: To Tutor or not to Tutor: That is the Question. In: Proc. of AIED. IOS Press, Amsterdam (2009b)

Sawyer, K. (ed.): The Cambridge Handbook of the Learning Sciences. Cambridge University Press, New York (2006)

Self, J.: The defining characteristics of intelligent tutoring systems research: ITSs care, precisely. IJAIED 10, 350–364 (1999)

VanLehn, K., Siler, S., Murray, C., Yamauchi, T., Baggett, W.B.: Why do only some events cause learning during human tutoring? Cognition and Instruction 21(3), 209–249 (2003)

VanLehn, K.: The behavior of tutoring systems. IJAIED 16(3), 227–265 (2006)

VanLehn, K.: Intelligent tutoring systems for continuous, embedded assessment. In: Dwyer, C. (ed.) The Future of Assessment: Shaping Teaching and Learning, Lawrence Erlbaum Associates, Mahwah (2007)

VanLehn, K.: What's in a step? Toward general, abstract representations of tutoring systems log data. In: Conati, C., McCoy, K., Paliouras, G. (eds.) UM 2007. LNCS (LNAI), vol. 4511, pp. 455–459. Springer, Heidelberg (2009)
VanLehn, K., Graesser, A.C., Jackson, G.T., Jordan, P., Olney, A., Rose, C.P.: When are tutorial dialogues more effective than reading? Cognitive Science 3(1), 3–62 (2007)
VanLehn, K.: The Interaction Plateau: Answer-Based Tutoring Step-Based Tutoring Natural Tutoring. In: Woolf, B.P., Aïmeur, E., Nkambou, R., Lajoie, S. (eds.) ITS 2008. LNCS, vol. 5091, p. 7. Springer, Heidelberg (2008)
Varma, S., McCandliss, B., Schwartz, D.: Scientific and Pragmatic Challenges for Bridging Education and Neuroscience. Educational Researcher Apr. 37, 140–152 (2008)
Vassileva, J., Wasson, B.: Instructional Planning Approaches: from Tutoring towards Free Learning. In: Proc. of EuroAIED 1996, Portugal, Lisbon (1996)
Virvou, M., Moundridou, M.: Adding an Instructor Modeling Component to the Architecture of ITS Authoring Tools. IJAIED 12, 185–211 (2001)
Wasson, B.: Identifying Coordination Agents for Collaborative Telelearning. IJAIED 9, 275–299 (1999)
Wenger, E.: Artificial Intelligence and Tutoring Systems: Computational Approaches to the Communication of Knowledge. Kaufman Publishers Inc., Los Altos (1987)
Woolf, B.: Building Intelligent Interactive Tutors: Student-centered Strategies for Revolutionizing E-learning. Morgan Kaufmann, Burlington (2009)
Zhang, X., Mostow, J., Beck, J.: A Case Study Empirical Comparison of Three Methods to Evaluate Tutorial Behaviors. In: Woolf, B., et al. (eds.) ITS 2008. LNCS, vol. 5091, pp. 122–131. Springer, Heidelberg (2008)

Chapter 8
Decision-Making in Cognitive Tutoring Systems

Daniel Dubois[1], Roger Nkambou[1], Jean-François Quintal[1], and François Savard[2]

[1] Computer Science Department, University of Quebec in Montreal, P.O. Box 8888, Centre-Ville Station, Montréal, Québec, H3C 3P8, Canada
{dubois.daniel,Nkambou.roger}@uqam.ca

[2] École Polytechnique de Montréal, C.P. 6079 Succursale Centre Ville Montreal, PQ, Montréal, QC, H3C 3A7, Canada
francois@fsavard.com

Abstract. Human teachers have capabilities that are still not completely uncovered and reproduced into artificial tutoring systems. Researchers have nevertheless developed many ingenious decision mechanisms which obtain valuable results. Some inroads into *natural* artificial intelligence have even been made, then abandoned for tutoring systems because of the complexity involved and the computational cost. These efforts toward naturalistic systems are noteworthy and still in general use. In this chapter, we describe how some of this AI is put to work in artificial tutoring systems to reach decisions on when and how to intervene. We then take a particular interest in pursuing the path of "*natural*" AI for tutoring systems, using human cognition as a model for artificial general intelligence. One tutoring agent built over a *cognitive architecture*, CTS, illustrates this direction. The chapter concludes on a brief look into what might be the future for artificial tutoring systems, *biologically-inspired cognitive architectures*.

8.1 Introduction

Human teachers have features and capabilities that remain beyond the reach of current computer-based tutoring systems, but artificial systems continue to improve. Many approaches have been explored and a number of artificial intelligence algorithms were developed which yield interesting results and often mimic human actions, even attitudes, quite well. But they still lag on the communicative aspects (perception, language), in transfer of abilities across domains, in sociability and emotions, and even autonomy. If the goal is to offer human-level capabilities, why not have systems work the way humans do? Well, for one thing, understanding humans is not so easy, as the multiplicity of psychological theories demonstrates. Also, at the neuronal level, the computational brute force required is impressive, with the brain's hundred billion neurons, each with thousands of connections to other neurons. We have yet to decipher how the neural "modules" operate and interact, but research delivers new insights by the week. Theories about

R. Nkambou et al. (Eds.): Advances in Intelligent Tutoring Systems, SCI 308, pp. 145–179.
springerlink.com

the mind are being proposed and refined, and incorporate more and more of the biological aspects.

In his 1987 book about artificial intelligence and tutoring systems, Wenger (1987) presents a surprising number of early but nonetheless advanced research efforts aimed at going beyond action-reaction tutoring. From the outset, they explored various aspects of the goal of rendering human cognition: presenting the appropriate content, conversing, storing factual knowledge, using procedural knowledge, diagnosing, and so on. Initial steps were taken by Carbonell in 1970 when he introduced the idea of *information-structure-oriented CAI* (Carbonell 1970). He applied Quillian's (1968) semantic networks to knowledge modeling in learning systems as an accurate model of how people store and access information. From this seminal idea, he derived SCHOLAR, in which domain knowledge was separate from tutorial capabilities. Collins collaborated with Carbonell to extend SCHOLAR with better inference and dialog capabilities (Collins et al. 1975), yielding WHY (Stevens and Collins 1977). That system implemented the human Socratic strategy (orienting the learner through a series of questions leading to general principles and contradictions) with production rules as a succession of simple decisions. Self, for his part, addressed the need for explicit modeling of the learner, and more importantly, the need to implement the model as a set of programs that would than play an active role in the tutor's decision-making (Self 1974). The two psychologists Donald Gentner and Donald Norman set out to build artificial tutors to further investigate the ramifications of their theories about memory organization as a network of schemata or prototype concepts (Norman et al. 1976). In 1979, O'Shea produced the QUADRATIC tutor (O'Shea 1979), after observing with Sleeman (O'Shea and Sleeman 1973) that intelligent tutoring systems should be able to self-improve their set tutorial strategies.

The stake was to achieve systems capable of adapted actions without having to prepare ahead of time for all combinations of constraints. Indeed, adapting to the student's inferred state of knowledge and presentation preferences, rapidly becomes a gigantic undertaking in frame-based systems, even without trying to adapt the interaction in line with a chosen pedagogical theory. Yet, it remains true that having learners build lasting knowledge can best be achieved when the context is appropriate, including high motivation. The teacher needs sensitivity to various factors, and the ability to decode, weigh and infer, and prioritize. Reproducing this requires much more than what is traditionally implemented in artificial systems. This is where human-inspired processing may really shine. Anderson, in his Adaptive Control of Thought – Rational (ACT-R) theory of cognition (Anderson 1976, 1993) and Newell with SOAR, his own version of the mind as a production-system (Newell 1973, 1990), went a step further with regard to cognition and production rules, both proposing unified theories of how the mind works. In their proposals, at a conceptual level and even at some physical level, things get processed and moved via the action of a large collection of simple rules. This is an instantiation of Fodor's physical symbol system hypothesis (Fodor 1975; 1983), which states that every general intelligence must be realized by a symbolic system. This principle offers a unified framework upon which to build complete cognitive systems. However, arbitrary processes and algorithms can result from these rule-based system, since rules are constrained only by practical considerations. But at least the fundamental unifying principle remains: rules do it all. At a high level of abstraction, they may correspond to the simple processes that form cognition (Baars 1997; Dennett 2001; Dennett and Akins 2008; Dehaene and Naccache 2001; Hofstadter 1995; Jackson 1987; Minsky 1985).

What justification is there for seeking to create cognitive architectures rather than being content with using just any artificial intelligence technique that works? Langley, Laird and Rogers (2009) express the opinion that "Research on cognitive architectures is important because it supports a central goal of artificial intelligence and cognitive science: the creation and understanding of synthetic agents that support the same capabilities as humans." In contrast to AI techniques such as *expert systems* that solve specific problems, cognitive architectures aim for breadth of coverage across a diverse set of tasks and domains. In the words of Newell (1990), "A cognitive architecture specifies the infrastructure for an intelligent system that remains constant across different domains and knowledge bases." The appropriate fundamental principle should sustain scalability and generalization – a seductive holy grail.

The word *architecture* implies that the system attempts to model not only the behavior, but also the structural properties of the modeled original. This definition goes deeper than Taatgen, Lebiere and Anderson's (2006) view that "A cognitive architecture is a computational modeling platform for cognitive tasks." The latter definition sees a "cognitive architecture" as a system that includes, and supports the creation of, a cognitive model of the task one wants to teach to a learner (able to reflect the learner's cognition to some extent); whereas the former requires that the structural organization, and its processing (not just the observable behaviours), mimic the original.

A quick lexical digression may be helpful here, to clarify a term that underlies this whole section: *cognition*. Like many words (such as intelligence, emotions, consciousness), this one is plagued with multiple definitions and excessive breadth of coverage. It may refer to knowledge building, decision making, information processing (the "cognitivist" view of mental processing), or all of these and more (Neisser 1967). In tutoring systems, one finds that a *cognitive* tutor may includes a model of the cognitive steps a learner may follow, or a tutor may adopt the label based on the "cognitive *architecture*" view, in which the underlying mechanisms mimic biological ones, at least at some level of abstraction.

We propose the following definition of an ideal artificial cognitive tutoring agent: *an agent built on an architecture that offers structures, features and functioning comparable to the human model so that it is similarly capable of adaptation, learning, generalization within and across domains, and action in complex situations encountered in tutoring learners*. This definition points in the direction of *general intelligence*, to which we will return in the final section.

8.2 Various Levels of Tutoring Decisions

Traditional computer-aided instruction (CAI) is not doomed. Thanks to simple knowledge structures, it is easily implemented and maintained by almost anyone.

It will keep being used in classrooms and niche applications for a long time. But such instruction isn't sufficient for complex domains, in higher education and for learners that are not highly autonomous. It cannot sustain strong adaptation and yet provide effective tutoring. Plain frame-oriented systems are not sufficient because flexible, adaptive tutoring systems have to *build* decisions, all the time, about lots of things. Some of their *intelligence* is geared toward the aim of adapting to the user, by means that may be simple or more elaborate. Ideally, as the user of an intelligent tutoring system (ITS), I would like human-level support with all its richness: seeing the tutor understanding what I really mean when I answer, interacting with me in the way I prefer, sensing from my voice and my face when I am struggling, showing empathy, adapting to my affective state, joking with me, guiding my efforts to resolve problems, finding out what keeps me motivated, inventing creative ways of helping me to learn. I would like the artificial agent not to respond solely on the basis of observed events, but to try and understand what is happening in my mind. These are goals for intelligent tutoring systems. And some of these call for human-inspired decision mechanisms.

But here again, the term "*intelligent* tutoring system" is widely applicable. As long as a system adapts in some significant way to the learner, to fall into this category. And any that use human-*inspired* mechanisms can be called *cognitive*. Real-world observation of tutors does not exert much pressure toward changing this "all-inclusive" categorization: looking at how some human tutors tackle their task, one may be surprised to discover how little inexperienced tutors use complex and involved approaches while still making a difference (Graesser et al. 2005). However, accomplished tutors perform better than naive ones (Merrill et al. 1995; Cohen et al. 1982). This observation underlies the subject of this chapter: architectures that can yield results on a par with those of skilled human tutors.

Although the idea of cognitive architectures may seem novel and strange to most people in the computer science field, quite a few research efforts for mingling cognition and AI have been ongoing at the same time as the work on ACT-R and SOAR , although they may not have achieved the same level of recognition (Wenger 1987). Recent years have seen the development of more such systems, often with common features, and each with its own merits. In an updated and extended version of their 2002 article (Langley et al. 2009), Langley, Laird and Rogers offer a recent and well documented account of the field. Wikipedia (http://en.wikipedia.org/wiki/Cognitive_architecture) adds some interesting items to their list. Giving an account of their similarities and specificities would exceed the scope of this chapter. We will limit ourselves to offering an illustration of such an architecture. This chapter proposes a view of the way some recent intelligent tutoring systems, cognitive and otherwise, make tutoring decisions, and how human-like architectures may support these processes. A sample system, CTS, is described in some detail. We also offer a glimpse of the emerging trend of biologically-inspired architectures that may close the gap between so-called intelligent tutoring systems and their human mentors.

8.3 An Overview of Some Tutoring Decision Mechanisms

ACT-R cognitive tutors and their student modeling techniques have shown their potential, and there has been an opportunity to explore their limitations. For example, a model-tracing mechanism cannot engage students in multi-turn dialogs and cannot be successfully applied to ill-defined domains. Researchers have started from these observations to work on extensions of the original ideas or pursue other paths. One such extension is exemplified by the *Ms. Lindquist* ITS, which adds a separate tutorial model to the central decision engine. *Ill-defined domains* (see Lynch et al. 2006), by definition, cannot be described in detail and necessitate another approach, such as constraint-based modeling (CBM; Ohlsson 1994; Mitrovic et al. 2001; you may also like to look up Chapters 4 and 5 in this book). As an example of an ITS for ill-defined domains, we will present *Rashi*, which uses an expert rulebase, rather than production rules, to deal flexibly with the more relaxed procedures of an enquiry. It also shows a way of monitoring progress in ill-defined domains, where the procedural knowledge cannot be fully covered or described as absolute sequential steps. ACT-R and constraint-based modeling tutors differ in that, whereas a model-tracing tutor is designed to represent the steps of procedures and use this model to trace the learner's operations step-by-step, an expert-rules system tries to match observations to expert principles described as constraints, then relies on separate mechanisms for giving advice and tutoring (Mitrovic et al. 2003). *Andes*, a mature tutoring system, exemplifies a less stringent way of using model tracing, compared to the line of ACT-R cognitive tutors which permit little variation in steps. Andes offers multiple levels and types of help, all on-demand except for the green or red coloring of the learner's equations. Its original reliance on bayesian estimates for plan recognition and estimation of the learner's knowledge has been replaced in Andes2 by simpler mechanisms: asking the learner to indicate his current goal and what principle should apply, and evaluating his equations rather than trying to derive them from canonical ones. Indeed, creating the solution graph was a heavyweight endeavor. Solutions remain tractable only for simple problems; the same is true of validating equations through derivation. *AutoTutor*, for its part, adopts an almost provocative view of tutoring, rejecting the need for planning and relying solely on dialog-driving to implement effective coaching.

In the following section, we will briefly describe these tutoring systems, as illustrations of some of the possible decision-making mechanisms utilizing cognitive constructs and AI techniques. We will begin with a presentation of Anderson's Cognitive Tutors, which introduced two fundamental cognitive mechanisms that are still widely used: model tracing and knowledge tracing.

8.3.1 Tutors Based on the ACT-R Theory of Cognition

Backed by a rich body of research and an impressive track record of educational successes, the line of intelligent tutoring systems grounded in the ACT production system (later extended and reorganized into ACT-R theory of cognition) are prime examples of *cognitive* tutors. The distinction between declarative and procedural knowledge – between merely knowing an algebraic rule and being able to apply it in a problem – is central to ACT-R. It postulates that procedural knowledge can only be acquired with progressive integration, through problem solving, of what is initially declarative knowledge (Anderson et al. 1995). This theory inspired a line of intelligent tutoring systems based on a **model-tracing** algorithm. The core idea is to try to *trace* a student's cognitive steps by the parallel application of a series of **production rules** to facts relating to the problems to be solved. As a student works his way through the problems, his mastery of each rule is inferred by another algorithm, **knowledge tracing**. We will present these two algorithms in a little more detail in the next section.

8.3.1.1 LISP/Geometry/Algebra Tutors: Tutoring through Model Tracing

Brief description. In the 1990s, most products and research in the cognitive field were based on the ACT-R software system (grounded in the ACT-R theory of cognition). The tutors were developed using the Tutor Development Kit (TDK), itself based on the Tertl production rule engine (Choksey and Heffernan 2003). Descriptions of some artificial tutors based on ACT-R can be found in (Anderson et al. 1995). From early 2000 onward, work on most new tutors in the line has been based on the JESS production rule engine, notably in the Cognitive Tutor Authoring Tools (CTAT) environment (Aleven et al. 2009). In both systems, however, production rules and model tracing are central.

Decision mechanisms
Model tracing over production rules
The model-tracing algorithm, as mentioned above, does its best to trace a student's cognitive steps in resolving a problem. It is given the following inputs:

- The student model, further divided into 1) the current state of working memory (composed of working memory elements) and 2) a set of relevant production rules
- The student's last input in the user interface

The working memory elements are facts related to a problem. Following an example in (Heffernan 2001), such facts can be named-quantities in a problem and operators linking them (e.g. "distance to shore", "boat speed", the division operation).

Production rules are *if-then* statements composed of actions to be taken when some conditions are met. The conditions refer to elements in working memory (including goal elements), while actions can, notably, alter those elements or cause a message (hint, error feedback) to be displayed. The model-tracing algorithm tries to find rules which can lead to the current student's input. For example, if the student altered an algebraic expression from "7+2*g" to "7+(2*g)", a production rule causing the addition of parentheses to a sub-expression would be returned by the algorithm (given that such a rule was input to the algorithm).

Importantly, "buggy" rules are used to model students' expected errors, and helpful error messages can be included in the "action" part of these rules. The tutor sends the student the error message when the system runs the rule identified by the algorithm. Additionally, hints can be associated with rules and be provided to the student upon request, based on rules that, as computed, could lead to the solution.

When the model-tracing algorithm finds a correct rule (or a chain of such rules) matching the student operation, it provides visual feedback in the user interface, highlighting the step as correct. In the alternative case of the step being matched by a buggy rule, the tutor instead displays an error message. Although the messages are contextual in the sense of referring to the cognitive operations performed by the student (provided the model-tracing model encompasses the specific path), there is no other consideration by the tutor beyond the fact that the student has utilized the buggy rule. For instance, help will be offered whether it has already been given or not, it will not be adapted to the learner profile of the student, and it will not support the learner's cognition beyond the superficial evidence of the faulty knowledge. Some of these limitations can be addressed by adding a completely separate set of tutorial rules embodying pedagogical and tactical knowledge to provide deeper remedial guidance, as explored by Heffernan in *Ms. Lindquist* (Heffernan 2001), which is the system we will examine next.

From a practical standpoint, it should be noted that the production rules must be created manually for the domain at hand (algebra, geometry, etc.) The difficulty of this task led to the creation of authoring environments such as CTAT and automated systems such as Demonstr8, which tries to infer rules from demonstrations by experts (Blessing 1997).

Knowledge tracing:

Starting with version 3.0 of ACT-R, new principles were added and others were modified. Such changes include modularization of the architecture, a set of "sub-symbolic" principles associating weights to rules and to (the equivalent of) working memory elements, and the compilation of new rules on-the-go, by the engine itself. However, it appears these features are *not* used in the TDK. Even though such capabilities might seem to be good candidates for modeling some further aspects of student learning, things did not evolve that way. Instead, knowledge tracing was pursued as a more efficient way (in terms of development effort and computationally) of implementing knowledge about the learner (Koedinger, personal communication, 2009). In knowledge tracing, each rule is assigned a probability of being known by the student. As new problems are worked out by the student, rules that are correctly used (or those that *should have been* used) have their probability adjusted through a bayesian learning formula. The tutor then uses the resulting estimated level of mastery for each rule to select which problem should be presented to the student (Koedinger and Aleven 2007).

Cognitive aspect. The way the word *cognitive* applies to these tutors may not be the obvious one for a newcomer to the field. They are not cognitive in the sense that they are built on a cognitive architecture (see our comment about cognitive *architectures* in the introduction). For reasons of efficiency, they leave aside many cognitive operations (sub-symbolic modeling in declarative memory, chunking, transfer of declarative knowledge to procedural knowledge in rules) in favour of a more efficient *computational modeling platform for cognitive tasks*. "Cognitive" refers here to the fact that these tutors use a cognitive modeling *of the domain* and of the mental operations, correct or faulty, someone is likely to use in solving the problems. The ITSs trace the *learner's cognition*, then apply the action specified in the appropriate part of the matched rule. Information processing is operated by the rules themselves. In stricter terms, the *tutoring* in itself does not rely on humanly cognitive processing.

8.3.1.2 Ms. Lindquist: Tutoring through Model Tracing Extended with Tutorial Rules

Brief description. Ms. Lindquist (Heffernan 2001; Heffernan et al. 2008) offers web-delivered tutoring on writing expressions for algebra "story" problems (symbolization). The behaviors it manifests are inspired by the experience of the real Ms. Lindquist and by Heffernan's own teaching. It offers dialog-based sessions in which the learner's inputs are examined for errors, and the observed difficulty is then broken down into simpler steps. Resolution is promoted by asking questions rather than directly giving hints and advice. The tutoring relies on an explicit cognitive model of the tutoring process, and allows for multi-turn tutorial strategies.

Decision mechanisms. Pursuing the approach taken in the ACT-R cognitive tutors, Ms. Lindquist uses the same model-tracing mechanism for its student model and primary diagnosis tool, and the same bayesian knowledge-tracing. However, it supplements these primary mechanisms by adding more involved tutoring capabilities, with tutorial behaviors arising from selection rules. Following the student's input, the model-tracing algorithm analyzes the content and transfers its conclusions to the tutorial model. The selection rules of the latter decide on the appropriate reaction, which may be to use a tutorial strategy, display a buggy message or provide a hint. Tutorial strategies are contained in plans called Knowledge-Construction Dialogs (KCD) and, in more specific ones called Knowledge-Remediation Dialogs (KRD). Selection rules try out the possible reactions, where relevant, in the following order: KRD, buggy message, KCD, hint. The heuristic behind this ordering is to offer the most contextual response possible (KRD and buggy messages), then use a tutorial strategy (KRD or KCD) before folding back to buggy messages or hints. In multi-step interventions (based on a KRD or a KCD), an *agenda* data structure keeps a memory of the dialog history for questions still awaiting a correct answer from the student, and the remaining steps in the plan (questions that the tutor is planning to ask later on). The steps chosen from the KCD or KRD by the selection rules are added to the agenda. The action at the top of the agenda dictates the next action, usually asking a question.

Cognitive aspect. Ms. Lindquist utilizes a cognitive model for its student modeling, based on ACT-R theory, and a cognitive model of pedagogical interventions for its decisions on tutoring; both operate on a rule-based engine. KRDs and KCDs, replicate questions human tutors ask themselves (their thinking) before posing acts toward the student.

8.3.2 *Tutors That Use an Engineering Approach to Decision Mechanisms*

The next systems adopt an "engineering" approach to problem-solving: what is efficient and useful is what is needed. These systems are thus not organized around any cognitive principle. If a resemblance to cognitive mechanisms is found, it was not intended or sought. But, of course, the mechanisms at work in our minds always influence what solution patterns we find (read "recognize"), at least unconsciously, and the decision mechanisms present in the ITSs can always relate at some level, in some form, to human cognition. So, the distinction we are making here is that the organizing principles of the ITSs we are about to discuss are not voluntarily constrained to reflect the human brain or mind.

8.3.2.1 Andes2: Process Critiquing Based on Model Tracing Plus On-Demand Help

Brief description. Andes (VanLehn et al. 2005) is a tutoring system for Newtonian physics, computer literacy, and critical thinking skills. Its philosophy is to maximize student initiative and freedom during the pedagogical interaction. Intended solely as a homework support, its user interface tries to stay close to the old paper and pencil environment, letting the student process aspects of the problem and enter information in any order. However, it goes beyond paper and pencil to offer multiple levels of help, providing immediate feedback by coloring correct entries green and incorrect ones red, responding to "What's wrong with that?" help requests and supporting "What should I do next?" queries. It can solve algebraically the equations that the student has entered, provided the student has entered enough correct ones. It implements these capabilities via some specific user tools: the Conceptual Helper displays mini lessons on defective knowledge, the Self-Explanation Coach guides the student through a solved physics problem, and the Procedural Helper responds to help requests while the student is solving problems. A more complete description is found in Chapter 21.

Decision mechanisms. Andes' tutoring decisions on when to produce tutoring interventions are straightforward: immediate feedback is given when the equations are entered, and nearly all other help is on-demand. However, several AI mechanisms are used to achieve *process critiquing* (VanLehn et al. 2004). In process critiquing, attention is not devoted to following consecutive steps of a procedural skill. And, while tutoring makes use of model tracing at some point, it isn't primarily a consequence of step matching. Emphasis is put on application of principles, that is, on the more global process, leaving some leeway to the learner. At the core of this tutoring, much of the previous version of Andes help relied chiefly on a bayesian network describing the domain in terms of physics knowledge (expert) rules, problem-specific facts and a solution graph of problem-solving steps. Andes2 has replaced the probabilistic evaluations (Conati et al. 2002) with simpler methods. The flat solution graph with over 600 inference rules has been reorganized around a hundred principles, into a "bubble graph" whose nodes represent physical quantities involved in the problem, and principle application nodes relating to these quantities. Also attached to each principle application node is a *method graph* containing the series of steps that can realize each principle application. To decide how good the learner's inputs and equations are and what help to offer, Andes2 uses multiple AI techniques, primarily revolving around the bubble graph. The reader is referred to Chapter 21 for details.

Cognitive aspect. Andes is a good example of a successful ITS involving purely *artificial AI* techniques with no reference to cognitive processing, either in the student's mind or in the tutor. Nonetheless, the principle of *error handlers* is reminiscent of *elementary processes* in many cognitive theories (see Baars', Crick and Koch's, Dehaene and Naccache's, Dennett's, John V. Jackson's), and the bubble graph is a form of semantic net, which effectively depicts the association of ideas and the thematic organization of knowledge.

8.3.2.2 Rashi: Rule-Based Domain-Independent Tutoring of Student's Exploration

Brief description. Rashi (Woolf et al. 2005) is an inquiry environment designed for domain-independent coaching of very active, hands-on learning in ill-defined domains. It offers an authentic setup in which the learner performs a proper investigation, using tools commonly employed by researchers. Examination room, field and image explorers, interview tools, inquiry notebook, argument editor, various objects such as a field guide and images of artifacts let the learner examine, analyze, find evidence, form hypotheses and reach conclusions, guided throughout the process by the tutor. A *Coach* is coupled to this environment, with rules that aim at global argument formation rather than at reaching specific answers or following process steps. In its various implementations (for forestry, geology, biology and art history), the tutor monitors the learner's unconstrained activities, examines its findings, compares them to expert knowledge and presents prompts and reminders that guide and motivate the student toward extracting sufficient evidence and reaching sound conclusions.

Decision mechanisms. The tutor's coaching is rule-based, utilizing an overlay of student inputs, context and current status on expert knowledge. Coaching currently intervenes only on demand. There is no immediate reaction as in model tracing. The coaching relies on a database containing two types of expert rules: those that serve in monitoring the learner's operations, either for supporting well-formed arguments or the adequate use of available tools; and those which orient the student's search for more material to support his arguments. The coach's reasoning amounts to comparing the learner's input (keyword search) and its context in Rashi to propositions contained in the expert knowledge base (a directed graph of weighted propositions). It looks up the Inquiry Notebook to see whether the student has support for an argument, then inspects the Argument Editor to verify how well these elements are connected to form a satisfactory support. In effect, the Coach inspects the underlying graph representing the learner's discoveries and inferences, counts the correct and incorrect support and refutation links between facts and hypotheses, and computes their weighted value to determine whether they amount to a critical mass of valid support. Then, these observations are compared to the expert rules database. A rule issues an advice or a question when it corresponds to some logical flaw, a lack of supporting data or a missing intermediate step, etc. Rules can also move the learner to another place (to suggest another tool, bring up artifacts, etc.). The generic decision process is subject to the author's (teacher's) settings, such as how many arguments are required to support a specific conclusion, whether such and such rule is desired in the domain, which of the five types of feedback he wants his students to receive, or the order in which they should be offered. Teachers can also influence how the Coach uses the expert knowledge by specifying the priority of content to be presented when needed.

Cognitive aspect. As in Andes, rules and algorithms implement expert and tutoring knowledge. While the system is effective in building knowledge in the learner (with some sort of student knowledge modeling), there is no cognitive processing in the tutor.

8.3.2.3 AutoTutor: Dialog-Based Tutoring

Brief description. AutoTutor (Graesser et al. 2008; Graesser et al. 1999) is a tutoring system concerned with computer literacy, qualitative understanding of Newtonian physics and critical thinking skills. Its aim is to support the learner in his construction of knowledge, helping him express what he knows about the aspect currently focused on. The dialog is central in this process of active learning (see Chapter 9). As an initial contribution, the learner is asked to compose a paragraph about his understanding of a problem and his proposed solution. A complete and correct answer is generally obtained after a series of corrections and additions, with the conversational partners taking turns in a mixed-initiative dialog. Supporting, guiding and encouraging the learner's self-expression on the subject is what AutoTutor is about.

Decision mechanisms. AutoTutor's decision process makes use of a number of cognitive and AI techniques and algorithms. Depending on the current state of knowledge demonstrated by the learner in the course of the conversation, and the latest textual input, AutoTutor provides feedback to the student on what the student types, pumps the student for more information, prompts him to fill in missing words, gives hints, fills in missing information with assertions, identifies and corrects misconceptions and erroneous ideas, answers the student's questions, or summarizes topics. To accomplish these functions, the consecutive outputs of the architecture's modules are used in a sequential chain of analyses, and then used jointly to decide on the tutor's dialog move (Fig. 8.1). The sequential processing is concerned with understanding and assessing the learner's utterance. For a more specific evaluation, the learner's input is segmented into main clauses by the Language Analyzer, which assigns to each of these a speech act classification. This information is organized into a structured *state object* and passed to the Assessor. The Assessor submits each clause to its Latent Semantic Analysis (LSA) algorithm for an evaluation of its similarity with sentences describing expected answers or misconceptions associated with the problem. It also gauges the likely effects of various dialog moves on the student's learning, and updates the Student Model (i.e., what the student knows about the expectations and misconceptions associated with the problem). All the new information is added to the state object. The Dialog Management module receives this updated, richer state object, including the dialog information from the previous state, notably the particular student speech act category. This rich information constitutes a context that is examined by a set of 15 fuzzy production rules that decide which dialog move category to pursue in AutoTutor's next utterance. They select a path in the Dialog Advancer Network (DAN) (Person et al. 2000) and adapt a plan to produce AutoTutor's next dialog move. The DAN is an augmented finite-state network which describes discourse pathways that may include one or a combination of the following components: Discourse markers (e.g., "Okay" or "Moving on"), AutoTutor dialog moves (e.g., Positive Feedback, Pump, or Assertion), Answers to questions, or Canned expressions (e.g., "That's a good question, but I can't answer that right now"). It is noteworthy that the dialog management relies on many collaborating sub-modules, all adding and passing information through a state object. This bears some similarity to the "blackboard" architecture (Erman et al. 1980).

AutoTutor refers to multiple pedagogical principles in deciding on the next aspect (expectation) to coach on. One of these is the frontier learning or zone of proximal development principle, selecting the next expectation that builds on what the student knows. It is applied here in selecting the expectation with the highest sub-threshold LSA value. For more information about dialog in AutoTutor, see Chapter 9.

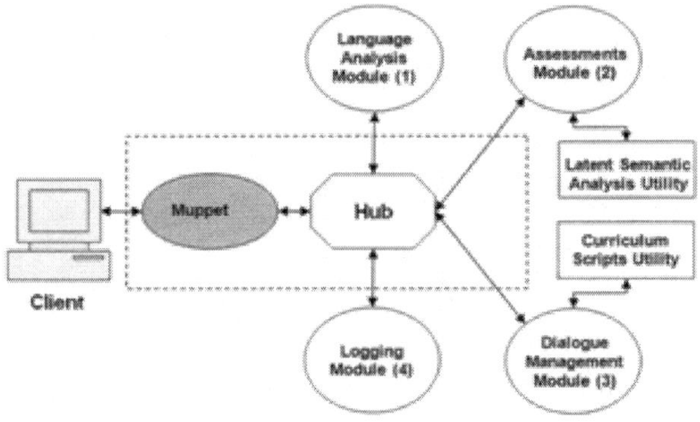

Fig. 8.1 AutoTutor relies on a number of mechanisms in analyzing the student's input and preparing a response. Source: Graesser et al. (2005)

Cognitive aspect. The central element in AutoTutor, its Dialog Management module, relies on inputs from many algorithms and mechanisms. Some of these (the state object that is collaboratively enriched and used by multiple modules; the production rules) do recall cognitive processes; others (those performing natural language interpretation) simply mimic their effects through *artificial AI* mechanisms.

8.3.2.4 Supportive AutoTutor and Shakeup AutoTutor: Decision with Consideration of the Learner's Emotions

Work on extending AutoTutor's reach with respect to the learner has been pursued in the past few years to include the learner's emotional and motivational states (D'Mello et al. 2009). While still under development, two emotion-sensitive versions of AutoTutor represent serious advances in these directions. It is interesting to see how the architecture could be stretched, very simply, to include the new parameters.

Brief description. The newer versions of AutoTutor, called Supportive AutoTutor and the Shakeup AutoTutor, receive not only textual inputs but also cues from complementary channels: posture features from a thin-film pressure pad laid on the chair, and facial feature tracking to monitor facial expressions coming through a camera. Dialog features have also been extended to allow inference of the learner's affect. Fusion of the multiple sensory channels should lead to a more reliable emotion classifier and more adept tutorial responses. The brief feedback AutoTutor provides is only intended to give an appraisal of the learner's latest textual input. "Good job" and "Well done" are examples of positive feedback, while "That is not right" and "You are on the wrong track" are examples of negative feedback. However, this feedback may provoke an emotional response in the learner in non-neutral cases, especially since it is accompanied by an appropriate facial expression from the character. The learner's emotional state is addressed in the tutorial action that follows the feedback, the tutorial move. At this point, the two tutoring variants diverge somewhat. What differentiates the two affect-aware versions is what is considered to be the cause of the emotions. Supportive AutoTutor attributes the source of the emotions to the material or to itself, whereas the Shakeup AutoTutor attributes responsibilities to the learner. This difference affects the wording of the tutorial moves and the conversational style. For instance, we may have "Some of this material can be confusing. Just keep going and I am sure you will get it" from the Supportive one, and "This material has got *you* confused, but I think you have the right idea. Try this..." from the Shakeup one. The way things are expressed will differ as well. In situations where boredom is detected, the Supportive tutor would offer, "Hang in there a bit longer. Things are about to get interesting", where the Shakeup version would instead say, "Geez this stuff sucks. I'd be bored too, but I gotta teach what they tell me".

Decision mechanisms. Attribution theory (Heider 1958; Batson et al. 1995; Weiner 1986), cognitive disequilibrium during learning theory (Piaget 1952; Craig et al. 2004; Festinger 1957) and experts recommendations have been synthesized into new production rules capable of processing emotional cues which reveal some of the emotions believed to exist in relation to learning: boredom, confusion and frustration. The original fuzzy production rules were sensitive to cognitive states of the learner, but not to his emotional states. The newly designed fuzzy production rules map dynamic assessments of the student's cognitive and affective states with tutor actions: feedback delivery (positive, negative, neutral), a host of dialogue moves (hints, pumps, prompts, assertions, and summaries), and facial expressions and speech modulation by AutoTutor's embodied pedagogical agent (EPA). These are triggered in the following order: (a) feedback for the current answer, (b) an empathetic and motivational statement, (c) the next dialog move, (d) an emotional display on the face of , and (e) emotional modulation of the voice produced by AutoTutor's text-to-speech engine. Five parameters in the learner model influence the decisions: (a) the current emotion detected, (b) the confidence level of that emotion classification, (c) the previous emotion detected, (d) a global measure of student ability (dynamically updated throughout the session), and (e) the conceptual quality of the student's immediate response.

Cognitive aspect. Adding emotions to an artificial system brings it a step closer to human information processing. The involved rules have an impact on cognition, and sometimes *are* cognition. Although one could question the way they are implemented, from a naturalistic point of view, the rules are founded on theories of emotions.

8.3.3 Summary of Existing Tutoring Agents

In the systems just described, with the exception of the ACT-R tutors, decision mechanisms are not based on a global theory of cognition – although it is interesting to note that many relate in some measure to high-level human cognition: diagnosing with multiple autonomous error handlers, reaching decisions through a chain of simple decisions, transferring and sharing information by means of a central state object. But these mechanisms are used for their convenience and efficiency, not as part of a global theory of cognition. This is fine, and efficient, as these systems have demonstrated. However, we would like to see how close to human performances a tutoring agent could get if it drew deeper inspiration from its mentor.

ACT-R tutors, for the most part, work within the confines of the ACT-R theory of cognition. The parallel holds for the agent called Steve (Rickel and Johnson 1998; not presented in this chapter), which embodies the Soar theory of cognition, a somewhat similar rule-based approach. Both have shown the feasibility and relevance of the endeavor. Both theories have inspired research, ideas and implementations since the late seventies. However, recent years have seen many proposals for alternative cognitive architectures, and many have been implemented computationally, as Samsonovich illustrates in his overview (Samsonovich 2010). We have singled out one of these, the Global Workspace (GW) theory, because it involves consciousness and offers a complete framework on cognition. We thought its implementation in IDA/LIDA could be extended and form a cognitive tutoring agent. We describe this agent in the coming pages.

8.4 CTS: Designing Cognitive Tutors on Human Bases

The name of the system we present here, CTS (Dubois 2007), stands for "Conscious Tutoring System". Simply put, the agent's processing is based on the conscious/unconscious distinction and on what is called the *access* consciousness, the consciousness phenomenon that renders a piece of information widely accessible to the whole brain. The reader may refer to Block (1995) and (2002) for a clarification of the many levels and types of consciousness. In our implementation, the reference to consciousness is at the functional level. There is no strong claim about CTS being "really" conscious in a human fashion. Not yet... Before embarking on a description of the agent and its architecture, let us see what impact Baars' view on cognition may have on information processing.

8.4.1 Cognition Based on Consciousness, and Some Functional Implications

It cannot be denied that the unconscious/conscious distinction exists. Simple experiences can demonstrate that there are things one can accomplish unconsciously, nearly effortlessly, whereas doing the same operations wilfully, while thinking about the steps, can wear us down pretty fast. Baars presents many such operations (Baars 1997). For instance, try reading a paragraph, then reading it again with the book turned upside down. Do you observe a difference? Of course you do! That shows the difference between doing something consciously and doing it mostly unconsciously. There is a tremendous gain in encapsulating all we can in automated, unconscious processes. However, highly efficient as they are, automated processes are specialized and remain limited in their adaptability. They can take charge only in known cases for which they have been grown. Conscious reflection allows solutions to be created for new situations, improving fitness (Baars 1997; Rolls 1999); it takes time and effort, but it can be done.

When one becomes *conscious* of something, when it "comes to one's mind", *all of one's mind* becomes aware of the information. It becomes available to all of the mind's processes as input for processing, and they can then respond and put forward elements that may be useful in finding a solution. Want it or not, biographical memory will bring back events, especially in fearful situations, and solutions used in situations that are related in some respect; semantic memory will also add "comprehension" of the situation by stringing related concepts to it. Analytic processes may spontaneously propose an overall structure for the event or scenario. Often, all this takes place before one has fully realized what is happening. With all this information at hand, one may decide what is important to address first, choose a multi-step procedure for doing so, and start performing the operations, under the control of the will, until something more important — such as satisfying the urge to sneeze — arises and requires attention.

Which part of the system actually responds is thus determined, and the response is shaped, by the context: current goal and plan, current mood and emotions, complementary information recovered by other systems such as memory or emotional systems, processes currently in the forefront, etc. The situation brought to consciousness is described by coalitions of processors presenting various aspects. Many such coalitions may try to have their information broadcast, but only one can have access to the broadcasting facility at a time, as only one situation can occupy the conscious "space" (one is conscious of only one idea, aspect, situation at a time). Thus, there is "competition" for access to this global workspace. Consciousness appears as a key element in "strong" adaptability.

Baars described this arrangement in his *Global Workspace* (GW) theory of the mind, which he first proposed in 1988. It provides a neuropsychological account and a high-level functional architecture that unifies many of the previous researchers' work on describing and modeling the human mind and consciousness. This theory proposes a general framework describing the essential roles of attention and consciousness in human beings.

Before examining how the agent makes decisions, we will now look at the architecture of CTS, which is rooted in that of IDA, extended to support tutorial decisions.

8.4.2 Conceptual Architecture of CTS

The conceptual architecture of CTS is founded on the ideas put forward by Professor Franklin and his team. Franklin has created an implementation of Bernard Baars' theory in the successive agents Conscious Mattie, IDA, and LIDA. CTS shares many of LIDA's features but is a reimplementation with some differences, some of which are extensions in order to achieve a tutoring agent. The architectures can be considered similar except where noted.

Resources are of two kinds: Codelets and modules (Fig. 8.2). Codelets are internal "micro-agents" that play a specific role in the architecture: perception, metacognitive observation, representing information, monitoring, etc. Those that represent information have an internal structure that allows them to store an item of information along with its classification. Here, we depict Codelets as clouds, since they may work alone or as a team. Modules, shown as rectangles, can be any piece of programming, agent or conventional, that receives information and returns some other element of information to Working Memory through an interface. Links (arrows) describe two kinds of information transfer between these resources: textual information and activation. "Textual information" refers to concepts, facts or sub-information, completed by a category (for instance "camera" + "left"). "Activation" refers to numbers that indicate how strongly individual Codelets or nodes are stimulated. We will come back to activation later on in our tour of the architecture, when we describe the Behavior Network.

All Codelets perform their function in the "unconscious" side of the architecture (the *audience* in the theatre metaphor of the Global Workspace theory), except for the information Codelets, which operate on both sides. Teaming of processors (Codelets) is an essential tenet of the theory: there is both collaboration and competition among the processors. Processors collaborate to describe the

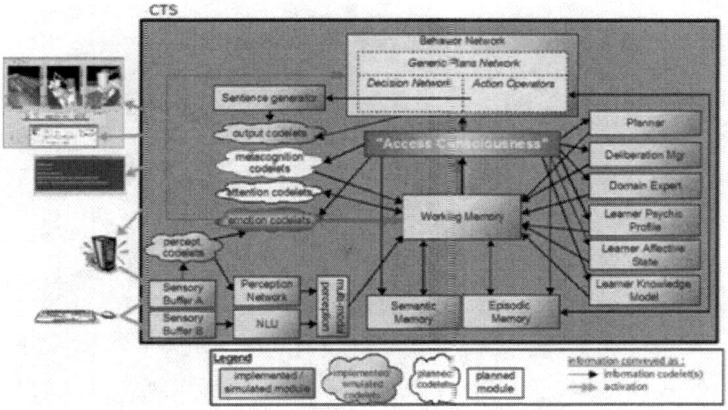

Fig. 8.2 Conceptual architecture for CTS

situation more fully, and then the resulting teams compete for access to broadcasting by consciousness. Central to the architecture are Working Memory and "Access Consciousness", which form the hub for most communications. These are a fundamental element of the architecture, and are placed in the center of the diagram. Working Memory receives input from all sources, and Access Consciousness copies the information selected by Attention to every processing resource.

Processing is organized by a cognitive cycle comprising eight steps (nine for LIDA), going from capturing external stimuli to performing an action. These steps specify when resources are solicited and allowed to contribute information (by sending it to Working Memory). In accordance with Baars' GW theory and experimental results indicating that performing a simple action normally requires about 200 ms, these steps are repeated 5 times per second in an endless cycle. However, since processing the information in a particular subsystem may take more or less time, depending on complexity or other factors, complete treatment of a piece of information may require idle cycling, at least with respect to that specific information (parallel processing of other information in overlapping cycles can, however, continue simultaneously).

Before going into more detail, it is appropriate to interrupt the presentation briefly to present the external environment and application CTS has been adapted for in its current instantiation. This will help in understanding some of the specific goals we have set.

8.4.2.1 CanadarmTutor to Support Astronaut Training

CTS has been developed in a collaborative project with the Canadian Space Agency to coach astronauts in need of training on manipulating the Canadarm2 (Nkambou et al. 2005). Since the robotic arm is a crane built with seven degrees of freedom, it is not easy to predict how a given part will actually move when a joint is activated. What makes matters even more difficult is that the Arm and the

Fig. 8.3 Astronauts have to deal with three camera views and a complementary textual source of information for figuring out Canadarm's situation. NASA

Fig. 8.4 A screenshot of the CanadarmTutor user interface. The red path is the course of action suggested by the path planner.

Space Station can be observed only via three computer monitors that show limited view-provided by cameras located at fixed positions (Fig. 8.3). Astronauts need precisely honed spatial awareness, because they have to operate in a setting very different from terrestrial operations, one in which "up" and "down" have little meaning. However, tutoring resources are scarce. An ITS would provide a welcome support in numerous respects, facilitating learning of concepts and procedures (system of axes, manipulation rules and safe procedures), suggesting appropriate exercises in spatial recognition, and providing the opportunity to practice manipulations with coaching.

Our first implementation of a tutoring system for the International Space Station and its Robotic Manipulator System, Canadarm2, has been designed to give primarily reactive feedback that helps enhance spatial and situational awareness (Roy et al. 2004). CanadarmTutor's tutoring capability initially relied on a path planner that serves as an expert to validate learners' actions. The planner automatically detects student errors in operating the manipulator, produces illustrations of correct and incorrect motions in training, and provides feedback and hints. The path planner we developed acts as a domain expert and can calculate the Arm's moves, consistent with the best available camera views, to avoid obstacles and achieve a given goal. The path planner enables CanadarmTutor to answer several types of student questions, such as "how to....", "what if...", "what's next", "why" and "why not". However, sometimes the solution paths supplied by the path planner prove too complex and difficult for students to follow. The path planner thus does not meet the two basic principles for tutoring agents in procedural tasks: it can't guide the user through an expert solution or recognize the student profile (novice, intermediate, or expert) to offer tailored feedback and assistance. To address these deficits, this capability has thus been extended with our "conscious" cognitive tutoring agent. We will refer to this simulated environment and its user interface (Fig. 8.4) for our coming descriptions and examples. We can now continue with our tour.

There are functional correspondences of modules with the human nervous system and, at a higher level, with functions it accomplishes. On the more biological side, we find Sensory Buffers and Perception, and Semantic and Episodic Memories. At a higher level of correspondence, we find Working Memory, the Learner Models and the Domain Expert. The Behavior Network fits somewhere in between. The inclusion of not-so-biological modules, especially the "peripheral" ones, in roles that may currently be played by external, independent agents, is a concession we are willing to make for two reasons. First, it makes the system easier for people from the ITS world to grasp. Second, it permits any currently operating module or agent to be accepted as an input or extension to CTS. And, if we need a third reason, this is a concession that will eventually be removed when the implementation of the ITS is complete. Then, all the necessary, psychologically and biologically plausible teams of Codelets will replace modules, collectively playing the same roles. Here are the roles Codelets and modules play in reaching decisions.

8.4.2.2 Resources Proper to the Architecture

Senses and Perception. CTS currently possesses two sensing channels for textual inputs, linked to two input buffers, one for the user interface and one for messages from the simulator. After the incoming message has been rearranged in a hierarchical tree by a syntactic parser, the perceptual Codelets (to be explained in the Behavior Network section below) scan the buffer and activate nodes in the Perceptual Network (PN) while simultaneously transferring to them the data they have recognized. These nodes represent the information as concepts the agent can recognize ("Canadarm2 manipulation", "user answer", etc.). Nodes already stimulated receive a new partial boost.

Working Memory (WM). In the GW theory, *consciousness is associated with a global workspace in the brain – a fleeting memory capacity whose focal contents are widely distributed ("broadcast") to many unconscious specialized networks* (Baars and Franklin 2003). The architecture depicts WM as a short-term "storage place" (figuratively speaking) where information Codelets from all sources meet, form associations and eventually become part of coalitions that may broadcast. Attention constantly hovers around WM and selects one of the competing coalitions. WM is an information hub between modules, which usually have no direct relation, and for Codelets that need to communicate information (in the form of an information Codelet that they create). Information that lands in WM is copied to Long-Term memories which, like other modules, return the coalition with information complements. The WM in CTS implements important aspects of the Blackboard principles, although modules do not pick what they find of interest to them, but only receive what has been selected by Attention.

Attention. Attention is primarily the mechanism that selects the most prominent (activated) information in WM and supplies it to Access Consciousness for global broadcasting. It corresponds to the *spotlight* used by Baars in his theater metaphor. This selection mechanism is involuntary and uncontrollable. Indeed, one cannot choose what comes to mind, except by making a conscious and difficult effort to *focus* attention, giving preeminence to some specific information. That voluntary focusing of attention is implemented in CTS with attention Codelets that add activation to information Codelets that correspond to their monitoring.

Access Consciousness. Access Consciousness "publishes" the information selected by the Attention mechanism to make it available to all (unconscious) modules. It allows all "actors" to become aware of the situation. During deliberations, this publishing returns to all actors the information contributed by one of them in WM. This mechanism is crucial for the collaboration of the parts in complex, conscious processing, for instance in elaborating a progressive diagnosis.

Transient Episodic Memory (TEM). TEM receives broadcasts and makes a record of them. It can then be probed for recent events by time of occurrence.

Associative Memory (AM). Implements relations between events at the conceptual level. Events can be retrieved via keyword. AM could also contain knowledge of the Domain Model but, for practical reasons, the DM is left as a separate module.

The Behavior Network (BN). Tutorial behavior is stored in a three-tiered Behavior Network. This is based on an original algorithm from Pattie Maes, which she called MASM (Maes' Action Selection Mechanism; Maes 1989), as reimplemented by Franklin and his team. We have reorganized it to include ideas expressed in BEETLE (Zin et al. 2002). It retains Maes' idea of a network with *competence modules*; there are pre- and post-conditions for each node, bidirectional links between them through their preconditions and effects, and both logical conditions and activation levels for deciding on the next behavior to select.

However, Maes had a network of competence nodes performing actions by themselves; Negatu and Franklin (2002) separated decision from action by adding *Codelets* (micro-agents; see further below) related to each node, which allow adapted actions without increasing the number of fixed competence modules. This also made it possible to automatize sequences of frequent actions in a biologically plausible, hebbian manner. Our version of the BN keeps these ideas. It also retains levels of planning and action similar to those in LIDA, but functions somewhat differently, with the help of other mechanisms: a "Planner" and deliberation, which are covered below.

CTS' behavioral knowledge is organized in three groups within the Behavior Network: 1) a Decision Network (DN), 2) a network of Generic Plans (GPN) and 3) a Bag of Actions (BA). The ultimate organization is done via CTS' Planner (4). When the context favors the emergence of a behavior in the DN, a way of implementing the general behavior contextually emerges in the GPN portion of the network, bringing into the Planner the specific steps (actions) that will implement the global behavior; finally, the action is proposed in Working Memory for adaptation and a last-call vote on its execution. The adaptation occurs via a deliberation that summons the various actors in the architecture for their contribution to the information payload.

BN-1) Decision Network: The Decision Network holds nodes representing high-level behaviors and contextually decides on which one to retain.

BN-2) Generic Plan Network: A generic plan contains ways (streams of lower-level actions) of realizing the behaviors. It is comprised of steps that necessarily follow each other, without alternative paths. Generic plans serve two general purposes. Some are high-level plans that orient behavior globally, giving general steps at a coarse level of specification. Others are closer to operations and no further expansion of their steps is required. The latter type either supply a more detailed account of the necessary steps, or present methods for fleshing out a high-level action found in the Decision Network. As an example of a coarse plan needing subspecification, consider the special case of the very generic plan for conducting a learning session, with the steps Beginning, Middle and Ending. Obviously, these steps need more detailed instructions. When the time comes, the Planner submits the step to Working Memory so that it can be elaborated with a more detailed generic plan. As examples of methods, we have the many generic plans describing ways for executing interventions – for instance, after an incorrect operation on the learner's part. Elementary actions are adapted to the context observed at the time the act is performed (specific relevant content, preferences of the current user with respect to optional steps). Deciding on the next thing to do depends partly on the level of activation of a node. The "strongest" node for which all of the *required* preconditions are satisfied is the one that pilots CTS next processing step.

BN-3) Bag of Actions: The Bag of Actions is the repertoire of individual, elementary actions in the form of a collection of Codelets that can be summoned for action. An action may be referred to in many places of the Decision Network, in different plans.

BN-4) Planner: The sub-elements of the BN are ultimately organized through CTS' Planner, which makes it possible to do some planning that is hard to achieve with Maes' algorithm. For instance, in some situations, it may be appropriate to focus temporarily on remediating the learner's faulty understanding, and then resume the presentation or whatever was going on, which in fact amounts to inserting a remediation plan into the presentation plan. Without the Planner, it would be uncertain at best (depending on the BN's concrete implementation) whether the current plan would be pursued when remediation was complete.

Despite the apparent implications of its name in computer terms, the Planner does not decide on anything, it does not *plan*. It simply toys with plans, storing steps and managing the addition, insertion and removal of steps. It implements that portion of Working Memory which holds the list of goals and steps we, as humans, keep alive in our minds.

Codelets. Elementary processing is a fundamental aspect of the architecture, as Baars hypothesized that the mind is a community of simple processes. Franklin used the Copycat concept of *codelet* (Hofstadter and Mitchell 1995) to implement these processes. A Codelet is a light, specialized agent with minimal autonomy. It receives information, discards what it does not recognize, and reacts according to its capabilities. Collectively, a team of Codelets may pursue a higher-level goal, such as interpreting a message in the Sensory Buffer. In our architecture, Codelets are grouped by type of specialty: information, attention, control, or action; or by where they reside or where they look for information: in Perception, Working Memory or the Behavior Network. Here is a description of the varieties:

C-1) Information Codelets: An information Codelet represents or contains information in a specific location, or serves as a transporter of data for communication between modules and Working Memory. Information Codelets may form associations with each other, each one containing a single element within coalitions that provide a richer content. They are short-lived, being created by modules or by some types of Codelets to represent states of fact in Working Memory. When a process becomes automated, they may not be needed anymore, being replaced by a direct communication link between connected actors ("unconscious" communication).

C-2) Action Codelets. Released by the BN's nodes, action Codelets implement the "voluntary" actions of CTS, both on the internal and on the "motor" (outer) side. They act on the environment that is external to CTS, or within CTS. Whereas a Codelet of most other types fulfills its responsibilities in a self-contained fashion and produces an information Codelet, action Codelets act as builders and modifiers in their environment. For instance, some send messages to other computers – say, to request video replay to the simulator; some others display messages in the user interface; etc.

C-3) Attention Codelets. This is a category that encompasses all of the processes that monitor other processes or communications. These Codelets serve in specialized roles: as reinforcers of information appearing in Working Memory (implementing voluntary attention), as processes overseeing the result of an operation in order to declare its success or failure (*expectation Codelets*), or as processes monitoring the global operations of CTS (implementing metacognitive processes).

C-4) Perception Codelets. Codelets in this group are collectively responsible for recognizing elements entering the Sensory Buffers and creating an interpretation by activating the information Codelets that form the perceptual network. They are hybrids, having the perceptual role of attention Codelets (they monitor Sensory Buffers in hope of recognizing something there), and the action role of action Codelets (they act on other Codelets, activating them).

C-5) Emotion Codelets. Collectively, they form CTS' *pseudo-amygdala* (the amygdala being an organ involved in learning emotional associations). Some of them are connected to Perception Codelets on one side, and to emotional motivator Nodes in the BN on the other. They evaluate the low-level features detected by the perception Codelets and stimulate accordingly the emotional motivator Codelets located in the BN (see below). Others are concerned with what comes to Working Memory, adding to relevant information Codelets activation with emotional tag.

C-6) State Codelets. State Codelets are hybrids similar to perception Codelets except that they exist within the Behavior Network, where they represent the current context as observed. They hear broadcasts from Access Consciousness and turn on when they recognize information. As a logical context, they form the required preconditions for behavior nodes to fire. As analog context, they supply activation to the single or multiple nodes they are attached to (which may form part of the DN, the GPN or the BA). Thus, the context determines what operation is most relevant, according to what Attention has selected for broadcasting. Some of the state Codelets are not absolute prerequisites for an action, but rather represent optional conditions: preferences, favorable circumstances, etc. These simply *add* activation to a node (or take away some, in the case of unwanted situations), increasing (lowering) the likeliness that this node will be favored.

C-7) Motivator Codelets. Motivators are the other source for activation supplied to nodes in the BN. Some Motivators implement values and principles dear to the agent; some connect to emotion Codelets and represent them in the BN. Like the optional state Codelets, they are part of a richer context and potentially add to the activation of a node, driving action selection "from the top", with respect to the agent's goals, and from situational analysis (emotions). They are variably stimulated by the perceived events or inferred beliefs they correspond to, and they pour activation into the Behavior nodes that connect to them. Motivator Codelets can be made more or less sensitive to excitators, and sets of parameters organized as profiles may form various personalities for CTS.

8.4.2.3 Resources External to the Architecture

External resources use privileged communication channels that make them virtually part of the architecture. An interface bidirectionally transforms information into information Codelets. These modules and external agents react to information they "hear" from the broadcasts just the same; they also volunteer information when they deem appropriate, eventually priming some “motivators” in the agent (see section C-7).

Domain Expert (DE). Knowledge of the domain and evaluation of the learner's actions has been delegated to an external agent based on MIACE (Mayers 1997) and developed by Fournier-Viger (see (Najjar et al. 2005) for details on how this agent encodes its knowledge). The Domain Expert receives Access Consciousness' broadcasts and then evaluates the learner's operations and answers. It communicates with CTS through an interface which does the bidirectional transformation to and from information Codelets.

Learner Model (LM). The learner model is a distributed one. Its static part, the Learner Profile (LP), contains psychological information, including the learner's learning style. Its dynamic part is twofold: the Learner Affective State (LAS) tracks the learner's mood and emotional state, while the Learner Knowledge Model (LKM) holds facts and the learning history, infers knowledge and trends, and computes statistics.

8.4.3 Global Decision Process in CTS

CTS makes decisions and operates on the basis of three fundamental processes: cognitive cycle, activation transfer and deliberation. They interact and create the global decision mechanism.

Activation. This is the first level of a systemic decision, the most elementary. Much in the architecture processing is activation-based, with decisions emerging both through voting during deliberations and through competition (in Working Memory and in the Behavior Network). Behavior nodes accumulate activation from state and motivator Codelets; information Codelets bring value to a coalition in WM; coalitions having enough activation occupy WM and compete for broadcasting (for becoming the "conscious" information) based on their global activation level. Information Codelets progressively lose their activation and eventually disappear from WM, and cannot anymore be part of an upcoming broadcasting.

Cognitive Cycle. The cognitive cycle is the next level of decision, where all resources get the chance to participate. In CTS/LIDA, it is a detailed version of the standard perceive-process-act cycle. Eight steps (nine in the case of LIDA) specify an order for actors to intervene and for the processing to take place in the Behavior Network (please look at Fig. 8.5): 1) Perception, 2) Percept becomes part of WM, 3) LT memories see the info and return related info, 4) Competition for consciousness of all coalitions in WM, 5) Broadcast of the strongest (coalition of) information, 6) Recruitement of behavioral resources (activation by state and motivator Codelets), 7) Selection (among all activated Behaviors), and 8) Acting (releasing action Codelets).

In one cycle, there is one and only one broadcast of information. But consecutive cycles may overlap, due to the highly distributed processing, at least in the human model. For instance, sensing may take place at the same time as memories are searching for information about a previous broadcast. In any case, a cycle is the elementary unit of conscious processing. However, processing information in a meaningful or useful manner may require more than one cycle, for iterative refinement of an idea or a plan, or for pondering alternatives. Thus, an *action* may not happen at every cycle, in the sense of taking an action on the exterior world, until the *deliberation* has ended.

Deliberation constitutes the most complex decision processing. Except in very special activities (dangerous situations, meditation), context is not a given from the external or the internal environment alone; ideas, points of view, hypotheses and preferences usually supplement the raw information from the outside to form a richer environment for human decisions. Deliberation in CTS is the gathering and analysis of information, including "opinions", through reflection loops made possible by Access Consciousness and its broadcasting of selected information to all actors in the architecture. In CTS, the Behavior Network is not the lone decision-maker that dictates what to do next. All the actors have input, sometimes signaling a situation that may need attention, sometimes opposing a suggested course of action, supplying alternatives, or helping to adapt the proposed intervention. When a proposition is opposed, new suggestions are sought, initiating a new round of deliberative cycles.

The three levels are interlinked, activation passing sustaining the selection of information toward broadcasting, and broadcasting allowing the multiple loops that form a deliberation.

Decisions in CTS may involve another aspect that we have not mentioned until now, a facet that is very humanly: emotions. They are the last component of CTS involved in decision-taking. We now turn our attention to this aspect.

8.4.4 Emotions in the Agent

CTS' architecture allows decisions to be made under the influence of emotions in a biologically plausible manner. Part of its apparatus mimics the involvement of the amygdala in learning and producing spontaneous emotional response. A good description of this segment can be found in (Faghihi et al. 2008). Another part memorizes emotional appraisal along with the factual content of an event, which has repercussions later, when memories are probed and brought back into Working Memory, influencing decision making. We will take a brief tour of these systems below, starting with the amygdala.

Amygdala's emotional involvementThe amygdala is known to play a role in emotional responses. Research in neurobiology suggests that there are two (or more) routes from perception to action (Rolls 1999). The first route is short and direct: information flows from the sensory thalamus directly to the amygdala, and from the amygdala to basal ganglia. Motor reaction is then rapid, even impulsive, as the received information is not interpreted by other brain structures (Squire and Kandel 1999). In the second route, information from the external environment is analyzed by various cortical areas (primary sensory cortex, unimodal associative cortex, polymodal associative cortex). It is then sent to the hippocampus, presumably for memory retrieval and temporary storage. All of this processing serves to give meaning to the external stimulus and link it to other events (through the hippocampus' episodic memory), before it returns to the amygdala for emotional appraisal and response.

These two routes are implemented in CTS and work in parallel. Emotions that become conscious as well as those that remain unconscious (follow the "short" route) excite *emotional nodes* located in the Behavior Network. The short route influences the selection of Behaviors (by priming some of them), even sometimes squarely causing a Behavior to fire before the longer, analytic process comes to a conclusion and selects the "logical" choice.

We'll look at the involvement of emotions in conscious action decisions before contrasting it to the unconscious influence of emotions. You may look at **Error! Reference source not found.** 8.6 to accompany the coming descriptions. It illustrates both the cognitive cycle in CTS and emotional processing. As we will describe, emotions get involved at many points in the cycle.

Emotions influencing *conscious* decisions in CTSIn step 1 of the cognitive cycle, the collective interpretation work made by the perception Codelets results in a percept, which is temporarily stored as the active nodes of the Perceptual Network. The percept becomes part of Working Memory as a single network of information Codelets (step 2), but is looked upon by the Coalition Manager as multiple possible coalitions of Codelets, each describing various aspects of the situation. During the Coalition Manager's processing, emotion Codelets inspect the informational content of each coalition and infuse it with a level of activation proportional to its emotional valuation. They also graft an information Codelet that tags them with the classification of the emotional energy. The activation supplement increases the likelihood of some coalitions to draw Attention to themselves. The emotionally-reinforced coalition may then become the information CTS considers in a forthcoming deliberation (if that coalition is selected and broadcast). In a nutshell, emotions influence the focus of attention. For example, imagine the situation where the learner does not acknowledge repeated prompts from CTS. The tutoring agent has learned in past occasions that this may be a sign of annoyance, so it should "feel" sorry for taking this course of action and tag this situation as negative.

When an emotionally-reinforced coalition is selected (step 4) and broadcast (step 5), its content is received by the Transient Episodic Memory which stores it with its emotional tag and value. When TEM is later probed for concepts or events relating to the current situation, emotional valuation heightens the salience of stored events, influencing which is restored to Working Memory (step 3). Here, too, emotions influence the decision making in CTS by modifying what is memorized and recalled.

The broadcast coalition is also received by the Behavior Network. Some emotional motivator Codelets should react to the emotional aspect of the information and infuse the Behavior nodes they are connected to with some activation, in this case inhibitory, lowering their probability of further being selected (step 6). This is a third involvement of emotions in CTS.

These direct (reaction to WM content) and indirect (memory recall) emotional interventions in WM is how the Amygdala gets involved in CTS' analytical long route.

Emotions causing reflex actions. In the short route (Fig. 8.5, ESR rectangles), emotional involvement occurs sooner, but based on a rough evaluation of the

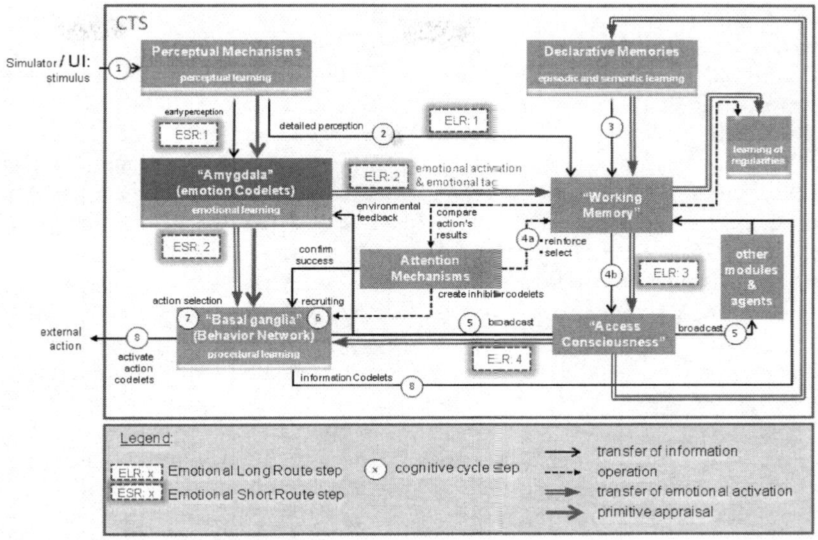

Fig. 8.5 CTS' cognitive cycle states when the various actors of the architecture may do their processing of information. Emotions impact CTS cognition at various points. This influence may happen unconsciously, either to promote the selection of a specific behavior, of to enhance memorization.

situation. The perception Codelets connect in parallel both to the PN and to emotion Codelets. Before their collective interpretation creates a rich and complete interpretation of the stimulus, their individual activation is copied directly to emotional motivator Codelets (themselves connected directly to acts in the BN) the first leg of the short route. This direct stimulation finds its application primarily in dangerous situations. When low-level basic information coming from the perception Codelets is recognized as highly dangerous, there is no time to think, and the emotional Codelets will likely force an action to fire in the Action Bag of the BN. This usually makes CTS act before it has had time to become "conscious" of the situation and consciously plan a course of action. This is the fourth emotional involvement. As a concrete example, say CTS learns that an input from the learner saying "I'm going to quit" after a series of failures shortly precedes his shutting off the session. This contradicts the tutor's goal of seeing the learner conclude an exercise successfully, and is interpreted as saddening. It is very much undesirable. This negative association, learned by an emotional Codelet in the past, will fire an immediate, simple reaction (a short message) allowing some time for CTS to reach a more involved, deliberated decision about what to do next. This implicit reaction corresponds to the process described by Squire and Kandel (1998).

But there is a fifth way emotions may have an impact: they modulate actions intensity. Although not implemented yet in CTS (the tutor currently has means of action limited to text boxes, canned demonstrations and sequences replay), we foresee motor actions being modified in their amplitude and rapidity by the emotional value of the coalition that sparked them.

Emotional reinforcement. Instinctive reactions do not block the longer analytical process; they merely step in first. Eventually, however, a better idea of the situation is worked out (following complete interpretation by Perception) and may come to consciousness, allowing normal action selection to take place. The instantaneous, emotional reflex initiated by the "short route" is then compared by expectation Codelet(s) to the action that has been selected in the BN in the "normal" way. If the two roughly correspond, the expectation Codelet does two things: it grafts an information Codelet with a strong negative valence onto the action proposition in WM (to prevent a repetition of the action); it also grafts a confirmation stating the correspondence, which will serve, when broadcast, as a reinforcer to the emotion Codelet(s) that was or were instrumental in setting off the reflex. In effect, this will cause CTS' "pseudo-amygdala" to reinforce its relevant "rule". However, when the initial reaction diverges from the behavior proposed by the more methodical analysis, some mechanism must be in place to control the incorrect reflex action. Indeed, according to Rolls (2000), the amygdala never unlearns a rule, always reacts to the same stimulus and needs cortical interventions to temper it (Morén 2002). From a neurological point of view, control over actions is the role of cortical areas. We implement the cortical controls here with *inhibition Codelets*, which are generated by the expectation Codelet(s) that spotted the discrepancy. These Codelets attach themselves to a behavior node in the action selection mechanism (BN) and constantly subtract activation from it.

8.4.5 Summary on CTS

It would be nice to have a tutoring agent capable of learning the domain on its own, adapting to various operational settings and offering appropriate tutoring for different types of knowledge, transferring relevant strategies and tutoring tactics to the new instructional constraints. CTS' proposes many of the necessary mechanisms (some of them not described here), and its architecture allows for extensions as well as adaptations to include biological features. The current prototype implements many principles of the GW theory and, while we keep adding and improving aspects of the agent, it already demonstrates that artificial cognition works. The separation unconscious vs. conscious and a functional consciousness offer a promising avenue of advancement.

8.5 Coming Trends: A Glimpse of the Future

In all of the systems presented, including CTS, the cognitive processing is accomplished at a high level of representation, which helps in the design and implementation aspects. However, artificial systems still have very limited capabilities compared to real-world requirements: their ability to interpret visual inputs (recognizing objects, interpreting scenes) is weak, they have little insight about focusing their attention on what counts, their linguistic capabilities remain wanting, they are still awaiting a reasonable foundation of knowledge for real-world "common-sense" reasoning and lack the framework for integrating diversified knowledge (Dutch et al. 2008; Samsonovich et al. 2008). They also fare poorly with regard to autonomy, in its widest ramifications, are pretty clumsy about emotional intelligence (Picard 2007) and remain very limited in their social ability, which may impede their potential for finding solutions and knowledge whose relevance has been tested and validated by other artificial agents. Fidelity to the human model remains remote, biologically speaking, and the explanatory power of the architectures is limited. A match closer to biology may be wished for. But does any of this really matter? Why insist on copying the way humans process information, and why even attempt to reproduce it at the lower levels of processing?

From a negative standpoint, applying non-human-readable artificial neuron networks to accomplish some chain of processing is much harder and not within anyone's reach, at least given the current state of authoring tools. The same may be true even for symbolic-level tools and architectures. On the positive side, having "living" models of biological theories is a good way of testing them, finding their limits and loopholes. In this way, research gains improved theories, and also becomes better able to predict the impact of surgical or psychological interventions. These are some of the reasons prompting the pursuit of biologically-inspired cognitive architectures, known increasingly under the acronym BICA.

We will now present a brief Q&A on the subject, and end by revisiting the question that is the focus of this chapter: "How will this impact an artificial intelligent tutor?"

How does BICA differ from current cognitive architectures? Well, it depends on where an architecture is positioned on the "biological fidelity scale". Research has produced, and is maintaining and developing, systems that use representation and computation means at varying degrees of "natural fidelity". Many existing architectures, such as ACT-R and LIDA/CTS, seek mind-level natural validity, with some functional correspondence to brain structures and mathematical reproduction of neural learning; others, such as CLARION (Sun 2004) and SOVEREIGN (Gnadt and Grossberg 2008), tend to add a lower-level biological *explanatory* level (with biologically-inspired structures and processing that cause and explain the results, with respect to biologically-inspired parameters and mathematical computations that justify the results). In other words, biologically-*inspired* architectures are not constrained to purely and essentially reproducing biological processing; they may incorporate "*artificial* AI" techniques, some being more difficult to justify in biological terms than others. What is really new is the interest for creating complete architectures that rely on low-level fidelity to biological processes ("complete architectures" meaning, here, "systems that cover all of the human nervous system"). Samsonovich proposes an excellent feature-by-feature overview of many cognitive architectures (online at http://members.cox.net/bica2009/cogarch/architectures.htm), showing, among other things, their correspondence with brain structures and their level of modeling. The reader is also referred to the article by Dutch *et al.* (2008) reviewing current symbolic cognitive architectures and emerging paradigms.

What is to be gained with BICA? One thing that has constantly eluded AI research efforts is artificial *general* intelligence (AGI), that is, the ability to adapt and react to unknown situations in unconstrained domains, by applying existing knowledge in new ways or acquiring new knowledge. AI has produced solutions that often outperform humans in specific tasks or on limited problems, but they are either not meant for general applicability or cannot transfer to other domains. What has set humans apart from other species is, for one, our ability to adapt, but more than that, our ability to learn and to set goals for this learning through self-monitoring, self-regulation and valuation of goals. Being able to judge its own lack of knowledge, to determine ways for acquiring that knowledge, to store it in a meaningful and usable manner, and have the motivation to do these things, would allow an artificial system to walk in the footsteps of a human baby and become better at whatever tasks it has been assigned, on its own. Researchers believe that strong ties to the human model, at the biological level, are currently our best bet and that we should pursue "strongly" biology-inspired cognitive architectures (Berg-Cross 2008).

What impact will it have on artificial teachers? One trait that is fundamental in a good teacher is his (her) ability to learn and better adapt to his (her) students. While this sounds simple enough, in reality it calls for a collection of abilities that are well integrated in good human teachers/tutors, both as high-level processes and as low-level mechanisms sustaining them. Reproducing their interplay requires an architecture capable of capturing and making sense of subtle signals, and bringing together various information sources, decision centers, and motivational mechanisms. Motivation fuels a good part of the process of taking action, and this naturally connects to values and emotions. Being interested in the learner's performance may be woven into the fabric of an ITS, but deciding to devote some attention to the learner's emotions and motivational state at some point, rather than solely to his knowledge state, is a more of a human game. And it is touchy. Striking the right balance in different situations, with a variety of learners, requires dynamic adaptation, and dynamic learning too, all done under the supervision of metacognition. Each situation has something to teach about the efficiency of attempts with respect to a specific learner (read: his personality profile). It is only through constant learning (and self-supervision) that an artificial teacher can improve. At some point, human designers can no longer be of much help. The agent has to see for itself, make attempts and gauge the results. Samsonovich and colleagues' Cognitive Constructor relies on these premises (Samsonovich et al. 2008).

We would love to see scenarios where the artificial teaching agent decides to take innovative actions to get out of a situation judged as an impasse. For instance, the agent might conclude that the learner just doesn't get it (results do not improve significantly) in spite of all the effort on both sides, and begin showing demotivation and anger. A clueless agent could keep repeating the same theory and exercises over and over again, or it could simply go on with the next subject after storing the poor grades in its database. An AGI teacher might, instead, detect distress and anger, evaluate the signals as deserving priority of attention, "feel" compassion at the sight of the effort the learner has expended and disappointment at the results, and become motivated ("feel" the urge) to devote more resources to finding a solution. In view of the global context (learner profile, past performances, valid goals), it might then try applying current abilities, such as communicating over the Internet, to a search for pedagogical solutions, or decide to consult other similar pedagogical agents about their proven solutions to similar situations. A reward system ("pleasure") would reinforce the tendency to use the newly created creative path or, conversely, increase the resistance to it. In the latter case (unsuccessful new solution path), the motivation for supporting the learner would remain highly activated and another plan for the creative application of current abilities would therefore be devised. In that scenario, we see the usefulness of emotions in orienting decisions and generating amplitude in the reaction. Besides endowing the agent with more social grace, emotions are a quick way of integrating multiple sources of information – somewhat akin to intuition – , reinforcing decisions and orienting the next action.

As we are coming to understand more and more, emotions are not irrelevant to performance; they are not simply something that's nice to have. They truly are part of intelligence. Plain logic and hard facts usually do not suffice to build lasting relationships among humans. At times, it may become irritating to deal with a senseless or otherwise poker-faced teacher. An ITS meant for long-term interventions (accompanying students year after year) needs to address learners' emotions. Empathy is sometimes sought, and a lack of it may demotivate the learner. Thus, the artificial agent should be able to detect, aptly deal with, and display emotions. It takes time, practice and fine observation to find what works in which situation, and it takes the appropriate memory structures to store emotions in a significant and useful way. Humans possess these, and they work quite well in most situations. Along with our perception and communication abilities, our metacognitive mechanisms and our goal-setting autonomy, they are yet another good reason to take a long, hard look at the human model.

Acknowledgements. We gratefully acknowledge the work and helping hand of many current and past members of the GDAC laboratory and other close collaborators in the realization of CTS and Canadarm Tutor: Patrick Hohmeyer, Pierre Poirier, Usef Faghihi, Mohamed Gaha, Philippe Fournier-Viger, Khaled Belghith.

References

Aleven, V., McLaren, B.M., Sewall, J., Koedinger, K.R.: A New Paradigm for Intelligent Tutoring Systems: Example-Tracing Tutors. International Journal of Artificial Intelligence in Education 19, 105–154 (2009)

Anderson, J.R.: Language, memory, and thought. Erlbaum, Hillsdale (1976)

Anderson, J.R.: Rules of the Mind. Erlbaum, Hillsdale (1993)

Anderson, J.R., Corbett, A., Koedinger, K.R., Pelletier, R.: Cognitive tutors: Lessons learned. J. Learn. Sci. 4(2), 167–207 (1995)

Anderson, J.R., Gluck, K.: What role do cognitive architectures play in intelligent tutoring systems? In: Klahr, D., Carver, S.M. (eds.) Cognition & Instruction: Twenty-five years of progress. Lawrence-Erlbaum Associates, Mahwah (2001)

Baars, B.J.: In the Theater of Consciousness. The Workspace of the Mind. Oxford University Press, New York (1997)

Baars, B.J., Franklin, S.: How conscious experience and working memory interact. Trends in Cognitive Sciences 7, 166–172 (2003)

Batson, C.D., Turk, C.L., Shaw, L.L., Klein, T.R.: Information Function of Empathic Emotion - Learning That We Value the Others Welfare. J. Personal Soc. Psychol. 68, 300–313 (1995)

Berg-Cross, G.: Introduction to Biological Inspiration for Intelligent Systems. In: Madhavan, R., Messina, E.R. (eds.) Performance Metrics for Intelligent Systems (PERMIS) Workshop (2008)

Blessing, S.B.: A Programming by demonstration authoring tool for model-tracing tutors. Int. J. Artif. Intell. Educ. 8, 233–261 (1997); reprinted in Murray, T., Blessing, S., Ainsworth, S. (eds) Authoring tools for advanced technology educational software. Kluwer Academic Publishers, Dordrecht, pp. 93–120 (2003)

Block, N.: On a Confusion About a Function of Consciousness. Behav. Brain Sci. 18, 227–247 (1995)

Block, N.: Some Concepts of Consciousness. In: Chalmers, D. (ed.) Philosophy of Mind: Classical and Contemporary Readings. Oxford University Press, New York (2002)

Carbonell, J.R.: AI in CAI: An Artificial Intelligence Approach to Computer-Aided Instruction. IEEE Trans. Man-Mach. Syst. 11, 190–202 (1970)

Carnegie Learning, Timeline of Cognitive Tutor History (2010), `http://ctat.pact.cs.cmu.edu/index.php?id=timeline` (accessed)

Choksey, S., Heffernan, N.: An Evaluation of the Run-Time Performance of the Model-Tracing Algorithm of Two Different Production Systems: JESS and TDK. In: Tech. Rep. WPI-CS-TR-03-31. Worcester Polytechnic Institute, Worcester, MA (2003)

Cohen, P.A., Kulik, J.A., Kulik, C.C.: Educational outcomes of tutoring: A meta-analysis of findings. Am. Educ. Res. J. 19, 237–248 (1982)

Collins, A., Warnock, E.H., Aiello, N., Miller, M.L.: Reasoning from incomplete knowledge. In: Bobrow, D., Collins, A. (eds.) Representation and Understanding: Studies in Cognitive Science. Academic Press, New York (1975)

Conati, C., Gertner, A., VanLehn, K.: Using Bayesian networks to manage uncertainly in stu-dent modeling. User Modeling & User-Adapted Interaction 12(4), 371–417 (2002)

Craig, S.D., Graesser, A.C., Sullins, J., Gholson, J.: Affect and learning: An exploratory look into the role of affect in leaning. J. Educ. Media 29, 241–250 (2004)

Crick, F., Koch, C.: Towards a neurobiological theory of consciousness. Seminars of the Neurosciences 2, 263–275 (1990)

Crick, F., Koch, C.: A framework for consciousness. Nature neuroscience 6(2) (2003)

Dehaene, S., Naccache, L.: Towards a cognitive neuroscience of consciousness: basic evidence and a workspace framework. Cogn. 79, 1–37 (2001)

Dennett, D.: Are we explaining consciousness yet? Cogn. 79, 221–237 (2001)

Dennett, D., Akins, K.: Multiple drafts model. Scholarpedia 3(4), 4321 (2008) (accessed 21, March 2010)

D'Mello, S.K., Craig, S.D., Fike, K., Graesser, A.C.: Responding to Learners' Cognitive-Affective States with Supportive and Shakeup Dialogues. In: Jacko, J.A. (ed.) Human-Computer Interaction. Ambient, Ubiquitous and Intelligent Interaction. Springer, Heidelberg (2009)

Dubois, D.: Constructing an agent equipped with an artificial consciousness: Application to an intelligent tutoring system. PhD Thesis, Université du Québec à Montréal (2007)

Duch, W., Oentaryo, R.J., Pasquier, M.: Cognitive architectures: where do we go from here? Frontiers in Artificial Intelligence and Applications 171, 122–136 (2008)

Erman, L.D., Hayes-Roth, F., Lesser, V.R., Reddy, D.R.: The Hearsay-II speech-understanding system: Integrating knowledge to resolve uncertainty. Computing Surveys 12(2), 213–253 (1980)

Faghihi, U., Poirier, P., Dubois, D., Gaha, M., Nkambou, R.: Implementation of Emotional Learning for Cognitive Tutoring Agents. In: Gelbukh, A., Morales, E.F. (eds.) MICAI 2008. LNCS (LNAI), vol. 5317. Springer, Heidelberg (2008)

Festinger, L.: A theory of cognitive dissonance. Stanford University Press, Stanford (1957)

Fodor, J.A.: The Language of Thought. Thomas Crowell, New York (1975)

Fodor, J.A.: The modularity of the mind. MIT Press/Bradford Books, Cambridge (1983)

Gnadt, W., Grossberg, S.: SOVEREIGN: An autonomous neural system for incrementally learning planned action sequences to navigate towards a rewarded goal: Technical Report CAS/CNS-TR-07-015. Neural Networks 21, 699–758 (2008)

Graesser, A.C., Millis, K., Chipman, P., Cai, Z., Wallace, P., Britt, A., Storey, J., Wiemer, K., Magliano, J., D'Mello, S.: A Demonstration of ITSs That Promote Scientific Inquiry Skills: Critical Thinking Tutor and ARIES. In: Proceedings of ITS 2008 Demo Session (2008)

Graesser, A.C., Olney, A., Haynes, B.C., Chipman, P.: AutoTutor: A cognitive system that simulates a tutor that facilitates learning through mixed-initiative dialogue. In: Forsythe, C., Bernard, M.L., Goldsmith, T.E. (eds.) Cognitive systems: Human cognitive models in systems design. Erlbaum, Mahwah (2005)

Graesser, A.C., Jeon, M., Dufty, D.: Agent technologies designed to facilitate interactive knowledge construction. Discourse Process 45, 298–322 (2008)

Graesser, A.C., Wiemer-Hastings, K., Wiemer-Hastings, P., Kreuz, R.: The TRG, Auto Tutor: A simulation of a human tutor. J. Cogn. Syst. Res. 1, 35–51 (1999)

Heffernan, N.T.: Intelligent tutoring systems have forgotten the tutor: adding a cognitive model of human tutors. Dissertation. School of Computer Science, Carnegie Mellon University, Pittsburgh, PA (2001)

Heffernan, N.T., Koedinger, K.R., Razzaq, L.: Expanding the model-tracing architecture: A 3rd generation intelligent tutor for Algebra symbolization. The International Journal of Artificial Intelligence in Education 18(2), 153–178 (2008)

Heider, F.: The Psychology of Interpersonal Relations. John Wiley & Sons, New York (1958)

Hofstadter, D.R., Mitchell, M.: The copycat project: A model of mental fluidity and analogy-making. In: Hofstadter, D. (ed.) The Fluid Analogies Research group. Fluid Concepts and Creative Analogies. Basic Books, New York (1995)

Jackson, J.V.: Idea for a Mind. Siggart Newsletter 181, 23–26 (1987)

Koedinger, K.R., Aleven, V.: Exploring the assistance dilemma in experiments with Cognitive Tutors. Educ. Psychol. Rev. 19(3), 239–264 (2007)

Langley, P., Laird, J.E., Rogers, S.: Cognitive architectures: Research issues and challenges. Cogn. Syst. Res. 10, 141–160 (2009)

Lynch, C., Ashley, K., Aleven, V., Pinkwart, N.: Defining Ill-Defined Domains; A literature survey. In: Aleven, V., Ashley, K., Lynch, C., Pinkwart, N. (eds.) Proceedings of the Workshop on Intelligent Tutoring Systems for Ill-Defined Domains at the 8th International Conference on Intelligent Tutoring Systems (2006)

Maes, P.: How to do the right thing. Connect. Sci. J. 1(3), 291–323 (1989)

Mayers, A.: Miace: Une architecture théorique et computationnelle de la cognition humaine pour étudier l'apprentissage. Thèse. Université de Montréal, Montréal (1997)

Merrill, D.C., Reiser, B.J., Merrill, S.K., Landes, S.: Tutoring: Guided Learning by Doing. Cogn. Instr. 13(3), 315–372 (1995)

Minsky, M.: The society of Mind. Simon & Schuster, New York (1985)

Mitrović, A., Mayo, M., Suraweera, P., Martin, B.: Constraint-Based Tutors: A Success Story. In: Monostori, L., Váncza, J., Ali, M. (eds.) IEA/AIE 2001. LNCS (LNAI), vol. 2070, p. 931. Springer, Heidelberg (2001)

Mitrovic, A., Koedinger, K.R., Martin, B.: A comparative analysis of cognitive tutoring and constraint-based modeling. In: Brusilovsky, P., Corbett, A.T., de Rosis, F. (eds.) UM 2003. LNCS, vol. 2702. Springer, Heidelberg (2003)

Morén, J.: Emotion and learning: A Computational Model of the Amygdala. In: Meyer, Berthoz, Floreano, Roitblat, Wilson (eds.) From Animals to Animats 6: Proceedings of the 6th International Conference on the Simulation of Adaptive Behaviour.. MIT Press, Cambridge (2000)

Najjar, M., Fournier-Viger, P., Mayers, A., Hallé, J.: Tools and Structures for Modelling Do-main/User Knowledge in Virtual Learning. In: Proceedings of the 16th AACE World Confer-ence on Educational Multimedia, Hypermedia & Telecommunications, ED-MEDIA 2005 (2005)

Negatu, A., Franklin, S.: An action selection mechanism for 'conscious' software agents. Cogn. Sci. Q. 2, 363–386 (2002)

Neisser, U.: Cognitive psychology. Prentice-Hall, Englewood Cliffs (1967)

Newell, A.: Production systems: Models of control structures. In: Chase, W.G. (ed.) Visual information processing. Academic Press, New York (1973)

Newell, A.: Unified theories of cognition. Harvard University Press, Cambridge (1990)

Nkambou, R., Belghith, K., Kabanza, F., Khan, M.: Supporting Training on Canadarm Simula-tor using a Flexible Path Planner. In: Proceedings of the 12th International Conference on Ar-tificial Intelligence in Education (AIED). IOS Press, Amsterdam (2005)

Norman, D.A., Gentner, D., Stevens, A.L.: Comments on learning: schemata and memory or-ganization. In: Klahr, D. (ed.) Cognition and Instruction. Lawrence Erlbaum Associates, Hillsdale (1976)

Ohlsson, S.: Constraint-based Student Modeling. In: Student Modeling: the Key to Individualized Knowledge-based Instruction. Springer, Berlin (1994)

O'Shea, T.: Self-improving Teaching Systems: an Application of Artificial Intelligence to Computer-aided instruction. Birkhauser Verlag, Basel (1979)

O'Shea, T., Sleeman, D.H.: A design for an adaptive self-improving teaching system. In: Rose, J. (ed.) Advances in Cybernetics. Gordon and Breach Publishers, London (1973)

Person, N.K., Bautista, L., Kreuz, R.J., Graesser, A.C., TRG: The Dialog Advancer Network: A Conversation Manager for AutoTutor. In: Proceedings of the workshop on modeling human teaching tactics and strategies at the Intelligent Tutoring Systems 2000 conference, University of Quebec, Montreal (2000)

Piaget, J.: The origins of intelligence. International University Press, New York (1952)

Picard, R.W.: Toward Machines with Emotional Intelligence. In: Matthews, G., Zeidner, M., Roberts, R.D. (eds.) The Science of Emotional Intelligence: Knowns and Unknowns. Oxford University Press, Oxford (2007)

Quillian, M.R.: Semantic memory. In: Minsky, M. (ed.) Semantic information Processing. MIT Press, Cambridge (1968)

Rickel, J., Johnson, W.L.: STEVE: A Pedagogical Agent for Virtual Reality, In: Proceedings of the Second International Conference on Autonomous Agents (1998)

Rolls, E.T.: The brain and emotion. Oxford University Press, New York (1999)

Rolls, E.T.: Neurophysiology and functions of the primate amygdala, and the neural basis of emotion. In: Aggleton, J.P. (ed.) The Amygdala: a Functional Analysis. Oxford Univ. Press, Oxford (2000)

Roy, J., Nkambou, R., Kabanza, F., Hartman, L., Jang, T.-S.: Supporting Spatial Awareness in Training on a Telemanipulator in Space. In: Lester, J.C., Vicari, R.M., Paraguaçu, F. (eds.) ITS 2004. LNCS, vol. 3220, pp. 860–863. Springer, Heidelberg (2004)

Samsonovich, A.V., Kitsantas, A., Dabbagh, N.: Cognitive Constructor: A biologically-inspired self-regulated learning partner. In: Samsonovich, A.V. (ed) Biologically Inspired Cog-nitive Architectures. AAAI Technical Report FS-08-04. AAAI Press, Menlo Park (2008)

Samsonovich, A.V.: Feature-by-Feature Comparison of Cognitive Architectures (2010), http://members.cox.net/bica2009/cogarch/architectures.htm (Accessed March 2, 2010)

Self, J.A.: Student models in computer-aided instruction. Int. J. Man-Mach. Stud. 6, 261–276 (1974)

Squire, L.R., Kandel, E.R.: Memory: From Mind to Molecules. WH Freeman & Co., NY (1999)

Stevens, A.L., Collins, A.: The goal structure of a Socratic tutor. In: Proceedings of the National ACM Conference, Seattle, WA (1977)

Sun, R.: The CLARION cognitive architecture: Extending cognitive modeling to social simulation. In: Sun, R. (ed.) Cognition and Multi-Agent Interaction. Cambridge University Press, New York (2004)

Taatgen, N.A., Lebiere, C., Anderson, J.R.: Modeling Paradigms in ACT-R. In: Sun, R. (ed.) Cognition and Multi-Agent Interaction: From Cognitive Modeling to Social Simulation. Cambridge University Press, Cambridge (2006)

VanLehn, K., Bhembe, D., Chi, M., Lynch, C., Schulze, K., Shelby, R., Taylor, L., Treacy, D., Weinstein, A., Wintersgill, M.: Implicit versus explicit learning of strategies in a non-procedural cognitive skill. In: Lester, J.C., Vicari, R.M., Paraguaçu, F. (eds.) ITS 2004. LNCS, vol. 3220, pp. 521–530. Springer, Heidelberg (2004)

VanLehn, K., Lynch, C., Schulze, K., Shapiro, J.A., Shelby, R., Taylor, L., Treacy, D., Weinstein, A., Win-tersgill, M.: The Andes Physics Tutoring System: Lessons Learned. Int. J. Artif. Intell. Educ. 15(3) (2005)

Wikipedia. Cognitive architecture, http://en.wikipedia.org/wiki/Cognitive_architecture (Accessed March 16, 2010)

Wenger, É.: Artificial intelligence and tutoring systems. In: Computational and Cognitive Approaches to the Communication of Knowledge. Morgan Kaufmann Publishers, San Francisco (1987)

Weiner, B.: An attributional theory of motivation and emotion. Springer, New York (1986)

Woolf, B.P., Murray, T., Marshall, D., Dragon, T., Kohler, K., Mattingly, M., Bruno, M., Murray, D., Sammons, J.: Critical Thinking Environments for Science Education. In: Proceedings of the 12th International Conference on AI and Education (2005)

Woolf, B.P.: Building intelligent interactive tutors. Student-centered strategies for revolutionizing e-learning. Elsevier, New York (2009)

Zinn, C., Morre, J.D., Core, M.G.: A 3-tier Planning Architecture for Managing Tutorial Dialogue. In: Cerri, S.A., Gouardéres, G., Paraguaçu, F. (eds.) ITS 2002. LNCS, vol. 2363, p. 574. Springer, Heidelberg (2002)

Chapter 9
Tutorial Dialog in Natural Language

Andrew M. Olney[1], Arthur C. Graesser[1], and Natalie K. Person[2]

[1] University of Memphis, Memphis TN 38152, USA
{a-graesser, aolney}@memphis.edu
[2] Rhodes College, Memphis TN 38112, USA
person@rhodes.edu

Abstract. This chapter reviews our past and ongoing investigations into conversational interaction during human tutoring and our attempts to build intelligent tutoring systems (ITS) to simulate this interaction. We have previously modeled the strategies, actions, and dialogue of novice tutors in an ITS, called AutoTutor, with learning gains comparable to novice tutors. There is evidence, however, that expert human tutors may foster exceptional learning gains beyond those reported for some categories of human tutors. We have undertaken a rigorous, large scale study of expert human tutors and are using these data to create Guru, an expert ITS for high school biology. Based on our analyses, expert human tutoring has several distinctive features which differ from novice human tutoring. These distinctive features have implications for the development of an expert ITS, and we briefly describe how these are being addressed in Guru.

9.1 Introduction

The empirical evidence that one-to-one human tutoring is extremely effective compared to classroom environments is well known (Bloom 1984; Cohen et al. 1982; Corbett 2001; Graesser and Person 1994). The effectiveness of one-to-one tutoring raises the question of what makes tutoring so powerful. Three different hypotheses, known as the tutor-centered, student-centered, and interaction hypotheses, have been proposed to answer this question (Chi et al. 2001; Chi et al. 2008). The tutor-centered hypothesis claims that effective tutoring stems primarily from the actions of the tutor, specifically, the tutor's pedagogical moves are tailored to a given student. In contrast, the student-centered hypothesis places the emphasis on the student, highlighting that students are active participants in the construction of their own knowledge rather than being mere information receptacles.

Interest in the interaction hypothesis is growing (Chi et al. 2001 ; Chi et al. 2008). However, the interaction hypothesis has long-standing roots in the tutoring literature. An early meta-analysis on a large sample of studies compared human-to-human tutoring with classroom environments and suitable comparison conditions (Cohen et al. 1982). The vast majority of the tutors in these studies were untrained in

R. Nkambou et al. (Eds.): Advances in Intelligent Tutoring Systems, SCI 308, pp. 181–206.
springerlink.com

tutoring skills and had moderate domain knowledge; they were peer tutors, cross-age tutors, or paraprofessionals, but rarely accomplished professionals.

These «unaccomplished" human tutors enhanced learning with an effect size of a .4 standard deviation unit (sigma), or approximately a half letter grade.

As one might expect, unskilled human tutors are not prone to implement sophisticated tutoring strategies that have been proposed in the fields of education, the learning sciences, and developers of ITSs (Graesser et al. 1995; Graesser et al. 2009; Person et al. 1995).

This chapter reviews our work on conversational interaction during human tutoring and our attempts to build intelligent tutoring systems (ITS) to simulate this interaction.To date, the bulk of our research addresses the strategies, actions, and dialogue of novice tutors (Graesser et al. 1994; Graesser et al. 1995; Person et al. 1994). We have implemented novice tutoring in an ITS, called AutoTutor, with learning gains comparable to novice tutors (Graesser2004; VanLehn et al. 2007). More recently, we have expanded our investigation to highly accomplished expert human tutors (Cade et al. 2008; D'Mello et al. in press). Our shift in emphasis is driven by a desire to understand what makes accomplished expert human tutors produce exceptional learning gains, as has been previously reported (Bloom 1984). We have undertaken a rigorous, large scale study of accomplished human tutors, and we are using these data to create Guru, an expert ITS for high school biology. In the following sections we further elaborate this contrast between novice and expert, both in terms of human tutoring and the ITS components required to mimic interaction with novice and expert human tutors.

9.2 Novice Human Tutoring

Two samples of novice tutoring were collected and analyzed (Graesser et al. 1994; Graesser et al. 1995). The first sample consisted of tutoring sessions on undergraduate research methods, the Research Methods Corpus (RMC). The 3 tutors in the RMC were graduate students who had never tutored for research methods, thus they were truly novice tutors in this domain. The 27 students receiving tutoring participated in two 1-hour sessions each with different tutors, for a total of 54 sessions. Each session was recorded; however, due to video quality, only 44 sessions could be transcribed. The second sample of novice tutoring was in the domain of 7th grade algebra, the Algebra Corpus (AC). The 10 tutors who participated were high school students with an average 9 hours of prior experience in tutoring. The 13 students receiving tutoring participated in 1-hour sessions for a total of 22 sessions. Thus in total the RMC and AC consist of 76 hours of tutoring for 40 students and 13 tutors.

Multiple codings schemes have been developed to analyze the RMC and AC along different dimensions including feedback, tutor examples, Gricean Maxims, student errors, and student questions (Graesser et al. 1994, Graesser et al. 1995; Person and Graesser 1999; Person et al. 1995). Tables 9.1 and 9.2 present the primary dialogue moves used by students and tutors across these analyses.

Table 9.1 Novice Student Dialogue Moves

Move Category	Description
Contribution Quality	
Complete	Student provides complete answer to tutor question.
Partial	Student provides partial answer to tutor question.
Vague	Student provides vague answer to tutor question.
Error-ridden	Student provides error-ridden answer to tutor question.
No Answer	Student fails to provide any answer to tutor question.
Asks Question	
Makes request	Student makes request unrelated to the problem/example.
Counter- clarification	Student needs clarification on tutor's previous statement.
Problem-related	Student asks question directly related to the problem/example.
Other	Any question not assigned to one of the other three question categories.
Misconception	Student states his or her own misconception. Reminding Example & Student comments on a similar example.
Meta-comment	Student comments on own ability or attribute of problem.
Acknowledgement	Student acknowledges tutor's contribution (e.g., Uh-huh).
Gripes	Student complains. Think aloud & Student thinks aloud.
Nonverbal	Student makes a nonverbal response (e.g., laughs).
Draw	Student draws on board.
Other	Any speech act not assigned to one of the other student categories.

Using the coding scheme in Tables 9.1 and 9.2, Graesser and Person's analyses of the RMC and AC uncovered three frequent dialogue structures (Graesser et al. 1994; Graesser et al. 1995; Graesser et al. 2005a). These same structures have featured prominently in the work of other researchers conducting fine-grained analyses of tutoring (Chi et al. 2001; Chi et al. 2004; Chi et al. 2008; Litman and Forbes-riley 2006; Shah et al. 2002). These three dialogue structures are:

1. 5-step Tutoring Frame
2. Expectation and Misconception Tailored (EMT) dialogue
3. Conversational Turn Management (which includes tutor pedagogical modes)

These three structures are multiply embedded: 3 is embedded in 2, which in turn is embedded in 1. There are two common features across all three of these structures. The first is that the tutor, rather than the student, tends to initiate and guide the conversational interaction. The second common feature is that all three of these structures exist at the level of the problem, rather than across larger spans of the tutorial discourse.

Table 9.2 Novice Tutor Dialogue Moves

Move Category	Description
Additional example	
Easier	Tutor provides student an easier example than the previous example.
Difficult	Tutor provides a more di_cult example than the previous example.
Equal	Tutor provides an example of equal difficulty with the previous example.
Asks question	
Error-repair	Tutor asks question speci_cally related to student error.
Directed-Activity	Tutor asks question in order to redirect student's activity.
Leading	Tutor asks question to expose student's misconception.
Counter-clarification	Tutor requests clari_cation of student's previous statement.

Table 9.2 (*continued*)

Pump	Tutor pumps student for additional information.
Assessment	Tutor assesses student's knowledge about a particular topic.
Global	Tutor globally assesses student's knowledge (e.g., Do you understand?)
Other	Any question not assigned to one of the other question categories.
Feedback	
Positive	Tutor gives positive feedback to student.
Negative	Tutor gives negative feedback to student.
Neutral	Tutor gives neutral feedback to student. Immediate Tutor provides immediate feedback for a student error.
Delayed	Tutor provides delayed feedback for a student error.
Reminding Example	Tutor comments on a similar example.
Specific Component	Tutor focuses on specific component of current problem/example.
General Level	Tutor discusses current example in more general terms. Hint Tutor provides the student with a hint.
Splice	Tutor splices in the correct answer.
Elaborates	Tutor elaborates current problem/example.
Answers	Tutor answers student question.

Table 9.2 (*continued*)

Rearticulates	
Solution	Tutor rearticulates the current problem's solution.
Representation	Tutor rearticulates the problem's representation.
Affective	
Own ability	Tutor comments on his or her own ability.
Student ability	Tutor comments on student's ability.
Problem	Tutor comments on the di_culty of the problem/example.
General	Tutor makes general empathetic comment.
Gripes	Tutor complains.
Directive	Tutor tells the student what to do.
Draw	Tutor draws on the board.
Nonverbal	Tutor makes some type of nonverbal response (e.g., laughs).
Other	Any speech act not assigned to one of the other tutor categories.

9.2.1 5-Step Tutoring Frame

The 5-Step Tutoring Frame begins once a problem has been introduced. As indicated by the name, the following five steps are enacted in order:

1. TUTOR asks a difficult question or presents a problem.
2. STUDENT gives an initial answer.
3. TUTOR gives short feedback on the quality of the answer.
4. TUTOR and STUDENT have a multi-turn dialogue to improve the answer.
5. TUTOR assesses whether the student understands the correct answer.

This 5-Step Tutoring Frame involves a great deal of conversational interaction. The structure of the 5-Step Tutoring Frame fosters both collaborative discussion and joint action as the tutor works with the student to iteratively construct a better answer.

The 5-Step Tutoring Frame can be better understood by contrasting it with the Initiate-Respond-Feedback (IRF) sequence typically used in classrooms (Sinclair and Cloutard 1975). The first three steps occur in classroom IRF, but the questions are easier short-answer questions. The classroom IRF sequence consists of the teacher initiating a question, the student giving a short-answer response, and the teacher giving a positive or negative feedback of the response. For example, consider the following IRF example for Newtonian physics.

TEACHER: According to Newton's second law, force equals mass times what?
STUDENT: acceleration
TEACHER: Right, mass times acceleration. Or
STUDENT: velocity
TEACHER: Wrong, it's not velocity, it is acceleration.

As the above example illustrates, IRF does not facilitate conversational interaction. The 5-Step Tutoring Frame goes beyond IRF by posing more difficult questions that stimulate the collaborative interactions found in step 4.

9.2.2 Expectation and Misconception Tailored (EMT) Dialogue

Novice human tutors maintain a basic representation of the correct answer to a problem (expectations) as well as some misconceptions that may arise. For example, expectations E1 and E2 and misconceptions M1 and M2 are relevant to the example physics problem below.

PHYSICS QUESTION: If a lightweight car and a massive truck have a head-on collision, upon which vehicle is the impact force greater? Which vehicle undergoes the greater change in its motion, and why?
E1. The magnitudes of the forces exerted by A and B on each other are equal.
E2. If A exerts a force on B, then B exerts a force on A in the opposite direction.
M1: A lighter/smaller object exerts no force on a heavier/larger object.
M2: Heavier objects accelerate faster for the same force than lighter objects

Expectations and misconceptions form a simple domain model which novice tutors use to select dialogue moves. Expectations are akin to the expert model of model tracing tutors, and misconceptions are likewise analogous to buggy libraries in model tracing tutors.

Novice tutors select dialogue moves based on the status of the current problem's expectations and misconceptions. Hints and prompts direct the student to articulate missing content words, phrases, and propositions. For example, a hint for expectation E1 might be "What about the forces exerted by the vehicles on each other?", which would ideally elicit the answer "The magnitudes of the forces are equal." A corresponding prompt to elicit "equal" would be "What are the magnitudes of the forces of the two vehicles on each other?" As the conversational interaction of the tutoring session unfolds, the student articulates the tutor's expectations in piecemeal (as in the examples given) or directly (for a high ability student). Novice tutors also have some awareness of common misconceptions associated with a problem. Thus when a student articulates a misconception, the tutor identifies the misconception and corrects it.

9.2.3 Conversational Turn Management

The preponderance of conversational interaction is tutor-led. Student led dialogue can occur when students ask questions, but it is well documented that students rarely ask questions, even in tutoring environments (Graesser et al. 1994; Graesser et al. 2005a). Tutor-led turns usually consist of three steps. The first step gives positive, neutral, or negative feedback on the student's last answer. The second step advances progress through the current problem, based on the expectations and misconceptions the student has already covered. Thus the second step may be instantiated with prompts for specific information, hints, assertions with correct information, or corrections of misconceptions. The third step signals the student that it is their turn to respond, i.e. via a question, rising intonation, or a gesture.

Novice human tutors use the 5-Step Tutoring Frame, EMT dialogue, and conversational turn management to present challenging problems or questions to the student, adaptively scaffold good answers through collaborative interactions, provide feedback when students express erroneous information, and answer occasional student questions. What is absent are sophisticated pedagogical strategies. According to our systematic analyses of the tutoring process (Graesseret al. 1995; Graesser et al. in press; Person et al. 1995), novice human tutoring is not characterized by sophisticated tutoring strategies that have been proposed in the fields of education, the learning sciences, and developers of ITS (Person and Graesser 2003). In particular, novice tutors rarely engage in pedagogical techniques such as bona fide Socratic tutoring strategies (Collins et al. 1975), modeling-scaffolding-fading (Rogoff and Gardner 1984), Reciprocal Teaching (Palinscar and Brown 1984), frontier learning (Sleeman and Brown 1982), building on prerequisites (Gagne 1985), or diagnosis/remediation of deep misconceptions (Lesgold et al. 1992). This is perhaps unsurprising because these strategies are complex and were not discovered for centuries.

9.3 AutoTutor

AutoTutor simulates a novice human tutor by holding a conversation with the learner in natural language. The pedagogical framework of AutoTutor was inspired by three bodies of theoretical, empirical, and applied research. These include explanation-based constructivist theories of learning (Aleven and Koedinger 2002; Chi et al. 2001; Chi et al. 1994; VanLehn et al. 1992), intelligent tutoring systems that adaptively respond to student knowledge (Anderson et al. 1995; VanLehn et al. 2002a), and empirical research that has documented the collaborative constructive activities that routinely occur during human tutoring (Chi et al. 2001; Fox 1993; Graesser et al. 1995; Moore 1994; Shah et al. 2002). The pedagogical strategies of AutoTutor are modeled on the novice human tutoring strategies described in Section 9.2, including the 5-Step Tutoring Frame, EMT dialogue, and conversational turn management.

AutoTutor implements the 5-Step Tutoring Frame by presenting a series of challenging questions or problems that require approximately a paragraph of information to answer correctly. An example question in conceptual physics is, «When a car without headrests on the seats is struck from behind, the passengers often suffer neck injuries. Why do passengers get neck injuries in this situation?" Although a perfect answer to this question is approximately 3-7 sentences in length, the initial answers by actual human learners are typically only 1 word to 2 sentences in length. The conversational interaction afforded by tutorial dialogue is particularly helpful when the student's answer is incomplete. AutoTutor uses the 5-Step Tutoring Frame to assist the learner in the evolution of an improved answer by drawing out more of the learner's knowledge that is relevant to the answer. The dialogue between AutoTutor and the learner typically lasts 50-200 turns (i.e., the learner expresses something, then the tutor, then the learner, and so on), which is on par with the interactivity in human tutoring.

AutoTutor uses expectations and misconceptions as an integral part of its domain model, and selects dialogue moves that elicit expectations and address misconceptions. More specifically, the goal of AutoTutor is to elicit the correct answer from the student. Since the correct answer is a paragraph of information, this goal reduces to eliciting each sentence, an expectation, in the correct answer paragraph. In order to elicit each expectation, AutoTutor generates tutorial dialogue moves including pumps, hints, prompts, and assertions:

Pumps. AutoTutor pumps the student for more information during the early stages of answering a particular question (or solving a problem). The pump signals the student to keep talking, for example using positive feedback (e.g., right, yeah, dramatic head nod), neutral back channel feedback (uh-huh, okay, subtle head nod), and explicit requests for more information (What else?, Tell me more). By encouraging the student to say more, pumping helps expose the student's knowledge while giving the student an opportunity to construct knowledge by herself.

Hints. When the student is having problems answering a question or solving a problem, the tutor gives hints by presenting a fact, asking a leading question, or reframing the problem. Hints cue the student to some relevant feature of the problem without revealing the role of that feature in answering the problem.

Prompts. AutoTutor supplies the student with a discourse context and prompts them to fill in a missing word or phrase. Prompting is a scaffolding device for students who are reluctant to supply information. Students are expected to supply more content and more difficult content as they progress in learning the domain knowledge.

Assertions. AutoTutor gives a summary to an expectation. This summary serves the function of succinctly codifying a lengthy, multi-turn, collaborative exchange when an expectation is covered or a problem step is completed.

It is worth noting the continuum of information provided by the tutor in different types of moves. Moves at the beginning of the list, i.e. pumps and hints, provide less information to the student than moves towards the end of the list, i.e. prompts and assertions. By only giving more information when the learner is floundering, AutoTutor promotes active construction of knowledge (Graesseret al. 1995 ; Chi et al. 2001). Analysis of AutoTutor experiments shows that deeper questions, i.e. pumps and hints, promote more learning than shallow dialogue moves such as prompts and assertions (Jackson et al. 2004).

AutoTutor assesses the student's answers to these dialogue moves using Latent Semantic Analysis (LSA), a vector space method capable of representing world knowledge (Deerwester et al. 1990; Dumais et al. 1991; Landauer and Dumais 1997; Landauer et al. 2007). In LSA, a word is represented by a fixed size vector of real numbers. A sentence or document is also represented by a fixed size vector, made by summing component word vectors. Words, sentences, and documents can all be compared to each other by comparing their vectors. AutoTutor uses LSA to compare the student's answer to the expectations by comparing the LSA vector of the student's answer to the vectors of the expectations. LSA vectors that are identical have a cosine of 1, but AutoTutor uses a lower threshold, e.g. 0.7, to allow the student some flexibility in their answer. In other words, LSA allows student answers with the same meaning, but different wording, to be recognized as correct answers.

AutoTutor uses conversational turn management to maintain a coherent conversational interaction with the student. The primary mechanisms in AutoTutor for conversational turn management are AutoTutor's speech act classifier and dialogue manager. Before AutoTutor responds to a student, the student's utterance is analyzed to determine its speech act (Olney et al. 2003). If a student asks an information-seeking question, AutoTutor launches a subdialogue to answer that question. This subdialogue can consist of multiple rounds of clarification, and even recursively nested subdialogues for more detailed questions (Graesser et al. 2005b). If the speech act is an answer or verification question, AutoTutor gives feedback based on the cosine between the student's LSA vector and the current expectation.

In the second and third steps of conversational turn management, AutoTutor selects and delivers a dialogue move. The specific dialogue move is selected based on the current context of the tutoring session, including the problem that the student is on, the current expectation, and the last dialogue move type generated (e.g. hint). AutoTutor loads this context into its state table, an information state (Larsson and Traum 2000), and then processes this state table through a dialogue manager (Graesser et al. 2005b). The dialogue manager is defined by a formal language for describing dialogues together with a corresponding interpreter to execute dialogues in this language. This approach has made it much easier to create new tutorial dialogue patterns than was possible with previous finite-state approaches (Graesser et al. 2001; McTear 1998). The dialogue manager's interpreter finds and returns a dialogue pattern, which is a plan that matches the current context. Recently this has been reimplemented using Prolog in GnuTutor, an open-source approximation of AutoTutor, which allows for more sophisticated backtracking (Olney 2009). The dialogue plan returned by the dialogue manager ends with a question, has a gesture, or has rising intonation to indicate to the student that the tutor expects them to respond.

The learning gains of AutoTutor have been evaluated in over 20 experiments conducted during the last 12 years. Assessments of AutoTutor on learning gains have shown effect sizes of approximately 0.8 standard deviation units in the areas of computer literacy (Graesser2004) and Newtonian physics (VanLehn et al. 2007). AutoTutor's learning gains have varied between 0 and 2.1 sigma (a mean of 0.8), depending on the learning performance measure, the comparison condition, the subject matter, and the version of AutoTutor.

9.4 Expert Human Tutoring

Most human tutoring studies that have been reported in peer-reviewed sources have primarily included untrained or "typical" tutors (Cohen et al. 1982). By comparison, expert tutoring studies are scarce, and such studies have included only a handful of expert tutors. This section reviews the expert tutoring studies most frequently cited in the literature and notes some problems that have contributed to our lack of expert tutoring knowledge.

First, several well-known studies do not mention the number of expert tutors included in the analyses (Aronson 2002; Fox 1993; Derry and Lajoie 1993; Lepperand and Woolverton 2002). Second, although some studies report five or six expert tutors (Derry and Potts 1998; Graesser et al. 2001; Lepper et al. 1990; Lepper et al. 1993; VanLehn et al. 2007), many included only one or two experts (Shah et al. 2002; Evens et al. 1993; Glass et al. 1999; Lajoie 2001; Jordan and Siller 2002).

Third, some of these studies have overlapping expert tutors. For example, the tutors included in (Graesser et al. 2001), (Jordan and Siller 2002), and (VanLehn et al. 2007) are the same five tutors. Fourth, not all studies on expert human tutoring investigate the same phenomena. A number of studies have focused on the motivational aspects of tutors instead of the cognitive and pedagogical features that contribute to student learning (e.g., the studies by Mark Lepper and colleagues). A fifth problem with these studies is that the credentials of the expert tutors are inconsistent. Some studies define expert tutors as Ph.D.s with extensive teaching or tutoring experience (Evens et al. 1993; Glass et al. 1999; Graesser et al. 2001; Jordan and Siller 2002), but other studies define expert tutors as graduate students working in tutoring centers (Fox 1993). Taken together, these problems raise uncertainty as to whether the findings generalize to all expert tutors. A more detailed analysis of these problems reveals two enduring themes. First, many studies suffer from small or unknown sample sizes. Secondly, it is difficult to aggregate findings across studies because of shared tutors, differing research goals, and inconsistent definitions of expertise.

To address these issues, we recently undertook a rigorous, large scale study of accomplished, expert human tutors. Our approach mirrors our previous study of novice human tutoring by collecting observations of naturalistic one-to-one tutoring.

Twelve expert math and science tutors were recruited to participate in the project. The expert tutors were recommended by academic support personnel from public and private schools in a large urban school district. All of the tutors have long-standing relationships with the academic support offices that recommend them to parents and students. The criteria for being an expert tutor in our project are as follows:

Have a minimum of five years of one-to-one tutoring experience (most of the tutors in our study have 10+ years of tutoring experience)
Have a secondary teaching license
Have a degree in the subject that they tutor
Have an outstanding reputation as a private tutor
Have an effective track record (i.e., students who work with these tutors show marked improvement in the subject areas for which they receive tutoring)

All of the students in our study were having difficulty in a science or math course and were either recommended for tutoring by school personnel or sought professional tutoring help.

We created our expert tutoring corpus by observing our expert tutors in one-on-one tutoring sessions. Fifty one-hour tutoring sessions were videotaped and transcribed. All of the videotapes were transcribed according to strict transcription guidelines and were verified for accuracy. To capture the complexity of what transpires during a tutoring interaction, three coding schemes were developed to classify every tutor and student dialogue move in the 50 hours of tutoring. In the analyses we conducted, a dialogue move was either a speech act (e.g., a tutor hint), an action (e.g., student reads aloud), or a qualitative contribution made by a student (e.g., partial or vague answer). Multiple dialogue moves could occur within one conversational turn.

Table 9.3 Tutor Motivational Moves

Move Category	Example
Attribution Acknowledgment	that's easy
Conversational Ok	alrighty
General Motivational Statement	cause you're such a good student I just enjoy . . .
Humor	so you're going to have kids and you go « oh I . . .
Negative Feedback	no no no no
Negative Feedback Elaborated	actually no you're gonna have some . . .
Neutral Feedback	not quite
Neutral Feedback Elaborated	mm you're thinking of vertical vertical angles . . .
Positive Feedback	very good alright
Positive Feedback Elaborated	very good because everything is on top
Repetition	negative 2
Solidarity Statement	let's do it

Two coding schemes were used to classify the tutor dialogue moves, the Tutor Pedagogical Scaffolding scheme and the Tutor Motivational Dialogue scheme. The Pedagogical Scaffolding scheme included 14 categories and was inspired by previous tutoring research on pedagogical strategies and dialogue moves (Graesser et al. 1995; Cromley and Azevedo 2005). The Tutor Motivational Dialogue scheme included 13 categories that were either reported previously in the literature or were extrapolated from the INSPIRE model (Lepper and Woolverton 2002). Each tutor dialogue move was classified as either pedagogical or motivational. The Tutor Motivational and Pedagogical Schemes are presented in Table 9.3 and Table 9.4.

A 16 category coding scheme was also developed to classify all student dialogue moves. Some of the student move categories captured the qualitative nature of a student dialogue move (e.g., Correct Answer, Partially Correct Answer, Error-ridden Answer), whereas others were used to classify types of questions, conversational acknowledgments, and student actions (e.g., reading aloud or solving a problem). The Student Dialogue Move Scheme is presented in Table 9.5. Four

Table 9.4 Tutor Pedagogical Moves

Move Category	Example
Counter Example	not multiply we'll add in the area of the bases . . .
Comprehension Gauging Question	you see what I'm saying
Direct Instruction/Explanation	so that's your lateral area
Example	so as a male you will undergo meiosis and your . . .
Forced Choice	so are we going bigger or smaller
Hint	but now we're not gonna add this many dots . . .
New Problem	let's look at this example here it's called . . .
Other	does he give you a time limit
Paraphrase	you take out an r squared and you'd have 4 . . .
Provide Correct Answer	first outer inner last
Preview	we're going to talk about how atoms ions . . .
Prompt	can we simplify the radical of 9 is simply
Pump	and then what do we do
Simplified Problem	what inside the cell would have an electrical . . .
Summary	so that's all there is to it so you got a circular . . .

Table 9.5 Student Dialogue Moves

Move Category	Example
Acknowledgment	yes, ma'am
Common Ground Question	the parasites?
Correct Answer	6 times 54
Error Ridden Answer	multiply
Gripe	I might as well not pay attention
Knowledge Deficit Question	well, what's a skeleton?
Metacomment	I don't know what I'm doing, hold on
Misconception	I thought you added two to it
No Answer	it will be, oh shoot it will be
Other	she didn't do that
Partial Answer	so I guess eliminate those 2
Read Aloud	first class levers are the most common type a pire of
Social Coordination Action	afternoon sunday? want to do it like sunday afternoon?
Student Works Silently	uh
Think Aloud	to the power of, no, x plus 1
Vague Answer	cause you, yeah times

trained judges coded the 50 transcripts on the three dialogue move schemes. Cohen's Kappas were computed to determine the reliability of their judgments. The Kappa scores were .96 for the Tutor Motivational Scheme, .88 for the Tutor Pedagogical Scheme, and .88 for the Student Move Scheme. Approximately 47,000 dialogue moves were coded.

In addition to these dialogue move coding schemes, we also developed a coding scheme for larger units of the tutoring session. We call these units *modes* (Cade et al. 2008). Two trained judges coded the 50 transcripts and found eight modes, including *Introduction, Lecture, Highlighting, Modeling, Scaffolding, Fading, Off-Topic,* and *Conclusion*, with Kappa above .80 for each mode. Each mode can be characterized by a specific kind of interaction:

Introduction. Expert tutoring sessions usually begin with an *Introduction* that contains greetings and establishes an agenda for the rest of the session.

Lecture. Approximately 20% of the sessions consist of direct instruction. We call these modes *Lecture*, but they are usually highly customized, just-in-time, and interactive, unlike traditional classroom lecture.

Highlighting. When a student encounters difficulty while problem solving, *Highlighting* draws attention to a problem solving step.

Modeling. In this mode, the tutor works a problem while the student watches.

Scaffolding. Expert human tutoring is dominated by *Scaffolding*, in which the tutor and student solve a problem together. Roughly 50% of all turns take place in this mode.

Fading. As the inverse to *Modeling*, the student predominantly solves a problem while the tutor watches in *Fading*.

Off-Topic. Non-tutoring related conversation, e.g. humor, infrequently occurs.

Conclusion. Expert tutoring sessions usually end in a characteristic fashion, similarly to how *Introduction* begins the session.

An individual mode can span dozens of turns, and so represents a major unit in the structure of a tutoring session. However, not all modes are equally prevalent or contain comparable numbers of turns. Approximately 70% of all turns are contained within *Lecture* and *Scaffolding* modes, with the remaining turns roughly divided amongst the remaining modes (Cade et al. 2008).

We are currently analyzing these 50 hours of expert human tutoring data, which we call the expert human tutoring corpus (EHTC). Our goal is to explore whether there are similar structures that we have found for novice tutoring, i.e. the 5-Step Tutoring Frame, EMT dialogue, and conversational turn management described in Section 9.2. However, there may be characteristics of expert human tutoring that are very different than those of novice human tutoring protocols.

Although our analyses are still underway, we believe that expert tutors are different from novice tutors in atleast six different ways (Person et al. 2007a; Person and Mallot 2007b; Person et al. 2007c). It is important to qualify these claims about expert tutors because with the exception of (Person et al. 2007c), there was never a systematic comparison of tutors with different expertise in any given study. Instead, the relative frequencies of tutor strategies and discourse moves were computed in the EHTC and compared with the relative frequencies of the same theoretical categories in published studies with unskilled tutors. At the same time, however, there is little evidence of EMT dialogue in the EHTC, and the patterns of dialogue are more complex than the typical instantiation of the 5-Step Tutoring Frame.

1. Expert tutors form more accurate student models that non-expert tutors. This is evidenced in the question asking analyses that we performed. Expert tutors ask proportionately more low specificity questions (e.g., So?) and more common ground questions (e.g., So, I use the Pythagorean Theorem?) than tutors and students in non-expert sessions. We interpret these findings to mean that expert tutors are more attuned to the needs of their students and have established considerable common ground. If this wasn't the case, low specificity questions (e.g., So?) would result in conversation breakdowns. We also found that students being tutored by experts ask fewer knowledge deficit questions (e.g., What do the ribosomes do?) than students working with non-expert tutors, indicating that knowledge deficit questions are less necessary when participants have established a high level of common ground.
2. Expert tutors are more dynamic in their instruction and do not rely on curriculum scripts. Experts typically begin the tutoring sessions by figuring out the

topics/problems that students are having difficulty with and by asking questions about the students' performance on quizzes, homework, and exams. After this data collection phase, the tutor decides where to begin the session and what material will be covered. Expert tutors do not begin a session with any pre-planned teaching agenda, but rather base their instruction on students' particular needs at the time of the tutoring session.
3. Expert tutors give more discriminating feedback than non-expert tutors. Non-experts are just as likely to give positive feedback to wrong answers as negative feedback (Person et al. 1994), but this is not true of expert tutors.
4. Expert tutors primarily rely on just-in-time direct instruction and evaluative feedback when responding to student dialog moves.
5. Expert tutors are task-oriented, direct, and do not appear to adhere to Lepper INSPIRE motivational model.
6. Particular tutoring modes (defined by tutor dialogue move frequencies and patterns) are evident in expert tutoring, including *Introduction, Lecture, Highlighting, Modeling, Scaffolding, Fading, Off-Topic*, and *Conclusion.*

We have recently used data mining techniques to discover significant patterns of dialogue moves in the EHTC (D'Mello et al. in press). Our basic approach is to consider two-step transitions, i.e. move to move, that significantly diverged from chance and whose effect sizes were greater than the median effect size. So far our analyses have focused on *Lecture*. In *Lecture*, only 34 transitions out of 1869 (43 x 43) are significant with effect sizes above the median. A visual inspection of these transitions revealed four meaningful clusters. In the first cluster, the information *transmission* cluster, the tutor mostly engages in direct instruction and only superficially monitors student attention and understanding. In the second cluster, the information *elicitation* cluster, the tutor elicits information from the student using direct questioning, e.g. forced choice, prompts, pumps, etc., the student tries to answer, and the tutor gives feedback on the student's answer. The information elicitation cluster is the *Lecture* cluster most like the IRF and 5-Step Tutoring Frame described in Section 9.2. The third cluster is the off-topic cluster, e.g. humor, consisting of just a few moves as opposed to the *Off-Topic* mode. The fourth and final cluster in *Lecture* is the questioning cluster that handles student-asked common ground questions and knowledge deficit questions. Each of these four clusters can be viewed as a subgraph of the larger *Lecture* graph or viewed as a subdialogue nested in the larger *Lecture* dialogue.

Our analyses of the EHTC have revealed a richer structure than has previously been reported for novice tutoring, though again, we stress that this might be confounded by the lack of novice tutors in our sample. The behavior of novice tutors, as described in Section 9.2, aligns fairly well with the distinction of inner and outer loop in the behavior of ITS (Vanlehn 2006). According to the inner/outer loop distinction, problem selection happens in the outer loop, and the actual working of the problem, step by step, happens in the inner loop. Under this analysis, the three features of novice tutoring described in Section 9.2 align quite well with the inner loop. The 5-Step Tutoring Frame provides an overall dialogue structure for the inner loop, with step 4 accounting for much of the individual steps, or expectations, in the problem. EMT dialogue contributes by structuring the content within step 4.

Finally, the conversational turn management further elaborates step 4. As stated in Section 9.2, these three structures are multiply embedded.

However, the multiple levels of structure found in our analysis of the EHTC, while embedded, are considerably more complex. Transition probabilities between modes indicate that sessions typically shift from *Introduction* to *Lecture* to *Scaffolding*. Because only some modes introduce problems, e.g. *Scaffolding*, mode transitions are a step above the outer loop of problem selection. Inner loops occur in clusters within modes, such as the information elicitation cluster in *Lecture*. Finally, there are the dialogue moves themselves. In contrast to the more straightforward multiple embedding of the 5-Step Tutoring Frame, EMT dialogue, and conversational turn management, the EHTC corpus is revealing a complex embedding in which many clusters exist in a single mode, and many dialogue moves exist within each cluster. In other words, the EHTC is revealing a web of embedded structure in expert human tutoring, as opposed to the simple nesting found in novice human tutoring.

9.5 Guru

Guru, like AutoTutor, is designed to simulate a human tutor by holding a conversation with the learner in natural language. However, Guru is design to simulate an *expert* human tutor rather than a novice human tutor. The characteristics of expert human tutors described in Section 9.4 are informative when considering the design of Guru, and how it should differ from AutoTutor.

First and foremost, our analyses revealed that expert tutors do not use curriculum scripts. However, curriculum scripts are a central element of AutoTutor. They contain all the EMT dialogue for a problem as well as the expectations which are used to track the student's progress and understanding. If curriculum scripts and EMT dialogue are not characteristic of expert human tutoring, then Guru requires a new way of tracking student understanding and organizing knowledge about the domain.

Second, in terms of dialogue structure, expert tutors rely a great deal on evaluative feedback and just-in-time direct instruction. Contrast this to the hints, prompts, and elaborations that constitute the bulk of AutoTutor's dialogue. Guru cannot soley rely on hints, prompts, and elaborations but rather must incorporate tutor dialogue moves into a new model for just-in-time direct instruction.

Third, experts are precise with their feedback. In AutoTutor, feedback is calculated by comparing the student's responses with the expectations from the curriculum script. Again, expert tutors do not appear to use such a script. Furthermore, the traditional way of comparing student answers with expectations in AutoTutor, LSA (Graesser2004; VanLehn et al. 2007; Graesser et al. 2005b), is relatively imprecise: to LSA, "do you want to drive me" and "do you want me to drive" mean the same thing. To model the precise feedback of expert human tutors, it is necessary to incorporate a more sensitive technique than LSA.

Fourth, expert tutors maintain highly accurate student models. In AutoTutor, the student model is simply the set of LSA comparisons of the student's input to each expectation in the curriculum script. Not only do expert tutors not use curriculum scripts, but LSA also doesn't have the precision to match an expert tutor. Therefore Guru should apply a different methodology for student modeling.

Fifth, expert tutors use a variety of tutoring modes and clusters within modes that have no clear correlates in AutoTutor. Contrasted with the linear hint-prompt-assertion cycle used in AutoTutor, the expert tutoring modes are both more numerous and more complex. Fortunately, the dialogue management used in AutoTutor is extremely powerful (Graesser et al. 2005), so a new approach to dialogue management per se for Guru is not required.

In summary, Guru needs a new way to model the domain, model the student, interpret student utterances, and generate direct instruction. We are working on a unified approach to all of these tasks, which is based on a single knowledge representation. Using a single knowledge representation for multiple purposes like these is not uncommon in an ITS. For example, overlay student models typically assume a domain decomposition in which chunks of content can be marked as understood by the student, rather like checking items off a list. An overlay student model is so called because it lays over the domain model in a rather transparent way, i.e. each element of the domain model is on the checklist for the overlay student model.

Clearly an overlay student model first requires a domain model. In the same way, interpretation of student input and the generation of direct instruction can also be yoked to a domain model. However, the creation of a domain model is sufficiently challenging to require special authoring tools and many man-hours to develop (Murray 1998; Corbett 2002; Aleven et al. 2006). Thus for Guru we have been particularly interested in unsupervised and semi-supervised knowledge representation techniques that can extract semantic representations from raw text. Although we still find LSA useful for some tasks, we have been developing a new technique for concept map extraction, which we believe holds promise for domain modeling, student modeling, interpretation of student utterances, and generation of direct instruction.

9.5.1 Concept Map Extraction

The term "concept map" has become largely associated with an educational practice in which students create a graph representation of ideas and the links between them (Novak and Canas 2006). However, similar notions to concept maps have been used in the education, artificial intelligence, and psychological communities for decades, and as a result there are dozens of different definitions of concept map (Fisher et al. 2000). Generally speaking, a concept map consists of a set of nodes (concepts) and edges (relations) describing a core concept or answering a core question (Novak and Canas 2006). We call a pair of nodes connected by an edge a *triple* because it consists of three elements: a start node, an edge relation, and an end node. Thus in general, relationships in concept maps are binary. This prevents or obfuscates the expression of some relationships such as a verb with three arguments, unless additional constraints are adopted which can convert a concept map into a first order logic (Sowa 2007, 2010).

Our unique concept map definition is a synthesis of previous work in both the psychology and education literatures (Graesser and Franklin 1990; Gordon et al. 1993 ; Fisher et al. 2000). The education literature, particularly relevant from an ITS point of view, has promoted relatively small, human-readable maps, such as the SemNet map (Fisher et al. 2000). The key feature that makes these concept maps easy to understand is that they are radial, with a core concept in the middle of the map and a single layer of links radiating from that concept. End nodes linked to the core concept can potentially be the centers of their own maps, but each map is coherent by itself. From the psychology literature, we adopt a limited set of edges linking two nodes in the concept map (Graesser and Franklin 1990; Gordon et al. 1993). Discrete sets of edges are also common in ontologies, e.g. *is-a* or *has-part*. For Guru, a salient advantage of having a restricted set of edges is that they facilitate both generating questions and answering questions from the map (Graesser and Franklin 1990; Gordon et al. 1993).

Recently, we developed a semi-supervised procedure for extracting concept maps with radial structure and discrete set of edges (Olney in press). The procedure operates on a textbook, using a semantic parser and post processing to transform the semantic parse into concept maps. More specifically, the LTH SRL Parser (Johansson and Nugues 2008) outputs a dependency parse annotated with semantic roles derived from Propbank (Palmer et al. 2005) and Nombank (Meyers et al. 2004) for each sentence in the textbook. For each syntactic or semantic relation found by the parser, we require that the start node be a key term in our domain. Key terms are defined as those terms existing in the glossary or index of the book. If the start node is a key term, a corresponding end term is found in the parse, and then the relation linking them is classified using a hand-built decision tree. Some relations are syntactic, e.g. *is-a* is determined by the presence of a "be" main verb as in "an abdomen is the posterior part of an arthropod's body." Other relations are semantic and are classified using the semantic information returned by Nombank or Propbank, e.g. *has-part* is determined by "body" in the example above because "body" is a Nombank predicate whose sense gloss is "partitive part." This process of concept map extraction is semi-supervised because the key terms and edge relations have been manually defined for our domain, but the rest of the procedure is unsupervised.

9.5.2 Domain and Student Modeling

The concept map extracted from Guru's biology textbook contains roughly 30,000 triples centered around 2,000 terms. Thus it is a fairly well elaborated model of the domain. The triples allow us to query particular properties of the key terms in our domain:

ABDOMEN is-a part
ARTHROPOD has-part ABDOMEN
ABDOMEN has-property posterior

It is fairly straightforward to build an overlay student model around this domain model. One can consider each key term as a chunk the student should master, and calculate a coverage score based on the number of triples a student has appeared to master. Although each chunk may be considered as a kind of expectation, or bundle of expectations, the overall structure of the concept map-based domain model is different from the script based model of EMT dialogue described in Section 9.2, in at least three ways. First, the concept map expectations are not attached to a particular problem, but instead are general to the domain. Second, rather than a limited set of expectations, the concept map (in theory) includes all of the salient relations in the biology textbook. Finally, the concept map relations, consisting of triples, are more structured than AutoTutor expectations, which are undifferentiated LSA vectors. In other words the concept map-based domain model appears to be more general, have broader coverage, and be more structured than curriculum script based EMT dialogue.

We are currently building richer links between the standards for high school biology instruction in our state and concept maps we've extracted from the state textbook. This will allow us to better focus the domain and student models of Guru to the content covered by state-wide standardized testing, which in turn will make it easier to integrate Guru into classroom activities.

9.5.3 Interpretation of Student Utterances

In Guru, and in an ITS generally, interpreting a student utterance means mapping that utterance to the domain model. In the case of Guru, which uses an overlay student model, such mapping facilitates both interpretation of the student utterance as well as assessment of the student's current understanding. The most straightforward way to accomplish this mapping is to use the concept map extraction technique from Section 9.5.1 on the student's utterances, and compare the resulting triples with those in the domain and student models. Intuitively, there are more ways to compare triples than monolithic LSA vectors. By definition, each triple has three components, and Guru's feedback can be differentially driven by the correctness of each component.

If only the start node of a student's triple is incorrect, we can hypothesize that the student has not adequately discriminated their start node from the actual start node. For example, if the student's utterance contains the triple **white blood cell** *has-consequence* **delivers oxygen**, then we can identify that this student knows something about red blood cells that is being incorrectly generalized to white blood cells. If only the edge relation of a student's triple is incorrect, then we can hypothesize that the student knows that the two concepts are related, but misunderstands the type of relation. For example, if the student's utterance contains the triple **red blood cell** *lacks* **delivers oxygen**, then we can target the *lacks* relation for remediation. Finally, if the end node only is incorrect, then we might hypothesize that the student lacks sufficient background knowledge. For example, the triple **red blood cell** *has-property* **found in plants** likely indicates that the student knows absolutely nothing about red blood cells, and some direct instruction is needed. These are just examples of possible strategies, but they illustrate how a concept map representation composed of triples can be used to make fine discriminations of the student's error and respond appropriately.

9.5.4 Generating Direct Instruction

Just-in-time direct instruction is, by definition, unplanned. As such it is impossible to render from a curriculum script, which is essentially pre-planned. Rather just-in-time direct instruction must be generated dynamically from an existing domain model. Concept maps have been previously used to generate text. Our concept maps use a fixed set of edge relations that can be set into correspondence with certain question types, e.g. definitional, causal consequent, and procedural, for both the purposes of answering questions (Graesser and Franklin 1990) as well as generating them (Gordon et al. 1993). For example **red blood cell** *has-consequence* **delivers oxygen** can be used to generate the questions «What causes oxygen to be delivered," "What does a red blood cell do," or "What can you say about a red blood cell and oxygen" depending on whether we want to query the start node, the end node, or the edge relation between them respectively. Of course, given the same triple, it is straightforward to create direct instruction like "a red blood cell delivers oxygen."

A similar approach is used in the Betty's Brain ITS (Biswas et al. 2005; Leelawong and Biswas 2008). In this "learning by teaching" system, students teach an animated agent named Betty, whose brain is visible as a causal concept map with additional hierarchical (i.e. is-a) and descriptive relations (i.e. has-property). Students teach Betty by explicitly creating linkages in the concept map "brain." Betty can "take" quizzes by applying a qualitative reasoning algorithm to the causal concept map. Moreover, Betty can describe her reasoning by reading off the relationships in the map, e.g. light increases algae growth which decreases oxygen in the water.

9.5.5 Limitations

In this section, we have outlined some of the major dimensions in which we believe expert human tutors differ from novice human tutors, and the implications for these differences on the design of an expert ITS. The major differences that we have emphasized, the domain model, the student model, interpretation of student utterances, and generation of direct instruction, appear to be well supported by a concept map knowledge representations. However, although our preliminary observations are plausible, these applications have yet to be rigorously evaluated.

9.6 Conclusion

This chapter described our previous and ongoing investigations into the conversational interaction that defines human tutoring. Both our analyses of novice and expert human tutors are corpus-based, driven by extensive collections of human tutoring dialogues. Our goal is to better understand the representations and processes of human tutoring by building computational models in the form of intelligent tutoring systems that embody our theory.

For novice human tutoring, we have identified three major dialogue structures, including the 5-Step Tutoring Frame, EMT dialogue, and conversational turn management. These three structures are nested such that each occurs within its preceding structure. These three structures are comprehensive enough that they can be used to specify the runtime of an ITS, and we have done so in the ITS AutoTutor. The 5-Step Tutoring Frame defines the overall structure of a problem, the EMT dialogue defines the components of a problem, and conversational turn management defines how each tutor turn is constructed in a conversationally appropriate way. In experimental evaluations of learning gains, AutoTutor yields an approximately .8 effect size increase relative to control conditions. Relative to the .4 effect size for novice human tutoring reported in a meta-analysis (Cohen et al. 1982), AutoTutor appears to be convincing as a model of novice human tutoring both in terms of its structure and its effectiveness.

Our recent collection and analysis of expert human tutoring has revealed some differences which may be attributable to the difference between expert and novice human tutors. The expert tutors in our study manifested very complex conversational interaction relative to novice human tutors, including dialogue modes, functional clusters of dialogue moves within modes, and finally the dialogue moves themselves. As discussed in Section 9.4, there is no clear correspondence between these dialogue structures and the structures associated with novice human tutoring. Moreover, our analyses of expert human tutoring suggest that expert human tutors utilize more precisely defined and well-organized domain and student models, are more precise in evaluating and responding to student answers, and utilize a just-in-time direct instruction that is highly adapted to the student's current knowledge state.

We have proposed a particular formulation of concept maps to address these four issues, and we have outlined how these concept maps can be extracted from a textbook, alleviating the burden of domain model authoring. Using the concept map representation, we have further proposed several strategies for addressing the four salient phenomena in our analysis of expert human tutoring, including the domain/student model, interpretation of student utterances, and generation of direct instruction. Though our current assessments are promising, these strategies await a more rigorous evaluation.

Moreover, since the goal of any ITS is to produce learning gains, the conclusive evaluation of the concept map representation and associated strategies is a learning outcome study. We are currently engaged in curriculum development, usability studies, and unit testing in preparation for a learning outcome study. If we have properly identified and represented the differences between expert and novice human tutors, then this should be reflected in a corresponding difference in learning gains.

References

Aleven, V., Koedinger, K.R.: An effective metacognitive strategy: Learning by doing and explaining with a computer-based cognitive tutor. Cognitive Science 26, 147–179 (2002)

Aleven, V., McLaren, B.M., Sewall, J., Koedinger, K.R.: The cognitive tutor authoring tools (CTAT): Preliminary evaluation of efficiency gains. Intelligent Tutoring Systems 2006, 61–70 (2006)

Anderson, J.R., Corbett, A.T., Koedinger, K.R., Pelletier, R.: Cognitive tutors: lessons learned. The Journal of the Learning Sciences 4(2), 167–207 (1995)

Aronson, J.: Improving academic achievement: Impact of physchological factors on education. Academic Press, San Diego (2002)

Biswas, G.: Designing learning by teaching agents: The Betty's Brain system. Int. J. Artif. Intell. Ed. 18(3), 181–208 (2008)

Bloom, B.S.: The 2 sigma problem: The search for methods of group instruction as effective as one-to-one tutoring. Educational Researcher 13(6), 4–16 (1984)

Cade, W.L., Copeland, J.L., Person, N.K., D'Mello, S.K.: Dialogue modes in expert tutoring. In: Woolf, B.P., Aïmeur, E., Nkambou, R., Lajoie, S. (eds.) ITS 2008. LNCS, vol. 5091, pp. 470–479. Springer, Heidelberg (2008)

Chi, M., Leeuw, N., Chiu, M., LaVancher, C.: Eliciting Self-Explanations Improves Understanding. Cognitive Science 18(3), 439–477 (1994)

Chi, M.T.H., Siler, S.A., Jeong, H., Yamauchi, T., Hausmann, R.G.: Learning from tutoring. Cognitive Science 25, 471–533 (2001)

Chi, M.T.H., Siler, S.A., Jeong, H.: Can tutors monitor students' understanding accurately? Cognition and Instruction 22(3), 363–387 (2004)

Chi, M., Roy, M., Hausmann, R.: Observing tutorial dialogues collaboratively: Insights about human tutoring effectiveness from vicarious learning. Cognitive science 32(2), 301–341 (2008)

Cohen, P.A., Kulik, J.A., Kulik, C.L.C.: Educational outcomes of tutoring: a meta analysis of _ndings. American Educational Research Journal 19, 237–248 (1982)

Collins, A., Warnock, E.H., Aiello, N., Miller, M.L.: Reasoning from incomplete knowledge. In: Bobrow, D., Collins, A. (eds.) Representation and Understanding: Studies in Cognitive Science. Academic Press, New York (1975)

Corbett, A.: Cognitive computer tutors: Solving the two-sigma problem. In: Bauer, M., Gmytrasiewicz, P.J., Vassileva, J. (eds.) UM 2001. LNCS (LNAI), vol. 2109, p. 137. Springer, Heidelberg (2001)

Corbett, A.: Cognitive tutor algebra I: Adaptive student modeling in widespread classroom use. In: Technology and Assessment: Thinking Ahead. Proceedings from a Workshop. National Academy Press, Washington (2002)

Cromley, J.G., Azevedo, R.: What do reading tutors do? a naturalistic study of more and less experienced tutors in reading. Discourse Processes 40(2), 83–113 (2005)

Deerwester, S.C., Dumais, S.T., Furnas, G.W., Landauer, T.K., Harshman, R.A.: Indexing by latent semantic analysis. Journal of the American Society of Information Science 41(6), 391–407 (1990)

Derry, S.J., Lajoie, S.P.: Computers as cognitive tools. Erlbaum, Hillsdale (1993)

Derry, S.J., Potts, M.K.: How tutors model students: A study of personal constructs in adaptive tutoring. American Educational Research Journal 35(1), 65–99 (1998)

D'Mello, S., Olney, A.M., Person, N.: Mining collaborative patterns in tutorial dialogues. Journal of Educational Data Mining (in press)

Dumais, S.: Improving the retrieval of information from external sources. Behavior Research Methods Instruments and Computers 23(2), 229–236 (1991)

Evens, M., Spitkovsky, J., Boyle, P., Michael, J., Rovick, A.: Synthesizing tutorial dialogues. In: Proceedings of the 15th Annual Conference of the Cognitive Science Society. Lawrence Erlbaum Associates, Mahwah (1993)

Fisher, K., Wandersee, J., Moody, D.: Mapping biology knowledge. Kluwer Academic Pub., Dordrecht (2000)

Fox, B.: The human tutoring dialogue project. Lawrence Erlbaum Associates, Inc., Hillsdale (1993)

Gagn, R.M.: The conditions of learning and theory of instruction. In: Holt, R., Winston Glass, M., Kim, J., Evens, M., Michael, J., Rovick, A. (eds.) Novice vs. expert tutors: A comparison of style. Midwest Artificial Intelligence and Cognitive Science Conference (1985)

Gordon, S., Schmierer, K., Gill, R.: Conceptual graph analysis: Knowledge acquisition for instructional system design. Human Factors: The Journal of the Human Factors and Ergonomics Society 35(3), 459–481 (1993)

Graesser, A.C., Franklin, S.P.: Quest: A cognitive model of question answering. Discourse Processes 13, 279–303 (1990)

Graesser, A.C., Person, N.K.: Question asking during tutoring. American Educational Research Journal 31(1), 104–137 (1994)

Graesser, A.C., Person, N.K., Magliano, J.P.: Collaborative dialogue patterns in naturalistic one-to-one tutoring. Applied Cognitive Psychology 9, 1–28 (1995)

Graesser, A.C., Lu, S., Jackson, G.T., Mitchell, H., Ventura, M., Olney, A., Louwerse, M.M.: AutoTutor: A tutor with dialogue in natural language. Behavioral Research Methods, Instruments, and Computers 36, 180–193 (2004)

Graesser, A.C., Person, N.K., Harter, D., et al.: Teaching tactics and dialog in AutoTutor. International Journal of Artificial Intelligence in Education 12(3), 257–279 (2001)

Graesser, A.C., Olney, A.M., Haynes, B.C., Chipman, P.: AutoTutor: A cognitive system that simulates a tutor that facilitates learning through mixed-initiative dialogue. In: Forsythe, C., Bernard, M.L., Goldsmith, T.E. (eds.) Cognitive Systems: Human Cognitive Models in Systems Design. Erlbaum, Mahwah (2005a)

Graesser, A.C., Chipman, P., Haynes, B., Olney, A.: AutoTutor: An intelligent tutoring system with mixed-initiative dialogue. IEEE Transactions on Education 48(4), 612–618 (2005b)

Graesser, A.C., DMello, S.K., Cade, W.L.: Instruction based on tutoring. In: Mayer, R., Alexander, P. (eds.) Handbook of Research on Learning and Instruction, Routledge, New York (in press)

Jackson, G.T., Person, N.K., Graesser, A.C.: Adaptive tutorial dialogue in autotutor. In: Proceedings of the workshop on Dialog-based Intelligent Tutoring Systems at the 7th International conference on Intelligent Tutoring Systems, Universidade Federal de Alagoas, Brazil, pp. 368–372 (2004)

Johansson, R., Nugues, P.: Dependency-based syntactic-semantic analysis with PropBank and NomBank. In: CoNLL 2008: Proceedings of the Twelfth Conference on Computational Natural Language Learning, Morristown, NJ (2008)

Jordan, P., Siler, S.: Student initiative and questioning strategies in computer mediated human tutoring dialogues. In: Proceedings of ITS, Workshop on Empirical Methods for Tutorial Dialogue Systems (2002)

Lajoie, S., Faremo, S., Wiseman, J.: Tutoring strategies for effective instruction in internal medicine. International Journal of Artificial Intelligence and Education 12, 293–309 (2001)

Landauer, T.K., Dumais, S.T.: A solution to plato's problem: The latent semantic analysis theory of the acquisition, induction, and representation of knowledge. Psychological Review 104, 211–240 (1997)

Landauer, T.K., McNamara, D.S., Dennis, S., Kintsch, W. (eds.): Handbook of Latent Semantic Analysis. Lawrence Erlbaum, Mahwah (2007)

Larsson, S., Traum, D.: Information state and dialogue management in the TRINDI Dialogue Move Engine Toolkit. Natural Language Engineering, 323–340 (2000)

Lepper, M.R., Aspinwall, L.G., Mumme, D.L., Chabay, R.W.: Self-perception and social-perception processes in tutoring: Subtle social control strategies of expert tutors. In: Olson, J.M., Zanna, M.P. (eds.) Self-inference processes: The Ontario Symposium, Erlbaum, Hillsdale (1990)

Lepper, M.R., Woolverton, M., Mumme, D., Gurtner, J.: Motivational techniques of expert human tutors: Lessons for the design of computer based tutors. In: Lajoie, S., Derry, S. (eds.) Computers as cognitive tools. Erlbaum, Hillsdale (1993)

Lepper, M.R., Woolverton, M.: The wisdom of practice: Lessons learned from the study of highly effective tutors. In: Improving academic achievement: Impact of psychological factors on education. Academic Press, San Diego (2002)

Lesgold, A., Lajoie, S., Bunzo, M., Eggan, G.: SHERLOCK: A coached practiceenvironment for an electronics troubleshooting job. In: Larkin, J.H., Chabay, R.W. (eds.) Computer assisted instruction and intelligent tutoring systems: Shared goals and complementary approaches, Lawrence Earlbaum, Hillsdale (1992)

Litman, D., Forbes-riley, K.: Correlations between dialogue acts and learning in spoken tutoring dialogues. Nat. Lang. Eng. 12(2), 161–176 (2006)

McTear, M.: Modelling spoken dialogues with state transition diagrams: experiences with the CSLU toolkit. In: Fifth International Conference on Spoken Language Processing, Sydney, Australia (1998)

Meyers, A., Reeves, R., Macleod, C., Szekely, R., Zielinska, V., Young, B., Grishman, R.: The NomBank project: An interim report. In: Meyers, A. (ed.) HLTNAACL 2004 Workshop, Frontiers in Corpus Annotation, Boston (2004)

Moore, J.D.: Participating in explanatory dialogues: Interpreting and responding to questions in context. The MIT Press, Cambridge (1994)

Murray, T.: Authoring Knowledge-Based tutors: Tools for content, instructional strategy, student model, and interface design. Journal of the Learning Sciences 7(1), 5 (1998)

Novak, J.D., Canas, A.J.: The theory underlying concept maps and how to construct them. Technical report, Institute for Human and Machine Cognition (2006)

Olney, A., Louwerse, M., Mathews, E., Marineau, J., Hite-Mitchell, H., Graesser, A.: Utterance classification in AutoTutor. In: Proceedings of the HLT-NAACL 2003 Workshop on Building Educational Applications Using Natural Language Processing, Philadelphia. Association for Computational Linguistics, pp. 1–8 (2003)

Olney, A.M.: Gnututor: An open source intelligent tutoring. system based on AutoTutor. In: Proceedings of the 2009 AAAI Fall Symposium on Cognitive and Metacognitive Educational Systems, Washington (2009)

Olney, A.M.: Extraction of concept maps from textbooks for domain modeling. In: Proceedings of the International Conference on Intelligent Tutoring Systems (in press)

Palinscar, A.S., Brown, A.: Reciprocal teaching of comprehension-fostering and comprehension-monitoring activities. Cognition and Instruction 1, 117–175 (1984)

Palmer, M., Gildea, D., Kingsbury, P.: The Proposition Bank: An annotated corpus of semantic roles. Comput. Linguist. 31(1), 71–106 (2005)

Person, N.K., Graesser, A.C., Magliano, J.P., Kreuz, R.J.: Inferring what the student knows in one-to-one tutoring: the role of student questions and answers. Learning and individual differences 6(2), 205–229 (1994)

Person, N.K., Kreuz, R.J., Zwaan, R., Graesser, A.C.: Pragmatics and pedagogy: Conversational rules and politeness strategies inhibit effective tutoring. Cognition and Instruction 13, 161–188 (1995)

Person, N., Graesser, A.: Evolution of discourse during cross-age tutoring. In: O'Donnell, A.M., King, A. (eds.) Cognitive perspectives on peer learning. Lawrence Erlbaum, Mahwah (1999)

Person, N., Graesser, A.: Fourteen facts about human tutoring: Food for thought for ITS developers. In: AI-ED 2003 Workshop Proceedings on Tutorial Dialogue Systems: With a View Toward the Classroom, pp. 335–344 (2003)

Person, N.K., Lehman, B., Ozbun, R.: Pedagogical and motivational dialogue moves used by expert tutors. Presented at the 17th Annual Meeting of the Society for Text and Discourse. Glasgow, Scotland (2007a)

Person, N.K., Mallott, L.: Question generation in expert tutoring. Presented at the 17th Annual Meeting of the Society for Text and Discourse, Glasgow, Scotland (2007b)

Person, N.K., Moses, J., Willson, W.: Question asking in expert and non-expert tutoring sessions. Presented at the 2007 AERA Annual Meeting, Chicago (2007c)

Rogoff, B., Gardner, W.: Adult guidance of cognitive development. In: Rogoff, B., Lave, J. (eds.) Everyday cognition: Its development in social context. Harvard University Press, Cambridge (1984)

Shah, F., Evens, M., Michael, J., Rovick, A.: Classifying Student Initiatives and Tutor Responses in Human Keyboard-to-Keyboard Tutoring Sessions. Discourse Processes 33(1), 23–52 (2002)

Sinclair, J., Coulthard, R.M.: Towards an Analysis of Discourse: the English used by teachers and pupils. Oxford University Press, London (in press)

Sleeman, D., Brown, J.S. (eds.): Intelligent tutoring systems. Academic Press, New York (1982)

Sowa, J.F.: Conceptual graphs. In: Van Harmelen, F., Lifschitz, V., Porter, B. (eds.) Handbook of knowledge representation. Elsevier Science, San Diego (2007)

Sowa, J.F.: Concept mapping. JFSowa.com (2010), `http://www.jfsowa.com/talks/cmapping.pdf` (Accessed May 15, 2010)

VanLehn, K., Jones, R., Chi, M.: A Model of the Self-Explanation Effect. The Journal of the Learning Sciences 2(1), 1–59 (1992)

VanLehn, K., Jordan, P., Rose, C., Bhembe, D., Bottner, M., Gaydos, A., Makatchev, M., Pappuswamy, U., Ringenberg, M., Roque, A., et al.: The architecture of Why2-Atlas: A coach for qualitative physics essay writing. In: Cerri, S.A., Gouardéres, G., Paraguaçu, F. (eds.) ITS 2002. LNCS, vol. 2363, pp. 158–167. Springer, Heidelberg (2002)

Vanlehn, K.: The behavior of tutoring systems. Int. J. Artif. Intell. Ed. 16(3), 227–265 (2006)

VanLehn, K., Graesser, A.C., Jackson, G.T., Jordan, P., Olney, A., Rose, C.: When are tutorial dialogues more effective than reading? Cognitive Science 31, 3–62 (2007)

Wiley, J., Jensen, M.: When three heads are better than two. In: Proceedings of the 28th annual conference of the cognitive science society, pp. 2375–2380 (2006)

Wiley, J., Bailey, J.: Effects of collaboration and argumentation on learning from web pages. In: Erkens, G. (ed.) Collaborative learning, reasoning, and technology. Lawrence Erlbaum, Mahwah (2006)

Chapter 10
Affective Tutors: Automatic Detection of and Response to Student Emotion

Beverly Park Woolf[1], Ivon Arroyo[1], David Cooper[1], Winslow Burleson[2], and Kasia Muldner[2]

[1] Center for Knowledge Communication, 140 Governor's Drive, University of Massachusetts Amherst, MA 01003-4610
[2] Arizona State University, PO Box 878709 Tempe, Az 85287-8709
{bev, ivon}@cs.umass.edu,
{winslow.burleson,Katarzyna.Muldner}@asu.edu

Abstract. This chapter describes the automatic recognition of and response to human emotion within intelligent tutors. Tutors can recognize student emotion with more than 80% accuracy compared to student self-reports, using wireless sensors that provide data about posture, movement, grip tension, facially expressed mental states and arousal. Pedagogical agents have been used that provide emotional or motivational feedback. Students using such agents increased their math value, self-concept and mastery orientation, with females reporting more confidence and less frustration. Low-achieving students—one third of whom have learning disabilities—report higher affective needs than their higher-achieving peers. After interacting with affective pedagogical agents, low-achieving students improved their affective outcomes and reported reduced frustration and anxiety.

10.1 Introduction

Affect is a central component of human cognition and strongly impacts student learning (McQuiggan et al. 2008; Goleman 1995; Efklides and Petkakim 2005; Brand et al. 2007). If computers are to interact naturally with humans, they must recognize affect and express social competencies. Affect has begun to play an important role in intelligent tutors (Conati and Maclaren 2004; D'Mello et al. 2007) and affective tutors seem to increase the effectiveness of tutorial interactions and, ultimately learning. The field of affective tutors investigates techniques for enabling computers to recognize, model, understand and respond to student emotion effectively. One obvious next frontier in computational instruction is to systematically examine the relationships between student affective state and learning outcomes (Shute 2008).

While early learning theories ignored the importance of emotion in learning, recent research has created a link between emotion and learning and the claim has been made that cognition, motivation and emotion are the three components of

R. Nkambou et al. (Eds.): Advances in Intelligent Tutoring Systems, SCI 308, pp. 207–227.
springerlink.com

learning (Snow et al. 1996; D'Mello et al. 2007). Various classroom studies have linked interpersonal relationships between teachers and students to increased student motivation over the long term (Wentzel and Asher 1995; Royer and Walles 2007).One goal of affective computers is to recognize affect or identify the affective state of people from a variety of physical cues that are produced in response to affective changes in the individual (Picard et al. 2004).

When humans use affect within one-to-one teaching relationships, the result is very powerful. For example, in their research on 'thin slices,' Ambady and Rosenthal demonstrated that based on a short segment of video, as little as six seconds of a teacher's first interactions with a student, participants could predict that teacher's effectiveness and student end-of-term grades based on the teacher's exhibited use of affect (Ambady and Rosenthal 1992). Wentzel (1997) has shown that caring bonds between middle schoolchildren and their teachers are predictive of learners' performance. This chapter looks at the role new technology plays in automatic recognition of and response to student affect. Affective interventions encourage learning, lessen student humiliation and provide support and motivation that outweighs or distracts from the unpleasant aspects of failure. Section 2 describes real-time automatic recognition of emotions exhibited during learning, while Section 3 describes attempts to automatically generate appropriate responses to student emotion. Section 4 describes ways to evaluate these recognition and response mechanisms and to integrate them into educational practice. This research is based on efforts at the University of Massachusetts, Arizona State University and the MITMedia Lab.

10.2 Automatic Recognition of Student Affect

Great interest exists in embedding affective support into tutoring applications and research has focused on automated detection of affective states as a first step towards this goal (Conati and Mclaren 2004; D'Mello and Graesser 2007; McQuiggan and Lester 2006; Graesser et al. 2007). Currently there is no gold standard for either labeling a person's emotional state or for responding to it. One approach to recognizing emotion is to triangulate among three different inputs: sensor data, student self-reports, and human observation of students. While we accept that there will never be definitive categorization of a human's emotional state, this triangulation has been used Arroyo et al., to identify clear examples of emotions (frustration, flow, etc.) that can be labeled using sensor information (Woolf et al. 2009).

Hardware sensors have the potential to provide information on students' physiological responses that have been linked to various affective states (D'Mello and Graesser 2007; Graesser et al. 2007; D'Mello et al. 2007). Research explores various sensors' potential for affect recognition, e.g., Burleson, 2006 developed a learning companion that depended on a sensor framework (incorporating a mouse, posture chair, video camera, skin conductance bracelet) to recognize and respond to student affect. Our sensor platform of four physiological sensors (Fig. 10.1) has been tested with more than 1,000 students in middle high and college classes. The platform is unobtrusive enough to be used by students in a typical setting and resource-conscious enough to run on average computer labs available to students

(Cooper et al., 2009). These sensors collect raw data about physical activity and state of a student and the challenge remains to map this data into models of emotional states and use this information productively.

Mental State Camera. We use a standard web-camera to obtain 30 frames per second at 320x240 pixels. This is based on an earlier facial expression recognition system used in Burleson's Affective Learning Companion (Mota and Picard 2003; Kapoor et al. 2007. The present system is coupled with El Kaliouby's MindReader applications (el Kaliouby 2005). We developed a Java Native Interface (JNI) wrapper around the MindReader library. The interface starts a version of the MindReader software, and can be queried at any time to get the most recent mental state values that have been computed by the library. In the version used in the experiments, only the six mental state features were available, but in future versions we can train it on new mental states.

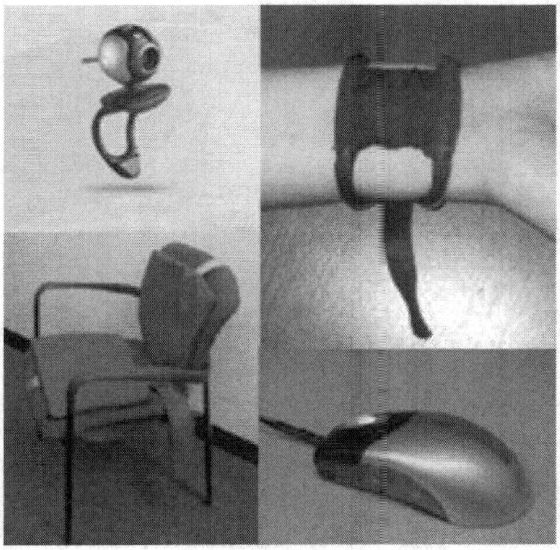

Fig. 10.1 Sensors used in the classroom (clockwise): mental state camera, skin conductance bracelet, pressure sensitive mouse, pressure sensitive chair.

Skin Conductance Bracelet. The Affective Computing Group at the MIT Media Lab has been advancing the development of wireless wearable skin conductance sensors for over a decade. Various implementations include the galvactivator, a glove that could illuminate an LED when its user had heightened levels of skin conductance (Picard and Scheirer 2001); HandWave which used a custom built Printed Circuit Board (PCB) and 9V battery to provide blue tooth wireless transmission of skin conductance data at rates up to5 Hz (Strauss et al. 2005).

The current system used in our research employs the next generation of Hand Wave electronics developed at MIT, providing greater reliability, lower power requirements through wireless RFID transmission, and a smaller form. This smaller form was redesigned to minimize the visual impact and increase the wearable aspects of previous versions. We integrated and tested these electronic components into a wearable package suitable for students in classrooms.

Pressure Sensitive Mouse. The pressure mouse was originally developed by the Affective Computing Group at MIT. It uses six pressure sensors embedded in the surface of the mouse to detect the tension in users' grip and has been used to infer elements of users' frustration(Qia and Picard 2002; Dennerlein et al. 2003). We replicated MIT's pressure mouse through a production of 30 units. The new design minimized the changes made to the physical appearances of the original mouse in order to maintain a visually non-invasive sensor state.

Pressure Sensitive Chair. We used a simplified posture state chair developed at ASU using a series of eight force sensitive resistors as pressure sensors dispersed throughout the seat and back of a readily available seat cover cushion. This posture chair sensor was developed at ASU.

In our framework, each feature source from each student is a separate stream of data. Hence we have streams of data that each report asynchronously and at different rates. In order to merge all of the data sources, an ID from each student, and a time of the report was needed from each source. We have a database table with a row for every time stamp and wrist ID pair, and a column for each reported sensor value and tutor data value. Each cell in a row represents the latest report of the data source.

10.3 Automatic Response to Student Affect

Once a student's emotion has been recognized, the next issue is to identify how to respond to improve student motivation and learning. Providing empathy or support strongly correlates with learning (Graham and Weiner 1996; Zimmerman 2000) and the presence of someone who cares, or at least appears to care, can be motivating. Various studies have linked interpersonal relationships between teachers and students to motivational outcomes (Wentzel and Asher 1995; Picard et al. 2004). Can this noted human relationship be reproduced, in part, by apparent empathy from a computer character? Apparently the answer is yes (Bickmore and Picard 2004). People seem to relate to computers in the same way they relate to humans and some relationships are identical to real social relationships (Reeves and Nass 1998). For example, students continue to engage in frustrating tasks on a computer significantly longer after an empathetic computational response (Klein et al. 2002), have immediately lowered stress level (via skin conductance) after empathy and after apology (Prendinger and Ishizuka 2005), and relational skills improve long-term ratings of caring, trust, respect, desire to keep working (Bickmore and Picard 2004). Computer agents impact student learning, affect and motivation based on gender, ethnicity and realism of the agent (Baylor 2005).

This is not to say that the inferences, movements and interventions of computer agents can exactly replicates those of people, nor can peer theories exactly map to

Fig. 10.2 Pedagogical Agents act out their emotion and talk with the student expressing full sentences of cognitive, meta-cognitive and emotional feedback.

the human peer-tutoring case; however, computer control does allow for careful testing of hypotheses about how to use virtual peer support for learning (Fig. 10.2) (Picard et al. 2004).

Peer learning companions can create adaptable vicarious experiences for students (Burleson and Picard 2008; Baylor 2005; Chan and Baskin 1990). Companions can create adaptable vicarious experiences that are difficult to create in classrooms and observation of peers succeeding may enhance the observing student's self-efficacy (McQuiggan et al. 2008). Verbal persuasion is a common motivational tool used by tutors both human and automated (Lepper et al. 1993). Companions that express confidence in a student's abilities can have profound effect on the student's own self-efficacy beliefs. The impact is determined by the value the student places on the persuader, so an established relationship between tutor and the student makes verbal persuasion all the more powerful (McQuiggan et al. 2008).

One goal of modeling and responding to student affect is to impact the student's affective state and subsequent changes in student physiology (Bandura 1997). Research has shown that strategies that guide students toward affective states with lower arousal levels will diminish the adverse effects of high-arousal physiological responses on student efficacy. Affect recognition can use pedagogical companions to take action when situations of arousal and low self-efficacy occur (McQuiggan et al. 2008).

10.3.1 Learning Companions

We describe gendered learning companions that provide support and encouragement, emphasizing the importance of perseverance, expressing emotions and offering strategies (e.g., "Use the help function"), see Fig. 10.3 (Arroyo et al 2009;

Cooper et al. 2009). These learning companions (LCs) are *empathetic* in that they visually reflect the last emotion reported by the student (queried within the system every five minutes); as long as that emotion is not negative, e.g., companions do not mirror frustration or boredom. Companions act out their emotion and talk with students expressing full sentences of meta-cognitive and emotional feedback. They are non-intrusive — they work on their own computer to solve the problem at hand, and react only after the student has answered the question. Agents respond with some of Carole Dweck's (2002) recommendations about disregarding success and valuing effort. This adds a new dimension to the traditional feedback regarding success/no-success generally given to students.

We measured the impact of LCs on student motivation and achievement and integrated controlled exploration of their communicative factors (facial expression and mirroring postures) as the student/agent relationship developed. Empirical studies show that students who use LCs increased their math value (e.g., questions such as "Mathematics is an important topic"), self-concept (e.g., "I am good in mathematics") and mastery orientation, see Sections 10.4.4-10.4.5. Students tend to become more bored (less interested) towards the end of any instructional session. Yet students using LCs maintain higher levels of interest and reduced boredom after 15 minutes of tutor use. They reported a higher mean confidence, interest and excitement. Despite the fact these results were not significant, this relative advantage for LCs indicates that they might alleviate students' boredom as the session progresses.

10.3.2 Automatics Affective Response

The learning companions used in Wayang (see Section 10.4) deliver approximately 50 different messages emphasizing the malleability of intelligence and the importance of effort and perseverance (Table 10.1). The messages also include meta-cognitive help related to effective strategies for solving math problems and effective use of Wayang's tools. Ultimately, the interventions will be tailored according to Wayang's affective student model. However, we are currently still validating the models and algorithms for deciding which intervention to provide and when, and thus relied on an effort model only to assign messages for this experiment. This section describes these interventions including *attribution* and *strategy* training, as well as *effort affirmation*.

The affective support was to train students motivationally, by emphasizing the importance of effort and perseverance and the idea that intelligence is malleable instead of a fixed trait (Dweck 2002). The characters provided this support by responding to the effort exerted by students rather than to the student's emotions. Characters were either unimpressed when effort was not exerted, or simply ignored that the student solved the problem. They also offered praise to students who exerted effort while problem-solving, even if their answers were wrong, highlighting that the goal is to lessen the importance of performance in favor of learning.

Table 10.1 Companions provided several responses based on student effort

Type	Sample message
Attribution (General)	I found out that people have myths about math, thinking that only some people are good in math. Truth is we can all be good in math if we try.
Attribution (Effort)	Keep in mind that when we are struggling with a new skill we are learning and becoming smarter!
Attribution (No Effort)	We will learn new skills only if we are persistent. If we are very stuck, let's call the teacher, or ask for a hint!
Attribution (Incorrect)	When we realize we don't know why the answer was wrong, it helps us understand better what we need to practice.
Effort Affirmation (Correct No-effort)	That was too easy for you. Let's hope the next one is more challenging so that we can learn something.
Effort Affirmation (Correct Effort)	Good job! See how taking your time to work through these questions can make you get the right answer?
Strategic (Incorrect)	Are we using a correct strategy to solve this? What are the different steps we have to carry out to solve this one?
Strategic (Correct)	We are making progress. Can you think of what we have learned in the last 5 problems?

The characters were highly positive, in the sense that they displayed encouraging gestures (e.g., excitement and confidence). Negative gestures (appearing frustrated or bored) were not effective and were eliminated by researchers. Characters behaviorally mimicked student self-reported emotions, which is a form of a non-verbal empathetic response (e.g., learning companions appeared excited in response to student excitement, see Fig. 10.3, right). In this experiment the companions occasionally expressed non-verbal behaviors of positive valence only, the underlying goal being to make them appear life-like and engaged and to impart some of their enthusiasm to the students. The next three types of interventions described are verbal messages tailored according to the tutors modeling of students' effort.

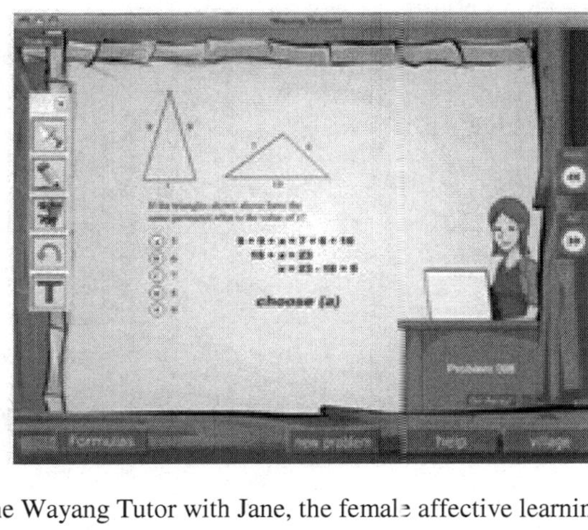

Fig. 10.3 The Wayang Tutor with Jane, the female affective learning companion

Attribution Interventions. Attribution theory proposes that students' motivation to learn is directly rooted in their beliefs about why they succeed or fail at tasks (Weiner 1972). If students can be taught to alter these beliefs, for instance to understand that failure is the result of a lack of effort instead of a lack of ability, then their motivation to learn and learning outcomes can be significantly improved (Robertson, 2000). For example:

- *General attribution* messages encourage students to reflect about myths and math learning in general;
- *Effort attribution* messages reinforce that effort is a necessary by-product of learning, and are specially tailored to situations where students are investing effort but are struggling;
- *No-effort attribution* messages are more emphatic than effort attributes and are designed to help students realize that effort is necessary to learn, and generated when students are not investing effort;
- *Incorrect attribution* messages are generated to motivate students after they provide an incorrect response, by re-formulating how they perceive errors.

Effort-Affirmation Interventions. In contrast to the effort-attribution messages described above, which aim to change students' attitude towards effort during problem solving and are generated before the student actually starts problem solving, *effort-affirmation* interventions acknowledge effort after students obtain a correct solution (see Table 10.1 for examples). These interventions include:

- *Correct no-effort interventions* are generated after a student invests no effort but obtains a correct solution, to make students realize that praise is not appropriate;
- *Correct-effort affirmations* are generated after a student both invests effort and obtains the correct solution, to acknowledge the student's effort.

Strategic Interventions. The final type of intervention focuses on meta-cognitive strategies, with the goal of both making students more effective problem solvers and motivating them for learning in general.

- *Incorrect strategic* messages are generated when students are not succeeding at problem solving, to motivate them to change their general problem-solving strategy, i.e., think about why they are not succeeding;
- C*orrect strategic* messages are generated when students are succeeding at problem solving, to encourage them to evaluate their progress.

10.4 Experiments with Affective Tutors

The affect recognition and response software described above are stand-alone; they can provide affective input to any tutor and can generate its responses for any tutor. We conducted several empirical evaluations with this software to directly connect several objectives of affect research. First we describe experiments that dynamically identified student emotion during learning and reliably identified these emotions through classifiers. Then we describe experiments to personalize tutor response based on student needs and to integrate this work into educational practice.

Currently we are using affect systems in tandem with Wayang Outpost, a multimedia tutoring system for high school geometry and algebra; see Figure 10.3 (Arroyo et al., 2007; 2009; Woolf et al., 2010). Problems are presented one at a time, each consisting of the problem statement with four or five solution options directly below it. Students select an answer and the tutor provides immediate visual feedback by coloring the answer green or red, for correct or incorrect respectively. Prior to or after selecting an answer, a student may ask for a hint, which Wayang displays in progression from general suggestions to the correct answer. In addition to this domain-based help, Wayang includes a wide range of metacognitive and affective support, delivered by learning companions; agents designed to act like peers who care about the student's progress, and offer support and advice on how to improve student learning strategies. Within each topic section, Wayang adjusts the difficulty of problems provided depending on past student performance.

Gendered and ethnically diverse companions allow exploration of how the gender and ethnicity of companions influences outcomes (e.g., learning, attitudes) (Arroyo et al. 2009). The learning companions' interventions are tailored to a given student's needs according to two models of affect and effort embedded in the tutor. The *effort* model uses interaction features to provide information on the degree of effort a student invests in generating a problem solution. An *affect* model assesses a student's emotional state; based on linear regression, this model is derived from data obtained from a series of studies described in (Arroyo et al. 2009; Cooper et al. 2009). Wayang has been used with thousands of students in the past and has demonstrated improved learning gains in state standard exams (Arroyo et al. 2008; 2009)

10.4.1 Methodology for the Affect Studies

We conducted several series of experiments involving the use of sensors, learning companions and Wayang Outpost (Arroyo et al., 2009; 2010). One study involved 35 students in a public high school (HS) in Massachusetts; another involved 29 students in the University of Massachusetts (UMASS); and the final study involved 29 undergraduates from Arizona State University (AZ). In the HS and UMASS studies, students used the software as part of their regular math class for 4-5 days and covered topics in the traditional curriculum. In the AZ lab study, students came into a lab for a single session. These three experiments yielded the results of 588 Emotional Queries from 80 students who were asked about their emotion, e.g., "How confident do you feel?" The response was a scale 1-5 and the queries separated into four emotion variables: 149 were about confidence/anxiety, 163 about excitement/depression, 135 about interest/boredom, and 141 about frustrated/not frustrated. 16 of the student responses gave no answer to the Emotional Query. Models were created to automatically infer student emotions from physiological data from the sensors. Students produced self-reports of emotions and all queries include valid data from at least one sensor.

Another set of studies quantitatively analyzed the benefit of learning companions on affective and cognitive outcomes. The subjects included one hundred and eight (108) students from two high schools 1 (one low and the other high achieving) in the state of Massachusetts and involved 9th and 10th graders. Two thirds of the students were assigned to a learning companion of a random gender, and one third to the no learning companion condition. We obtained complete data (pre and posttest survey and math test) for a smaller subset of subjects. Students took a mathematics pretest before starting, and completed a survey that assessed their general attitude towards mathematics.[1]The pretest covered general attitudes towards math and learning, such as likes/dislikes of math, how much was math valued as important, and how students felt when they solved math problems (anxiety, confidence, frustration, boredom, excitement). Four questions asked about student feelings towards problem solving before they began to work with the tutor, including interest/boredom, frustration, confidence/anxiety, excitement (e.g. how frustrated do you get when solving math problems). For the next three days, students used the Wayang instead of their regular mathematics class. Approximately every five minutes, students were asked to provide information on one of the four target emotions (e.g. how frustrated do you feel?). At the start of a student's interaction with Wayang, learning companions introduced themselves and when students needed help during problem solving, the companions reminded students about the "help button," which provided multimedia based support in the form of animations with sound. Characters spoke out the messages as described in the previous section, occasionally at the beginning of a new problem or after a correct or incorrect attempt to solve the problem. After students used the tutoring module for three days, they took a mathematics post-test, and answered the same questionnaire they had received prior to using the tutor. In addition, the post-survey included five questions about the student's perceptions of the Wayang tutoring system (*Did you learn? Liked it? Helpful? Concerned? Friendly?*). Several student behaviors were logged, e.g., success at problem solving and use of tools and help. Students' self-report of their emotions within the tutor were logged, as well as students behavior, e.g., muting the characters (using a mute button), and whether they abused help or quick-guessed.

10.4.2 Automatic Affect Recognition Empirical Studies

Using the four sensors described in Section 10.2 and placed on each student's chair, mouse, monitor, and wrist, information was conveyed to the tutor about student posture, movement, grip tension, arousal, and facially expressed mental states. Experiments showed that when sensor data supplemented a user model based on tutor logs, the model reflects a larger percentage of the students'

[1] The pre-test included 3 items for self-concept in math ability, e.g., students compared themselves to other students in their math ability and compared mathematics to other subjects; 3 items to address subjective mathematics liking/value).

self-concept than does a user model based on the tutor logs alone. The models were further expanded to classify four ranges of emotional self-concept including frustration, interest, confidence, and excitement with over 78% accuracy. We used stepwise regression analysis with each of the emotions as the dependent variable, and tutor and sensor features as the independent variables. Results from the regression show that the best models for the emotions confidence, frustration, and excitement came from the subset of examples where all of the sensor data was available, and the best model for interested came from the subset of examples with mouse data available.

Table 10.2 shows that the best classifier of each emotion in terms of Accuracy ranges from 78% to 87.5%. By using Stepwise Regression we have isolated key features for predicting user emotional responses to four categories of emotion. These results are supported by cross validation, and show improvement using a very basic classifier.

Table 10.2 This table shows the results of the best classifier of each emotional response. Accuracy of no classifier is a prediction that the emotional state is not high. Values in parentheses include the middle values in the testing set as negative examples.

Classifier	True Pos.	False Pos.	True Neg.	False Neg.	Accuracy (%)	Accuracy (%) No Classifier
Confident All	28(28)	5(24)	10(16)	1(1)	86.36(63.77)	34.09(57.97)
Frustrated All	3(3)	0(0)	46(58)	7(7)	87.5(89.7)	82.14(85.29)
Excited Wrist	25(25)	9(37)	25(40)	5(5)	78.1(60.7)	53.12(71.96)
Interested Mouse	24(25)	4(19)	28(53)	7(7)	82.54(74.76)	50.79(69.90)

This affect recognition evaluation made several important contributions to the field of sensor recognition of student affect in intelligent tutors. We showed that students' self-reports of emotion can be automatically inferred from physiological data that is streamed to the tutoring software for students in real educational settings. Summaries of this physiological activity, in particular data streams from facial detection software, can help tutors predict more than 78% of the variance of students emotional states, which is much better than when these sensors are not used (Table 10.2). We analyzed how students feel and behave while solving mathematics problems in a public school setting and identified state-based fluctuating student emotions through student's self-reports. These fluctuating student reports were related to longer-term affective variables (e.g., value mathematics and self-concept) and these latter variables, in turn, are known to predict long-term success in mathematics, e.g., students who value mathematics and have a positive self-concept of their mathematics ability perform better in mathematics classes (Zimmerman 2000). An opportunity exists for tutoring systems to optimize not only learning, but also long-term attitudes related to students' emotions while using the software. By modifying the "context" of the tutoring system including students' perceived emotion around mathematics, a tutor might optimize and improve their mathematics attitudes.

10.4.3 Automatic Response to Affect Results

The sensor data described above provides emotional predictions that are a first step for intelligent tutor systems to create sensor based personalized feedback for each student in a classroom environment. While many researchers have created affective agents, e.g., (Baylor 2005; Lester et al. 1999), evaluation of their impact on learning has not been conclusive.

Our studies targeted several population demographics. First we evaluated the differences between male and female approaches to learning with the tutor. Second we focused on low–achieving students particularly students with learning disabilities. For each of these populations, we evaluated students' affect and cognition both before and after using the tutor. Within each population, we additionally examined high school (HS) and undergraduate (UG) populations, which can be summarized as follows. HS students had less math incoming ability than UG. Students in the HS study were more "pessimistic" than the UG study, both in pre-test surveys and self-reports of emotions, while UG students were not generally frustrated, HS students reported more frustration, and less interest, confidence and excitement. The combination of both populations provided an interesting mix of students with different feelings and math abilities. Both populations learned based on post-pre test tests; they improved an average 10% in math performance (25% proportional learning gain). The details of our gender studies and evaluation of students who are low achieving continues below.

Table 10.3 Significant Post-Tutor Outcomes: Main and interaction effects for Affective and Cognitive Outcomes. Key: H-A—High-Achieving students; L-A –Low-Achieving students; LC —Learning Companions;∅— No significant difference across conditions; ∅ MathAbility —No significant MathAbility effect or MathAbilityxLC interaction effect

	Overall Effect	**Targeted and Differential Effects by Gender and Achievement Level**
Learning	Students learned in all conditions (paired samples t-test, *t(99) = 2.4, p = .019), but no significant effect for LC	L-A students improved more than H-A in all conditions *F(99,1) = 5.3, p = 0.02
Perceptions of the Tutor	∅	Females using LCs have higher perception of the tutor **F(50,1)=7.5, p=.009; Males not using LCs have a higher perception of the tutor **F(94,1)=10.5, p=.002
		H-A students perceive the tutor better than L-A when LCs are absent, LCxMathAbility **F(96,1)=6.84, p = 0.01
Liking of Mathematics	Students receiving Female LC demonstrated higher math liking. *F(93,2) = 3.7, p = 0.03	∅ MathAbility
Math Ability Self-concept	Students receiving Jane showed higher posttest self-concept. *F(94,2) =3.6, p = 0.03	When LCs are absent, H-A students had higher increase in self-concept than L-A. LCxMathAbility: ⁺F(94,3) = 2.3, p = .08

We carried out Analyses of Covariance (ANCOVA) for each of the affective and behavioral dependent variables (post-tutor and within tutor) shown in Table 10.2.

Table 10.4 Significant emotion within and after using the tutor. Key: **LC**–Learning Companions; H-A—High Achieving; L-A –Low Achieving; ∅—No significant difference across conditions; **∅MathAbilityxLC**—No significant MathAbilityxLC interaction effect/MathAbility effect.

	Overall Effect	**Targeted Effect by Gender and Achievement Level**	**DifferentialEffect by Gender and Achievement Level**
Frustration	Students with female companions reported less overall frustration **F(213,2) = 6.1, p =.003	Females using the female companion have less frustration, within tutor: ***F(99,2) = 8.2, p = .001 After tutor $^+$F(49,1) = 3.1, p = 0.09 L-A students have lower post-tutor frustration in the LC condition than no-LC. $^+$F(58,1) = 3.4, p=.07	When LCs are absent, L-A students have higher post-tutor frustration than H-A. LC x MathAbility $^+$F(93,3) = 2.4, p = .08
Confidence	Students using LCs have higher overall confidence, within tutor: *F(204,1) = 5.3, p = .02	Females using the female companion have more confidence, within tutor: **F(96,1) = 5.6, p = .01 L-A students in the LC condition have higher confidence. Within Tutor LC effect: **F(108,1)= 7.3, p =.008 Post-tutor LC effect: *F(56,1)= 3.8, p =.05 and	H-A students have higher confidence than L-A students (but esp. when companions are absent) MathAbility effect within: *F(204,1)= 4.1, p = .05 MathAbility effect posttutor: *F(91,1) = 5.8, p = .02
Interest	Students in the LC condition have higher overall interest at posttest time. LC main effect: $^+$F(94,1) = 3.4, p = .07	L-A students in the LC condition report marginally more post-tutor interest. LC main effect: $^+$F(58,1) = 2.7, p =.1	L-A students report more boredom than H-A students across all conditions MathAbility effect $^+$F(219,1) = 2.9, p = .09
Excitement	∅	Females using the female companion report more excitement. After tutor: $^+$F(53,1) =3.2, p = 0.08	Females report less excitement than males with no LC, within tutor, GenderxLC: *F(200,1) = 6.1,p = .02 Post-tutor, GenderxLC: *F(67, 1) = 5.3, p = .02 H-A students report less excitement when LCs are absent, no difference when LCs are present. MathAbilityxLC within: *F(200,1) = 5.2, p=.02
Productive behavior: time in hint problems	∅	L-A students spend more time in hinted problems with LCs. $^+$F(67, 1) = 2.9, p = 0.095	Females spent more time than males on "helped problems" in the LC condition, within tutor, Gender xLC: $^+$F(109,1) = 2.78, p = 0.09
Gaming Behavior; Quick-guess. Help abuse	∅	Females using companions made fewer quick guesses by females in LC condition. Mean guesses per student: **F(55,1) = 7.4, p = 0.009	Females abused help marginally less, across all conditions. Gender: $^+$F(110,1) = 2.9, p = 0.09 Females made fewer quick-guesses with LCs; males made more quick guesses with LC: GenderxLC **F(109,1) = 9.03,p = 0.003 L-A students quick-guess more than do H-A students MathAbility effect: **F(109,1) = 5.9, p = 0.017 No MathAbilityxLC interaction effect

Table 10.3 shows the results for general post-tutor outcomes, while Table 10.4 presents the results for affect-related and other variables measured within the tutor. As far as emotions, we include findings both on students' self-reported emotions within the tutor, and post test differences in survey responses (note that in Table 10.2, we reported how students were feeling *before* they interacted with Wayang, while Tables 10.3 and 10.4 look at how interaction with Wayang influenced these feelings).

Our covariates consisted of the corresponding pretest baseline variable (e.g., we accounted for students' pretest baseline confidence when analyzing confidence while using the tutor or afterwards). Independent variables corresponded to *condition*, specifically learning companion (LC) present vs. absent and LC type (Female (Jane) vs. Male (Jake) vs. no-LC). We analyzed both main effects and interactions for achievement level (MathAbility) and conditions over all student data (see second and last columns of Tables 10.3 and Table 10.4). In addition, because of the special affective needs of the targeted group (e.g., females or low-achieving), we repeated the ANCOVAs for that population only, for a "targeted effect," Table 10.4 (third column). Results showed that all students demonstrated math learning after working with Wayang, with low-achieving students learning more than high achieving students across all conditions (Table 10.3). Learning companions did not affect student learning directly, but successfully induced positive student behaviors that have been correlated to learning, specifically, students spent more time on hinted problems (Arroyo and Woolf 2005) (see "Productive behavior" row, Table 10.4). The beneficial effect of learning companions was mainly on affective outcomes, particularly on confidence (see "Confidence" row, Table 10.4). Low-achieving students who received learning companions improved their confidence while using the tutor and at post test time more than students with no learning companions, while their counterparts in the no-LC condition tended to decrease their confidence (Table 10.4).

10.4.4 Gender Studies

While learning companions afford affective advantages for all students, several significant effects in the ANCOVAs indicated a higher benefit of learning companions for female students. In the case of the emotional outcomes just mentioned (confidence and frustration, in particular), the effects are stronger for females than for males (i.e. while all students improved their confidence and reduced their frustration, the third column of Table 10.3 shows stronger significance for females alone). Last column of Table 10.3 also shows that females' confidence is improved but not confidence for males. It is important to note that these gender effects on emotions (within or after the tutor) are not due to females starting out feeling worse, as our analyses account for that baseline pretest emotion as a covariate.

Females especially perceived the learning experience with Wayang significantly better when learning companions were present, while the opposite happened for males, who actually reported worse perceptions of learning when learning companions were present. Female students in the LC condition also had more productive behaviors in the tutor: they spent more time than did males on "helped problems" compared to females in the no-LC condition; they "gamed" less when characters were present (a significant interaction effect revealed that the opposite happens for males).

10.4.5 Behavior of Students with Low Achievement

Currently, students with learning disabilities (LD) who require extra resources comprise 13% percent of students in USA (NCES 2009). These students show improved performance with certain classroom interventions (e.g., providing extra time on tasks, peer tutoring). However these interventions are difficult or impossible to sustain in classrooms without additional instructional support, something that schools are increasingly unable to provide due to budgetary constraints. To the extent that these students are not being educated to their full potential, there is a large negative impact not only in the lives of these students but on society at large.

The under-achievement of students with LD in math does appear to have a biological basis, and there is evidence that many of these students have difficulties with working memory, executive control and procedural knowledge (Geary et al. 1999; 2007). As a result, many students with LD may persist in using counting strategies (e.g., finger counting) long after their typically achieving peers have switched to retrieving answers from memory (Fletcher et al. 2007), taking longer to solve math problems and performing poorly in math class and high-stake tests (Olson 2005). Students with LD develop more negative feelings towards math, choose less advanced math classes in high school and are later under-prepared for science and math careers. LD is a complex multi-factor problem and most educational institutions do not have the tools needed to provide cost-effective instruction tailored to each individual.

Table 10.5 Affective self-reports of high-achieving vs.low-achieving students prior to tutoring

Affective Criterion	**Means, standard deviations and between-subjects test** Low-achieving: N=64; High-achieving: N=43
<u>Self-concept</u> of math ability (in comparison to other students, other subjects, 3 items)	Low-achieving: M=3.2 SD=1.1 High-achieving: M=4.1 SD=1.0 *****F(106,1)=18.2, p=.000**
How <u>confident</u> do you feel when solving math problems?	Low-achieving: M=3.1 SD=1.3 High-achieving: M=4.0 SD=1.3 *****F(105,1)=11.5, p=.001**
How <u>frustrating</u> is it to solve math problems?	Low-achieving: M=3.6 SD=1.2 High-achieving: M=3.0 SD=1.1 **** F(106,1)=7.6, p=.007**
How <u>exciting</u> is it to solve math problems?	Low-achieving: M=2.2 SD=1.2 High-achieving: M=2.7 SD=1.4 ***F(106,1)=3.64, p=0.05**

Low-achieving students were defined as those who scored lower than median grade on the math pretest. One third of these low-achieving students had been previously diagnosed as having a specific learning disability in math or reading and had an Individualized Education Plan (IEP), a document that identifies a student's academic, physical, social and emotional needs. Most students with IEPs (95%) are part of this low-achieving group. Table 10.5 shows that low-achieving students disliked math more, valued it less, had worse perception of their math ability, and reported feeling worse when solving math problems. Since low achieving students (both with and without disabilities) struggle with math, our conjecture was that *all* low achievers could require additional affective support. Thus, the first goal of the study was to examine the affective needs of both low achieving and learning disability students in our data (15% of subjects).

10.4.6 Discussion of Results for Low-Achieving Students

Learning companions had a positive impact for all students on some measures, e.g., all students receiving the female companion (Jane) improved math liking and self-concept of their math ability. This was not the case for the male learning companion (Jake), which was muted by students twice as much as Jane, making it too similar to the control version.

Some differential effects (last column Table 10.4) suggest that learning companions are essential for low-achieving students' affect. When LCs are present, low achieving students report positive affect nearly as much as do high-achieving students and it is only when learning companions are absent that a large gap exists between these student groups. This affective gap reduces when learning companions are present. This result is found for several outcome variables: self-concept, perceptions of learning, frustration, excitement.

However, learning companions did not manage to change some negative feelings and behaviors: low-achieving students did quick-guess more across all conditions than high achieving students; low achievement students reported less interest than high achieving in all conditions. We did see an increase in productive behaviors that lead to learning (Arroyo and Woolf 2005), low-achieving students spent more time in problems where help is requested (i.e. students pay more attention to hints).

General implications for tutors include the possibility of defining features and tool sets that support low-achieving students differentially from the rest. In future studies we will analyze separately the impact of companions on a large population of students with learning disabilities, compared to students without learning disabilities.

10.4.7 Discussion of Gender Studies Results

Overall effects in Tables 10.3 and 10.4, second column, suggest a general advantage of learning companions (both Jane and Jake) for some affective outcomes. Table 10.3 shows that students reported significantly <u>less frustration and more interest</u> (less boredom) when learning companions were used compared to the no learning companion condition. At the same time, Table 10.4 shows that students receiving the female learning companion reported significantly higher self-concept and liking of math at posttest time. Students receiving Jane also reported higher confidence towards problem solving and in post-tutor surveys. One reason why Jake was at a disadvantage compared to Jane might be the fact that the male character was muted twice as much as was the female character. If students mute the characters, then the experimental condition turns out to be highly similar to the control condition (no learning companion) thus diminishing its effect. While significant results are limited to affective outcomes —learning companions did not impact learning—we are impressed given the short exposure of students to the tutoring system.

10.5 Discussion and Future Work

This chapter described both automatic recognition and automatic response to student affect within intelligent tutors and provided examples of such systems that contribute to the growing body of work on affective reasoning for intelligent tutoring systems. This research represents a first step towards a computational theory of affect that can be leveraged to increase student motivation and learning. For example, bringing sensors to our children's schools addresses real problems of students' relationship to mathematics as they learn the subject and supports adaptive feedback based on an individual student's affective states. Within tutor environments, such sensors and animated pedagogical agents have the potential to support students by engaging them through social interaction.

As an example of automatics affect recognition, we described wireless sensors to recognize student emotion, along with a user model framework that predicted emotional self-concept. The framework was used in classrooms of up to 25 students with four sensors per student. By using stepwise regression we isolated key features for predicting user emotional responses to four categories of emotion. This was backed up by cross validation, and shows a small improvement using a very basic classier. This data has demonstrated that intelligent tutoring systems can provide adaptive feedback based on an individual student's affective state. As an example of automatic affect response, we described the evaluation of emotional embodied animated agents and their impact on student motivation and achievement.

There are a number of places for improvement in affective tutors. The effect of specific pedagogical actions on student learning should be investigated and perhaps used to quantitatively gauge the influence of competing tutorial strategies on learning. Additionally, summary information of all sensor values was used in the experiment described above. We may find better results by considering the time series of each of these sensors. In addition, the software in the Mental State Camera can be trained for new mental states or we might look at individual differences in the sensors. Creating a baseline for emotional detection before using the tutor system could help us to better interpret the sensor features.

Emotional predictions from sensors and agents are only a first step towards personalized feedback for students in classroom environments. We propose that tutors will ultimately identify desirable (e.g. flow) and non-desirable (e.g. boredom) student states. Different interventions will be tested in an attempt to keep students in desirable states as long as possible (e.g. a confused student might be invited to slow down, reread the problem and ask for a hint). Part of this approach includes embedding user models into tutors to provide instructional recommendations. Interventions algorithms are being developed based on tutor predictions, e.g. mirror student emotion, support student effort, provide more immediate feedback on student progress, and allow students increased control of their experience.

Modeling gender and student achievement level are potentially powerful as they can enrich the predictive power of the student model and improve teaching power at a very low cost. The importance of including gender and achievement level in a user model is not a mere hypothesis, but is based on extensive research, for examples on gender differences and learning at the K-12 level (Sax 2005; Beal 1994) Some research suggests that girls and boys have different approaches to problem solving (Fenneman et al. 1998; Carr and Jessup 1997) and even that they should be taught differently (Sax 2005). While this literature involves gender differences in the classroom, we have found empirical evidence over the years that gender differences exist when males and females use tutoring systems at the K-12 level (Arroyo and Woolf 2005).

Research on affective tutors may ultimately lead to delicate recommendations about the type of support to provide for individual students. Should male students receive affective support at all? Should all females be provided with learning companions? Should students with learning disabilities use learning companions? These are harder questions to answer from these experimental results. While our results suggest that high school females will affectively benefit more than high school males when exposed to learning companions, we cannot conclude that males in general should not receive affective learning companions. We might suggest that low achieving students (males and females) will highly benefit from affective learning companions. It was only high achieving males who clearly did not benefit from affective learning companions, though our data set is not large enough to provide statistically significant results on the impact of learning companions for a combination of math ability and gender characteristics of students. Further studies with larger number of students might result in more nuanced recommendations about how to modulate the feedback to individualize instruction in affective tutors.

Acknowledgement. This research was funded by two awards (1) National Science Foundation, #0705554, IIS/HCC *Affective Learning Companions: Modeling and supporting emotion during teaching*, Woolf and Burleson (PIs) with Arroyo, Barto, and Fisher; and (2) U.S. Department of Education to Woolf, B. P. (PI) with Arroyo, Maloy and the Center for Applied Special Technology (CAST), *Teaching Every Student: Using Intelligent Tutoring and Universal Design To Customize The Mathematics Curriculum.* Any opinions, findings, conclusions or recommendations expressed in this material are those of the authors and do not necessarily reflect the views of the funding agencies.

References

Arroyo, I., Cooper, D.G., Burleson, W., Woolf, B.P., Muldner, K., Christopherson, R.: Emotion Sensors Go To School. In: International Conference on Artificial Intelligence and Education, pp. 17–24. IOS Press, Amsterdam (2009)

Arroyo, I., Ferguson, K., Johns, J., Dragon, T., Mehranian, H., Fisher, D., Barto, A., Mahadevan, S., Woolf, B.: Repairing Disengagement With Non Invasive Interventions. In: International Conference on Artificial Intelligence in Education. IOS Press, Marina del Rey (2007)

Arroyo, I., Woolf, B., Royer, J.M., Tai, M.: Affective Gendered Learning Companions. In: Inter. Conference on Artificial Intelligence in Education. IOS Press, England (2009)

Arroyo, I., Woolf, B.P.: Inferring learning and attitudes from a Bayesian Network of log file data, pp. 33–40. IOS Press AIED, Amsterdam (2005)

Baker, R.S., Corbett, A.T., Koedinger, K.R., Wagner, A.Z.: Off-Task Behavior in the Cognitive Tutor Classroom: When Students Game The System. In: Proceedings of ACM CHI 2004: Computer-Human Interaction, pp. 383–390 (2004)

Bandura, A.: Self-efficacy: The Exercise of Control. Freeman, New York (1997)

Baylor, A.: The Impact of Pedagogical Agent Image on Affective Outcomes. In: Proceedings of Workshop on Affective Interactions: Computers in the Affective Loop, San Diego (2005)

Beal, C.: Boys and girls: The development of gender roles. McGraw-Hill, New York (1994)

Bickmore, T., Picard, R.W.: Establishing and Maintaining Long-Term Human-Computer Relationships. Transactions on Computer-Human Interaction 12, 293–327 (2004)

Brand, S., Reimer, T., Opwis, K.: How do we learn in a negative mood? Effect of a negative mood on transfer and learning. Learning and Instruction 17, 1–16 (2007)

Burleson, W.: Affective Learning Companions: Strategies for Empathetic Agents with Real-Time Multimodal Affective Sensing to Foster Meta-Cognitive Approaches to Learning, Motivation, and Perseverance. MIT PhD thesis (2006), `http://affect.media.mit.edu/pdfs/06.burleson-phd.pdf` (Acesses May 14, 2010)

Burleson, W., Picard, R.W.: Gender-specific approaches to developing emotionally intelligent learning companions. IEEE Intelligent Systems 22, 62–69 (2007)

Carr, M., Jessup, D.: Gender Differences in First-Grade Mathematics Strategy Use: Social and Metacognitive Influences. Journal of Educational Psychology 89(2), 318–328 (1997)

Chan, T.W., Baskin, A.: Learning companion system. In: Frasson, C., Gauthier, G. (eds.) Intelligent Tutoring Systems. Ablex Publishing, Norwood (1990)

Conati, C., Mclaren, H.: Evaluating A Probabilistic Model of Student Affect. In: Lester, J.C., Vicari, R.M., Paraguaçu, F. (eds.) ITS 2004. LNCS, vol. 3220, pp. 55–66. Springer, Heidelberg (2004)

Cooper, D.G., Arroyo, I., Woolf, B.P., Muldner, K., Burleson, W., Christopherson, R.: Sensors Model Student Self Concept in the Classroom. In: International Conference on User Modeling and Adaptive Presentation, pp. 30–41 (2009)

D'Mello, S.K., Picard, R.W., Graesser, A.C.: Towards an Affect-Sensitive AutoTutor. Special issue on Intelligent Educational Systems IEEE Intelligent Systems 22(4), 53–61 (2007)

D'Mello, S.K., Graesser, A.: Mind and Body: Dialogue and Posture for Affect Detection in Learning Environments. In: Frontiers in Artificial Intelligence and Applications Conference (2007)

Dennerlein, J., Becker, T., Johnson, P., Reynolds, C., Picard, R.W.: Frustrating computer users increases exposure to physical factors. In: Proceedings of International Ergonomics Association, Seoul, Korea, pp. 24–29 (2003)

Dweck, C.: Messages that motivate: How praise molds students' beliefs, motivation, and performance (In Surprising Ways). In: Aronson, J. (ed.) Improving academic achievement. Academic Press, New York (2002)

Efklides, A., Petkakim, C.: Effects of Mood on students' metacogntiive experience. Learning and Instruction 15, 415–431 (2005)

El Kaliouby, R.: Mind-reading Machines: the automated inference of complex mental states from video. PhD thesis, University of Cambridge (2005)

Fennema, E., Carpenter, T., Jacobs, V., Franke, M., Levi, L.: A Longitudinal Study of Gender Differences in Young Children's Mathematical Thinking. Educational Researcher 27(5), 6–11 (1998)

Fletcher, J., Lyon, G., Fuchs, L., Barnes, M.: Learning disabilities: From identification to intervention. NY (2007)

Geary, D., Hoard, M., Byrd-Craven, J., Nugent, L., Numtee, C.: Cognitive Mechanisms Underlying Achievement Deficits in Children with Mathematical Learning Disability. Child Development 78(4), 1343–1359 (2007)

Geary, D., Hoard, M., Hamson, C.: Numerical and arithmeticalcognition: Patterns of functions and deficits in children at risk for a mathematicaldisability. Journal of Experimental ChildPsychology 74, 213–239 (1999)

Goleman, D.: Emotional Intelligence. Bantam Books, New York (1995)

Graesser, A., Chipman, P., King, B., McDaniel, B., D'Mello, S.: Emotions and Learning with AutoTutor. In: Luckin, R., Koedinger, K., Greer, J. (eds.) 13th International Conference on Artificial Intelligence in Education. IOS Press, Amsterdam (2007)

Graham, S., Weiner, B.: Theories and principles of motivation. In: Berliner, D., Calfee, R. (eds.) Handbook of Educational Psychology. Macmillan, New York (1996)

Kapoor, A., Burleson, W., Picard, R.: Automatic prediction of frustration. International Journal of Human-Computer Studies 65(8), 724–736 (2007)

Klein, J., Moon, Y., Picard, R.: This Computer Responds to User Frustration: Theory, Design, Results, and Implications. Interacting with Computers 14(2), 119–140 (2002)

Lepper, M., Woolverton, M., Mumme, D., Gurtner, J.: Motivational techniques of expert human tutors: lessons for the design of computer-based tutors. In: Lajoie, S., Derry, S. (eds.) Computers as Cognitive Tools. Erlbaum, Mahwah (1993)

McQuiggan, S., Lester, J.: Diagnosing Self-Efficacy in Intelligent Tutoring Systems: An Empirical Study. In: Ikeda, M., Ashley, K., Chan, T. (eds.) ITS 2006. LNCS, vol. 4053, pp. 565–574. Springer, Heidelberg (2006)

McQuiggan, S., Mott, B., Lester, J.: Modeling self efficacy in intelligent tutoring systems: An inductive approach. User Modeling and User-Adapted Interaction 18(1), 81–123 (2008)

Mota, S., Picard, R.: Automated posture analysis for detecting learner's interest level. In: Computer Vision and Pattern Recognition Workshop (2003)

NCES, National Center for Educational Statistics, Digest of Educational Statistics, ch. 2 (2009), `http://nces.ed.gov/fastfacts/display.asp?id=64`

Picard, R., Papert, S., Bender, W., Blumberg, B., Breazeal, C., Cavallo, D., Machover, T., Resnick, M., Roy, D., Strohecker, C.: Affective Learning–A Manifesto. BT Journal 2(4), 253–269 (2004)

Picard, R., Scheirer, J.: The galvactivator: A glove that senses and communicates skin conductivity. In: 9th International Conference on Human-Computer Interaction, pp. 1538–1542 (2001)

Prendinger, H., Ishizuka, M.: The Empathic Companion: A Character-Based Interface that Addresses Users' Affective States. Applied Artificial Intelligence 19(3-4), 267–285 (2005)

Qi, Y., Picard, R.: Context-sensitive bayesian classifiers and application to mouse pressure pattern classification. Pattern Recognition 3, 448–451 (2002)

Reeves, B., Nass, C.: The media equation: How people treat computers, television and new media like real people and places. CSLI, New York (1998)

Robertson, J.: Is Attribution Training a Worthwhile Classroom Intervention For K–12 Students with Learning Difficulties? Education Psychology Review 12(1), 111–134 (2000)

Royer, J., Walles, R.: Influences of gender, motivation and socioeconomic status on mathematics performance. In: Berch, D., Mazzocco, M. (eds.) Why is math so hard for some children. Paul H. Brookes Publishing Co., MD (2007)

Sax, L.: Why Gender Matters: What Parents and Teachers Need to Know about the Emerging Science of Sex Differences. Doubleday (2005)

Snow, R., Corno, L., Jackson, D.: Individual differences in affective and cognitive functions. In: Berliner, D., Calfee, R. (eds.) Handbook of Educational Psychology. McMillan, New York (1996)

Strauss, M., Reynolds, C., Hughes, S., Park, K., McDarby, G., Picard, R.: The handwave bluetooth skin conductance sensor. Affective Computing and Intelligent Interaction 2005, 699–706 (2005)

Weiner, B.: An attributional theory of motivation and emotion. Springer, New York (1986)

Bernard, W.: Attribution Theory, Achievement Motivation, and the Educational. Review of Educational Research 42(2), 203–215 (1972)

Wentzel, K., Asher, S.: Academic lives of neglected, rejected, popular, and controversial children. Child Development 66, 754–763 (1995)

Woolf, B., Burleson, W., Arroyo, I., Dragon, T., Picard, R.: Emotional Intelligence for Computer Tutors. In: el Kaliouby, R., Craig, S. (eds.) Special Issue on Modeling and Scaffolding Affective Experiences to Impact Learning, International Journal of Learning Technology (2009)

Woolf, B., Arroyo, I., Muldner, K., Burleson, W., Cooper, D., Dolan, R., Christopherson, R.: The Effect of Motivational Learning Companions on Low-Achieving Students and Students with Learning Disabilities. In: International Conference on Intelligent Tutoring Systems, Pittsburgh (2010)

Zimmerman, B.: Self-Efficacy: An Essential Motive to Learn. Contemporary Educational Psychology 25, 82–91 (2000)

Chapter 11
Ontology-Based Formal Modeling of the Pedagogical World: Tutor Modeling

Riichiro Mizoguchi[1], Yusuke Hayashi[1], and Jacqueline Bourdeau[2]

[1] The Institute of Scientific and Industrial Research (I.S.I.R), Osaka University
8-1, Mihogaoka, Ibaraki, Osaka, Japan, 5670047
{miz,hayashi}@ei.sanken.osaka-u.ac.jp

[2] LICEF Research Center, TÉLUQ-UQAM
100, Sherbrooke Street West, Montréal, QC, Canada H2X 3P2
jacqueline.bourdeau@licef.ca

Abstract. This chapter discusses an ontological approach to tutoring actions design as a special case of target-world modeling. Although a lot of research on the learner model has been done to improve the adaptivity of intelligent tutoring systems (ITSs), the modeling of tutoring actions has not been sufficiently investigated. The authors have been performing ontological modeling of learning/instructional theories to remedy this situation. Intelligent tutoring systems must have a good number of primitive actions to generate intelligent actions. Paying close attention to the importance of modeling tutoring actions, we have developed an ontology of learning/instructional theories, named OMNIBUS, in the ITS domain. Drawing on our long experience in ontological engineering research, this chapter discusses the modeling of tutoring actions as well as target-world modeling per se, using an example of learning/instructional actions from the OMNIBUS/SMARTIES project.

11.1 Introduction

Advanced information processing technology, and especially knowledge processing, has influenced the design of learning support systems. Advances that have played a central role in the Intelligent Tutoring System (ITS) area include applications of artificial intelligence (AI) technology to learner modeling and the generation of tutoring actions that can be adapted to the state of the learner, based on the learner model. In both of these applications, inference techniques are the key technology. However, knowledge processing has not been focused on as the main technology, despite the fact that inference cannot work without knowledge.

Intelligence is attributed to adaptive actions of a system. Such actions can only be generated by a system that has been designed to change its actions adaptively to a situation. Researchers in ITS have placed greater emphasis on building the learner model than on modeling the tutor to perform intelligent tutoring actions.

R. Nkambou et al. (Eds.): Advances in Intelligent Tutoring Systems, SCI 308, pp. 229–247.
springerlink.com

This is natural, considering that no adaptation is possible without knowing the learner's state to which the system must adapt. After considerable research on learner modeling, we now have more results on that aspect than on the modeling of tutoring actions. We believe, as we have been claiming since our vision paper (Mizoguchi and Bourdeau, 2000), that it is time to pay more attention to the importance of modeling tutoring actions.

People usually concentrate on the adaptivity of a tutoring system from the inference point of view—which is not incorrect, as far as it goes. However, researchers need to be aware that a good number of primitive actions must be prepared in advance in order to perform such adaptive actions, because systems cannot generate intelligent actions from nothing. Although it seems obvious, the reality is that quite a few researchers do not take this simple fact into consideration. Our aim is to *model* the target world, including the primitive actions that are expected to be the infrastructure underlying intelligent actions. We believe that what has been missing in ITS research is not only the modeling of tutoring actions but also *target-world modeling* per se.

We have been conducting ontological engineering in the ITS domain for years with such *target-world modeling* in mind. In fact, our claim for the importance of ontological engineering in ITS research grows out of efforts to overcome the difficulties of expert system technology and knowledge engineering, and is in the same vein as the idea of *target-world modeling*. Expert system technology played a role in redirecting AI research away from its preoccupation with form-oriented matters such as inference/reasoning, logic, and knowledge representation for toy problems, toward content-oriented research that deals seriously with real-world problems and requires sophisticated knowledge base (KB) building technology, together with practical inference on the KBs. An AI boom resulted, and a tremendous number of expert systems were built and used in industry. Despite this success, the AI boom has subsided now because of the difficulty of KB maintenance, as well as poor intelligence resulting from a shallow understanding of the domain.

Although expert system technology enables us to develop a KB specially tuned to solve a particular problem, we cannot claim that such a system "understands" the domain and the problem. Knowledge-processing techniques in expert system technology are designed for building KBs that contain rule-coded expert heuristics solely for solving particular problems, without any justification for the individual rules in the KB or any explanation about the nature of the problem and of the domain itself. This is far from the idea of *target-world modeling*.

Ontological engineering has emerged to overcome difficulties like those encountered by expert system technology. Ontological engineering helps people build domain models and expert problem-solving activities within the domain—in the real world. It helps to reveal the essential properties and structure of the target domains and describe them in a manner understandable by computers. This is the theory and technology that can be expected to achieve *target-world modeling*.

This chapter discusses tutoring action design as a special case of *target-world modeling*. Ontological engineering is exploited in order to demonstrate its power to solve this problem, taking our research experience in the OMNIBUS/SMARTIES project as an example. The theoretical details have already been published in

(Hayashi et al., 2009c). Here, we explain the philosophy underlying the modeling of learning actions as well as our policies for ontology building. The next section presents the background of this study, followed by a brief overview of our project in Section 3. Section 4 gives a detailed explanation of the process of ontology building in OMNIBUS, with eight policies. In Section 5, we introduce a next-generation authoring tool named SMARTIES in the context of these policies. Section 6 discusses how SMARTIES works, demonstrating its unique architecture (which is different from that of conventional expert systems), and ends with some concluding remarks.

11.2 Background

The aim of the OMNIBUS/SMARTIES project includes the design of learning and tutoring actions for an intelligent authoring system. OMNIBUS is an ontology of learning and instructional theories, which models learning and tutoring (instructional) actions implicitly or explicitly.

One might think that expert system technology would be a promising approach when someone is asked to build such an intelligent authoring system. He/she would try to interview experts in instructional design and/or learning theories to uncover how they design learning/instructional scenarios and on what model of the learning/instructional world these scenarios are based. After a series of successful interviews, he/she would build a rule base that nicely simulates the behavior of experts in authoring learning/instructional scenarios.

Now, we would like to pose a question about how we can evaluate such a rule-based intelligent authoring system. It will show reasonable performance when the KB is built well. But the issue here is whether we can say that the system is wise enough to understand learning and instruction or not. We believe that the wisdom lies not with the system but still with those experts who gave their knowledge to the system. This is because the system does not "know" what learning/instruction is, and because it is not built based on a good model of learning actions.

Building a truly intelligent authoring system is a really challenging goal, one worthy of the attempt. We have tackled this problem by investigating learning/instructional theories rather than expert heuristics, motivated by the strong belief that ontological engineering is more powerful than expert system technology. We have set the following seven goals for our long-term research project:

1. Building a system that "understands" a good number of learning/instructional theories;
2. Devising a function to support the generation of a learning/instructional scenario justified by learning/instructional theories;
3. Capturing prescriptive aspects of learning/instructional theories in a declarative manner;
4. Avoiding the use of expert heuristics;
5. Avoiding the use of procedural modeling in any circumstances;
6. Being able to explain the theories it has and the rationale for the learning/instructional scenarios it generates;
7. Being able to add new theories easily.

It is apparent that form-oriented AI research would not be useful here. The issue is not the reasoning but the content modeling, which conventional AI research has not yet tackled seriously. However, the situation is improving currently, thanks to the emergence of ontological engineering, which provides theories and techniques to build a fundamental model of the target world. What we must address is the modeling of learning and tutoring/instructional actions, as *target-world modeling*, that is necessary in order for a system to perform inference. Ontology engineering provides us with theories and techniques for modeling the world of interest. Technologically, its central feature is its use of a declarative description to build a model, in a computer-understandable manner, as objectively as possible. This is why we have decided to adopt ontological engineering in our project. However, while ontological engineering thus appears to be a powerful methodology that can be applied to our research, an issue remains: the modeling of prescriptive aspects of instructional theories that relies heavily on the model of learning and tutoring/instructional actions. We have decided not to use procedural modeling. How, then, can we tackle this problem?

11.3 Overview of the OMNIBUS Project

The OMNIBUS project was motivated partially by the desire to build an innovative authoring system aware of learning/instructional theories. It is also expected to be able to produce IMS-LD-compliant learning/instructional scenarios. Named SMARTIES, it satisfies all of these aforementioned requirements as well as being able to connect Learning Objects (LO) available through GLOBE[1] with the generated scenarios.

The heart of SMARTIES is *Instructional_Learning event (I_L event)* decomposition, described in 4.5. OMNIBUS prepares all concepts necessary for performing I_L event decomposition interactively with the author (designer). The prescriptive aspects of learning/instructional theories are organized as unit operations of *I_L event* decomposition. The unit should be understood as a unit of instructional strategy. *I_L event* decomposition is done by proposing candidates for decomposition by the system, and then having authors perform a selection among the candidates. The authors could also choose to input their own strategies, in which case SMARTIES proposes tutoring actions and learner's states as terms it understands by which authors can describe their strategies. This allows SMARTIES to understand the author-input strategies. In selecting among system-provided candidates, authors can blend theories to select a candidate derived from theories other than those that prompted past selections. Note here, however, that justification for such blending is not supported by theories. It must be done at the author's risk.[2]

---

[1] The Global Learning Objects Brokered Exchange (GLOBE), http://www.globe-info.org/

[2] A description of how SMARTIES works can be found at http://www.ei.sanken.osaka-u.ac.jp/pub/miz/AIED07 WS_Invited.pps

11.4 Building the OMNIBUS Ontology

11.4.1 Overview

The following are some of the difficulties that must be overcome in building an ontology of learning/instructional theories:

1. Defining what exactly an ontology of theories is.
2. Reconciling the, long-lasting conflicts among theoretical paradigms: Behaviorism, Cognitivism, Constructivism and Socio-constructivism have been competing for long years. It seems impossible to find a common ground on which to describe them.
3. Modeling the variety of instructional strategies in a declarative form.
4. Selecting terms/concepts to include in the ontology.

As these difficulties suggest, the project was a real challenge. We could say that capturing prescriptive aspects of theories was the hardest part of our whole enterprise. After struggling with this issue, we were fortunately able to come up with a powerful solution. The idea is a good example of knowledge transfer across domains, since the solution was devised by using a technique developed for capturing the functional structure of engineering artifacts (Kitamura, 2004). This technique will be discussed in detail below, because it has a strong potential to become a standard way of capturing human problem-solving actions in general in a declarative manner.

For the moment, let us move on to the discussion of how to build an ontology. Contrary to what most people believe, building an ontology does not mean building an *is-a* hierarchy of the target concept. What an ontology should be is an *is-a* hierarchy of all the related concepts, to capture the target concept. An *is-a* hierarchy of the target concept is just one of them. To put it in terms of our problem, the OMNIBUS ontology should comprise an *is-a* hierarchy not only of theories but also of *actions, states, events* and *strategies.*

The next biggest difficulty is reconciling the learning paradigms. We should not create new theories; we have to respect the existing theories. But this does not mean that we should blindly follow what experts claim. We as ontologists need to keep a domain-neutral attitude so as to capture reality with the highest fidelity, in order to get rid of possible non-moderate views originating in domain-specificity. In addition, we have to be empowered by sound ontological theories.

While the notion of "learning" is not shared by theorists across paradigms, we firmly believe it has to be. This is the starting point and the solid ground on which we can build an ontology of learning and instructional theories. To justify this view, let us consider a situation where we deploy learning/instructional theories in a classroom instruction situation. With what degree of accuracy can we ensure that each learner enjoys the learning conditions a particular theory requires in such a situation? Each learner is in a different state of comprehension, motivation, etc. We have to apply one instructional strategy uniformly to all the learners; that is the reality of doing education in a classroom. This clearly suggests that we have to accept <u>approximation</u> in deploying theories. In other words, when our interest is in how to use theories in real-world education, we can realistically expect to capture them in terms of a less precise vocabulary than those used by theorists to describe them, even though this must be unsatisfactory for the theoreticians.

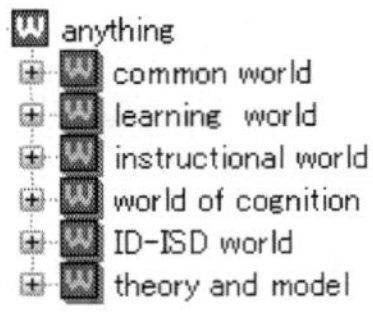

Fig. 11.1 Top-level structure of OMNIBUS

Now, let us move on to the OMNIBUS ontology we have built. Figure 11.1 shows its very top-level concepts: *common world, learning world, instructional world, instructional (system) design (ID-ISD) world, world of cognition* and *theory and model.* Although it is not very compliant with the common upper ontology (BFO,[3] DOLCE[4]), it captures the learning/instructional world. In fact, we encountered the following dilemma here: If we follow the popular upper ontology for the top-level structure, then the major characteristics of the learning/instructional world disappear. If we try to make the target world more visible, then the upper-level categories become less compliant with the popular upper ontologies. We finally adopted our current approach and achieved a reasonable upper-level structure in the *common world*, in which all the common concepts are organized according to the top-level ontology YATO (Mizoguchi, 2009).[5]

Figure 11.2 shows the ontology in a bit more detail. The three worlds of *learning, instructional* and *instructional (system) design* share a common conceptual structure. *Theory and model* is organized as an independent world in which theories and models are organized in several *is-a*

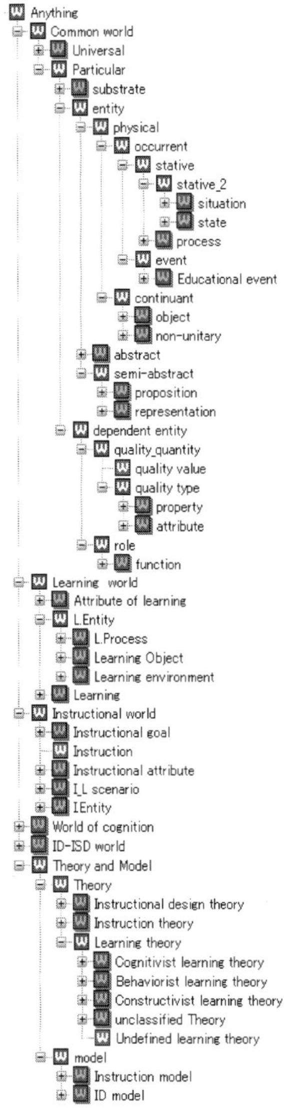

Fig. 11.2 More details of Omnibus

[3] Basic Formal Ontology (BFO), http://www.ifomis.org/bfo/home.

[4] DOLCE: a Descriptive Ontology for Linguistic and Cognitive Engineering, http://www.loa-cnr.it/DOLCE.html

[5] The new updated version is called YAMATO. See: http://www.ei.sanken.osaka-u.ac.jp/hozo/onto library/upperOnto.htm

hierarchies. *Common world* includes action (process), state, event, etc. These three constitute the core of our conceptualization of the learning/instructional world and will be explained in detail in the next subsection

11.4.2 State, Action and Event

Following the conventions of AI, we represent *actions* as *state changes* between the time before and the time after the execution of the action. This is why *states* play the key role in our ontology. *Stative* is a technical term from philosophy, by which we mean "state-like thing". A *process,* which includes *actions* as its subclasses, is a kind of *stative*, but *events* are not. We have to omit the philosophical details, but *process* and *event* are clearly differentiated in philosophy (see Galton and Mizoguchi, 2009, for details). *Events* are composed of *processes* as material. The relation between the two is analogous to that between vase and clay. Like a vase, an event is a unitary whole in the temporal space, whereas a process, like an amount of clay, is not a unitary whole. Furthermore, an *action* usually plays a role in an *event* as a *context*. For example, an utterance can play an explanatory or instructional role in the context of education. This observation suggests that we should model all *processes* (actions) as *state changes* objectively (context-independently), and that they can be used to represent a context as an event.

The next topic is how to model *states*, and, in particular, how many states to model and with what degree of granularity. At first glance, it looks hard to decide the right scope and granularity for state modeling. However, the fact that our goal is to capture theories tells us that we need only model as many states as the theories require, with the requisite granularity. This policy applies to the modeling of actions as well. If we notice we need more states and/or actions when dealing with new theories, we simply add them as needed.

The above discussion can be summarized in two policies, as follows:

Policy 1: All occurrents except events are modeled in terms of state change.

Policy 2: Events are modeled using processes as material. An event is a unitary whole in the temporal space. Actions are modeled in the minimalist way to be used for event description.

State modeling is very important in OMNIBUS for the following two reasons:

1. It provides infrastructure to enable us to model all the theories on a common ground.
2. It allows us to express all phenomena occurring during learning and instructional processes, and hence it guarantees that all the application programs based on OMNIBUS work in a uniform way.

Of course, we are aware that there are quite a few people who oppose these claims. In particular, many of the theorists would like to raise counterarguments such as, "It seems almost impossible for you to implement my theory in a computer without losing the deep understanding about learning that I have put into it." We are ready to agree with such a theoretician. But, we would like to reply, "That is not the issue." What we have been trying to do is not to represent theories as accurately as possible but rather to represent them so that the computer can understand them and use them in a manner as close as possible to the way experts use and apply them in real-world education. In other words, we have been modeling theories not from the viewpoint of "making" them but from that of "using" them. Engineers seek <u>not best but better</u>. Their number one priority is utility rather than perfection. They make steady advances toward improving real life by using a better solution, instead of waiting for the best solution. Now, we are ready to introduce the notion of *"engineering approximation"* for modeling theories, as follows:

> *Policy 3: In order to model theories in a realistic manner, we introduce engineering approximation.*

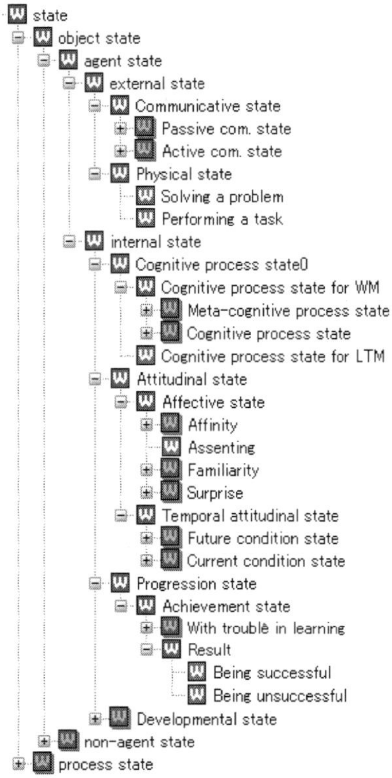

Fig. 11.3 *Is-a* hierarchy of state

Figure 11.3 shows an *is-a* hierarchy of *state*. *Agent state*, which represents the *state* of learner, is divided into internal and external states. The former is further divided into *cognitive process state, attitudinal state* and *progression state* and the latter into *communicative state* and *physical state*. By *communicative state*, we mean states such as *being told, having said* and so forth.

11.4.3 I_L event

In order to capture prescriptive aspects of theories, we need a model of actions, and to capture actions we need a model of states. We now have both actions and states in the ontology. The next issue is how we can build models of theories in terms of actions and states. It is quite possible that such a fine-grained set of primitives would enable us to capture all the possible phenomena occurring in the course of learning/instruction, and hence it would be a good set to use. However, there remains a concern that it might be a bit too fine. If the granularity is finer than necessary, the result will be an overly complicated model. In our case, learning and instructional actions are independent of each other, so they allow us to model every possible interaction between learning and instructional actions, which yields an overly complicated model of the behavior of an authoring system. What we need is a model that can explain learning and instructional theories and use them to help build theory-compliant scenarios. In the light of this goal, the granularity should be fine enough to explain theories and coarse enough to produce easy understandable and manageable behavior on the part of the authoring system.

When we pay attention to the kinds of events, we find two primitive types: *learning event* and *instructional event*. In the former, a learner performs a *learning action* and changes his/her *state*, under some conditions which the learning theories take care of. In the latter type, on the other hand, an instructor performs an instructional action to facilitate a learning action, in order to cause a planned change in the learner. When we view these two kinds of events as a unified entity, we come up with a composite event, which we call *I_L event*, in which "I" stands for Instruction and "L" for Learning. An *I_L event* thus consists of a triple of *<instructional action, learning action, state change>*. If we can represent all the possible event sequences occurring in the course of learning/instruction, then *I_L event* can be thought of as the core of the model of prescriptive aspects of theories. The above discussion is summarized in Policy 4:

Policy 4: Try to find event units with maximal granularity, under the condition of keeping the capability of modeling all the possible phenomena under consideration.

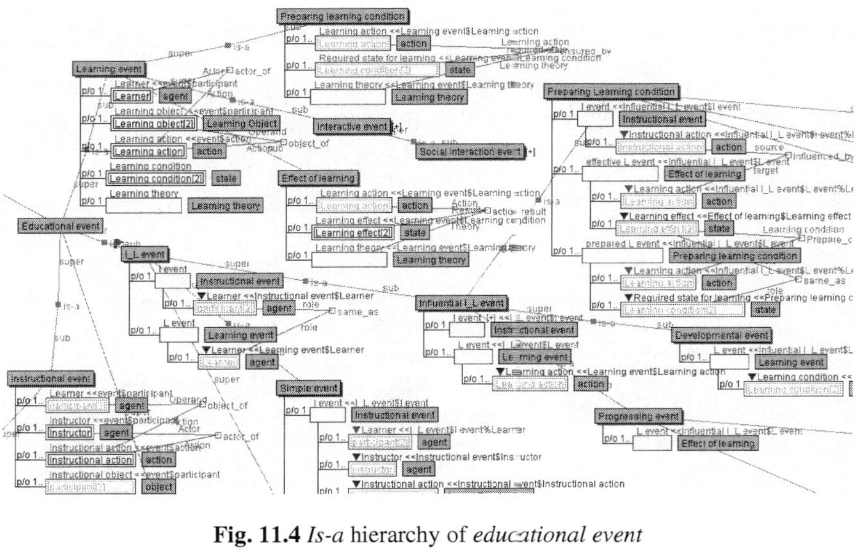

Fig. 11.4 *Is-a* hierarchy of *educational event*

What is important here can be summarized as follows: Learning theories tell us under which conditions learning happens and how learners change (what learning outcome they gain) as a result; instructional theories tell us how to facilitate such learning to maximize the learning outcome; and this nested structure of instructional and learning action is captured by the *I_L event*.

Policy 4 works well for tackling the modeling of procedural knowledge in general. It is apparent that state-based modeling works well when the modeling of actions is the main interest. As we noted above, however, we have to be careful about the granularity issue. It should be neither too small nor too large, but just the right size to represent the phenomena under consideration with maximal understandability.

Educational event is divided into several sub-events, as shown in Figure 11.4, in which definitions of *I_L event* and *learning event* are also shown. *I_L event* is further divided into *simple event, reciprocal I_L event* and *influential I_L event. Preparing learning condition*, a subclass of *influential I_L event,* consists of three slots: *I event, effective L event,* and *prepared L event. I event* influences *learning action*, which causes *learning effect*, which in turn serves as a precondition for the next learning action, defined in the *required state of learning* slot of *prepared L event.*

11.4.4 Function Decomposition

We are now ready to talk about how to model the prescriptive aspects of theories. We believe most people will agree with us that building learning/instructional scenarios is a kind of design process, called Instructional Design or Learning Design. Let us see what is happening in the mechanical design community, because engineering design is the most typical form of design and has a long research history. Artifacts are composed of parts each of which contributes to the function of the whole by performing its own function collaboratively. This applies perfectly to a scenario which is composed of several sub-scenarios which are ordered in a sequence and perform their individual functions to achieve the function of the whole scenario. Bearing this in mind, let us see what has been achieved toward the development of a functional ontology.

One of the authors has been intensively involved for more than 15 years in the development of a functional ontology and its application to representing the functional structure of artifacts (Kitamura et al., 2006). The research objectives include uncovering the essential properties of a function and identifying how many kinds of functions and how many functional concepts exist out there. As a result, we have come up with about 90 concepts and a reference ontology of functions to explain the functional world. Furthermore, the idea of functional decomposition has been proposed to represent the functional structure of any artifact. There are two kinds of decomposition: in terms of what and how, and in terms of granularity. These are discussed below.

11.4.4.1 Decomposition in Terms of What and How

We decided to apply the idea of functional decomposition to our problem. One of the key techniques in the functional ontology is the first type of decomposition, in terms of *what to achieve* and *how to achieve* it. Let us take *to weld* as an example.

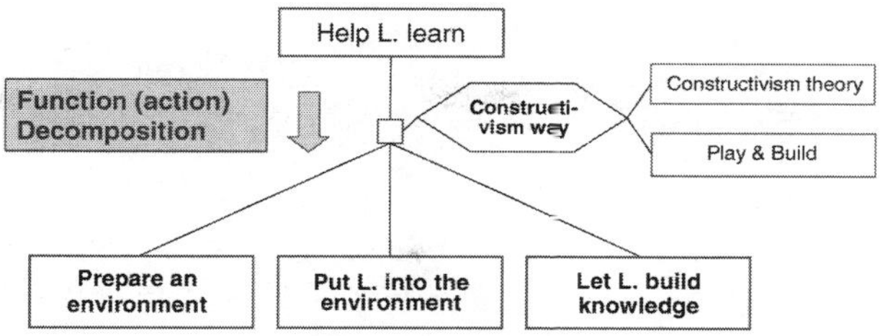

Fig. 11.5 Decomposition of *I_L event*

Although people believe *to weld* is a function, it is not correct from our theory. It is a composite concept of *to join* and *fusion way*. *To join* is achieved by not only *fusion way* but also by *bolt&nut way* or *glue way*. What is important here is that *to join* is a concept, independent of the way it is implemented. It even possesses high domain-independence. We call *to join* a part of *what to achieve* and *fusion way a* part of *how to achieve* it. *To cut* is not a function either. It is a composite of *to separate* and *sharp tool way*. *To wash* is another example that is not a function. It is a composite of *to clean* and *water way*. We have thus defined function as the *what to achieve* part, detaching the *how to achieve* part from the original concept. This decomposition purifies the functional concept and we end up with only about 90 functional concepts.

Conceptualization of *how to achieve* is also of value, and we introduce the notion of *way* for it. This is not a conceptualization of a sequence of sub-functions such as *<put them together, melt, cool>*, the sequence usually considered as an implementation for achieving *to weld.* The conceptualization of this sequence of actions as a "sequence of sub-functions" is not interesting at all because it is still a process. What we have done is different. We conceptualize it as a *principle* of why the sequence of sub-functions makes sense, rather than as a sequence. In terms of ontology, we conceptualize *how to achieve* as a relation between the original functional concept and the derived sequence of sub-functions, e.g., the relation between *to weld* and *<put them together, melt, cool>* in our case.

11.4.4.2 Decomposition in Terms of Granularity

Let us come back to the topic of learning with the notions of function and way in mind. *What to achieve* corresponds to *Help learners learn*, and *how to achieve* corresponds to, say, *constructivism way*, etc. Figure 11.5 shows a diagrammatic explanation of this relation. To achieve the goal by the *constructivism way*, when we read the diagram in a top-down manner, it says "I_L event of *Help learner learn* is decomposed into three sub-events: *prepare an environment, put learner into the environment* and *let learner build knowledge"*. This is the second type of decomposition in terms of granularity, which is explained in 4.6.

This implies that learning/instructional theories are understood as how to achieve, that is, a way for achieving learning goals in our conceptualization. In fact, as Figure 11.5 suggests, the same goal "Help learners learn" can be achieved by the behaviorism way, the cognitivism way, the constructivism way, etc. This is what we want to do with learning/instructional theories, that is, they are modeled in the system as alternatives to one another rather than competing and conflicting with one another. We can summarize the above discussion in Policy 5 below.

Policy 5: In the case of capturing actions and processes, detach how to achieve *from them to obtain* what to achieve, *and organize each separately. The latter should be conceptualized as "purified action" and the former as "way".*

Of course, the quality of the goal achieved is influenced by the *way* which has been applied to it. If two sheets of metal are welded, they cannot be separated, whereas if the bolt&nut way is used, they are detachable. The strength of the joint must be different from the *way* used. This would be the very point that experts are concerned with. That is, the difference would be their justification for loudly proclaiming the distinction between theories. Note, however, that we do not intend to neglect such differentiation between theories/*ways*. What we would like to do is to find a common background for theories so that users have a chance to access theories and compare them in order to choose the best one for their goals. So, we intentionally leave freedom of choice to the users, and the system would propose possible theories usable in the situation the users are in.

11.4.5 Modeling Procedures

Our preparations are now complete and we are ready to introduce our main idea of capturing procedures declaratively. The second type of decomposition in terms of granularity can be repeated until it reaches sub-functions of satisfactorily fine granularity to specify explicitly enough what should be done and in what way. Figure 11.6 depicts an example of two-way decomposition. Note that all four nodes represent *I_L events*, each of which is composed of *instructional action, learning action* and *state change*. The top node to be decomposed says "the instructor wants the learners to recognize what they are going to learn," "the learners recognize it" and the resulting state is "have recognized". There can be alternative ways to achieve the *state change* or perform the function: One is *Way1*, which explains *what to learn* and then how to learn it, and the other is *Way2*, which merely displays examples without any explanation. *Way1* is taken from Gagne and Briggs (Gagne and Briggs 1979) theory and *Way2* from Collins (Collins et al. 1989). In OMNIBUS such *ways* derived from theories are called *way-knowledge*. Because what is obtained by decomposition is also an *I_L event*, the decomposition operation can be carried out further on them. We call the resulting tree the *I_L event decomposition tree*. When you select a *way*, you have alternatives, so the links from a node to *ways* are in *"or"* relation. After the selection, on the other hand, links from the way to nodes are in *"and"* relation, as shown in Figure 11.6. Usually, authors end up with a tree whose *ways* are all determined, so picking up all the leaves (*I_L events* at the bottom) of the tree from left to right will generate an instructional/learning scenario. Furthermore, the upper structure of the tree

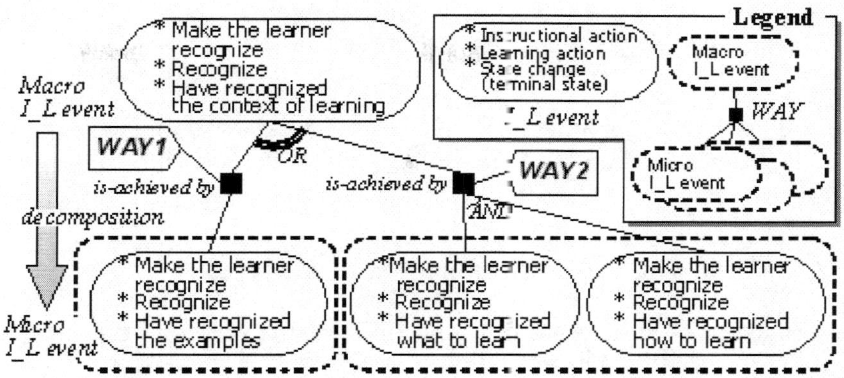

Fig. 11.6 A portion of an I_L event decomposition tree

composed by intermediate nodes will give you the design rationale of the scenario, since each *I_L event* in it represents the goal of the corresponding *actions/changes*.

In functional ontology research, it is well known that accumulation of *ways*, organized in a good structure, is beneficial for understanding a design methodology. In fact, while purified functions embody domain- and implementation-independent goals, *ways* contain rich domain-specific information which is worth analyzing. Ways contain the justification of the decomposition as well as a sequence of sub-actions. In the case of the *I_L event* decomposition tree, each *way* is a compact description of instructional/learning strategy elaborated in theories. Therefore, analyzing *ways* extracted from theories contributes to understanding them from innovative perspectives. In fact, we can organize these pieces of *way-knowledge* in an *is-a* hierarchy and each of them is defined by referring to states and actions which are already defined in the ontology (Hayashi et al., 2009b). Such an organization of *ways* gives us very different perspectives from a taxonomy of theories. What is most interesting is that each of the *ways* is interpreted as a unit of strategy for instructional activity and can be directly used to build learning/instructional scenarios by decomposing the starting *I_L event*.

Imagine a situation where a lot of *ways* are stored and are used to author a learning/instructional scenario. As explained thus far, all the *ways* are described in terms of *I_L events* that are defined by referring to *states* and *actions* defined in the ontology. Therefore, the computer can easily find applicable *ways* to decompose the target *I_L event* by simple pattern-matching, and then it proposes all the candidates to the user. The user selects one he/she likes from among them and decomposes the *I_L event* to get finer-grained *I_L events*. This decomposition process is continued until executable actions are reached, to obtain a learning/instructional scenario. The scenarios thus obtained are necessarily justified by theories, and hence they should be reliable, generic and articulate. The above discussion is summarized in three policies:

Policy 6: Build action-decomposition trees using basic conceptual units obtained following Policy 4, the purified action obtained from Policy 5, and its ways.

Policy 7: Extract ways from sources of procedures, and organize them in an is-a hierarchy.

Policy 8: Constitute a problem-solving engine by considering the extension (decomposition) operation of the I_L event decomposition tree as a kind of inference.

11.4.6 From Theories to Strategies and vice versa

Ways are educational strategies that are derived from theories (theory-based), inspired by practice (empirical) or both. In OMNIBUS, a theory is specified as being a hypothesis that is supported by evidence. Learning theories are attempts to explain the learning phenomenon through a learning mechanism, and consequently through states, actions and events. A general level of mechanism is by paradigm: behaviorists see learning as an association, cognitivists as information processing, constructivists as construction through interaction, and socio-constructivists as social interaction. A more specific level is by theory: Piaget's view is construction by accommodation and assimilation, etc. Educational strategies are derived from these theories (Bourdeau et al., 2007). In OMNIBUS, each *way* has a property called *theory of reference*. This is how SMARTIES allows authors (instructional designers) to make design decisions that are explicitly linked to a theory, and to reflect on each decision they make while building a learning scenario, at the macro- or the micro-level. Several examples have been designed to illustrate the power of the OMNIBUS-SMARTIES system. The example in Figure 11.7 was inspired by Reigeluth's book *Instructional Theories in Action* (Reigeluth, 1987), which provides a variation of scenarios, based on different theories, for the same lesson in optics (concerning lenses and microscopes). In our example, our system allows a designer to select a theory on which to build a complete scenario with fine-grained learning activities, using learning objects such as a virtual microscope and a simulation in optics. The OMNIBUS-SMARTIES system welcomes all theories without any preference. The strength of a theory in education is, as in any domain, its power of explanation and prediction, and the evidence that supports it.

11.5 The Design of a Learning Activity

SMARTIES has been designed and implemented based on OMNIBUS, in line with the aforementioned policies. We will now discuss how these policies work in building SMARTIES. OMNIBUS was designed with the notion of *target-world modeling* in mind. It provides us with all the necessary conceptual building blocks to model learning/instructional activities and captures the prescriptive aspects of theories in the form of *ways,* the *way* being a unit of *I_L event* decomposition.

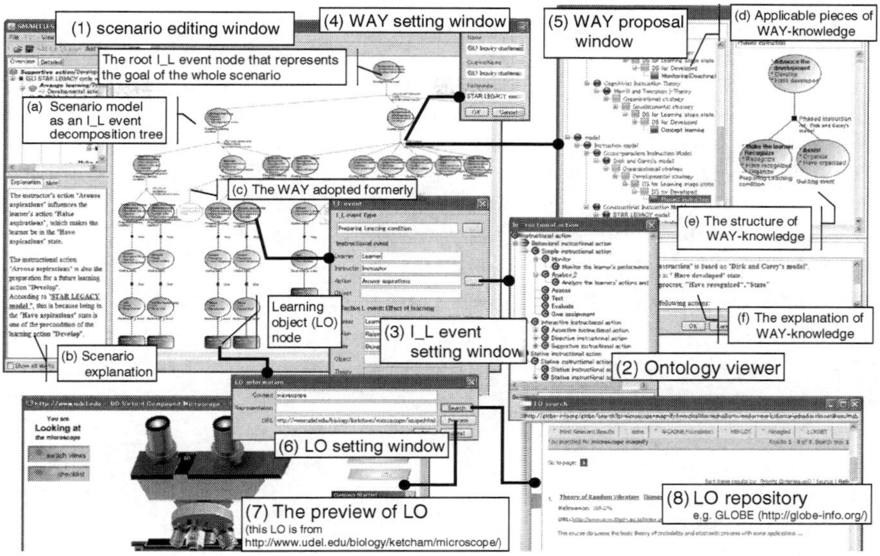

Fig. 11.7 Design of learning/instructional activity in SMARTIES.

Figure 11.7 shows a SMARTIES window configuration. The largest, main window (1) is the *scenario editing window*, in which authors can build a scenario model in terms of an *I_L event* decomposition tree, on the basis of the models of learning/instructional actions in OMNIBUS. SMARTIES can also help users to get leaf *I_L events* connected with LOs.

Following *Policy 1*, authors can define their own *I_L events* and *ways* in terms of *states* and *actions* defined in OMNIBUS (Fig. 11.7 (3), (4)). Then, those *I_L events* are understood by SMARTIES, which can use them in its future behavior. Although authors can define *I_L events* and *ways* in their own terms, SMARTIES cannot utilize them because it cannot understand the meaning of such terms. It can find applicable pieces of *way-knowledge* in the *way-knowledge* base, which stores *ways* defined in accordance with *policies 2* and *3*. Fig. 11.7 (5) is the *way-proposal* window, which shows applicable pieces of *way-knowledge*, sorted in order of types of paradigms, theories and strategies, for decomposing the current *I_L event*. In any situation, authors can query SMARTIES for an explanation of the object they indicate. Thanks to the proposals and explanations given by SMARTIES, authors can easily build theory-guaranteed learning/instructional scenarios as *I_L event* decompositions. Such activity constitutes the design of learning/instructional actions by SMARTIES. When LOs are connected to an *I_L event*, a learning/instructional model has been completed. *Policies 5* to *8* appear to enable SMARTIES's authoring behavior.

11.6 Concluding Remarks

We have thus far discussed the ontological approach to *target-world modeling*, taking an example of learning/instructional actions from the OMNIBUS/SMARTIES project. We summarized our experience in building the OMNIBUS ontology into eight *policies* on the basis of our long experience in ontological engineering research.

As discussed above, the difficulties we had to overcome were reconciling the gap between paradigms and modeling prescriptive aspects of theories. Let us summarize how we coped with these to build SMARTIES/OMNIBUS.

11.6.1 SMARTIES Is Not an Expert System

Viewing SMARTIES in terms of its performance, it looks like an expert system. From the viewpoint of its system architecture, however, it is not. It has no rule base, no inference engine and no heuristics. SMARTIES performs only simple operations, merely reading and writing concepts defined by OMNIBUS, and performing pattern matching of *I_L events* between one to be decomposed in a scenario model and others described as the macro events in the pieces of *way-knowledge*. The pieces of *way-knowledge* that have the macro event agreeing with the *I_L event* to be decomposed are applicable ones, that is to say, applicable learning/instructional strategies from theories.

The reason SMARTIES can provide versatile functions with such simple operation is that it has been designed in accordance with the seven requirements discussed in Section 2. Roughly speaking, OMNIBUS corresponds to the knowledge base of an expert system. However, unlike expert systems which model the problem space according to the heuristics of human experts, OMNIBUS models the problem space by *target-world modeling*, that is, the problem space is modeled as objectively as possible to represent what theories say about the target world. In other words, while expert systems have tried to capture how human experts solve particular problems, OMNIBUS/SMARTIES tries to capture the problem domain as a whole by utilizing the *target-world modeling* strategy, which matches very well with the philosophy of ontology building. Needless to say, there is no idea of ontology building in conventional expert systems, which hide the underlying reasons why their rules make sense and are applicable to the situations. On the other hand, prescriptive aspects of theories are successfully captured in the OMNIBUS ontology and hence, SMARTIES can explain its own behavior and the knowledge used in most cases.

11.6.2 Qualitative Evaluation of the Model of Learning/Instructional Theories

In building OMNIBUS, we did not attempt to model theories with total precision, which is inherently impossible because of the nature of representation in a computer. We aimed, rather, at developing a common background for theories based on *engineering approximation*. Success factors here include employing state-based representation and viewing theories from the standpoint of their use rather than of what they are. Although it is true that theories are very different from paradigm to paradigm, they are captured in OMNIBUS in terms of the learner's state, which is assumed by theories. As a result, we have confirmed that our model of learning/instructional theories has two remarkable characteristics:

(1) Classification of states roughly matches that of theories in terms of paradigm. For example, it is understood that cognitivism theories refer mainly to cognitive states, while constructivism theories refer to states related to meta-cognitive activities. Statistical analysis of the states referred to by *ways* extracted from theories clearly reflects that understanding. That is, references to cognitive states by cognitivist theories represent the highest percentage when compared either with other states referred to by cognitivist theories or with references by other theories to cognitive states. The same applies to constructivism theories and states related to meta-cognitive actions. In summary, we confirmed a high correlation between state classification and theory classification.

(2) Identification of overlapping characteristics among theories: Theories in each paradigm have different features by which they are differentiated. Nevertheless, they are not completely different, since they all are about learning/instruction. In spite of this simple fact, there is no convincing explanation for their degree of similarity to one another What we have found in the detailed analysis of *ways* extracted from theories is the existence of a reasonable amount of overlap among the states referred to by theories in different paradigms. This should be easily inferred from the fact that the percentages of the number of states referred to by constructivism theories are 6.3, 36.7, 41.4, 4.7, 0.8, 11.2 for learning stages, cognitive process state, meta-cognitive process state, attitudinal state, developmental state and external state, respectively. For details, see Table 4 in (Hayashi et al., 2009c). A similar scattering tendency is definitely found for the cognitivism case as well. These facts suggest that theories in different paradigms are not so isolated or conflicting as we would think. They could be even blended if authors wish. At least, theories in the same paradigm have a much higher blending potential. Although blending theories is an interesting option, there is no theory justifying it. Consequently, it must be done at the author's own risk.

11.6.3 Future Work

Analyzing theories in terms of ways and states is interesting because it provides us with a new insight about theories. We have started to develop a support function and environment in SMARTIES for analyzing theories in terms of strategy, action and state. The first results are available at (Hayashi et al. 2009c).

Another thing that needs doing is integration of one-to-one/many tutoring and collaborative learning. OMNIBUS/SMARTIES collects theories essentially for one learner and no theory for collaborative learning is included in it. One of the authors has been involved in ontology building for collaborative learning and its use for group formation and authoring learning materials for collaborative learning (Isotani et al. 2009; Isotani et al. 2010). A good thing is that both ontologies are based on the *I_L event*, so that they semantically interoperate and can be integrated into a unified framework. The key issue here is that interaction in collaborative learning can be viewed as any participant learns through interaction, and that view allows us to see that the other participant who gives a stimulus to the participant who learns is playing the role of "instructor" in the broad sense, whether or not he/she intends to do so. This is the reason why *I_L event* can be the core of an ontology of collaborative learning. The first result of this integration is available at (Hayashi et al., 2009a). We believe that this challenge will help to open up a new world of learning/instructional activity modeling.

References

Bourdeau, J., Mizoguchi, R., Hayashi, Y., Psyche, V., Nkambou, R.: When the Domain of the Ontology is Education. In: Proc. of LORNET 2007, User Centered Knowledge Environments: from theory to practice, The fourth annual LORNET conference I2LOR 2007 of the LORNET Research Network (2007), `http://www.licef.ca/bourdeau`

Collins, A., Brown, J.S., Newman, S.E.: Cognitive apprenticeship: Teaching the crafts of reading, writing & mathematics. In: Resnick, L.B. (ed.) Knowing, learning, & instruction: Essays in honor of Robert Glaser. Lawrence Erlbaum Associates, Hillsdale (1989)

Gagne, R.M., Briggs, L.J.: Principles of Instructional Design, 2nd edn. Holt, Rinehart & Winston, New York (1979)

Galton, A., Mizoguchi, R.: The water falls but the waterfall does not fall: New perspectives on objects, processes and events. Journal of Applied Ontology 4(2), 71–107 (2009)

Hayashi, Y., Bourdeau, J., Mizoguchi, R.: Toward a Learning/Instruction Process Model for Facilitating Instructional Design Cycle. In: Proc. of 9th IFIP World Conference on Computers in Education (WCCE 2009), pp. 138–147 (2009a)

Hayashi, Y., Bourdeau, J., Mizoguchi, R.: Towards Better Understanding of Learning/Instructional Design Knowledge with Strategy-centered Structuring. In: Proc. of ICCE 2009, pp. 91–98 (2009b)

Hayashi, Y., Isotani, S., Bourdeau, J., Mizoguchi, R.: Using Ontological Engineering to Organize Learning/Instructional Theories and Build a Theory-Aware Authoring System. International Journal of Artificial Intelligence in Education 19(2), 211–252 (2009c)

Hozo: ontology editor, `http://www.hozo.jp/`

Isotani, S., Inaba, A., Ikeda, M., Mizoguchi, R.: An Ontology Engineering Approach to the Realization of Theory-Driven Group Formation. International Journal of Computer-Supported Collaborative Learning 4(4), 445–478 (2009)

Isotani, S., Mizoguchi, R., Inaba, A., Ikeda, M.: The foundations of a theory-aware authoring tool for CSCL design. International Journal of Computers and Education 54(4), 809–834 (2010)

Kitamura, Y., Koji, Y., Mizoguchi, R.: An Ontological Model of Device Function: Industrial Deployment and Lessons Learned. Journal of Applied Ontology 1(3-4), 237–262 (2006)

Mizoguchi, R., Bourdeau, J.: Using Ontological Engineering to Overcome Common AI-ED Problems. International Journal of Artificial Intelligence in Education 11(2), 107–121 (2000)

Mizoguchi, R., Hayashi, Y., Bourdeau, J.: Inside Theory-Aware & Standards-Compliant Authoring System. In: Proc. of SWEL 2007, pp. 1–18 (2007)

Mizoguchi, R.: Yet Another Top-level Ontology YATO. In: Proc. of the Second Interdisciplinary Ontology Meeting, pp. 91–101 (2009)

Murray, T., Blessing, S., Ainsworth, S.: Authoring Tools for Advanced Technology Learning Environments: Toward Cost-Effective Adaptive, Interactive & Intelligent Educational Software. Springer, Heidelberg (2003)

Reigeluth, C.: Instructional Design Theories in Action. Erlbaum Associates, Hillsdale (1987)

Chapter 12
Using Drawings in Knowledge Modeling and Simulation for Science Teaching

Wouter R. van Joolingen, Lars Bollen, and Frank A.J. Leenaars

University of Twente, Faculty of Behavioral Sciences, P.O. Box 217, 7500 AE Enschede, The Netherlands
{w.r.vanjoolingen,l.bollen}@utwente.nl,
f.a.j.leenaars@student.utwente.nl

Abstract. Modeling knowledge in simulation-based inquiry learning requires a model of the domain that is executable, as well as a model of the learners' knowledge about the domain. An intermediate level is formed by models of the domain that are created by students as is done in modeling environments. An approach is presented for generating student created models from drawings. This approach requires drawing segmentation, shape recognition and model generation, which is done based on density-based clustering, elementary shape recognition combined with a shape ontology and model fragment composition respectively. The final result is an executable model that can be used to generate simulation outcomes based on learners' conceptions. The role of such a system is discussed, especially with respect to the diagnosis of misconceptions and the generation of tutoring interventions based on confronting learners with the consequences of their conceptions.

12.1 Introduction

In inquiry learning with the help of simulations and modeling, knowledge is modeled at three levels. First there is the level of the *authored simulation* (van Joolingen and de Jong 2003) in the form of an executable model that drives the simulation as a main resource for inquiry. The second level is that of *models created by students*, in the form of concept maps, system dynamics models or stated hypotheses and conclusions (Novak 1998; Penner 2001; Schwarz et al. 2007; Wilensky and Reisman 2006; Wilensky and Resnick 1999). The third level is that of *models the system makes of learners' knowledge*. Although these levels of knowledge modeling serve different purposes and therefore need to satisfy different requirements, they also have much in common as they rely on similar representations representing relations between variables in the domain. In many cases representations at all three levels need to be simulated. At the level of the simulation this is obvious. It has also been known that for the level of learner created models a simulation based on a learner generated model can have a beneficial effect on the learning process (Alessi 2000; Bliss 1994; Ergazaki et al. 2007;

R. Nkambou et al. (Eds.): Advances in Intelligent Tutoring Systems, SCI 308, pp. 249–264.
springerlink.com

Feurzeig and Roberts 1999; Sins et al. 2005). At the level of the model of learners' knowledge simulation of a model can help to identify conflicts between learners' hypotheses and predictions on one side and the model that the learner is studying on the other. Differences in models generate difference in simulation results which are opportunities to confront and discuss with learners in the knowledge building process.

In the current chapter we focus on the middle level, the representation of models by learners. As noted, learners can use many different kinds of representation to express their own models. Some of those representations, such as system dynamics models allow to be simulated right away, but have the drawback of being quite formal and requiring prior knowledge of the variables and relations in the domain, as well as knowledge about the notation and syntax of system dynamics. Other representations such as concept maps can be helpful, but cannot be simulated. Moreover, concept maps are good for building conceptual structures, but they are not really geared towards representing computational operations. Finally, concept maps also impose a kind of formalism on the kind of representations to use by students. Learning environments such as Cool Modes (Bollen et al. 2002) try to combine different visual languages like system dynamics, UML diagrams, freehand drawing etc. in one workspace, but these languages rarely interoperate, and especially freehand-drawings are only integrated on a visual level.

We try to address the problem for representing learners' models by letting learners make drawings representing their understanding of the domain. Using freehand drawings and sketches provides the most representational freedom, but it usually lacks any form of operational semantics. Recent sketch recognition systems try to include this kind of modeling support to drawings, e.g., for drawing logic diagrams (Alvarado and Lazzareschi 2007) or for recognizing mathematical expressions (LaViola and Zeleznik 2004), but they also inherit the limitations and restriction from the domain they are trying to support and from the language they try to recognize.

The approach presented in this paper brings together representational freedom and operational semantics. This approach allows learners to externalize and visualize their ideas on a phenomenon by using freehand drawings, which can be used to intelligently support the creation of a quantitative model by means of segmentation support to recognize coherent components in a drawing, sketch recognition for detecting basic shapes (e.g. arrows, links between components) and labeling to provide a means to the user to identify and tag relevant characteristics and properties of sketch components.

12.2 Modeling with Inaccurate Drawings

In this section we will describe the main properties of the system to generate models from drawings. We start with describing the context and rationale, and proceed with describing the necessary steps to move from drawing to model. In subsequent sections, the first results of implementing the approach will be presented.

When creating a model, many people, including experts, start by making a drawing that is a more or less schematic representation of the system that is being modeled. Drawings help identifying the main components that need to be included in a model and be represented as one or more variables (Van Meter and Garner 2005). Therefore, drawings can form a bridge between initial ideas and a formal model of a system. An example may make this clearer.

Suppose learners study the water cycle, that represents the origin of rainfall (water evaporates from the sea, condensates, flows to land where it forms raindrops and it starts to rain). Rain water comes together in rivers that feed back to the sea. A drawing of such a system could look like Fig. 12.1. Such a drawing by no means qualifies as a computational model as it is inaccurate and ambiguous (for instance there are several arrows that all have different meanings, and important concepts, such as the temperature of the water, are not represented. However, the drawing does represent relevant components of the system – sea, water in various states, wind, the sun as energy source – as well as processes (evaporation, flow, condensation) and could help in arriving at a model such as the system dynamics model represented in Fig. 12.2, that can be used to simulate the water cycle and investigate the influence of parameters such as the intensity of the Sun's radiation. On the other hand, the drawing conveys concepts and details that may not be included in a formal system dynamics model, like spatial relations between components (e.g. mountain, river, sea).

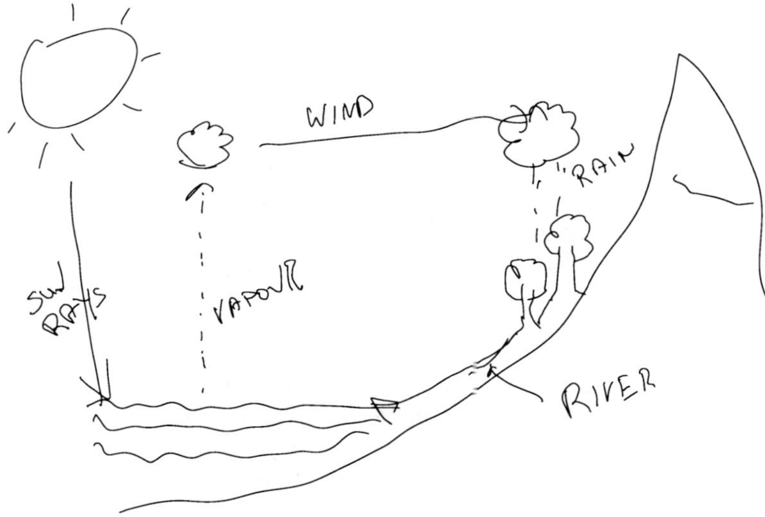

Fig. 12.1 Possible drawing to start modeling the water cycle.

The basic idea of our approach is to support learners in transferring their ideas, as expressed in a drawing, into a formal model; either by translating the drawing into a formal language such as system dynamics, or by adding information to the drawing in such a way that it becomes a computational model.

In the first case, a model such as the one presented in Fig. 12.2 would be created using the drawing as a means of support for creating the model elements, in

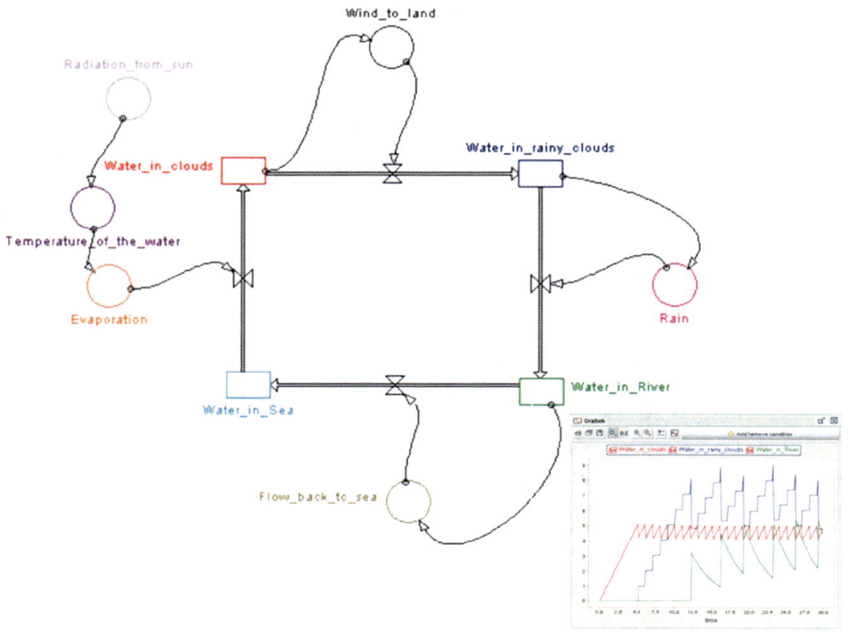

Fig. 12.2 Possible system dynamics model of the water cycle (created with Co-Lab (van Joolingen et al. 2005))

the second case, the drawing itself could be the model, meaning for instance that the drawing elements could be animated, based on the state of the model. For instance, the cloud could grow as more water evaporates and it could move onto the land area.

In order to make this work, the drawing must be used to identify objects, variables and processes that will be components of the model. In our drawing there are objects recognizable as containers of water (the sea, clouds above the sea, clouds above the land, the river). Each of these objects has a pictorial representation. Moreover, there are processes such as heating the sea water, transporting water through wind or flow, and rain. These processes are represented as arrows (such as the wind) or icons (such as the raindrops). An executable model would require formulation in terms of *variables* and *relations* (such as computational functions or differential equations). In the system dynamics formalism, this would come down to *stocks* (variables that represent a state of a container (e.g. the amount of water in a cloud), a *flow*, representing the rate of change of one or two states (e.g. water evaporating from the sea influences the amount of water in the sea as well as the amount of water in the clouds), or an *influence*, such as the temperature of the sea water influencing the evaporation rate.

Understanding learners' drawings can be supported by a system that understands the drawing to a certain extent. For such support to work it is necessary that the system (1) recognizes elements in the drawing, (2) enters a dialogue with the learner about the meaning of these elements and helps formalizing them into a model and (3) supports the simulation based on the model. The purpose of such a support system would be to bring modeling approaches within the reach of students that have not been trained in formal modeling techniques. Also drawings can support beginning modelers in learning such a formalism and potentially also help experienced modelers in dealing with models of complex systems. The next section will elaborate on the details of this support.

12.2.1 Converting a Drawing into a Model

To achieve the kind of support mentioned in the previous chapter, and to integrate sketching and modeling, a number of different approaches are being implemented and evaluated. As we aim for a generic, domain-independent modeling support, there will be no single, ideal solution, but a number of different approaches will act together to provide flexible, yet powerful assistance to the learner. In the following, various approaches are introduced and discussed. Together they form a step-by-step plan to generate a model out of a learner's drawing.

12.2.2 Segmentation and Grouping

To identify distinct objects in a drawing, clustering algorithms or Bayes classification constitute appropriate techniques. Taking into account information on time, location, color and thickness of each stroke, collections of strokes that belong together can be automatically detected.

This can be regarded as a preparatory step and support for the learner to identify relevant objects in a sketch. First experiences, which are described and illustrated in detail in the next chapter, showed that the results of automatic segmentation approaches are promising, but that this approach is also dependent on the learner's individual drawing style and needs to allow for users' interaction and intervention, e.g. by querying for ambiguous parts of a drawing or by providing manual segmentation and grouping features.

12.2.3 Sketch Recognition and Labeling

A number of sketch recognition systems have been developed and are used in learning and teaching settings) (Alvarado and Lazzareschi 2007; Hammond and Davis 2003; LaViola and Zeleznik 2004), and the increased spreading of pen-based devices, like Tablet PCs, PDAs and smartphones, recently promoted their usage. However, the available applications either focus on recognizing elements from specific domains, like logic diagrams (Alvarado and Lazzareschi 2007), letter recognition (Koile et al. 2007) or they aim at the beautification of strokes as in (Paulson and Hammond 2008).

When dealing with arbitrary sketches of varying domains, no currently available approach would be able to recognize elements of domain-independent drawings. For example, the trees, the sea and the clouds in Fig. 12.1 are not identifiable with reasonable effort. Still, sketch recognition approaches as proposed in (Tracy Hammond & Davis, 2003) are expected to be helpful to find basic and typical elements like arrows, geometric shapes, connecting lines, etc. that can be used in combination with grouping and labeling approaches as described in the next sections.

Labeling is a manual way to add descriptive tags to elements of a drawing, as shown in Fig. 12.3 below. Adding labels serves two purposes: (1) It helps a learner to externalize his ideas and to think about the meaning and characteristics of his sketch, and (2) it may be used to automatically deduce initial, draft models from a drawing, using the labels as variables and parameters. A leaner may be instructed to use labels in a specific way, as mentioned above.

12.2.4 Model Generation

Once recognized, either through automatic recognition or manual labeling by the learner, the various components need to be converted into elements of a computational model, and be connected in such a way that an executable model emerges. Such an approach has been already used in the qualitative modeling system GARP (Bouwer and Bredeweg 2001) that employs the idea of connecting *model fragments* into complete models. Using a model fragment for each of the identified sketch components, qualitative or quantitative attributes can be assigned to each of them. For instance, to the component representing the sea in Fig. 12.1, a temperature and an evaporation rate can be assigned. Arrows in the figure can be used to identify the way the fragments need to be linked together.

12.2.5 Integration of Modeling and Drawing

The techniques mentioned in the previous sections aim at facilitating a tight integration of drawing and modeling. Elements from a learner's drawing are used to create an initial model and serve as a starting point for the learner's modeling activities. In reverse, a drawing illustrates a model and is able to provide additional information that cannot be expressed in a formal modeling language. Furthermore, first study results indicate that creating a drawing in addition to modeling or text-writing activities can particularly activate learners' prior knowledge.

Fig. 12.1 and Fig. 12.2 intuitively illustrate this argumentation: The drawing, as an external representation of a learner's knowledge, depicts a complex phenomenon and "tells a story". It could be used in a presentation, for own recollection or to explain the issue to peers. Even more, the drawing contains elements that are relevant for the water cycle and weather in general, e.g. the trees and the mountains, but that are lacking in the model. In the system dynamics model, though, other (quantitative and temporal) aspects are represented, e.g. the simulation and the graphs explain why it is raining periodically and not permanently, as could be guessed from the drawing.

A complementary approach to the one outlined above is CogSketch by Forbus (Forbus & Usher, 2002). In contrast to the approach proposed here, in CogSketch the learner explicitly indicates the individual objects in the drawing by creating so-called "glyphs". A glyph can then be labeled and assigned to pre-defined objects from a given knowledge-base. As there is no sketch-recognition feature in CogSketch, it is possible to have large discrepancies between the sketched glyph and the assigned object, that is the glyph does not have to look like the object represented.

CogSketch, however, provides interesting capabilities in terms of reasoning with spatial relationships between drawing elements. Such relationships can be exploited in identifying relations between objects that need to be included in the model. As a consequence, it is possible to provide tutorial support in the form of prompts based on spatial features of the sketch. In a nutshell, we expect that a strong integration of drawing and modeling is beneficial for prior knowledge activation, knowledge externalization and modeling activities.

12.3 Supportive Technologies and First Results

In the current section, the above mentioned supportive technologies for segmentation, labeling, sketch recognition, and model integration will be picked up again and first results will be described.

12.3.1 Grouping and Labeling

For grouping and labeling we performed an exploratory study to gain insight into the way students represent real world systems as freehand drawings. The study deals specifically with drawings created for the purpose of answering questions about certain variables and relations in the described system. Results of this study lead to the following questions:

- How are described systems commonly represented by students?
- How do students use labels in their drawings and how could these labels be used by the system?
- Can these drawings be automatically segmented?

12.3.1.1 Study Setup

Ten participants used a graphics tablet to create their drawings. They worked in a specifically created sketch collection environment, which integrates drawing, labeling and simulation tools. The simulation tool was based on SimQuest (van Joolingen and de Jong 2003; van Joolingen et al. 1997). Descriptions of real world systems were presented to participants as short case texts along with a simulation that allowed participants to manipulate a number of variables. Extensive logs were kept of all the participants' actions in this application. This environment also offered tutorials to familiarize participants with both the graphics tablet and the software. Figure 12.3 shows a screenshot of this environment.

When participants felt comfortable working in this environment, they were asked to create sketch representations of two real world systems. The first system consisted of a toy car with a small engine that was connected to a table with a rubber band. The second system concerned a house, leaking energy to its environment, heated by a radiator that was controlled by a thermostat. Participants were then asked to add labels to those parts of their drawing that they believed had an effect on a specified variable. An English translation of the first case text is given below. The second case text was similar in length and amount of detail.

"A toy car is connected to a table leg with a rubber band. The car contains a small engine that produces a constant forward force on the car. The engine is switched on and the car starts to move away from the table leg, causing the rubber band to be pulled tight. Because the rubber band pulls the car backwards, the car may start to oscillate, but it is also possible that it slowly comes to a halt without oscillating. Make a sketch of this situation and use labels to identify those parts of your drawing that have an effect on whether or not the car will oscillate. Decide whether the variables mentioned below are relevant, and try to think of as many other relevant variables as possible."

An implementation of the locally scaled density based clustering (LSDBC) algorithm, as described in (Biçici and Yuret 2007), was used to locate clusters in the participants' drawings. Each point in the drawing was defined by its X, Y and time coordinates.

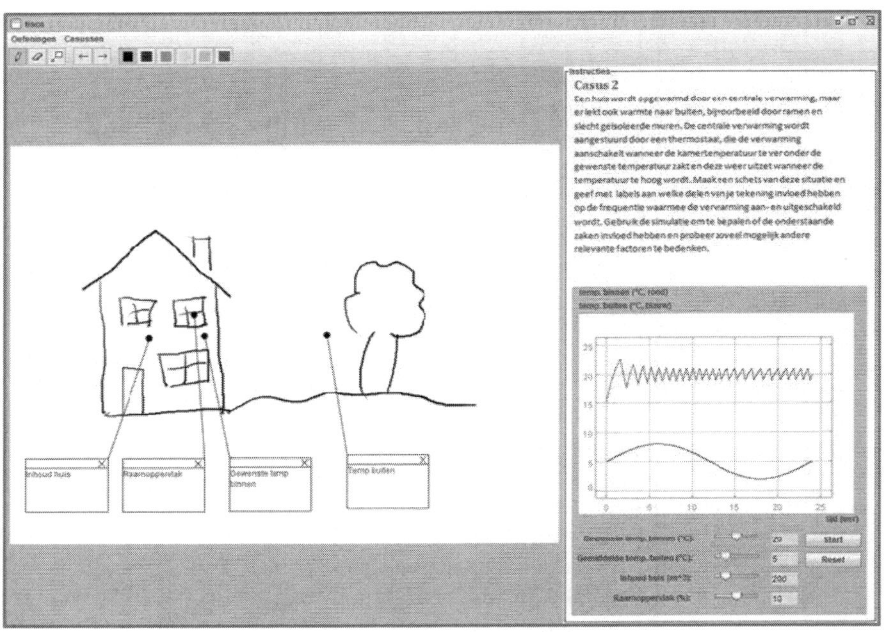

Fig. 12.3 Screenshot of the sketch collection environment. The right side of the picture shows a short case description and a simulation, the left side shows a labeled sketch that was drawn based on this information

12.3.1.2 Results

First results show that although there was quite some variety in the way participants represented the described systems, there were many similarities as well. About half the participants drew a solid house, while the other half drew a house that was 'see-through', so they could show the thermostat and radiator. Another interesting pattern was that all participants represented the table / rubber band / toy car system with the table on the left side and the car driving away to the right. While interesting in itself, this is the kind of result that could be very useful for distinguishing and identifying objects in an automated system.

Participants used labels to indicate which parts of their drawings they believed had an influence on a specified variable. Labels were not used to identify other objects that were drawn. For example the tree in Figure 12.3 is not labeled because it is not believed to influence the temperature in the house, but the window (specifically the surface area of the window) is labeled. During the course of the study participants were asked to give more information in the labels. E.g. instead of just mentioning that the surface area of the windows affects the temperature in the house, participants were later asked to write in more detail about the direction of this effect. When labels are used to identify as many objects as possible in the drawing, this can be of great help during clustering and sketch recognition. During the clustering phase, the labels can serve as seed points for the clustering algorithm and the text in the labels can be scanned for domain specific words to help in the sketch recognition phase. For instance, if a label containing the string 'window' is used to identify a rectangular shape; this is quite strong evidence that this shape in fact represents a window. A drawback of this approach is that it would require domain specific lexicons containing description strings of the objects to be recognized.

Other interesting results have already been found by the clustering algorithm, which was at times able to very accurately detect different parts of the drawing. Figure 12.4 shows the results of applying the LSDBC algorithm to four different drawings of a toy car connected to a table with a rubber band. The clustering algorithm accurately detected different parts of the sketch in all but the bottom-right drawing. The algorithm was able to distinguish the rubber band from the table by using time information. While this leads to good results when the drawing order is table – car – rubber band, it fails when the drawing order is table – rubber band – car, as it was in the bottom-right drawing. The current clustering algorithm could be further improved by using color and stroke information.

Recent developments used the participants' drawings as training data for a Naïve Bayes Kernel Distribution approach, using the RapidMiner / YALE libraries (Mierswa et al. 2006). This approach turned out to be more accurate than the LSDBC clustering approach to identify distinct objects in a sketch.

12.3.2 Sketch Recognition

Once partitioned into separate segments, the objects that have been found need to be identified as components of a model. The approach used for this is based on

Fig. 12.4 Four drawings of a toy car connected to a table by a rubber band. The colors were added by the clustering algorithm, with each color representing a different cluster.

creating a large library of model elements that can cover a number of domains. In the examples listed above, elements could be, for instance, *house*, *car*, or *cloud*. Apart from manual labeling by students, there are two ways in which shapes can be recognized. The first is based on the sketch recognition algorithm as devised by Hammond (Hammond & Davis, 2005; Paulson & Hammond, 2008). This algorithm identifies elementary shape elements such as lines, ellipses and spirals. These elements can be combined into more complex shapes such as squares, arrows and more by specifying rules in the LADDER language ("Language for Describing Drawing, Display, and Editing for use in sketch Recognition"). LADDER rules use concepts such as above/below, perpendicularity of lines, joint points etc. These complex shapes can again be reused to define even more complex shapes. For instance, a *house* can be defined in terms of a *rectangle,* with a *triangle* on top of it. A rectangle in its turn is defined as four lines, that meet in four points under straight angles. As we are dealing with freehand drawings, all rules include an amount of tolerance. However, as a trade-off, the more objects you define, the less tolerant your rules have to be to avoid ambiguity. As a consequence, a drawing has to be very accurate to be still recognizable.

A second approach for sketch identification would be using a database of drawings that are manually classified and, using data mining techniques, match the characteristics of a learner's drawing to that database. For this, basic shapes such as horizontal and vertical strokes, ellipses, etc. still need to be classified, but there is no need for defining the complex shapes. It may be expected that data mining techniques prove less sensitive to variation in details of the drawing (for instance to the exact shape of a cloud) and more tolerant to mistakes than those based on pure shape recognition.

12.3.3 Converting a Drawing into a Model

To achieve the kind of support mentioned in the previous chapter, and to integrate sketching and modeling, a number of different approaches are being implemented and evaluated. As we aim for a generic, domain-independent modeling support, there will be no single, ideal solution, but a number of different approaches will act together to provide flexible, yet powerful assistance to the learner. In the following, various approaches are introduced and discussed. Together they form a step-by-step plan to generate a model out of a learner's drawing.

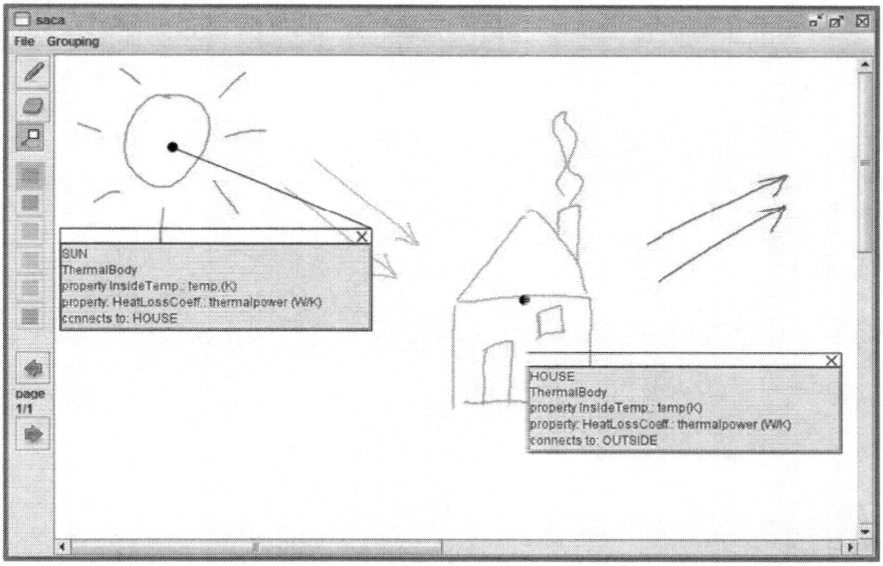

Fig. 12.5 A simple drawing in the domain of heating a house. The coloring originates from automatic grouping, the labels stem from automatic shape recognition

In a trial with data collected from the students in the study described above, a LADDER-based algorithm was capable of positively identifying shapes such as *house*, *sun* and *arrow*. The main addition to the standard LADDER algorithm was that the recognition took place after partitioning the drawing into distinct, smaller objects. The advantage of this approach is that interference of different shapes is avoided, and that parts of the shape (e.g. a window in the house) that are not part of the definition can be included in the shape. This allows for a 'loose' definition of shapes, e.g. the house reducing to just four lines that indicate two walls and a roof. This results in a very accurate recognition of shapes, as is shown in Figure 12.5. The robustness of this approach needs to be tested on larger shape libraries with more shapes.

12.3.4 Model Generation

The final stage in creating an executable model out of the drawing is to link the shapes, identify partial models and to create causal and computational links between the components. In this approach we assume that the system as a whole can be modeled in ordinary differential equations (ODE). ODEs are suitable to model many systems, and modeling systems such as System Dynamics base themselves on ODEs. This approach will not cover all systems (in particular discrete systems such as cellular automata would be excluded) but we! believe similar approaches could be found for such systems as well.

```
ThermalBody
canConnectTo  neighbouringBody : ThermalBody (0..N)
properties
public property       Temperature : temperature (K)
internal property     HeatCapacity : thermalcapacity (J/K)
internal property     ThermalEnery : energy (J)
connection(neighbouringBody) property HeatLossCoefficient :
   thermalpower (W/K)
equations
Temperature = ThermalEnergy / HeatCapacity
Var TotalHeatLoss =
SUM-OVER-ALL(neighbouringBodies) {
    (self<->neighbouringBody).HeatlossCoefficient *
       neighbouringBody.Temperature - Temperature
}
dThermalEnergy/dt = TotalHeatLoss
end ThermalBody
```

Fig. 12.6 Model fragment identifying a thermal body

The basic idea behind the model generation is the use of model fragments and connectors. For the example of the house heating this will be explained. First of all, an object such as *house* as identified by the sketch recognition mechanism is hardly useful for modeling. In order to make it useful, physical properties have to be added. This is done by creating a hierarchical ontology that can classify a house as a *thermal-body*, that associates with a model fragment that identifies the properties of that physical object as well as the connections it can have to other objects. These connections in their turn can have properties of their own. In Fig. 12.6 the model fragment for a thermal body is given. Note statements that specify that a thermal body can connect to any number of other thermal bodies, representing the exchange of heat. This adds flexibility compared to standard functional block approaches in which the number of connections is fixed for each object. Also each connection can have its unique properties, in this case the heat loss coefficient that determines how much heat is transferred over the link per unit of time.

Based on the layout of the drawing, model fragments can be combined. For instance, let us assume that we model a system consisting of a house and its environment. From the drawing a connection between the environment (ENV) and the house (HOUSE) can be inferred. This means that two thermal object model fragments can be instantiated, including a link between them. This instantiation leads to the set of equations that is presented in Fig. 12.7. This is a set of two ODE's and four algebraic equations that can be fed into a simulation engine such as available in – for instance – the SimQuest simulation system (van Joolingen & de Jong, 2003).

```
HOUSE.Temperature = HOUSE.ThermalEnergy / HOUSE.HeatCapacity
HOUSE.TotalHeatLoss = (HOUSE<->ENV).HeatlossCoefficient
    *(ENV.neighbouringBody.Temperature – HOUSE.Temperature)
dHOUSE.ThermalEbergy/dt = HOUSE.TotalHeatLoss
ENV.InsideTemperature = HOUSE.ThermalEnergy / HOUSE.HeatCapacity
ENV.TotalHeatLoss = (ENV<->HOUSE).HeatlossCoefficient *
    (HOUSE.neighbouringBody.Temperature - ENV.Temperature)
dENV.ThermalEbergy/dt = ENV.TotalHeatLoss
```

Fig. 12.7 Set of equations that can be generated from a drawing

This – relatively simple – example shows how the final step from drawing to model can be made. For more complex systems, of course the number of elements and links will grow, but given the complexity of an average drawing, we do not expect computational problems.

12.4 Conclusion and Outlook

In the preceding sections we presented a means of deriving a computational model from a more or less systematic drawing made by a learner. The foreseen benefits are twofold, aiming at the different levels of knowledge modeling addressed in this paper. For the level in which the learner models his own knowledge, generating a simulation based on that knowledge can have a positive effect on the learning process. Learners will be confronted with the consequences of their own ideas, and will be able to adapt these ideas based on this confrontation.

The approach we sketch here extends the possibility of generating computational models beyond using standard modeling languages such as system dynamics, LOGO or NetLogo. Instead of asking learners to learn a representational formalism, we create a system that generates formal models from the representations that learners create spontaneously. The advantages are that (1) in this way the idea of modeling by learners can have a wider application beyond that within dedicated modeling systems and (2) the system can make more extended use of the information it retrieves from learner generated data such as diagrams and models.

An example of the latter would be that the system simulates a learner's model and compares the results with the data that learners obtain from the simulation.

Presenting both simulation results and indicating the differences can result in a dialogue with the learner. Agents in the learning environment can use this information to adapt the learning environment, and for instance suggest experiments that provoke thought on their misconceptions, in the line of Posner and colleagues (Posner et al. 1982). This can best be illustrated by an example. A common misconception in astronomy is that the seasons are caused by the earth being closer to the sun in summers than in winters. This a misconception that can easily be picked up from a drawing. Agents can then respond in several ways:

- Tell the learner that the reality is different.
- Ask questions about the misconception and especially what this would mean for the times of seasons for the northern and southern hemispheres.
- Generate a simulation and run it to show that this would also mean that the summer would become shorter and that the temperature on the whole earth would rise and fall.

The option that would be chosen would be dependent on the pedagogical strategy, and on the level of the learner, but it is clear how the sketch-based modeling would enlarge the repertoire of ITSs, by extending the possibilities for diagnosis and feedback, including simulation-based feedback.

The work presented here is in progress. The segmentation part is up to a level that it can be used in practice. The recognition step can be covered for simple shapes, but there is a need for building and analyzing a large corpus of drawings to extract rules for more complex shapes than houses and suns. Once the corpus has been built, data mining techniques can be used to create and test classification rules. In the long run this should lead to a large library and ontology of learner generated shapes that can be used for identification of drawings, and become a learning system when it is also fed with the results of manual labeling by learners and experts.

The model generation part, finally, is rather straightforward once the shapes and their relations have been identified. Equations like the ones generated in the example presented above can be processed with existing simulation engines such as those available for SimQuest (van Joolingen and de Jong 2003). However, identifying relations is tricky, and strongly dependent on the way learners represent them. So far we need to rely on pre-stored relations and drawn arrows by learners.

In the long run, the outlook for drawing based modeling systems is that they can form an integrated part of inquiry-based environments that offer intelligent support for the learner. A traditional issue in Intelligent Tutoring Systems, including those based on inquiry learning, such as Co-Lab (van Joolingen et al. 2005) and SCY (de Jong et al. in press) is that of estimating the learners' knowledge level. The way we use drawings as described in this chapter provides a natural input for the knowledge modeling systems Drawings can form an excellent means of sharing and communicating knowledge, with fellow learners as well as with the tutoring system. If drawings can be augmented by a simulation, the learning environment can become a partner in the learning process, by detecting and acting upon misconceptions, similarities and differences between drawings by several learners. As such we believe that drawing-based modeling will provide a valuable addition to the repertoire of ITS design.

References

Alessi, S.M.: The Application of System Dynamics Modeling in Elementary and Secondary School Curricula. Paper presented at the Conference of the Red Iberoamericana de Informática Educativa, RIBIE, Vina del Mar, Chile (2000)

Alvarado, C., Lazzareschi, M.: Properties of Real-World Digital Logic Diagrams. Paper presented at the Proceedings of 1st International Workshop on Pen-based Learning Technologies, PLT 2007, Catania, Italy (2007)

Biçici, E., Yuret, D.: Locally Scaled Density Based Clustering. In: Beliczynski, B., Dzielinski, A., Iwanowski, M., Ribeiro, B. (eds.) ICANNGA 2007. LNCS, vol. 4431, pp. 739–748. Springer, Heidelberg (2007)

Bliss, J.: From mental models to modelling. In: Mellar, H., Bliss, J., Boohan, R., Ogborn, J., Tompsett, C. (eds.) Learning with Artificial Worlds: Computer Based Modelling in the Curriculum. The Falmer Press, London (1994)

Bollen, L., Hoppe, H.U., Milrad, M., Pinkwart, N. Collaborative Modeling in Group Learning Environments. Paper presented at the Proceedings of the XXth International Conference of the System Dynamics Society, Palermoy, Italy (2002)

Bouwer, A., Bredeweg, B.: VisiGarp: Graphical Representation of Qualitative Simulation Models. Paper presented at the Proceedings of 10th International Conference on Artificial Intelligence in Education, AI-ED 2001, San Antonio, Texas (2001)

de Jong, T., van Joolingen, W.R., Anjewierden, A., Bollen, L., d'Ham, C., Dolonen, J., et al.: Learning by creating and exchanging objects: the SCY experience. British Journal of Educational Technology (in press)

Ergazaki, M., Zogza, V., Komis, V.: Analysing Students' Shared Activity while Modeling a Biological Process in a Computer-Supported Educational Environment. Journal of Computer Assisted Learning 23(2), 158–168 (2007)

Feurzeig, W., Roberts, N.: Modeling and simulation in science and mathematics. Springer, New York (1999)

Forbus, K.D., Usher, J.: Sketching for Knowledge Capture: A Progress Report. Paper presented at the IU 2002, San Francisco (2002)

Hammond, T., Davis, R.: LADDER: A Language to Describe Drawing, Display, and Editing in Sketch Recognition. Paper presented at the Proceedings of International Joint Conference on Artificial Intelligence, Hyderabad, India (2003)

Hammond, T., Davis, R.: Ladder, a Sketching Language for User Interface Developers. Computers & Graphics 29, 518–532 (2005)

Koile, K., Chevalier, K., Low, C., Pal, S., Rogal, A. Singer, D., et al.: Supporting Pen-Based Classroom Interaction: New Findings and Functionality for Classroom Learning Partner. Paper presented at the 1st International Workshop on Pen-Based Learning Technologies, PLT 2007, Catania, Italy (2007)

LaViola, J., Zeleznik, R.: MathPad: A System for the Creation and Exploration of Mathematical Sketches. ACM Transactions on Graphics (Proc. SIGGRAPH) 23(3) (2004)

Mierswa, I., Wurst, M., Klinkenberg, R., Scholz, M., Euler, T.: YALE: Rapid Prototyping for Complex Data Mining Tasks. Paper presented at the 12th ACM SIGKDD International Conference on Knowledge Discovery and Data Mining, KDD 2006 (2006)

Novak, J.D.: Learning, creating and using knowledge: Concept map TM as facilitative tools in schools and corporations. Lawrence Erlbaum Associates Inc., Mahwah (1998)

Paulson, B., Hammond, T.: Accurate Primitive Sketch Recognition and Beautification. Paper presented at the Proceedings of the International Conference on Intelligent User Interfaces, IUI 2008, Canary Islands, Spain (2008)

Penner, D.E.: Cognition, Computers, and Synthetic Science: Building knowledge and meaning through modelling. Review of Research in Education 25, 1–37 (2001)

Posner, G.J., Strike, K.A., Hewson, P.J., Gertzog, W.A.: Accomodation of a scientific conception: Towards a theory of conceptual change. Science Education 66(2), 211–227 (1982)

Schwarz, C.V., Meyer, J., Sharma, A.: Technology, Pedagogy, and Epistemology: Opportunities and Challenges of Using Computer Modeling and Simulation Tools in Elementary Science Methods. Journal of Science Teacher Education 18(2), 243–269 (2007)

Sins, P.H.M., Savelsbergh, E.R., van Joolingen, W.R.: The Difficult Process of Scientific Modelling: An analysis of novices' reasoning during computer-based modelling. International Journal of Science Education 27(14), 1695–1721 (2005)

van Joolingen, W.R., de Jong, T.: SimQuest: Authoring educational simulations. In: Murray, T., Blessing, S., Ainsworth, S. (eds.) Authoring tools for advanced technology educational software: Toward cost-effective production of adaptive, interactive, and intelligent educational software. Kluwer Academic Publishers, Dordrecht (2003)

van Joolingen, W.R., de Jong, T., Lazonder, A.W., Savelsbergh, E.R., Manlove, S.: Co-Lab: research and development of an online learning environment for collaborative scientific discovery learning. Computers in Human Behavior 21(4), 671–688 (2005)

van Joolingen, W.R., King, S., de Jong, T.: The SimQuest Authoring System for Simulation-Based Discovery Learning. Paper presented at the Proceedings of the International Conference on Artificial Intelligence and Education, AIED 1997 (1997)

Van Meter, P., Garner, J.: The Promise and Practice of Learner-Generated Drawing: Literature Review and Synthesis. Educational Psychology Review 17(4), 285–325 (2005)

Wilensky, U., Reisman, K.: Thinking Like a Wolf, a Sheep, or a Firefly: Learning Biology Through Constructing and Testing Computational Theories — An Embodied Modeling Approach. Cognition and Instruction 24(2), 171–209 (2006)

Wilensky, U., Resnick, M.: Thinking in Levels: A Dynamic Systems Approach to Making Sense of the World. Journal of Science Education and Technology 8(1), 3–19 (1999)

Part III
Student Modeling

Chapter 13
Student Modeling

Beverly Park Woolf

Center for Knowledge Communication, 140 Governor's Drive,
University of Massachusetts Amherst, MA 01003-4610
bev@cs.umass.edu

Abstract. This chapter describes how to build student models for intelligent tutors and indicates how knowledge is represented, updated, and used to improve tutor performance. It provides examples of how to represent domain content and describes evaluation methodologies. Several future scenarios for student models are discussed. For example, we envision that student models will support assessment for both formative issues (the degree to which the student has learned how to learn – for the purposes of improving learning capacity and effectiveness) and summative considerations (what is learned– for purposes of accountability and promotion). We envision that student models will track when and how skills were learned and what pedagogies worked best for each learner. Moreover, they will include information on the cultural preferences of learners, their personal interests, learning goals, and personal characteristics. Ultimately, student model servers will separate student models from tutors and will be a part of wide area networks, serving more than one application instance at a time.

13.1 Introduction

Student models in intelligent tutoring systems represent student competencies and learning achievements. Modeling may involve techniques to represent content skills (e.g., mathematics, art history), knowledge about learning (e.g., metacognitive knowledge), and affective characteristics (e.g., emotional state). Although students' general knowledge might be determined quickly from quiz results, their learning style, attitudes, and emotions are less easily determined and need to be inferred from long-term observations. Models may be used for assessment by measuring changes in the student in any or all three of these areas. Student models generally represent inferences about users (e.g. their level of knowledge, misconceptions, goals, plans, preferences, beliefs), relevant characteristics of users (stereotypes) and users' records, particularly past interactions with the system.

A student model in an intelligent tutor observes student behavior and creates a qualitative representation of her cognitive and affective knowledge. This model partially accounts for student performance (time on task, observed errors) and reasons about adjusting feedback to the student. By itself, the student model achieves very little; its purpose is to provide knowledge that is used to determine the

R. Nkambou et al. (Eds.): Advances in Intelligent Tutoring Systems, SCI 308, pp. 267–279.
springerlink.com

conditions for adjusting feedback and it supplies data to other tutor modules, particularly the teaching module. One long-term goal of the field of AI and Education is to support learning for students with a range of abilities, disabilities, interests, backgrounds, and other characteristics (Shute et al. 2005).

This chapter describes how to build student models for intelligent tutors and indicates how knowledge is represented, updated, and used to improve tutor performance. The chapter first clarifies how to build a student model and then positions student models in the context of tutor system research. The chapter concludes with a series of open questions about the future of student models.

13.2 Motivation for Building Student Models

Human teachers learn about students through years of experience. Master teachers often use secondary learning features, (e.g., a student's facial expressions, body language, and tone of voice) to augment their understanding of students' learning. They also learn which pedagogical strategies work best with which students. Classroom teachers support student learning in many ways, e.g., by patiently repeating material, recognizing misunderstandings, and adapting feedback. Teachers adjust teaching strategies and customize their responses to an individual's learning needs. Interactions between students and teachers provide critical data about student goals, skills, motivation, and interests. Similarly, intelligent tutors make inferences about presumed student knowledge and store them the student model. A primary reason to build a student model is to ensure that the system has principled knowledge about each student so it can respond effectively, engage students ' interest and promote learning. Customized feedback is pivotal to producing learning. Instruction tailored to students' preferred learning style increases their interest in learning and enhances learning, in part, because tutors can support weak students' knowledge and develop strong students' strengths. Master human teachers are particularly astute at adapting material to students' cognitive and motivational characteristics. In mathematics, for example, using more effective supplemental material strongly affects learning at the critical transition from arithmetic to algebra and achievement of traditionally underperforming students (Beal 1994). Students show a surprising variety of preferred media; given a choice, they select many approaches to learning (Yacci 1994). Certain personal characteristics (gender and spatial ability) are known to correlate with learning indicators such as mathematics achievement (Arroyo et al. 2004) and learning methods (Burleson 2006). Characteristics such as proficiency with abstract reasoning also predict responses to different interventions. Thus, adding more detailed student models of cognitive characteristics may greatly increase tutor effectiveness.

Student models typically represent student behavior, which includes student answers (to questions or problems), actions (writing a program), results of actions (written programs), intermediate results (scratch work), and verbal protocols. Student behavior is assumed to reflect student knowledge as well as common misconceptions. Student models are typically qualitative (neither numeric nor physical); they describe objects and processes in terms of spatial, temporal, or causal relations (Clancey 1986; Sison and Shimura 1998). These models are also approximate and possibly partial (not fully accounting for all aspects of student behavior). In other words, tutor development focuses on computational utility rather than on cognitive fidelity (Self 1994). A more accurate or complete student model is not necessarily better, because the computational effort needed to improve accuracy or completeness might not be justified by any extra pedagogical leverage obtained. Two issues need to be considered when building student models: representing and updating student knowledge.

13.2.1 Representing Student Knowledge

Resenting student knowledge takes many forms, from simple numeric rankings about student mastery to complex plans or networks explaining student knowledge (Brusilovsky 1994; Eliot 1996) and the models may encode many types of knowledge (topics, misconceptions and bugs, affective characteristics, student experience, and stereotypes). Knowledge representation is foundational to artificial intelligence. It is a methodology for encoding concepts (e.g., objects, procedures) within a computer and providing efficient operations on these concepts so computers can reason about concepts and begin to appear intelligent (Brachman and Levesque 2004). A closely related knowledge category is misconceptions, which includes well-understood errors, or incorrect or inconsistent facts, procedures, concepts, principles, schemata, or strategies that result in behavioral errors (Sison and Shimura 1998). Not every error in student behavior is due to incorrect or inconsistent knowledge; behavioral errors can also result from insufficient knowledge.

Affective characteristics, e.g., student emotions and attitudes, are also represented in student models (see chapters 10 and 17 in this book). Confusion, frustration, excitement, boredom also motivation, self-confidence, and fatigue have been represented. Affective computing typically involves emotion detection or measuring student emotion, using both hardware (pressure mouse, face recognition camera, and posture sensing devices) and software technology (e.g., machine learning), and then providing interventions to address negative affect. Stereotypes have been used, specifically collections of default characteristics about groups of students that satisfy the most typical description of a student from a particular class or group (Kay 1994). For example, default characteristics may include physical traits, social background, or computer experience. Stereotypes may represent naïve, intermediate, and expert students (Rich 1983).

13.2.2 Updating Student Knowledge

Updating student knowledge is the second issue to consider in building student models. Updating is used to infer the student's current knowledge and frequently rules are used to compare the student's answers with comparable expert answers or sequences of actions. Student knowledge, as initially represented in the student model, is not usually equivalent to knowledge of the domain as represented in the domain model. The hope is that students' knowledge improves from that of a naïve student toward that of an expert over several sessions. Conceptually these two data structures are distinct, but practically they may be very similar. Student models typically miss some knowledge contained in domain models or have additional knowledge in terms of misconceptions. Comparison methods are used to update knowledge in the two models, assuming an overlay model. In some cases, the tutor generates models of faulty behavior by using slightly changed rules (malrules) to reproduce the results produced by a student with misconceptions. Student knowledge can be updated by plan recognition or machine learning techniques, which uses data from the problem domain and algorithms to solve problems given to the student. Plan recognition might be used to determine the task on which a student is currently working, for example by predicting student behavior, refined stereotypes, and using plan recognition techniques to recognize which planning behaviors were relevant for updating the student model. If the student is pursuing plans or a recognizable set of tasks, plan recognition techniques construct the student model and compare student behavior to expert procedures to indicate on which plan the student is working. For example, the Andes tutor used updated Bayesian belief networks to infer which new topics the student might know but had not yet demonstrated (see chapter 14 in this part of the book).

13.3 Alternative Methods for Building Student Models

Knowledge in student models can take many forms, from simple numeric rankings about student mastery to complex plans or networks. Different techniques work better or worse for different academic disciplines. Given this variety of knowledge, many techniques are needed to update student knowledge. This section describes techniques based on either cognitive science or artificial intelligence methods used to represent student knowledge. Cognitive science techniques include model-tracing and constraint-based methods, where as AI techniques include formal logic, expert systems, plan recognition and Bayesian belief networks. This classification is not meant to be exclusionary; techniques from one category might be used in conjunction with those from the other (e.g., adding a Bayesian belief network to a model-tracing tutor).

In some cases, a student model can be quite different from the domain model based on topics. For example, affective knowledge of a student (engagement, frustration and boredom) is independent of the domain knowledge and is typically inferred and recorded in a student model. Alternatively, a procedural model might use a specially developed expert model and compare the expert solution at a finer level of granularity, at the level of subtopic or subgoals. Such models have stronger diagnostic capabilities than overlay models. Procedural models overlap with generative bug models when they use algorithms divided into stand-alone portions, corresponding to pieces of knowledge that might be performed by students (Self 1994).

Student knowledge can be updated by *plan recognition* or *machine learning* techniques, which use data from the problem domain and algorithms to solve problems given to the student. Analysis involves structuring the problem into actions to be considered. Plan recognition might be used to determine the task on which a student is currently working. If the student is pursuing plans or a recognizable set of tasks, plan recognition techniques construct the student model and compare student behavior to expert procedures to indicate on which plan the student is working. The Andes tutor used updated Bayesian belief networks to infer which new topics the student might know but had not yet demonstrated (ref). The student model in Wayang Outpost used a Bayesian belief network to infer hidden affective characteristics (Woolf 2008). Open Student Models reflect the student's right to inspect and control the student model and participate in its creation and management. The aim of open modeling (also called overt, inspectable, participative, cooperative, collaborative, open learner, and learner-controlled modeling) is to improve the student modeling enterprise. Open user model refers to the full set of tutor beliefs about the user, including modeling student knowledge as well as preferences and other attributes. One aim of such a model is to prompt students to reflect on their knowledge (including lack of knowledge and misconceptions) and to encourage them to take greater responsibility for their learning. Learners are believed to enjoy comparing their knowledge to that of their peers or to the instructor's expectations for the current stage of their course. See chapter 15 in this part of the book for more details about open learner models.

13.3.1 Cognitive Science Methods

Two cognitive science techniques, model-tracing and constraint satisfaction, are briefly described in this section. The reader can found more details in chapters 3 and 4 of this book. Each technique is based on viewing learning as a computational process. For example, model-tracing tutors (or Cognitive Tutors) are grounded in cognitive psychology theory based on ACT-R cognitive model and assumes that human learning processes can be modeled by methods similar to information processing, e.g., rules or topics that will be learned by students. These tutors provide an underlying model of the domain to interpret students' actions and follow their solution path through the problem space. Model tracing assumes that student's actions—e.g., steps in solving mathematics problems, can be identified and explicitly coded through topics, steps, or rules (if-then rules in the case of a cognitive tutor) (Anderson and Reiser 1985). The student model uses these rules or steps to represent the knowledge. The tutor then traces students' implicit execution of these rules, assuming that students' mental model state (or knowledge level) is available from their actions (Anderson et al. 1987). The tutor is mostly silent, working in the background, yet when help is needed, it reasons about student knowledge and infers whether they traveled down a path encoded by the production rules. Comparison of student actions with execution by the domain model yields error diagnoses. After a student's action, the tutor might suggest which rule or set of rules the student used to solve the problem. The Andes physics student model is an example of a system that modeled students' steps while solving physics problems, was on the edge between simple and complex knowledge (references). It required a separate model for each physics problem (several hundred problems over two semesters) and inferred a new Bayesian network for each student. Building such models required a great deal of work, even with mature authoring tools.

The operative principle for model-tracing tutors is that humans (while learning) and student models (while processing student actions) are input-output equivalents of similar processes (i.e., both humans and the model have functionally identical architectures). Cognitive methods place a premium on the empirical fit of student actions to psychological data. At each opportunity for students to employ a cognitive rule, simple learning and performance assumptions are employed (encoded as production rules) to update an estimate of the probability that the student has learned the rule (Corbett and Anderson 1995). Cognitive tutors for algebra and geometry tutors have had a large impact on educational practice and are used by more than 500,000 students in over 1400 school districts across the U.S.

Whereas model-tracing assumes that human learning processes can be accurately modeled by computer techniques, constraint-based modeling (CBM) assumes the opposite. CBM methods are based on the assumption that learning cannot be fully recorded and only errors (breaking constraints) can be recognized by a computational system. For example, constraints in the field of adding fractions might check that all denominators in the problem and solution are equal before a student adds fractions. Thus the tutor checks that in adding fractions, students submit answers where the numerator equals the sum of operand numerators, and the constraint is satisfied when all denominators (a, d, and n) are equal. CBM methods are particularly powerful for intractable domains, in which students' knowledge cannot be exactly articulated, student approaches cannot be sufficiently described, and misconceptions cannot be fully specified. Many disciplines are intractable, e.g., programming languages, music composition, legal reasoning. In many domains, student input is limited to a few steps (graph lines, variable names, and equations) and this input might include a great deal of noise (student actions unrelated to learning because of a lack of concentration or tiredness) (Mitrovic 1998). Constraint-based methods are based on a psychological theory of learning that asserts that procedural learning occurs primarily when students catch themselves (or are caught by a third party) making mistakes (Ohlsson 1996, 1994; Self 1990). Students often make errors even though they know what to do because their minds are overloaded with many things, hindering them from making the correct decision. In other words, they may already have the necessary declarative knowledge, but a given situation presents too many possibilities to consider when determining which one currently applies (Martin 2001). Thus, merely learning the appropriate declarative knowledge is not enough; students must internalize that knowledge and know how to apply it before they can master the chosen domain.

Constraints represent the application of declarative knowledge to a current situation. Each constraint is an ordered pair of conditions that reduce the solution space. These conditions are the relevance condition (relevant declarative knowledge) and satisfaction condition (when relevant knowledge has been correctly applied): IF *relevance condition* is true THEN *satisfaction condition* will also be true. The *relevance condition* is the set of problem states for which the constraint is relevant, and the *satisfaction condition* is the subset of states in which the constraint is satisfied. If the constraint is violated, the student does not know this concept and requires remedial action.

When constraints are violated, an error is signaled that translates into a student's incomplete or incorrect knowledge. CBM reduces student modeling to pattern matching or finding actions in the domain model that correspond to students' correct or incorrect actions. In the example above the tutor might say: Do you know that denominators must be equal in order to add numerators? If the denominators are not equivalent, you must make them equal. Would you like to know how to do that?

Constraint-based models are radically different from model-based tutors in both underlying theory and resulting modeling systems. Although the underlying theories of model-tracing (Anderson 1983) and of Ohlsson's performance errors (Ohlsson 1996) may be fundamentally different in terms of implementing intelligent tutors, the key difference is level of focus. Model-tracing tutors focus on the procedures carried out and faithfully model procedures to be learned, whereas performance error–based tutors are concerned only with pedagogical states and domain constraints and represent just the pedagogical states the student should satisfy, completely ignoring the path involved (Martin 2001). CBM tutors represent only basic domain principles, through constraints, not all domain knowledge (Mitrovic 1998; Ohlsson 1994). They detect and correct student errors and do model the whole domain. Thus no expert model or a bug library is needed; the process is computationally efficient and neutral with respect to pedagogy (Mitrovic et al. 2004).

Several constraint-based intelligent tutors were developed for university students learning database programming (Mitrovic 1998; Mitrovic and Ohlsson 1999). One system teaches structured query language (SQL, pronounced "Seguel"), and another teaches database techniques, e.g., design and databases (Mitrovic 1998). Web-enabled versions of the Database Tutors have been available at DatabasePlace since 2003, and tens of thousands of students have used them. In addition to the SQL-Tutor, two other database tutors were evaluated: NORMIT, a tutor to normalize a database, and Entity-Relationship (EER), a tutor on database design for generating a database schema. Remote students have used the database tutors on the Internet completely independently from the courses in which they were enrolled. These students demonstrated equivalent learning even though they had no human teacher in the loop.

13.3.2 Artificial Intelligence Methods

Whereas cognitive science methods for building student models are based on the assumption that human learning (or errors) can be modeled, artificial intelligence (AI) methods are agnostic with regard to this assumption. These methods, e.g., including plan recognition and machine learning techniques, are not based on any attempt to model human learning. They are simply techniques that are successful at reasoning about knowledge. Using such methods, intelligent tutors have been able to induce student models, extend their knowledge and infer students' learning strategies. For example, machine learning (ML) paradigms enable intelligent tutors to extend themselves by learning new knowledge rather than by being programmed with that knowledge. Many ML methods have been used in tutors, e.g., Bayesian belief networks, reinforcement learning, hidden Markov models, fuzzy logic and decision theory, see Woolf (2009) for a description. This section first describes reasoning under uncertainty, which is an underlying principle for many ML techniques and then describes, Bayesian belief networks.

ML techniques describe the probability of an event occurring. *Probability theory* is used to reason about student knowledge and to predict future action by use of data techniques based on prior or current data. ML techniques enable tutors to reason about the *probability* of events as a way to draw useful conclusions about students.

We describe Bayesian belief networks as used in intelligent tutors. Bayesian theory can roughly be boiled down to one principle: To see the future, one must look at the past (Leonhardt 2001). Bayesian methods reason about the probability of future events, given their past and current probabilities. They are based on the understanding that the world is rife with uncertainty and often not suited to clean statistical tests. Bayesian belief networks (BBNs) enable computers to combine new data with prior beliefs about data, make subjective decisions about how strongly to weigh prior beliefs, and provide a policy for keeping new information in the proper perspective (Leonhardt 2001). They provide a graphical method to design probabilistic models based on conditional probabilities and the Bayes formula.

BBNs are used in intelligent tutors to support classification and prediction, model student knowledge, predict student behavior, make tutoring decisions, and (combined with data about student's proficiencies) determine on which steps students will need help and their probable method for solving problems (Mayo and Mitrovic 2001). They represent curriculum sequencing, e.g., skills in a domain. Tutors decide among alternatives, within a probabilistic model of student knowledge and goals, which problem to present next. They seek out propositions that are both part of a solution path and ones that students are likely to know. A basic formulation of BBNs represents causal networks among hidden skills and observed actions. Building a BBN in an intelligent tutor begins by recognizing that human teachers have uncertain knowledge of students and learning and they have only explicit knowledge about observed student actions (problems solved, equations submitted, or questions answered). Like most ML techniques, BBNs begin with observed actions and infer the probability of unobserved (hidden) skills (e.g., topics that students know).

Defining the structure of a BBN begins with a statement of probability:

$$P\,(student\_knows\,(S) \mid student\_knows\,(R))\,\_\,.95$$

which says that most students who know S also know R. is a graphical representation of the probability

$$P\,(Answer\ Problem044 \mid Skilla\,)\,\_\,.95$$

and means that the probability is high that people who know Skill a are very likely to answer Problem 044 correctly. The BBN represents the observed variable as well as the unobserved variable (Skill a). An arc lists the probability that one variable can be inferred from another (e.g., skill from answer). If an arc joins two nodes, it means the probability of all possible values for the pointed-at-node depends on the value of the previous node. If no arc joins two nodes, it means that the values for these nodes do not influence each other. Bayesian belief networks involve supervised learning techniques and rely on the basic probability theory and data methods. Graphical models are directed acyclic graphs with only one path through each (Pearl 1988). In intelligent tutors, such networks often represent relationships between prepositions about the student's knowledge and tutoring decisions. Nodes often represent the state of the teaching world (topics, level of skill, or correctness of questions).

Many BBNs have been developed to model student knowledge in intelligent tutors. One category of student model BBN is *expert-centric* or networks and conditional probabilities specified either directly or indirectly by experts (Mayo and Mitrovic 2001). Experts create the network structure or topology, draw the arcs, and defi ne the conditional probability of the arcs. Consider a naïve representation of student knowledge of physics, which states that student success on Problem 023 indicates understanding of Newton's law and that understanding Newton's law may result from reading the text.

A second example expert-centric student model is HYDRIVE (Mislevy and Gitomer 1996), which used a highly abstract method similar to Andes. HYDRIVE trained personnel to troubleshoot aircraft hydraulics involved in flight control, landing gear, and aerial refueling. It simulated features of troubleshooting by presenting students with a video sequence problem in which a pilot described the aircraft malfunction to technicians (e.g., " The rudders do not move during the preflight check"). The student performed troubleshooting procedures by accessing video images of aircraft components.

A second category of student model BBNs is *data-centric* models or networks that are specified from real-world data (e.g., log data) (Mayo and Mitrovic 2001). The structure and conditional probabilities are learned primarily from data collected from real-world evaluations of the tutor. This involves mining data from previous users of the system and classifying relationships (e.g., between solution of a problem and inferred student skill). Confirmation that a student really has a particular skill might come from evaluating student performance on simple problems that require only this skill. Several tutors used a data-centric approach. In one tutor, the hidden parameters of a Bayesian network were inferred using expectation maxima, an ML technique that deals with missing data (Ferguson et al. 2006). In another system, much of the student model was inferred from data (Johns and Woolf 2006).

13.4 Discussion and Future Research on Student Models

Student model have been enormously successful, enabling tutors to predict student performance in a number of fields. A variety of student models have been built including open and affective models. This chapter outlined the nature of student models and described how to build and update these models.

Many improvements are needed to fully represent and reason about students. We clearly need socio-technical solutions, recognizing the need for solutions that have large social components. In this section, we suggest some technology areas that provide promising developments. Such technology predictions are based on current research trends in student models and speculative at best. No one can know the future of student models nor accurately specify solutions for their development due, in part to the rapidly changing nature of software, languages, networks and hardware. Yet, we list considerations about likely future capabilities for student models and identify technologies that seem promising for their future development.

We envision that future student models will be *complex*, not only representing what students know, probably do and have abilities for, but other factors too. For instance, student models will probably track when and how skills were learned and what pedagogies worked best for each learner (Bredeweg et al. 2009). Moreover, student models might include information on the cultural preferences of learners, their personal interests, learning goals, and personal characteristics and will be able to select the optimal mix of learning environments, pedagogy, visualizations, and contexts that maximize engagement, motivation and learning outcomes for each individual. When the learner is part of a group, the model should provide the best estimate among the individuals who are part of the group, to attribute authorship.

We also envision that future student models will *support assessment* for both formative issues (the degree to which the student has learned how to learn – for the purposes of improving learning capacity and effectiveness) and summative considerations (what is learned–for purposes of accountability and promotion). In this regard, approaches to student modeling are needed that lead to valid and reliable inferences about student learning that are both diagnostic and predictive. Such a perspective concurs with the view that assessment should be dynamic over time.

We expect that in the future *privacy issues* in educational student models will be adequately addressed. Student privacy concerns and national and international privacy legislation have a considerable impact on what education applications may do. Strict privacy enhancing software tools and Internet services are needed. Generic student modeling systems will facilitate compliance with such regulations, as well as support privacy-enhancing services.

Currently student modeling techniques are developed and encoded into each individual educational program. For example, to measure a specific construct (e.g., algebra skills, persistence, help-seeking behavior) requires a substantial amount of effort to construct the relevant conceptual and statistical models (Shute et al. 2009). The construction cost of such models is about one year's time for a graduate student. The current approach does not scale to the increasing numbers of electronic learning environments that should have student models. We suggest that future student models will be developed as *shells* that exist independent of the instructional software and attached to the software only after they have been activated (Kobsa 2007). The term "shell" is borrowed from the field of expert systems and describes environments containing the basic components of expert systems (Nii 2009). Associated with each shell is a prescribed method for building a student model by configuring and instantiating encoded components. Shells support construction of knowledge bases through use of inference engines. Instead of building a student model for each instructional system, generic models will define their basic functionality and then be further constructed during development time.

These generic student models will serve as separate components and include a representation system for expressing the particular domain knowledge (e.g., logic formalism, rules, or simple attribute-value pairs) and a reasoning mechanism for deriving assumptions about users from existing models.

Most likely, future *student model servers* will be readily available for education. Servers are similar to generic student models in that they are separate from the application and will not run as part of it (Kobsa 2007). Student model servers will probably be part of local or wide area networks and serve more than one application instance at a time.

It is likely that educational data mining (EDM) and machine learning (ML) techniques will play larger role in augmenting student models automatically. These new approaches are presented in chapter 16 of this book. ML refers to a system's ability to acquire and integrate new knowledge through observations of users and to improve and extend itself by learning rather than by being programmed with knowledge (Shapiro 1992). These techniques organize existing knowledge and acquire new knowledge by intelligently recording and reasoning about data. For example, observations of students' past behavior will be used to provide training examples that will form a model designed to predict future actions (Webb et al. 2001). These techniques have been used to acquire models of individual students interacting with educational software and group them into communities or stereotypes with common interests. ML techniques are promising in cases where very large sets of usage data are available, like educational software on the Web (Kobsa 2007). These techniques improve teaching by repeatedly observing how students react and generalizing rules about the domain or student. These paradigms enable tutors to adapt to new environments, use past experience to inform present decisions, and infer or deduce new knowledge. Teaching environments will use ML techniques to acquire new knowledge about students and predict their learning (Arroyo and Woolf 2005; Johns and Woolf 2006).

One last prediction is that student models will probably adapt to new student populations. Obviously students have a variety of learning needs (e.g., exceptional students learn beyond their age group, special needs students require accommodations). Yet educational software is often built for the average student, expecting all students to be at the same academic level and be ready to learn. This is not so. Students are at various levels, have different skills and different learning abilities. No single instructional method works for all students and all disciplines. ML techniques can help enable software to acquire knowledge about distinct student groups and add that information to the tutor. Techniques can make decisions based on experience with prior populations and enable software to reason "outside" the original variables that made up the system.

References

Anderson, J.R.: The Architecture of Cognition. Harvard University Press, Cambridge (1983)

Anderson, J.R., Reiser, B.: The Lisp Tutor. BYTE 10, 159–175 (1985)

Anderson, J.R., Boyle, C.F., Farrell, R., Reiser, B.J.: Cognitive principles in the design of computer tutors. In: Morris, P. (ed.) Modelling Cognition. Wiley, Chichester (1987)

Arroyo, I., Woolf, B.: Inferring Learning and Attitudes from a Bayesian Network of Log File Data. In: Looi, C.K., McCalla, G., Bredeweg, B. Breuker, J. (eds.) Twelfth International Conference on Artificial Intelligence in Education, Amsterdam (2005)

Arroyo, I., Beal, C.R., Murray, T., Walles, R., Woolf, B.P.: Web-Based Intelligent Multimedia Tutoring for High Stakes Achievement Tests. In: James, R.M.V., Lester, C., Paraguaçu, F. (eds.) ITS 2004. LNCS, vol. 3220, pp. 468–477. Springer, Heidelberg (2004)

Beal, C.R.: Boys and Girls: The Development of Gender Roles. McGraw Hill, New York (1994)

Brachman, R., Levesque, H.: Knowledge Representation and Reasoning. Morgan Kaufmann (part of Elsevier's Science and Technology Division) (2004)

Bredeweg, B., Arroyo, I., Carney, C., Mavrikis, M., Timms, M.: Intelligent Environments. In: Global Resources for Online Education Workshop, Brighton, UK (2009)

Brusilovsky, P.: The Construction and Application of Student Models in Intelligent Tutoring Systems. Journal of computer and systems sciences international 32(1), 70–89 (1994)

Burleson, W.: Affective Learning Companions: Strategies for Empathetic Agents with Real-Time Multimodal Affective Sensing to Foster Meta-Cognitive and Meta-Affective Approaches to Learning, Motivation, and Perseverance. MIT PhD Thesis. (2006), http://affect.media.mit.edu/publications.php

Clancey, W.: Qualitative Student Models. Annual Review of Computer Science 1, 381–450 (1986)

Corbett, A.T., Anderson, J.R.: Knowledge tracing: Modeling the acquisition of procedural knowledge. User Modeling and User-Adapted Interaction 4, 253–278 (1995)

Eliot, C.: An Intelligent Tutoring System Based Upon Adaptive Simulation, Computer Science Department. University of Massachusetts, Amherst (1996)

Ferguson, K., Arroyo, I., Mahadevan, S., Woolf, B., Barto, A.: Improving Intelligent Tutoring Systems: Using EM to Learn Student Skill Levels. In: Ikeda, M., Ashley, K.D., Chan, T.-W. (eds.) ITS 2006. LNCS, vol. 4053, pp. 453–462. Springer, Heidelberg (2006)

IJAIED 2009 special issue on Authoring IEEE-TLT 2009 (2009)

Johns, J., Woolf, B.: A Dynamic Mixture Model to Detect Student Motivation and Proficiency. In: Conference on Artificial Intelligence (AAAI 2006), pp. 2–8. AAAI Press, Menlo Park (2006)

Kay, J.: Lies, Damned Lies and Stereotypes: Pragmatic Approximations of Users. In: Proceedings of the Fourth International Conference on User Modeling, pp. 175–184 (1994)

Kobsa, A. (ed.): Adaptive Web 2007. LNCS, vol. 4321. Springer, Heidelberg (2007)

Koedinger, K., Corbett, A.: Cognitive Tutors. In: Sawyer, K. (ed.) The Cambridge Handbook of the Learning Sciences. Cambridge University Press, Cambridge (2006)

Leonhardt, D.: Adding Art to the Rigor of Statistical Science. The New York Times, New York (2001)

Martin, B.: Intelligent Tutoring Systems: The Practical Implementations of Constraint-Based Modeling. Computer Science. University of Cantebury, Christchurch (2001), http://coscweb2.cosc.canterbury.ac.nz/research/reports/PhdTheses/2003/phd_0301.pdf

Mayo, M., Mitrovic, A.: Optimizing ITS Behavior with Bayesian Networks and Decision Theory. International Journal of Artificial Intelligence in Education 12, 124–153 (2001)

Mislevy, R.J., Gitomer, D.H.: The role of probability-based inference in an intelligent tutoring system. User Modeling and User Adapted Interaction 5(3-4), 253–282 (1996)

Mitrovic, A.: Experience in Implementing Constraint-Based Modeling in SQL-Tutor. In: Goettl, B.P., Halff, H.M., Redfield, C.L., Shute, V. (eds.) ITS 1998. LNCS, vol. 1452, pp. 414–423. Springer, Heidelberg (1998)

Mitrovic, A., Ohlsson, S.: Evaluation of a Constraint-Based Tutor for a Database Language. Artificial Intelligence in Education 10(3-4), 238–256 (1999)

Mitrovic, A., Suraweera, P., Martin, B., Weerasinghe, A.: DB-suite: Experiences with Three Intelligent, Web-based Database Tutors. Journal of Interactive Learning Research (JILR) 15(4), 409–432 (2004)

Nii, H.P.: Expert Systems Building Tools: Definitions (2009), `http://www.wtec.org/loyola/kb/c3_s2.htm`

Ohlsson, S.: Constraint-Based Student Modeling. In: Greer, J.E., McCalla, G. (eds.) Student Modeling: The Key to Individualized Knowledge-Based Instruction, pp. 167–189 (1994)

Ohlsson, S.: Learning from performance errors. Psychological Review 103, 241–262 (1996)

Pearl, J.: Probabilistic Reasoning in Intelligent Systems Networks of Plausible Inference. Morgan Kaufmann, San Francisco (1988)

Razzaq, L., Patvarczki, J., Almeida, S., Vartak, M., Feng, M., Heffernan, N., Koedinger, K.: The ASSISTment Builder: Supporting the Life Cycle of Tutoring System Creation. IEEE Transaction on Learning Technologies 2(2), 157–166 (2009)

Rich: Users are Individuals: Individualizing User Models. International Journal of Man-Machine Studies 18, 199–214 (1983)

Self, J.A.: Bypassing the intractable problem of student modelling. In: Frasson, C., Gauthier, G. (eds.) Intelligent tutoring systems: at the crossroads of artificial intelligence and education. Ablex Publishing, Norwood (1990)

Self, J.A.: Formal Approaches to Student Modeling. In: McCalla, G., Greer, J.E. (eds.) Student Models: The Key to Individual Educational Systems. Springer, New York (1994)

Shapiro, S.: Encyclopedia of Artificial Intelligence, 2nd edn. John Wiley & Sons, Chichester (1992)

Shute, V.J., Graf, E.A., Hansen, E.: Designing adaptive, diagnostic math assessments for individuals with and without visual disabilities. In: PytlikZillig, L., Bruning, R., Bodvarsson, M. (eds.) Technology-based education: Bringing researchers and practitioners. Information Age Publishing, Greenwich (2005)

Shute, V.J., Zapata, D., Kuntz, D., Levy, R., Baker, R., Beck, J., Christopher, R.: Assessment: A Vision, Global Resources for Online Education (GROE), Tempe Arizona (2009)

Sison, R., Shimura, M.: Student Modeling and Machine Learning. International Journal of Artificial Intelligence in Education 9, 128–158 (1998)

Webb, G., Pazzani, M., Billsus, D.: Machine Learning for User Modeling in User Modeling and User-Adapted Interaction, Netherlands 11, 19, 29 (2001)

Woolf, B.: Building Intelligent Interactive Tutors: Student-centered strategies for revolutionizing e-learning, vol. 480. Morgan Kaufmann, Burlington (2008)

Yacci: A Grounded Theory of Student Choice in Information-Rich Learning Environments. Journal of Educational Multimedia and Hypermedia 3(3-4), 327–350 (1994)

Chapter 14
Bayesian Student Modeling

Cristina Conati

Department of Computer Science, University of British Columbia,
2366 Main Mall, Vancouver, BC, V6G3C1
conati@cs.ubc.ca

Abstract. Bayesian networks are a formalism for reasoning under uncertainty that has been widely adopted in Artificial Intelligence (AI). Student modeling, i.e., the process of having an ITS build a model of relevant student's traits/states during interaction, is a task permeated with uncertainty, which naturally calls for probabilistic approaches. In this chapter, I will describe techniques and issues involved in building probabilistic student models based on Bayesian networks and their extensions. I will describe pros and cons of this approach, and discuss examples from existing Intelligent Tutoring Systems that rely on Bayesian student models

14.1 Introduction

One of the distinguishing features of an Intelligent Tutoring System (ITS) is that it is capable of adapting its instruction to the specific needs of each individual student, as good human tutors do. Adaptation can be performed at different levels of sophistication, from responding to student observable performance (e.g., errors), to targeting student assessed knowledge (or lack thereof), to helping students achieve specific goals (e.g., generate a given portion of a problem solution), to reacting to student emotions, to scaffolding meta-cognitive abilities (e.g., self-monitoring).

The more an ITS needs to know about its student to provide the desired level of adaptation, the more challenging it is for the ITS to build an accurate *student model* (see chapter by Beverly Woolf) based on the information explicitly available during interaction, because this information usually provides only a partial window on the desired student states. In other words, student modeling can be plagued by a great deal of uncertainty. In this chapter, I will illustrate an approach to handle this uncertainty that relies on the sound foundations of probability theory: Bayesian networks (Pearl 1988). Since the late eighties, Bayesian networks have been arguably the most successful approach for reasoning under uncertainty in AI, and have been widely used for both user modeling and student modeling. The rest of this chapter starts by providing some basic definitions. Next, it introduces *Dynamic Bayesian networks*, an extension of Bayesian networks to handle temporal information, and provides case studies to illustrate when and how to use static vs. dynamic networks in student modeling. The last part of the chapter discusses two

R. Nkambou et al. (Eds.): Advances in Intelligent Tutoring Systems, SCI 308, pp. 281–299.
springerlink.com

main challenges in using Bayesian networks in practice: how to choose the network structure and how to specify the network parameters. For each of these challenges, the chapter illustrates a variety of solutions and provides examples of how they have been used in applications to student modeling.

14.2 Bayesian Networks in a Nutshell

Bayesian networks are graphical models designed to explicitly represent conditional independence among random variables of interest, and exploit this information to reduce the complexity of probabilistic inference (Pearl 1988). Formally, a Bayesian network is a directed acyclic graph where nodes represent random variables and links represent direct dependencies among these variables. If we associate to each node X_i in the network a conditional probability table (CPT) that specifies the probability distribution of the associated random variable given its immediate parent nodes *parents*(X_i), then the Bayesian network provides a compact representation of the Joint Probability Distribution (JPD) over all the variables in the network.

$$\mathbf{P}(X_1, \ldots, X_n) = \prod\nolimits^{n}_{i=1} \boldsymbol{P}\,(X_i \mid Parents(X_i)) \tag{1}$$

This equation holds assuming that the network has been constructed so that each node is conditionally independent of all its non-descendant nodes given its parents (see (Russel and Norvig 2010) for more details).

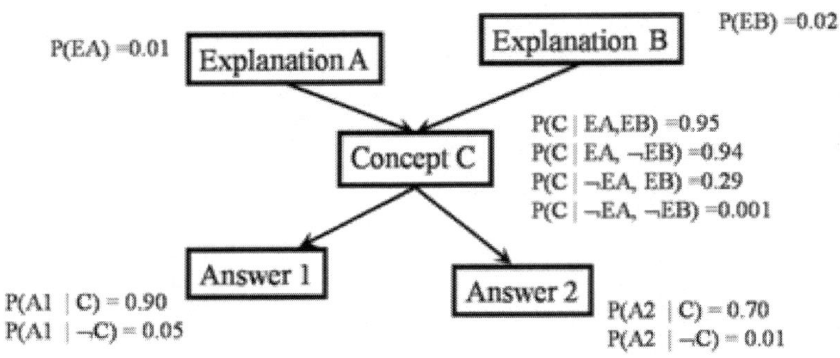

Fig. 14.1 Sample Bayesian network

Figure 14.1 shows a simple Bayesian network representing the following domain: the nodes *Explanation A* and *Explanation B* (indicated as *EA* and *EB* in the relevant CPTs) are binary variables each representing the probability that a student receives a corresponding explanation of concept C. The two explanations are provided independently, e.g., one at school by a teacher and one at home by a parent. The node *Concept C* (indicated as *C* in the relevant CPTs) is a binary variable representing the probability that a student understands the corresponding concept. The nodes *Answer 1* and *Answer 2* (indicated as *A1* and *A2* in the relevant CPTs) are binary variables each representing the probability that a student responds correctly to two different test questions related to concept C. The links and conditional probabilities in the network represent the probabilistic dependencies between receiving each of the two possible explanations for the concept, understanding it and then being able to answer related test questions correctly.

14.3 Static vs. Dynamic Bayesian Networks

The Bayesian network in Fig. 14.1 is *static*, i.e., it is suitable to perform probabilistic inference over variables with values that don't change over time. What changes and is tracked by a static Bayesian network is the belief over the state of these variables as new evidence is collected, i.e., the posterior probability distribution of the variables given the evidence.

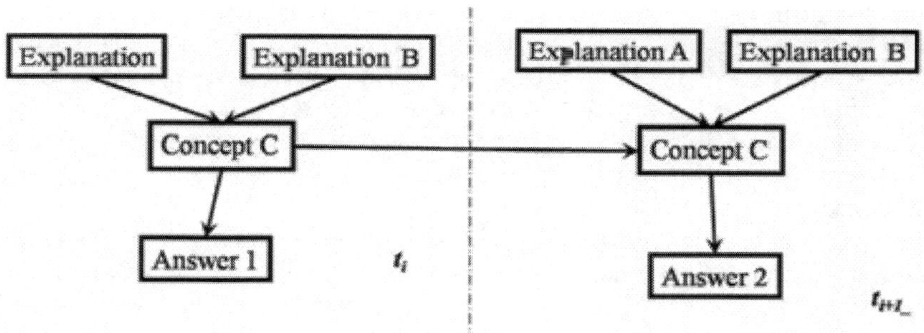

Fig. 14.2 Example DBN

Dynamic Bayesian networks (Dean and Kanazawa 1989), on the other hand, track the posterior probability of variables whose value change overtime given sequences of relevant observations. A Dynamic Bayesian networks (DBN from now on) consists of *time slices* representing relevant temporal states in the process to be modelled. For instance, Fig. 14.2 shows two time slices of a dynamic version of the network in Fig. 14.1. The first slice to the left represents the state of the variables *Concept C*, *Explanation A* and *Explanation B* from Fig. 14.1 after observing a student's answer to the first test at a given time t_i. The second slice represents the state of the same variables after observing a student's answer to the second test at a successive time t_{i+1}. The link between the variables for *Concept C* at times t_i and t_{i+1} models the influence of time on knowledge of this concept. It can be used, for instance, to model forgetting by adding to the CPT for *Concept C* at time t_{i+1} a non-zero probability that the student does not know *concept C* at that time given that she knew it at time t_i.

A key difference between the static network in Fig. 14.1and the dynamic network in Fig. 14.2 is in how evidence on student test answers is taken into account to update the posterior probability of *Concept C*. In Fig. 14.1, two subsequent observations on *Answer 1* and *Answer 2* would have the same weight in updating the probability of

Concept C, which makes sense if the true value of that variable does not change as the observations are gathered. In Fig. 14.2, the effect of having observed *Answer 1* at t_i on the probability of *Concept C* at t_{i+1} is mediated by the probability of *Concept C* at time t_i, while having observed *Answer 2* at t_{i+1} has a direct effect. This makes sense if the true value of *Concept C* can change overtime, because more recent observations are better reflections of the current state of a dynamic process than older ones.

14.3.1 Sample Applications to Student Modelling

Static networks can be used in student modeling as assessment tools under the assumption that the variables to be assessed (e.g., knowledge) are not changing as new evidence (e.g., test results) comes in. For instance, Mislevy (1995) describes a Bayesian student model used by the HYDRIVE tutoring system to assess a variety of skills and knowledge related to troubleshooting an aircraft hydraulics system. Martin and Vanlehn (1995) use a Bayesian student model for off-line assessment of student physics knowledge from evidence on completed problem solutions. Arroyo and Woolf (2005) describe a Bayesian network that assesses student attitudes toward learning with Wayang Outpost, an ITS for math (e.g., whether the student liked the system, found it helpful, learned from it) from statistics on the student interaction with the system (e.g., time spent per problem, time spent per action, average incorrect actions).

One of the first examples of using DBNs in student modeling is the *knowledge tracing* mechanism implement in the CMU Cognitive Tutors (Corbett and Anderson 1995). This mechanism uses Bayes theorem to compute the probability of mastering a rule at time t_{i+1} as a function of both the probability of knowing the rule at time t_i and observations of student problem solving steps pertaining to that rule at time t_{i+1}. While the original formulation of this mechanism was not in terms of DBNs, Reye (Reye 1998)has shown that it can be formulated as a DBN with the same basic behavior. One limitation of this *knowledge tracing* mechanism is that it requires knowing exactly which domain rule the current student solution step refers to. In order to eliminate possible ambiguities in mapping student solution steps with domain rules, this approach requires that students follow one specific solution defined a priory in the student model. For the same reason, it requires that students explicitly show all their solutions steps, i.e., it does not allow students to combine solution steps in their heads and generate actions that are the results of these mental computations. These requirements result in fairly constrained interaction that may become frustrating for some students. Finally, in *knowledge tracing* probabilistic update is limited to one rule at the time, i.e., this mechanism does not exploit the dependencies among the different rules involved in creating a complete problem solution.

The student model of the Andes tutoring system for physics (Conati et al. 2002) extends the approach proposed by Martin and Vanlehn (1995) to address the above limitations of knowledge tracing. For each problem solved by a student, Andes builds a static Bayesian network whose nodes and links represent how the various steps in a problem solution derive from previous steps and physics rules

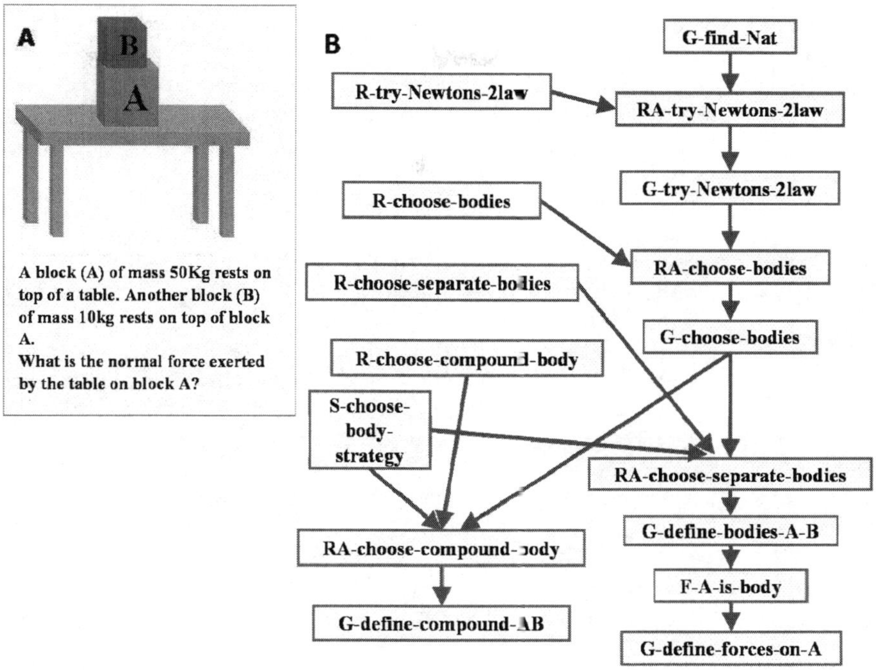

Fig. 14.3 A physics problem and a segment of the corresponding Bayesian network in the Andes tutoring system

(*task-specific network* from now on). For instance, Fig. 14.3B shows a (simplified) section of the task specific network for the problem in Fig. 14.3A, involving the application of Newton's second law to find the value of a normal force. Nodes in this network represent (i) facts corresponding to explicit solution steps (nodes labeled with a *F*- prefix in, Fig. 14.3B); (ii) problem solving goals (nodes labeled with a *G*- prefix); (iii) physics rules (nodes labeled with a *R*- prefix) that generate these facts and goals when applied to preceding facts and goals in the solution. Specific rule applications are indicated by nodes labeled with a *RA*- prefix in Fig. 14.3B. Alternative ways to solve a problem are represented as alternative paths to one or more solution steps in the network. Students can perform problem solving steps in their heads as they desire. When a problem solving step is entered in the Andes interface, Andes retrieves the corresponding fact node in the current task-specific Bayesian network, sets its value to *true* and computes the posterior probability of the other nodes in the network given this new evidence. All nodes involved in generating this step (i.e., all ancestors of the corresponding fact node) may be influenced by this update, with strength dictated by the probabilistic dependencies defined in the network's CPTs. Essentially, the task-specific Bayesian network allows Andes to guess which implicit reasoning has generated a given step, with the accuracy of the guess being influenced by how many steps the student has kept in her head and how many alternative ways to generate each step are represented in the network.

It should be noted that the task-specific network that Andes uses to track how a student solves a specific problem is not dynamic. Instead, Andes uses a form of dynamic network to track the evolution of student knowledge from one solved problem to the next. In particular, Andes maintains a *long-term student model* that encodes the posterior probability of each physics rule known by the system given all the solutions that a student has generated to so far. When the student starts a new problem, Andes generates the task-specific network for that problem as in Fig. 14.3, and initializes the prior probabilities of the rule nodes in the network using the posterior probabilities of the corresponding rules in the long-term model. As soon as the student terminates the problem, Andes discards its task-specific network, but saves the posterior probability of each of the network's rule nodes in the domain-general student model. This probability will then become the prior of the rule node in the task-specific network for the next problem that uses that rule. This process essentially corresponds to having a DBN where each time slice contains a rule node for each rule in the Andes' knowledge base; a new time slice is created when the student opens a new problem, and spans the time it takes the student to terminate problem solving. Removing a time slice when problem solving is over and saving rule posteriors to be used as priors in the next time slice is a form of *recursive filtering* (or *roll-up*). This process allows for maintaining at most two time slices in memory, as opposed to all the time slices tracked (Russel and Norvig 2010).

An alternative to the approach used in Andes is to create a new time slice every time a student generates a new action. We did not adopt this approach in Andes because the roll-up mechanism can be computationally expensive when performed after every student action on networks as large as Andes'. While this approximation may prevent Andes from precisely tracking learning that happens in between solution steps, it did not prevent Andes and its student model to perform well in empirical evaluations (Conati et al. 2002). Because, in the worst-case scenario, probabilistic update in Bayesian networks is intractable, simplifications like the one discussed here must often be made to ensure that the networks are usable in practice, and their impact/acceptability must be verified empirically. (Murray et al. 2004) describe an approach that does create a new slice after every student actions in networks comparable to Andes'. Despite adopting techniques to make network structure and CPTs more compact, performance testing based on simulated student actions showed that exact inference on the resulting models was not feasible. Using algorithms for approximate inference (Russel and Norvig 2010) improved performance, but still resulted in delayed response times on the larger networks tested.

An example of a DBN-based student model that creates time slices after every action and that has been used is practice is found in Prime Climb, an educational game to help students learn number factorization.

In Prime Climb students in 6th and 7th grade practice number factorization by pairing up to climb a series of mountains. Each mountain is divided into numbered sectors (see Figure 4), and each player can only move to a number that does not

Fig. 14.4 The Prime Climb Interface

share any common factors with her partner's number, otherwise s/he falls. To help students with the climbing task, Prime Climb includes a pedagogical agent (see Figure 4) for each player, that provides individualized support, both on demand and unsolicited, when the student does not seem to be learning from the game. To provide well-timed and appropriate interventions, the agent must have an accurate model of student learning, but maintaining such model is hard because performance tends to be a fairly unreliable reflection of student knowledge in educational games. PrimeClimb uses DBNs to handle the uncertainty involved in this modelling task. More specifically, there is a DBN for each mountain that a student climbs (the *short-term student model*). This DBN assesses the evolution of a student's number factorization knowledge during game play, based on the student's game actions. Each time slice in the DBN includes a *Factorization node* F_x for each number that is relevant to make correct moves on the current mountain (i.e., there is a node for each number on the mountain and for each of its factors). Each of these factorization nodes represents whether the student has mastered the factorization of that number. A new time slice is created after every new student action, e.g., after the student clicks on a number x to move there. The one-slice-per-action approach is feasible in Prime Climb because each time slice rarely contains more than a few dozens nodes. We will provide more details of the nature of the Prime Climb's DBNs in a later section.

14.4 Using Bayesian Networks in Practice

There are several advantages in using Bayesian networks for reasoning under uncertainty in general, and for student modeling in particular.

- They provide a more compact representation of the joint probability distribution (JPD) over the variables of interest. To fully specify the JDP $P(X_1, \ldots, X_n)$ over variables $X_1, \ldots, X_n$, it is necessary to specify the probability of each possible combination of variable's values (e.g., m^n numbers in the case of n m-valued variables). To fully specify the same distribution expressed via a

Bayesian network, it is sufficient to specify for each node with k parents the m^k entries of the associated CPT. If $k << n$, i.e., if the variables to be represented are sparsely connected, then the Bayesian network brings a substantial saving in the number of parameters that need to be specified.

- Algorithms have been developed that exploit the network's structure for computing the posterior probability of a variable given the available evidence on any other variable in the network. While the worse case complexity of probabilistic inference in Bayesian networks is still exponential in the number of nodes, in practice it is often possible to obtain performances that are suitable for real-world applications.
- The intuitive nature of the graphical representation facilitates knowledge engineering. It helps developers focus on identifying and characterizing the dependencies that are important to represent in the target domain. Even when dependencies are left out to reduce computational complexity, these decisions are easy to track, record and revise based on network structure, facilitating an iterative design-and-evaluation approach to model construction.
- Similarly, the underlying network structure facilitates the process of generating automatic explanations of the results of probabilistic inference, making Bayesian networks very well suited for applications in which it is important that the user understands the rational underling the system behavior, as it is often the case for Intelligent Tutoring systems (e.g., Zapata-Rivera and Greer 2004).
- Finally, Bayesian networks lend themselves well to support decision making approaches that rely on the sound foundations of decision theory. This means that selection of tutorial actions can be formalized as finding the action with maximum expected utility given a probability distributions over the outcomes of each possible action and a function describing the utility (desirability) of these outcomes (e.g., Murray et al. 2004; Mayo and Mitrovic 2001).

As is the case for any representation and reasoning paradigm, however, the benefits brought by Bayesian networks come with challenges. The two that arguably have the highest impact on the effort required by adopting this technology are: how to select a suitable structure and how to set the necessary network parameters. The next section discusses these two challenges and solutions proposed in the context of using Bayesian networks in student modeling.

14.5 Choosing Network Structure and Parameters: Examples from Student Modeling

14.5.1 Network Structure

14.5.1.1 Structure Defined Based on Knowledge

One common misconception related to structure definition in Bayesian networks is that the direction of the link between two variables must represent causality. In reality, the only constraint on structure is that every variable be (or can be reasonably assumed to be) independent of all its non-descendant nodes in the network, given its parent nodes. What is true is that structuring the network in the direction of causality makes it easier to satisfy the above constraint, because effects are independent of any previous influence given their immediate causes. In the domain represented in Fig. 14.1, for instance, whether the student understands or not concept C fully defines the probability that the student be able to answer questions about that concept, regardless of which explanation, if any, the student received.

Furthermore, defining links in the causal direction generally results in a more sparsely connected network. In our example, because understanding the concept fully specifies the probability of each answer, there is no direct dependency between the answers and thus there is no need for a link between the corresponding nodes. There is also no need for a direct link between the two *explanation* nodes, because we said they are provided independently.

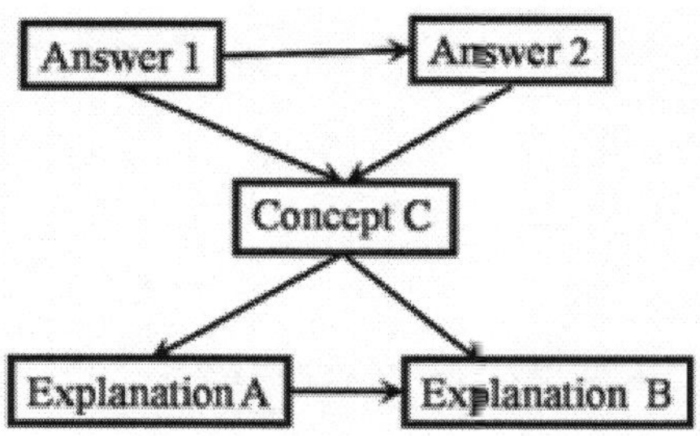

Fig. 14.5 Alternative structure for the Bayesian network in Figure 1

On the other hand, if we define the network as in Fig. 14.5, things change. We need a direct link between the two answer nodes because, given no other information, the belief that a student can generate a correct answer to a test is affected by whether or not the student can generate a correct answer to a different test that taps the same knowledge. Similarly, we need a direct link between the two explanation nodes because they are dependent if we know the true state of the student understanding of concept B. For instance, knowing that the student understands the concept and did not receive explanation A should increase the probability that the student received explanation B. This relationship between explanation A, explanation B and the understanding of concept C is fully captured by the structure in Fig. 14.1, but needs the extra arc between EA and EB in Fig 14.5. Still, the two networks in Fig 14.1 and Fig. 14.5 are equivalent if their CPTs are specified so that they represent the same JPD over the five variables involved. Which structure to select depends mostly on how much effort is required to specify the necessary network parameters (i.e., probabilities in the CPTs). Sparser networks include fewer parameters, but it is also important to consider how easy it is to quantify the needed probabilities.

In Andes, for instance, network structure is purely causal, capturing the following basic relation between knowledge of physics principles and problem solving steps: in order to perform a given problem solving step, a student needs to know the related physics rule and the preconditions for applying the rule. If a step can be derived from different rules, the student needs to apply at least one of them. As we will see in more detail in the next section, this causal structure yields very intuitive CPTs that can be specified via a limited number of parameters.

Matters are bit more complicated with the student model for the Prime Climb educational game. As we mentioned in a previous section, the student's progress on a Prime Climb mountain is tracked by a DBN that includes factorization nodes F_x for all the numbers on that mountain and their factors. Click nodes C_x are introduced in the model when the corresponding actions occur, and are set to either *true* or *false* depending upon whether the move was correct or not. Fig. 14.6 illustrates the structure used in the model to represent the relations between factorization and click nodes. The action of clicking on number *x* when the partner is on number *k* is represented by adding a click node C_x with parent nodes F_x and F_k (see Fig. 14.6b).

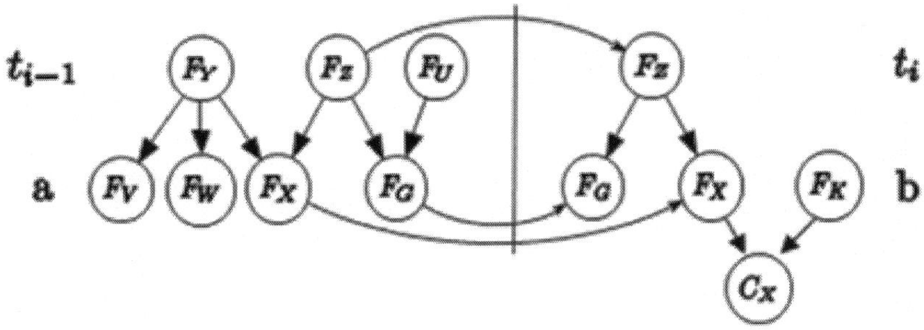

Fig. 14.6 Factorization nodes in the Prime Climb student model, where Z=X*G and Y=V*W*X; b: Click action

This structure represents the causal relationship between factorization knowledge and game actions that depend on it, which is intuitive to formalize: the correctness of a click is influenced by whether the student knows the factorization of the two numbers involved. The probability should be very high if the student knows both numbers, lower if the student knows only one number, and close to 0 if the student knows neither. Less obvious is how to choose the structure that represents the relationship between the factorization knowledge of a number and the factorization knowledge of its factors, because this relationship is not strictly causal. The rationale underlying the structure that was chosen for the Prime Climb network was derived based on discussion with mathematics teachers: knowing the prime factorization of a number influences the probability of knowing the factorization of its factors, while the opposite is not true. It is hard to predict if a student knows a number's factorization given that s/he knows how to factorize its non-prime factors. To represent this rationale, factorization nodes are linked as parents to nodes representing their non-prime factors. The conditional probability table (CPT) for each non-root factorization node (e.g. F_x in

Fig. 14.6a) is defined so that the probability of the node being known is high when all parent factorization nodes are true, and decreases proportionally with the number of unknown parents.

14.5.1.2 Structure Defined Based on Data

So far we have discussed how to define network structure based on existing knowledge of the dependencies among the relevant variables, but this approach is not feasible when the variables involved are not as clearly related as the ones in Andes and Prime Climb. The alternative is to define the structure based on data. Existing algorithms (e.g., Buntine 1996; Moore and Wong 2003) perform some form of heuristics search over the space of possible structures. The heuristics used to evaluate points in the search space generally rely on either statistical measures of correlation to verify whether the dependencies implicit in a given structure reflect the dependencies in the data, or measures related to the model's log likelihood *P(data|model)*, i.e., how well a given model explains the available data. These algorithms, however, require substantial amount of data to learn complex networks, which has limited their adoption in student modelling so far. To deal with limited data availability, existing work on learning the structure of Bayesian student models has combined ideas from these algorithms with heuristics based on knowledge of the target domain. Zhou and Conati (2003) for instance, have used a data-based approach to define the structure of a Bayesian student model that combines information on student personality and interaction patterns to assess student goals while playing Prime Climb. Using expert knowledge to define the structure of this DBN was not possible. While there are theories in psychology that can be used to relate personality to goals users may be pursuing while playing an educational game (e.g,, learn vs. having fun), these theories are too high-level to allow defining specific dependencies among these variables (see for instance Costa and McRae (1992)). Similarly, while it is intuitive that interaction behaviours should are in general affected by user goals, there is limited knowledge on how goals actually impact interaction behaviours in novel environments such as Prime Climb.

To learn the structure of the goal assessment network from data, Zhou and Conati (2003) run a user study during which the interaction patterns of students playing Prime Climb were logged and questionnaires were used to collect data on user personality and interaction goals. Because the amount of data collected was not sufficient to reliably apply existing algorithms to learn the complete network structure, this work used a greedy variation that separately builds and then combines different subparts of the network. The dependencies to be represented in each subpart are selected by running a correlation analysis over the relevant variables and choosing only those correlations that are statistically significant and above a given threshold for strength. The choice among the alternative structures that can represent the selected dependencies is made based on measures of log likelihood, and by using intuition to choose between structures with similar scores. Although this approach is not sound because the log marginal likelihood measure is not additive over network subparts, the resulting network (shown in Fig. 14.7) showed to be effective in assessing student goals when inserted in a larger model

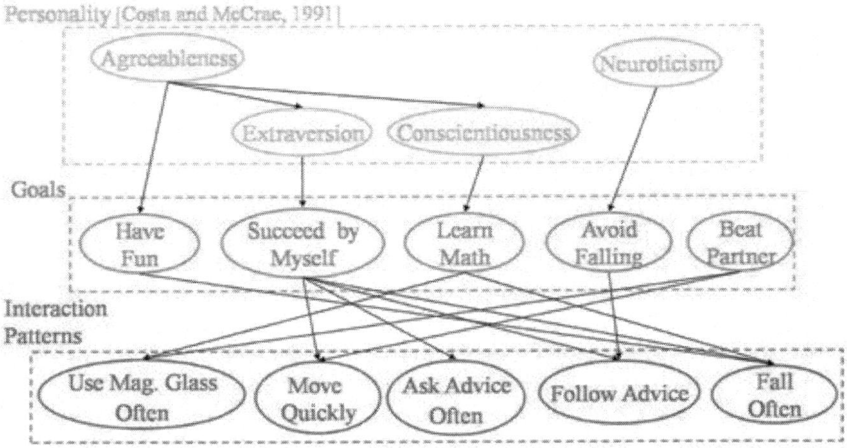

Fig. 14.7 Fragment of the goal assessment network in (Zhou and Conati 2003)

that relies on these goals as one of the elements to infer student emotions (Conati and Maclaren 2009). Arroyo and Wolf (2005) use a similar approach to learn the structure of the Bayesian network that relates interaction behaviors to user attitudes, mentioned in section 14.3.1.

14.5.2 Network Parameters

"Where do the parameters come from?" is arguably the first and most common objection that is raised in research that applies Bayesian networks to real world problems. As is the case for structure, the two main approaches to parameter specification are learning the parameters from data, or relying on domain experts to estimate them. Relying on expert judgment is costly and error prone. It is difficult for humans to commit to numbers their intuitions over given probabilistic dependencies. There has been substantial research on techniques that support the probability elicitation process (e.g., Keeney and von Winterfeldt 1991), but these techniques usually involve rather lengthy elicitation procedures and thus tend to be impractical when expert availability is limited. Still, when data is not available, relying on experts is the only viable approach and having conditional probabilities that are intuitive to specify can greatly facilitate parameter elicitation. In this section, will discuss one technique that can facilitate parameter specification by reducing the number of parameters to be specified, and two techniques for learning parameters from data

14.5.2.1 Parameters Reduction

One approach that can help reduce the effort of parameter specification is to reduce the number of parameters by approximating the necessary conditional probabilities via probabilistic variations of standard logic gates. This is the approach used by Andes to define the conditional probabilities in its task-specific networks.

Recall from section 14.3.1 that a task-specific network in Andes represents one or more solutions to a problem in terms of how each solution element derives from a physics rule and from the solution elements that are preconditions for rule application. Solution elements are either physics facts or problem solving goals, (collectively identified for convenience as *propositions* nodes PROP- in Fig. 14.8). Specific rule applications are represented in the network by rule application nodes (Rule-Appl nodes in Fig. 14.8).

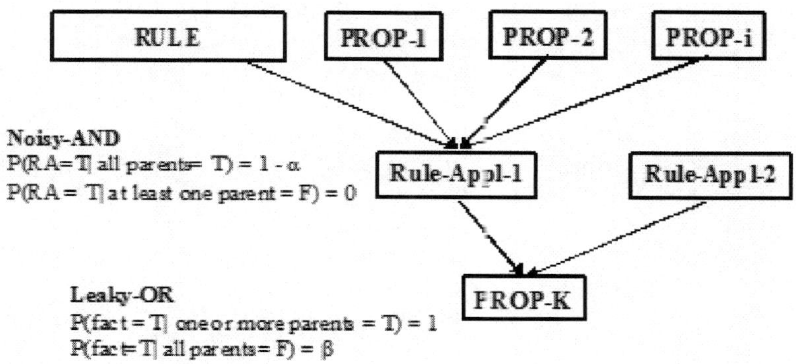

Fig. 14.8 probabilistic relations among rules, rule applications and their effects in Andes's task specific network

The parents of each Rule-application node include exactly rule, and a number of Proposition nodes corresponding to the rule's preconditions (see Fig. 14.8). A Rule-application node's value is *true* if the student has applied or can apply the corresponding rule to the propositions representing its preconditions, *false* otherwise. The probabilistic relationship between a Rule-application node and its parents is a Noisy-AND probabilistic gate (Henrion 1989). Here the Noisy-AND models the assumption that, in order to apply a rule, a student needs to know the rule and all its preconditions, although there is a non-zero probability α (the *noise* in the Noisy-AND), that the student will fail to apply the rule when s/he can, because of an error of distraction or some other form of *slip*. Thus, the α in Andes' Noisy-AND gates is an estimate of how likely it is that a student commits a slip, and it is the only parameter that needs to be specified to define the CPTs of rule-application nodes, regardless of how many parents they have.

Proposition nodes have as many parents (rule-application nodes) as there are ways to derive them. Thus, if there are two different rule applications that lead to the same solution element, then the corresponding Proposition node will have two parents (see Fig. 14.8). In Andes, the conditional probabilities between Proposition nodes and their parents are described by a Leaky-OR relationship (Henrion 1989), as shown in the lower part of Fig. 14.8. In a Leaky-OR relationship, a node is true if at least one of its parents is true, although there is a non-zero probability β of a "leak," that the node is true even when all the parents are false. This leak represents in Andes the probability that a student can derive a step via guessing or in some other way not represented in the network, and it is the only parameter that needs to be specified to define the CPTs of proposition nodes, regardless of how many alternative ways to derive a step the network encodes.

While the use of probabilistic logic gates in Andes greatly reduces the number of parameters that need to be specified, assessing the probability of a slip for each rule application and the probability of a guess for each solution element can still be a daunting task. The approach used in Andes follows a strategy that is often helpful when using Bayesian networks: make one or more simplifying assumptions that facilitate model definition and verify empirically whether the resulting model still yields an acceptable performance. The simplifying assumption made in Andes with respect to network parameters is that all slip and guess parameters are the same in the task-specific networks. Model adequacy was verified indirectly via empirical evaluations of the complete Andes system. The most extensive evaluation involved an experimental condition with 140 students using Andes for homework activities over the course of several weeks, and a control condition with 135 students doing homework without Andes. Students in the Andes condition scored significantly higher on a midterm exam covering relevant material. The accuracy of the Andes model was also analyzed directly by studying its performance in assessing the knowledge profile of simulated student (VanLehn and Niu 2001). This evaluation focused on performing *sensitivity analysis* on the Andes models to identify the factors that most impact model performance. The analysis revealed that the factor with the highest impact is, not surprisingly, the number of solution steps available as evidence to the model. I contrast, varying slip and guess parameters showed to have little effect on accuracy, confirming that the assumption of uniform slip and guess parameters was an acceptable one to make in light of the savings that it brings in effort for model specification.

14.5.2.2 Learning Parameters from Data

When all nodes in a Bayesian networks are observable, the entries for the network's CPTs can be learned via maximum-likelihood parameter estimation from frequency data (Russel and Norvig 2010). Unfortunately, in student modelling it is often the case that the variables of interest are not observable (e.g. student knowledge). Even when the variables are in theory observable (e.g., student goals, emotional states), in practice it can be very difficult to collect data on them,. Still, learning parameters from data is desirable because it eliminates the need to resort to the subjective judgment of experts. This judgement is not only hard to obtain and possibly fallacious, it can also be altogether unavailable when trying to model novel phenomena such as the relationships between student interaction with an ITS and student emotional states.

For this reason, there has been increasing interest in investigating how to exploit data-based techniques for parameters definition in student modeling. One approach, pioneered by Mayo and Mitrovic (2001), is to include in the student model only variables that are easily observable from interaction events with the tutoring system. In (Mayo and Mitrovic 2001) these variables model success or failure in using a variety of skills involved in correctly punctuating sentences. In particular, for each relevant skill, the Bayesian network in (Mayo and Mitrovic 2001) includes a variable representing the probability that a student will apply the skill correctly the next time it is relevant, given the outcome of the student's last attempt to apply the skill. The CPTs in the network were learned from log files of students solving punctuation problems in CAPIT, a tutoring system to help students practice punctuation skills. The network predictions are then used by CAPIT to automatically select new exercises for students, based on the criterion that a good exercise should involve several skills that the student has mastered and one or two skills that the student may still apply incorrectly. The idea of including in the student model only variables that are easily observable from interaction events obviously constraints the depth and sophistication of the inferences that an ITS can make about its students. However, Mayo and Mitrovic (2001) show that this approach is suitable and effective when the target instructional domain and interactions are of limited complexity.

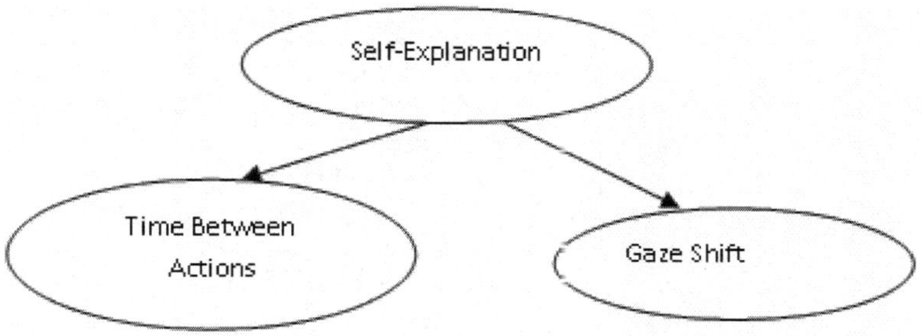

Fig. 14.9 Simple Bayesian network to predict self-explanation from action latency and gaze patterns in ACE

A second approach to learning the parameters of a student model from data relies on conducting empirical studies designed ad hoc to collect data on variables not observable from basic interaction events (we'll call these variables "hard-to-observe", to distinguish them from truly unobservable variables such as knowledge). For instance, Conati et al. (2005) conducted a study to collect data for a DBN that assesses student self-explanation behaviour from action latency and gaze patterns while the student is using an interactive simulation of mathematical functions. Self-explanation is the process of clarifying and elaborating instructional material to oneself, and it generally has a strong impact on learning (Chi 2000). In the context of studying interactive simulations, self-explanation relates to the effort a student makes to explain the effects of the manipulations performed on simulation parameters. The ACE system (Bunt et al. 2001; Conati and Merten 2007) aims to track student effort in self-explanation and provide adaptive interventions to increase this effort when needed. The study in (Conati et al. 2005) collected verbal protocols of students interacting with ACE, and analyzed these protocols to identify both episodes in which students generated self-explanations as well as episodes in which students failed to reason about the behaviour of the interactive simulation. These episodes where then matched to both log data on latency between student actions, as well as attention patterns over salient elements on the ACE interface, tracked via an eye-tracker. Frequencies from this dataset where then used to set the CPT of the simple Bayesian network shown in Fig. 14.9 (also known as a Naive Bayes classifier). A follow-up study showed that, when added to a more complex model of student learning, the network in Fig. 14.9 reliably supports the assessment of both student self-explanation and learning during interaction with ACE (Conati and Merten 2007). D'Mello et al. (2008) and Conati and Maclaren (2009) have adopted similar approaches relying on sophisticated data collection to build student models that assess student emotions from a variety of evidence sources.

In the research described above, it was not known upfront which observable factors could be good predictors of the hard-to-observe variables. Under these circumstances, in order to create a Bayesian student model researchers need to first find these predictors, which requires setting up experiments to collect data on the hard-to-observe variables. Once the data is collected and predictors are identified, everything is in place to apply standard maximum-likelihood parameter estimation. On the other hand, if there is an established dependency between the target hard-to-observe variables and a set of observable predictors, then the network parameters can be learned using EM (Dempster, et al. 1977). EM (which stands for Expectation-Maximization) is a class of algorithms that learn the parameters of a model with hidden variables by successive approximations based on two steps: the *expectation* step generates expected values of hidden variables from the current version of the model with approximated parameters; the *maximization* step refines the model parameters by performing maximum-likelihood parameter estimation using the expected values as if they were observed values. Thus, using EM removes the need for setting up complex studies to get values for hard-to-observe variables, when the dependency structure between these variables and a battery of observable variables is already known. Fergusson et al. (2006), for instance, used EM to learn the parameters of a Bayesian network that models knowledge of 12 geometry skills. In particular, EM was used to learn the dependencies between the variables representing this knowledge, and observable variables representing test questions designed specifically to assess the 12 target skills. The data for this work comes from a test that students in a Massachusetts high school had to take as part of a field study to evaluate the Wayang Outpost ITS for math.

Collecting sufficient amounts of data is the bottleneck in using any form of machine learning to specify a student model. It often requires setting up strong relationships with schools so that the necessary data can be collected as part of school activities involving whole classrooms. This process is generally very laborious. Schools, however, are becoming more and more willing to participate in these initiatives as the ITS field matures and produces concrete evidence of the benefits of having intelligent tutors available in the classroom, as it is shown by the increasing number of large scale school studies reported in ITS-related publications.

14.6 Discussion and Conclusions

Building a reliable picture of a student's relevant cognitive and affective states during learning is a task permeated with uncertainty that can be challenging even for experienced human tutors. Bayesian networks is a formalism for reasoning under uncertainty that has been successfully used in many AI applications, and that has been extensively used in student modeling, and user modeling in general. Critics of this approach mention the difficulty of reliably defining the model parameters (conditional probabilities) as one of its main drawbacks. An alternative approach for building a model of relevant student states would be to specify heuristic rules to define how available evidence should be integrated to assess the states. Defining these rules, however, still requires quantifying at some point complex probabilistic dependencies, because not explicitly using probabilities does not magically get rid of the uncertainty inherent to the modeling task. The advantage of a formal probabilistic approach is that the model only needs to quantify local dependencies among variables. The sound foundations of probability theory define how these dependencies are processed and affect the other variables in the model. In contrast, heuristic approaches require defining both the dependencies and ways to process them. This task is not necessarily simpler that defining conditional probabilities and entails a higher risk of building a model that generates unsound inferences. Furthermore, the Bayesian network graphical representation provides a compact and clear description of *all* the dependencies that exist in the domain, given the direct conditional dependencies encoded in the model. This helps to both verify that the postulated conditional dependencies define a coherent model, as well as debug the model when it generates inaccurate assessments. Similarly, the underlying network structure facilitates the process of generating automatic explanations of the results of probabilistic inference, making Bayesian networks very well suited for applications in which it is important that the user understands the rational underling the system behavior, as it is often the case for ITS (e.g., Zapata-Rivera and Greer 2004). Finally, Bayesian networks lend themselves well to support decision making approaches that rely on the sound foundations of decision theory. While decision theoretic approaches can still be too computationally expensive for handling complex tutorial interactions (e.g., Murray et al. 2004), researchers have shown their feasibility for dealing with particular pedagogical decisions in simpler domains, such as problem selection in sentence punctuation tasks (Mayo and Mitrovic 2001). Furthermore, continuous advances in research on decision theoretic planning suggest that more and more real world problems will be solvable with these approaches (see for instance the proceedings of *POMDP Practitioners Workshop: solving real-world* at http://users.isr.ist.utl.pt/~mtjspaan/POMDPPractioners/), including problems related to complex student modeling.

References

Arroyo, I., Woolf, B.: Inferring learning and attitudes from a Bayesian Network of log file data. In: 12th International Conference on Artificial Intelligence in Education, AIED 2005 (2005)

Bunt, A., Conati, C., Hugget, M., Muldner, K.: On Improving the Effectiveness of Open Learning Environments through Tailored Support for Exploration. In: 10th World Conference of Artificial Intelligence and Education, AIED 2001 (2001)

Buntine, W.: A Guide to the Literature on Learning Probabilistic Networks from Data. IEEE Transactions on Knowledge and Data Engineering 8(2), 195–210 (1996)

Chi, M.: Self-explaining: The dual processes of generating inference and repairing mental models. In: Glaser, R. (ed.) Advances in instructional psychology: Educational design and cognitive science, vol. (5), pp. 161–238. Lawrence Erlbaum Associates, Mahwah (2000)

Conati, C., Maclaren, H.: Empirically Building and Evaluating a Probabilistic Model of User Affect. Modeling and User-Adapted Interaction 19(3), 267–303 (2009)

Conati, C., Merten, C.: Eye-Tracking for User Modeling in Exploratory Learning Environments: an Empirical Evaluation. Knowledge Based Systems 20(6), 557–574 (2007)

Conati, C., Gertner, A., VanLehn, K.: Using Bayesian Networks to Manage Uncertainty in Student Modeling. Journal of User Modeling and User-Adapted Interaction 12(4), 371–417 (2002)

Conati, C., Merten, C., Muldner, K., Ternes, D.: Exploring Eye Tracking to Increase Bandwidth in User Modeling. In: Ardissono, L., Brna, P., Mitrović, A. (eds.) UM 2005. LNCS (LNAI), vol. 3538, pp. 357–366. Springer, Heidelberg (2005)

Corbett, A.T., Anderson, J.R.: Knowledge tracing: Modeling the acquisition of procedural knowledge. User Modeling and User-Adapted Interaction 4(4), 253–278 (1995)

Costa, P., McRae, R.: Four ways five factors are. Personality and Individual Differences 13, 653–665 (1992)

Dean, T., Kanazawa, K.: A Model for REasoning About Persistence and Causation. Computational Intelligence 5(3), 142–150 (1989)

Dempster, A., Laird, N., Rubin, D.: Maximization-likelihood from Incomplete Data via the EM Algorithm. Journal of Royal Statistical Society, Series B (1977)

D'Mello, S., Craig, S., Witherspoon, A., McDaniel, B., Graesser, A.: Automatic detection of learner's affect from conversational cues. User Modeling and User-Adapted Interaction, 45–80 (2008)

Ferguson, K., Arroyo, Y., Mahadevan, S., Park Woolf, B., Barto, A.: Improving Intelligent Tutoring Systems: Using Expectation Maximization to Learn Student Skill Levels. In: Ikeda, M., Ashley, K.D., Chan, T.-W. (eds.) ITS 2006. LNCS, vol. 4053, pp. 453–462. Springer, Heidelberg (2006)

Henrion, M.: Some practical issues in constructing belief networks. In: 3rd Conference on Uncertainty in Artificial Intelligence, pp. 161–173 (1989)

Keeney, R.L., von Winterfeldt, D.: Eliciting probabilities from experts in complex technical problems. IEEE Transactions on Engineering Management 38, 191–201 (1991)

Martin, J., VanLehn, K.: Student assessment using Bayesian nets. International Journal of Human-Computer Studies 42, 575–591 (1995)

Mayo, M., Mitrovic, T.: Optimising ITS Behaviour with Bayesian Networks and Decision Theory. International Journal of Artificial Intelligence in Education 12, 124–153 (2001)

Mislevy, R.: Probability-based inference in cognitive diagnosis. In: Nichols, P., Chipman, S., Brennan, R. (eds.) Cognitive Diagnostic Assessment, pp. 43–71. Erlbaum, Hillsdale (1995)

Moore, A., Wong, W.: Optimal Reinsertion: A New Search Operator for Accelerated and More Accurate Bayesian Network Structure Learning. In: ICML 2003, pp. 552–559 (2003)

Murray, C., VanLehn, K., Mostov, J.: Looking Ahead to Select Tutorial Actions: A Decision-Theoretic Approach. International Journal of Artificial Intelligence in Education 14(3-4), 235–278 (2004)

Pearl, J.: Probabilistic Reasoning in Intelligent Systems. Morgan Kaufmann, San Mateo (1988)

Reye, J.: Two-phase updating of student models based on dynamic belief networks. In: Goettl, B.P., Halff, H.M., Redfield, C.L., Shute, V.J. (eds.) ITS 1998. LNCS, vol. 1452, pp. 274–283. Springer, Heidelberg (1998)

Russel, S., Norvig, P.: Artificial Intelligence - A Modern Approach, 3rd edn. Prentice Hall, Englewood Cliffs (2010)

VanLehn, K., Niu, Z.: Bayesian student modeling, user interfaces and feedback: A sensitivity analysis. International Journal of Artificial Intelligence in Education 12, 154–184 (2001)

Zapata-Rivera, D., Greer, J.: Interacting with Inspectable Bayesian Student Models. International Journal of Artificial Intelligence in Education 14(2), 127–163 (2004)

Zhou, X., Conati, C.: Inferring User Goals from Personality and Behavior in a Causal Model of User Affect. In: UI 2003, International Conference on Intelligent User Interfaces, pp. 211–281. ACM Press, New York (2003)

Chapter 15
Open Learner Models

Susan Bull[1] and Judy Kay[2]

[1] Electronic, Electrical and Computer Engineering, University of Birmingham, B15 2TT, UK
[2] School of Information Technologies, University of Sydney, NSW 2006, Australia
bull@bham.ac.uk, judy@it.usyd.edu.au

Abstract. An Open Learner Model makes a machines' representation of the learner available as an important means of support for learning. This means that a suitable interface is created for use by learners, and in some cases for others who aid their learning, including peers, parents and teachers. The chapter describes the range of purposes that Open Learner Models can serve, illustrating these with diverse examples of the ways that they have been made available in several research systems. We then discuss the closely related issues of openness and learner control and the ways that have been explored to support learning by making the learner model available to people other than the learner. This chapter provides a foundation for understanding the range of ways that Open Learner Models have already been used to support learning as well as directions yet to be explored.

15.1 Introduction

Open learner models are learner models that can be viewed or accessed in some way by the learner, or by other users (e.g. teachers, peers, parents). Thus, in addition to the standard purpose of the learner model of maintaining data to enable adaptation to the individual according to their current learning needs, the learner model contents can also be of direct use to the user.

There are a variety of ways in which a learner model might be helpful to the learner, identified in the SMILI☺ (Student Models that Invite the Learner In) Open Learner Modelling Framework (Bull et al. 2007) as:

- Promoting metacognitive activities such as reflection, planning and self-monitoring;
- Allowing the learner to take greater control and responsibility over their learning, encouraging learner independence;
- Prompting or supporting collaborative and/or competitive interactions amongst groups of students;
- Facilitating interaction between learners and peers, teachers and parents;
- Facilitating navigation to materials, exercises, problems or tasks, etc., where links are available from the learner model;

R. Nkambou et al. (Eds.): Advances in Intelligent Tutoring Systems, SCI 308, pp. 301–322.
springerlink.com

- Supporting assessment - in particular providing formative assessment opportunities for students, but also enabling the learner model to be used as a summative assessment;
- Increasing the accuracy of the learner model data if the user is allowed to contribute additional or corrective information, to enable a more precise adaptive interaction to follow;
- Increasing learner trust in an adaptive educational environment by showing the system's inferences about their knowledge;
- The (non-educational) issue of people having the right to access electronic data about themselves.

A learner model that is inferred using any learner modelling technique could potentially be opened to the learner, if viewing the information in their learner model may be of benefit to the user. For example, learner models have been opened in simple weighted numerical models (Bull et al. 2009), fuzzy models (Mohanarajah et al. 2005), constraint-based models (Mitrovic and Martin 2007), Bayesian models (Zapata-Riviera and Greer 2004), transferable belief models (VanLabeke et al. 2007), knowledge tracing (Corbett and Bhatnagar 1997), and models constructed using conceptual graphs (Dimitrova 2003). The modelling technique does not necessarily determine the form in which the learner model will be presented to the user, or the level of interactivity the learner can have with their learner model.

Similarly, learner models can be opened in a variety of domains. As with adaptive learning environments that do not open the learner model to the user, they are often in programming (Weber and Brusilovsky 2001), mathematical (VanLabeke et al. 2007), scientific (Zapata-Riviera and Greer 2004) or second language (Bull and Pain 1995) domains. However, in line with increasing interest in less traditional domains in the field of artificial intelligence in education, open learner models are also becoming available in a broader selection of subjects. Examples include music theory (Johnson and Bull 2009) and historical text comprehension (Grigoriadou et al. 2003). Domain-independent open learner models have also been developed (Brusilovsky and Sosnovsky 2005; Bull et al. 2008; Mazza and Dimitrova 2004; Rueda et al. 2003; Kay and Lum 2005; Czarkowski et al. 2005). Furthermore, learner model attributes presented to the learner are not only cognitive, but may also include, for example, affective or social attributes (Chen et al. 2007).

In this chapter we give an overview of the presentation of open learner models, including common and less-traditional learner model externalisations; user/system control of the learner model and learning; and learner models open to other users such as peers, instructors and parents.

15.2 Presentation of Open Learner Models

Opening the learner model generally involves more than simply showing the learner the representations from the underlying system's model of their knowledge (or other attributes modelled), as these representations are not usually designed for interpretation by humans. In particular, learner models may not be designed for interpretation by those who are still learning a subject. While there is a case for ensuring that the underlying representation is designed explicitly to support the valuable roles that the open learner model can provide, this has generally not been the approach adopted. Regardless of the internal representation, a key challenge in opening a model to serve one of the purposes listed above, is to create an effective interface for presenting the model and supporting interaction with it.

As noted above, the method of model presentation does not necessarily match the underlying complexity of the model. For example, simple learner model overviews in the form of skill meters have been used to display a learner's knowledge level in a constraint-based model (Mitrovic and Martin 2007) (Fig. 15.1), and in a simple weighted numerical model (Ahmad and Bull 2009) (left of Fig. 15.2). Moreover, the complexity of a view of the learner model may differ within a system. Figure 15.2 shows both skill meters and a structured view of the learner model data available to the user in a single system (colour of nodes shows level of knowledge in the structured view) (Ahmad and Bull 2009).

Skill meters are the most commonly used simple overviews of the learner model contents, with a meter assigned to each topic or concept, which may include separate skill meters for sub-topics. (The latter allows simple structuring in the model presentation - e.g. Weber and Brusilovsky 2001.) Most skill meters show the level of the user's knowledge, understanding or skill as a subset of expert knowledge (Weber and Brusilovsky 2001), (Papanikolaou et al. 2003). The two examples in Figures 15.1 and 15.2 additionally show: (i) level of understanding as a proportion of areas covered (Mitrovic and Martin 2007); and (ii) the proportion

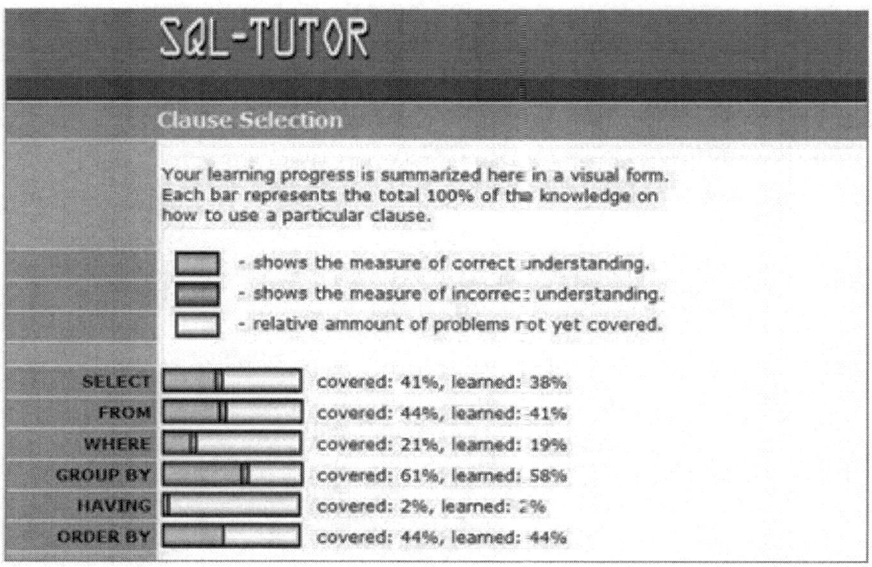

Fig. 15.1 Skill meters in SQL Tutor (Mitrovic and Martin 2007)

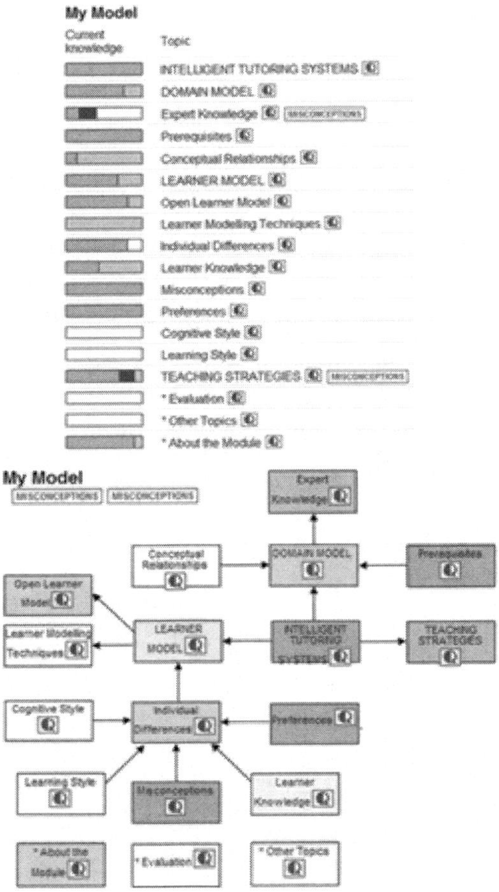

Fig. 15.2 Skill meters and structured view in OLMlets (Ahmad and Bull 2009)

of areas of difficulty that can be attributed to specific misconceptions (dark shading – clicking one of the misconceptions boxes presents a text statement of the specific misconception) (Bull et al. 2008). Other extensions to the standard skill meter are possible, e.g. varying the length of the meters to reflect the relative size of each topic (Bull et al. 2003); the skill meter showing the probability that the user knows a concept (Corbett and Bhatnagar 1997); or the level of knowledge of a learner contrasted against the combined knowledge of other groups (Linton and Schaefer 2000). Other representations that show similar information to the overviews displayed by skill meters for topics and concepts include: number of arrows in a target (Brusilovsky and Sosnovsky 2005); amount of liquid in a cup/container (Papanikolaou et al. 2003); smiley faces (Bull and McKay 2004); and images of the growth of trees (Lee and Bull 2008) (Fig. 15.3).

Fig. 15.3 Arrows in a target (Brusilovsky and Sosnovsky 2005); level of liquid (Papanikolaou et al. 2003); smiley faces (Bull and McKay 2004); and trees (Lee and Bull 2008)

The most common type of structured open learner model is probably the concept map (e.g. Rueda et al. 2003; Perez-Marin et al. 2007; Mabbott and Bull 2004; Kumar and Maries 2007), illustrated in Figure 15.4 by Willow (Perez-Marin et al. 2007) and Figure 15.5 by Flexi-OLM (Mabbott and Bull 2006).

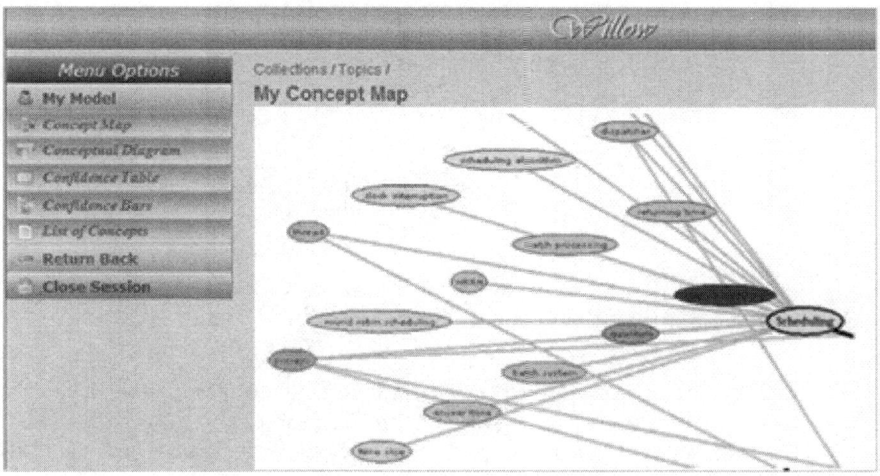

Fig. 15.4 The Willow concept map (Perez-Marin et al. 2007)

Concept maps can be pre-structured to reflect the domain, with nodes indicating the strength of knowledge or understanding of a concept (Mabbott and Bull 2004); they may reflect the learner's own conceptual structure or model, as inferred by the system (Perez-Marin et al. 2007); or they may be constructed by the learner (Cimolino et al. 2004; Mabbott and Bull 2007). Other detailed open learner model structures include tree structures, illustrated in Figure 15.6 with hierarchical structures in UM (Kay 1997) and Flexi-OLM Mabbott and Bull 2006); and individual instances of other types of structural relationship, e.g. Zapata-Riviera and Greer 2004; VanLabeke et al. 2007; Dimitrova 2003 (Fig. 15.7).

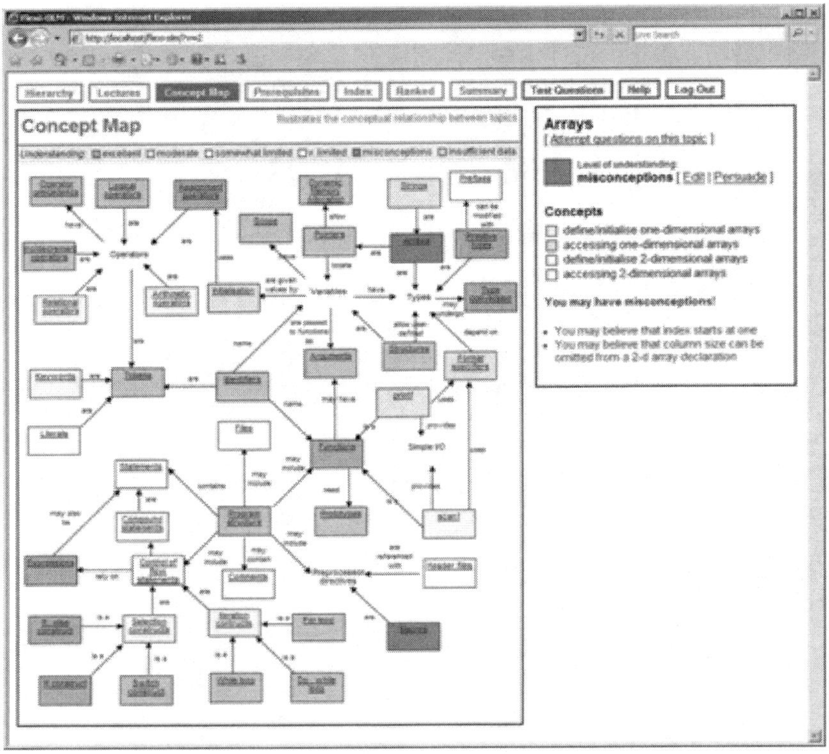

Fig. 15.5 The Flexi-OLM concept map (Mabbott and Bull 2006)

Consideration needs to be given to cases where detailed complex or structured learner model information is to be presented to the learner. UM (Kay 1997) (left of Fig. 15.6) allows the learner to expand and contract nodes to accommodate this requirement; SIV (Kay and Lum 2005) uses size, position and colour of text to display a large amount of learner model data (left-hand panel of Fig. 15.8). (Other uses of text in open learner models include Mohanarajah et al. 2005; Bull and Pain 1995; Tchetagni et al. 2007.)

Work has also investigated less traditional learner model presentations, including:

- audio and domain-specific representations for open learner models – e.g. MusicaLM (Johnson and Bull 2009) provides learner model information to the learner using a text description of learner beliefs, but also audio (music notes) and music notation (Fig. 15.9);
- haptic feedback in an open learner model, where strength of knowledge is portrayed by the 'hardness' or 'softness' of a sphere representing each area of the domain, and misconceptions feeling 'soft and sticky' (Lloyd and Bull 2006) (Fig. 15.10);

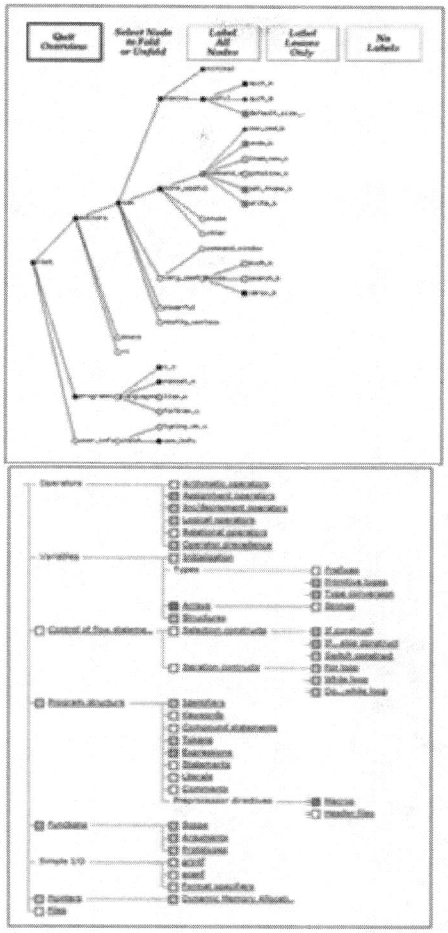

Fig. 15.6 Tree structures in UM (Cimolino et al. 2004) and Flexi-OLM (Mabbott and Bull 2004)

- open learner models for simulation tasks, such as the operation of a pole and cart device simulator (Morales et al. 2001) (Fig. 15.11);
- animations of a learner's misconceptions for comparison to the correct domain concepts, currently implemented for programming and chemistry (Johan and Bull 2009) (Fig. 15.12).

In addition to learner models that contain a single view of the learner model data, systems can offer multiple views of the same (or similar) model data for learners to explore (as many as seven views have been available in a system, with

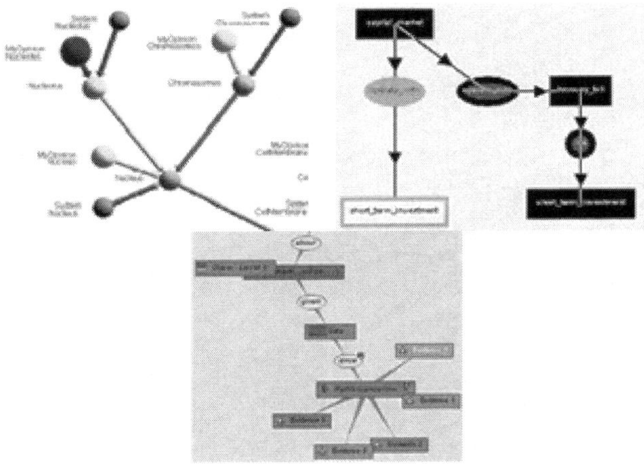

Fig. 15.7 Other structured open learner model presentations (. Zapata-Riviera and Greer 2004; VanLabeke et al. 2007; Dimitrova 2003)

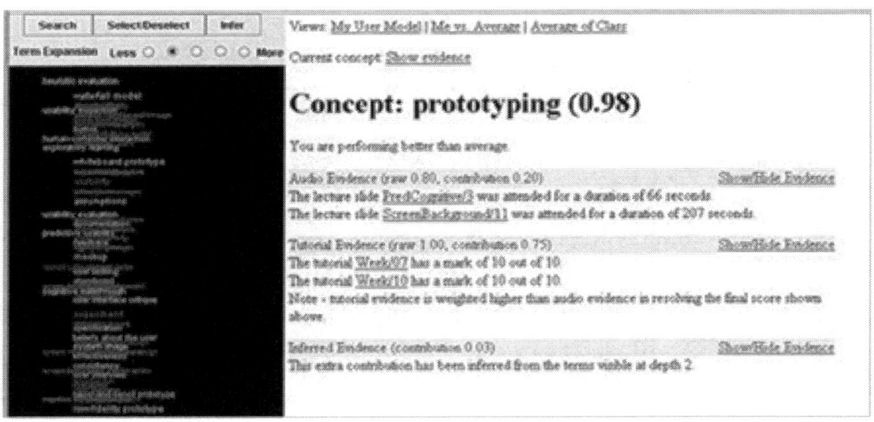

Fig. 15.8 Textual learner model presentation in SIV (Kay and Lum 2005)

users able to select their preferred views without difficulty (Mabbott and Bull 2006)); or multiple views with different aspects of the model information (e.g. VanLabeke et al. 2007)). Furthermore, the presentation method of the learner model may be adapted to the individual (Mazzola and Mazza 2009). Both user choice of view to access, and adaptive model presentations, are feasible – the most appropriate likely depending on the context of use.

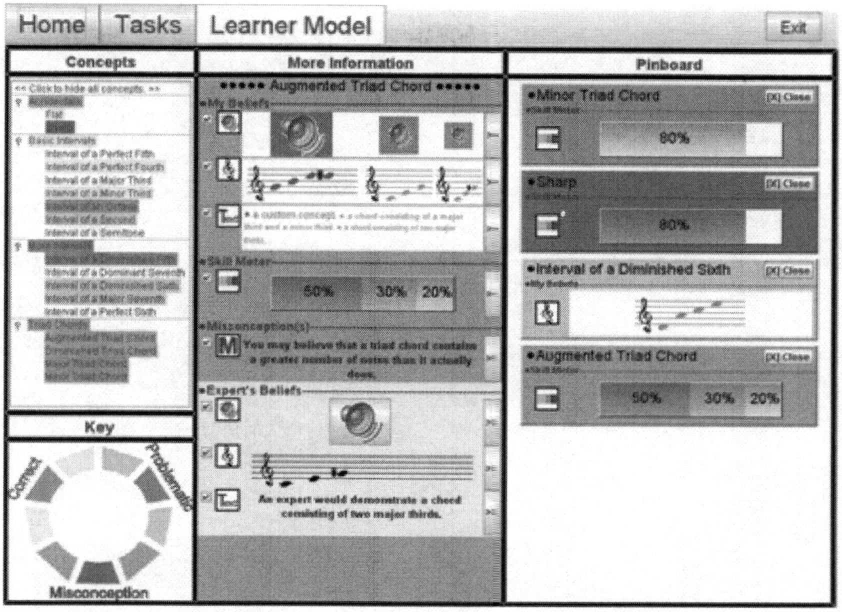

Fig. 15.9 Non-traditional open learner model presentation I: music notation and audio (extended from Johnson and Bull 2009)

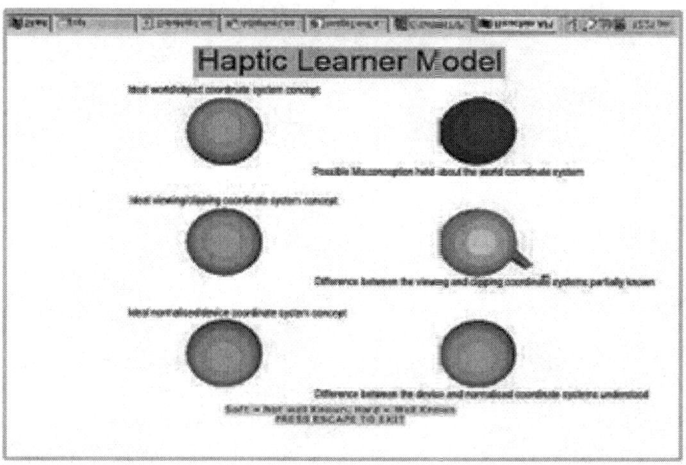

Fig. 15.10 Non-traditional open learner model presentation II: haptic learner model feedback (Lloyd and Bull 2006)

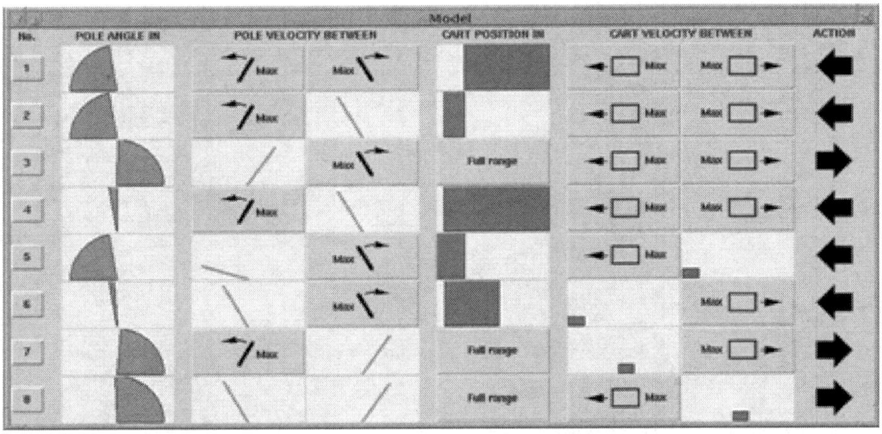

Fig. 15.11 Non-traditional open learner model presentation III: simulation task (operation of pole and cart device) (Morales et al. 2001))

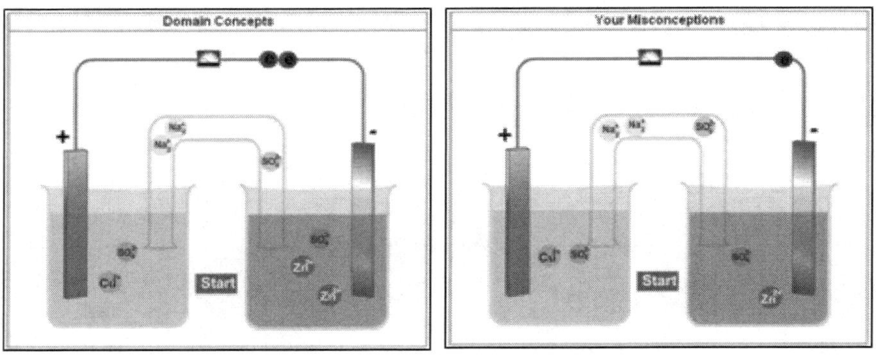

Fig. 15.12 Non-traditional open learner model presentation IV: animating learner beliefs (extended from Johan and Bull 2009)

15.3 Learner/System Control

As well as different methods of presentation of open learner models, the type of interaction a learner may have with their model and the level of control they have over the model contents, may differ. The question of learner/system control in open learner modelling is closely related to issues of metacognition in this context (Bull and Kay 2008).

15.3.1 System Control over the Learner Model Contents

The most straightforward form of open learner model is where the model is available for user viewing, but with no additional interactivity possible. These are known as *inspectable* learner models, and are completely under the control of the system - i.e. although the learner may view (or inspect) their learner model, they cannot make or suggest any changes to the system inferences about their understanding (or other attributes modelled). Most open learner models are in this category, including many of those introduced above. Inspectable learner models address the requirements for people to be allowed access to electronic data about themselves; but they also have the function of raising learner awareness of their knowledge, and prompting reflection, planning and formative assessment. Inspectable learner models are likely also to influence user trust in a system by confronting users directly with the system's inferences about their understanding.

15.3.2 Learner Control over the Learner Model Contents

In contrast to the above, *editable* learner models are learner models that are available for user viewing (as are inspectable models), but where the learner is able to change, or edit the contents. This is illustrated in Figure 15.13 by Flexi-OLM (Mabbott and Bull 2006). While the system may offer evidence or information to demonstrate where it disagrees with the learner's viewpoint (as is the case in Flexi-OLM, which can provide evidence in the form of the learner's responses to recent questions), the learner can in all instances override this if they wish, and so ensure that their desired changes to the model are effected.

Other examples help the learner to understand why and how the model was adapted to them, as illustrated in Figure 15.14 by SASY-unix, where the right-hand panel summarises the reasons for adaptation (e.g. "you want to get more than a pass grade", "you know the unix file system but haven't passed the quiz") (Czarkowski et al. 2005). Users of any SASY-based system can click the *why?* to see the reasons that the system believes this part of the learner model has its current value. They can inform the system of their own assessment of this part of the model. They can see if this will directly alter the value, as is the case for the belief "*you want to get more than a pass grade*". Or, if they cannot alter it directly, as in the case of "*you know the unix file system but haven't passed the quiz*", they can see that doing the quiz could change the value. The upper right of the display shows how the page was adapted; in this case, 2 items were removed and 5 added, because of the current learner model components listed. The learner can click this to see just how the presented content was personalised, which parts added or omitted. Each of these presents details of the parts of the learner model that controlled that personalisation.

Therefore the learner can scrutinise the personalisation to see what has been personalised. They can also see the precise parts of the learner model that caused this, as well as the reasons that the learner model has its current value.

Similar to learner model editing is an approach where the learner may *challenge* their model, and justify the changes they make to the model. The system will still accept the user's changes (and therefore the learner can still be said to be editing their model), but this additional information from the learner can feed back into the learner modelling process (VanLabeke et al. 2007).

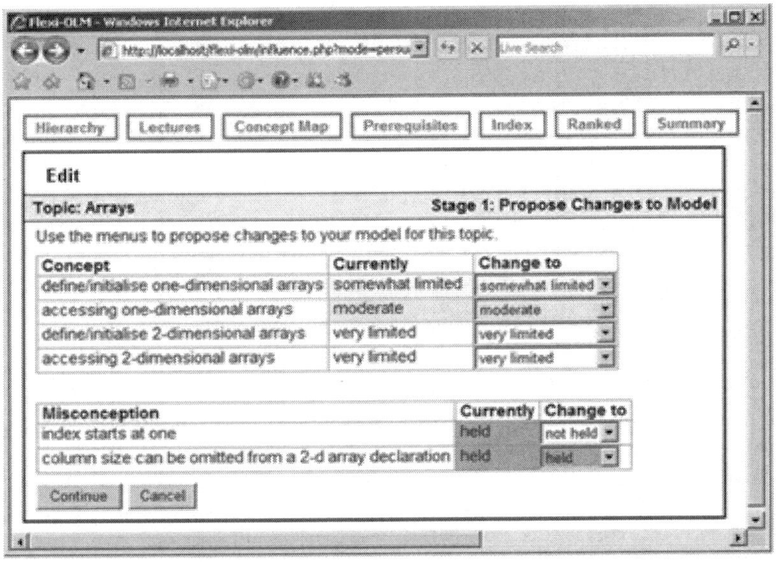

Fig. 15.13 Editing the Learner Model in Flexi-OLM (Mabbott and Bull 2006)

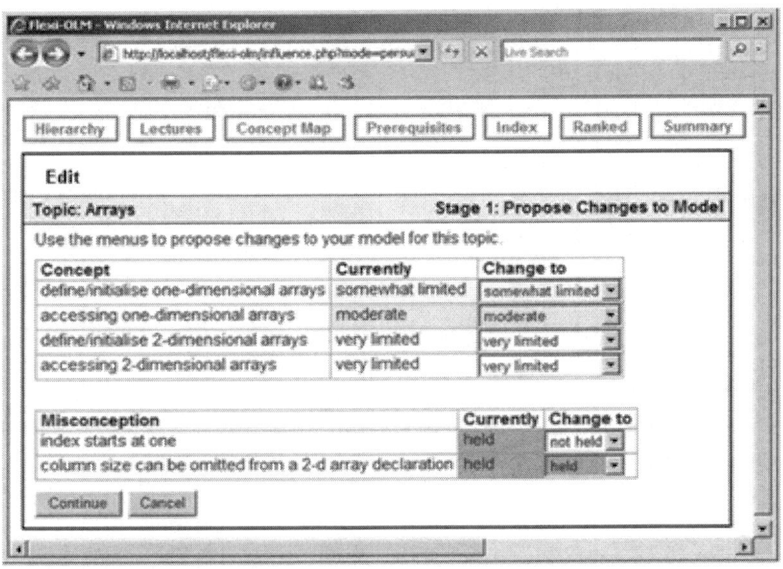

Fig. 15.14 Explaining the Learner Model in SASY-unix (Czarkowski et al. 2005)

Reasons for allowing learners to edit their model include enabling them to inform the system directly if: (i) their knowledge has increased (e.g. after a lecture or their own reading); (ii) they have forgotten information or techniques (to allow the system to consider or recommend appropriate revision); (iii) they were guessing (to allow the system to revaluate the learner model contents accordingly). Editable learner models place the learner in full control, inviting them to take responsibility for their learning interactions. The fact that a learner will need to be sure of any permanent changes they make to their learner model should also lead to them considering the options carefully, and so promote metacognitive activities as is the case with inspectable models. It may also increase user trust in a system as the user can control the data that determines adaptations in the interaction, if they wish.

15.3.3 Mixed Control over the Learner Model Contents

Interactive open learner models that allow an intermediate level of learner control include *co-operative* models (Beck et al. 1997), where the user and system provide different and complementary information for the learner model; *persuasion*, where the learner aims to change the model data (as in editable models), but the system will only subsequently alter the model if the learner can demonstrate that their self-assessment is accurate (e.g. by an additional short test) (Mabbott and Bull 2006); where additional information is provided for the learner model, to be considered alongside system inferences - i.e. the user may *add evidence* (Cimolino et al. 2004); and a truly balanced distribution of control of the learner model contents between the user and the system, where a negotiation process aims to achieve an agreed learner model. If this is not possible, any conflicts between the learner's and the system's representations are maintained in the model. This is illustrated in Figure 15.15 with Mr Collins (in the domain of pronoun placement in European Portuguese) (Bull and Pain 1995).

Negotiation can be undertaken in a variety of ways, for example: by menu selection (Bull and Pain 1995); dialogue games (Dimitrova 2003); and chatbots (Kerly and Bull 2008). Whatever technique is used, the key point is that the system and the user have equal rights to initiate and end discussions in negotiated learner models; and have the same negotiation moves available (e.g. offer information, request information, confirm, accept, critique, challenge, refute, justify).

Intermediate levels of interactivity in learner model maintenance may also require users to think carefully about any evidence they wish to add or remove, or discuss with the system. Therefore opening the learner model has a strong reflective element at various levels/types of interaction. While the learner is not able to edit the model to increase its accuracy, they do have the opportunity to directly influence the contents (though any proposed changes may not be agreed by the system). This increased involvement in the modeling processes may also affect user trust. Indeed, it has been suggested that learners may have greater trust in an environment where there is mixed control over the learner model, than full user control (Ahmad and Bull 2008). The most appropriate approach will likely depend on the context in which the open learner model is to be used.

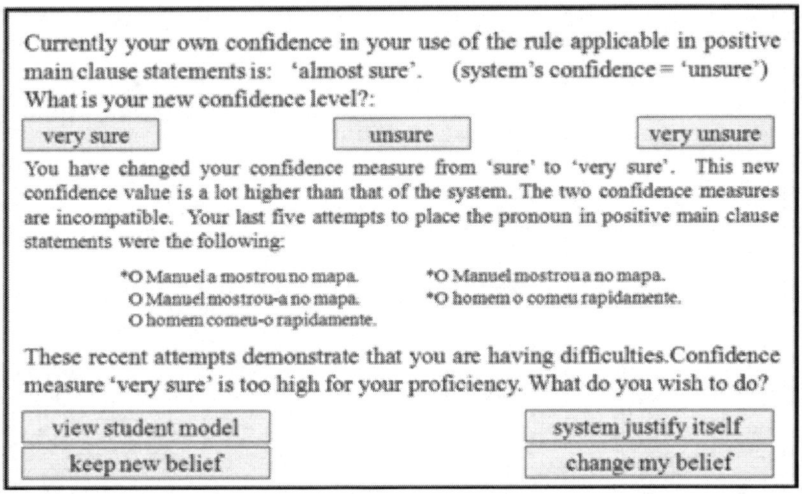

Fig. 15.15 Negotiating the Learner Model in Mr Collins (Bull and Pain 1995)

15.3.4 Independent Open Learner Models

Open learner models are usually integrated into an adaptive learning environment for consultation by the learner during their personalised interaction. This takes various forms, but well-known examples include skill meters for display alongside problem-solving and other tutorial support in an intelligent tutoring system (Mitrovic and Martin 2007) and in adaptive educational hypermedia providing navigation support (Weber and Brusilovsky 2001).

Whereas traditionally, adaptive learning environments (including those with open learner models) often have greater control over an interaction, independent open learner models (Bull and Mabbott 2008) are *independent* of a larger tutoring system. The learner model is constructed as in any system, but there is no additional (or only limited) tutoring or system guidance based on the learner model. The control and responsibility for decisions in their learning rest entirely with the learner: their planning and activity choices are facilitated by the contents of their learner model (we are therefore not concerned in this sub-section with control over the learner model contents, though the above distinctions are also applicable in independent open learner models). Although most open learner models are part of a larger system, independent open learner models have been successfully trialled in a variety of contexts (e.g. with trainee pilots (Gakhal and Bull 2008), language learners (Shahrour and Bull 2008), university students (Bull and Mabbott 2008) and schoolchildren (Kerly and Bull 2008)). Figures 15.2 (OLMlets) and 15.5 (Flexi-OLM) show examples of independent open learner models.

15.4 Open Learner Models for Other Users

Open learner models are applicable not only to individual learning scenarios, but can also be used in group learning. This may include individual models released by the learner to other group members – i.e. instructors or peers will see the specific information that the learner releases to them, as in OLMlets (Bull and Britland 2007) (Fig. 15.16). This enables learners to find collaborators (e.g. to seek help or to jointly work on an area where both individuals have weak knowledge; to compete with others; or simply to compare their progress against that of other users while preferring to work alone). Individuals may also be able to compare their own knowledge of a topic or concept (indicated by a star) on a five-point scale of very low to very high, to the knowledge of the group. Instructors can benefit from information about the progress of the group they are teaching, or

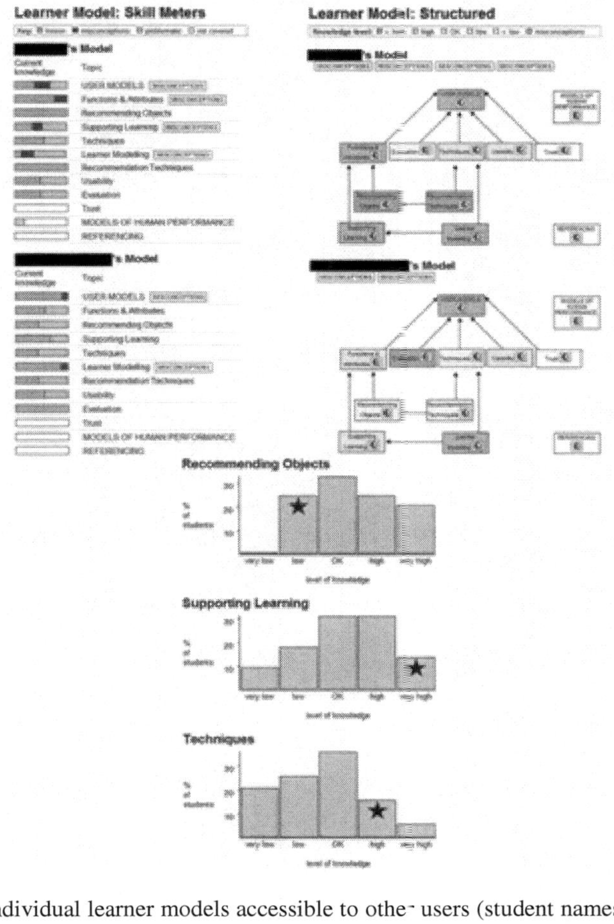

Fig. 15.16 Individual learner models accessible to other users (student names hidden in this image), and group model in OLMlets (Ahmad and Bull 2009; Bull and Britland 2007)

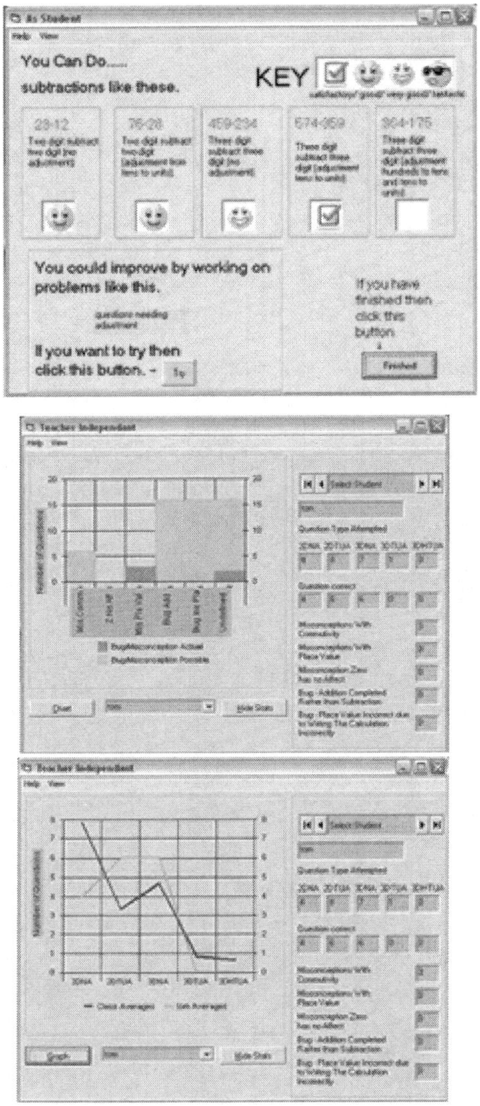

Fig. 15.17 An open learner model for children (upper) and teachers (lower) in Subtraction Master (Bull and McKay 2004)

from information about an individual's specific knowledge (or misconceptions). The learner model can be accessed in the format of choice of the viewer of the model (i.e. the same views are available to all users, but a peer or instructor who has been given access to an individual's learner model may select how to view it).

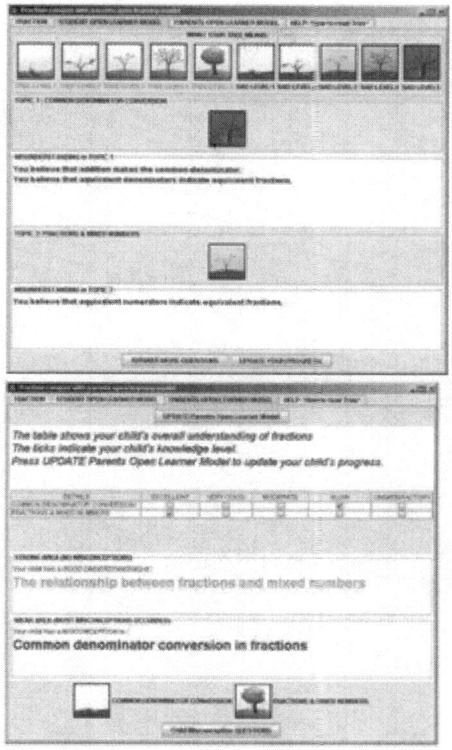

Fig. 15.18 An open learner model for children (left) and additional information for parents (right) in Fraction Helper (Lee and Bull 2008)

Similar approaches also allow the teacher to access information about an individual or groups of students, for example, using concept map, table, bar chart and text presentations for individuals or groups (Perez-Marin et al. 2007b). Systems may use different model presentation formats for different categories of user. For example, in DynMap (Rueda et al. 2003), the student may view the concept map of their current beliefs in a simpler format than does the teacher, who can follow more detailed representations of the evolution of the student's knowledge. In Subtraction Master (Bull and McKay 2004), the child views their learner model as a series of simple smiley faces representing the extent of their subtraction skills at different levels of difficulty. The teacher has more detailed information available, to help them to help a child individually. (This includes not only misconceptions identified, but also cases where misconceptions could have been demonstrated but were not, based on the subtraction questions that the child attempted.) This is illustrated in Figure 15.17. Similarly, as shown in Figure 15.18, Fraction Helper (Lee and Bull 2008) provides additional information to parents, over the learner

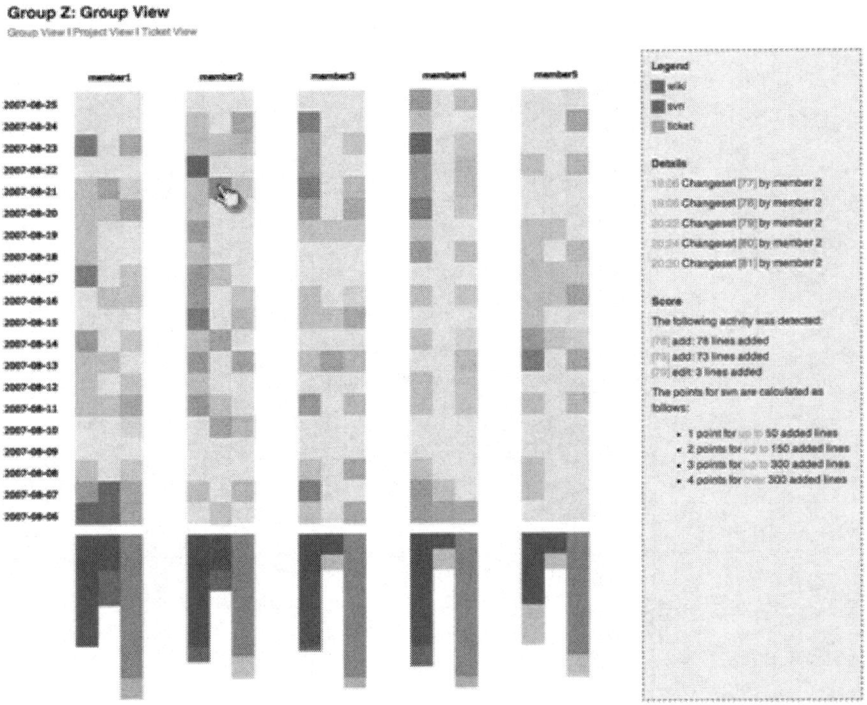

Fig. 15.19 Group activity model in Narcissus (Upton and Kay 2009)

model representations seen by their child. Nevertheless, the information for parents is still quite simple, as it is not assumed that all parents will have sufficiently advanced skills to interpret complex information. (Adapting the learner model presentation for parents is an issue that may benefit from further research.)

Other open learner models that permit instructor (and sometimes other user) access include REPro (Eyssautier-Bavay et al. 2009), CosyQTI (Lazarinis et al. 2007), CourseVis (Mazza and Dimitrova 2004), INSPIRE (Papanikolaou et al. 2008), Logic-ITA (Yacef 2005), MIM (Zapata-Riviera et al. 2007), PDinamet (Gaudioso et al. 2009).

Open learner models may also reflect group learning. A simple example was shown in Figure 15.16, but group open learner models can have much greater complexity. For instance, supporting long term group work based on evidence from group work tools, with the aim of facilitating effective group functioning. Figure 15.19 shows individual contributions to the group over time (each column represents one group member, with the brightness of a square showing the level of activity at that time) (Upton and Kay 2009) but as it uses very simple measures, it also links directly to the evidence and allows the learner to control the system decisions about when to make a square brighter. Evaluations of this and an earlier version (Kay et al. 2007) indicates that it helped team members identify group work problems and negotiate solutions.

It was also used by the group facilitator (tutor) to help with group problems and it served as a way to navigate the information space, a new role for an open learner model.

15.5 Summary

This chapter has provided an overview of some of the central issues in open learner modelling, with a particular focus on the ways in which open learner models are presented to the user; differences in user/system control over the learner model data and their learning; and learner models that are accessible to other users (e.g. peers, instructors, parents). While there are now many systems containing open learner models, there remain a range of directions for research in this area, including those areas presented in the introduction (from the SMILI open learner modelling framework (Bull and Kay 2007)), but new areas will likely also emerge.

Acknowledgements. We thank the following for their screen shots: N. Ahmad, P. Brusilovsky, M. Czarkowski, V. Dimitrova, R. Johan, M. Johnson, J.H.S. Lee, T. Lloyd, A. Mabbott, M. McKay, R. Morales, A. Mitrovic, K. Papanikolaou, D. Perez-Marin, K. Upton, N. VanLabeke, J.D. Zapata-Rivera.

References

Ahmad, N., Bull, S.: Do students trust their open learner models? In: Nejdl, W., Kay, J., Pu, P., Herder, E. (eds.) AH 2008. LNCS, vol. 5149, pp. 255–258. Springer, Heidelberg (2008)

Ahmad, N., Bull, S.: Learner Trust in Learner Model Externalisations. In: Dimitrova, V., Mizoguchi, R., du Boulay, B., Graesser, A. (eds.) Artificial Intelligence in Education. IOS Press, Amsterdam (2009)

Beck, J., Stern, M., Woolf, B.P.: Cooperative Student Models. In: du Boulay, B., Mizoguchi, R. (eds.) Artificial Intelligence in Education. IOS Press, Amsterdam (1997)

Bull, S., Britland, M.: Group Interaction Prompted by a Simple Assessed Open Learner Model that can be Optionally Released to Peers. In: Brusilovsky, P., Papanikolaou, K., Grigoriadou, M. (eds.) Proceedings of Workshop on Personalisation in E-Learning Environments at Individual and Group Level (PING), User Modeling (2007)

Bull, S., Kay, J.: Student Models that Invite the Learner In: The SMILI Open Learner Modelling Framework. International Journal of Artificial Intelligence in Education 17(2), 89–120 (2007)

Bull, S., Kay, J.: Metacognition and Open Learner Models. In: Roll, I., Aleven, V. (eds.) Proceedings of Workshop on Metacognition and Self-Regulated Learning in Educational Technologies, Intelligent Tutoring Systems, pp. 7–20 (2008)

Bull, S., Gardner, P.: Highlighting Learning Across a Degree with an Independent Open Learner Model. In: Dimitrova, V., Mizoguchi, R., du Boulay, B., Graesser, A. (eds.) Artificial Intelligence in Education. IOS Press, Amsterdam (2009)

Bull, S., McEvoy, A.T.: An Intelligent Learning Environment with an Open Learner Model for the Desktop PC and Pocket PC. In: Hoppe, U., Verdejo, F., Kay, J. (eds.) Artificial Intelligence in Education. IOS Press, Amsterdam (2003)

Bull, S., McKay, M.: An Open Learner Model for Children and Teachers: Inspecting Knowledge Level of Individuals and Peers. In: Lester, J.C., Vicari, R.M., Paraguaçu, F. (eds.) ITS 2004. LNCS, vol. 3220, pp. 646–655. Springer, Heidelberg (2004)

Bull, S., Pain, H.: Did I say what I think I said, and do you agree with me?: Inspecting and Questioning the Student Model. In: Greer, J. (ed.) Proceedings of World Conference on Artificial Intelligence and Education, Association for the Advancement of Computing in Education, Charlottesville, VA, pp. 501–508 (1995)

Bull, S., Mabbott, A., Gardner, P., Jackson, T., Lancaster, M., Quigley, S., Childs, P.A.: Supporting Interaction Preferences and Recognition of Misconceptions with Independent Open Learner Models. In: Nejdl, W., Kay, J., Pu, P., Herder, E. (eds.) AH 2008. LNCS, vol. 5149, pp. 62–72. Springer, Heidelberg (2008)

Brusilovsky, P., Sosnovsky, S.: Engaging Students to Work with Self-Assessment Questions: A study of two approaches. In: Proceedings of 10th Annual Conference on Innovation and Technology in Computer Science Education, pp. 251–255. ACM Press, New York (2005)

Chen, Z.-H., Chou, C.-Y., Deng, Y.-C., Chan, T.-W.: Active Open Learner Models as Animal Companions: Motivating Children to Learn through Interacting with My-Pet and Our-Pet. International Journal of Artificial Intelligence in Education 17(2), 145–167 (2007)

Cimolino, L., Kay, J., Miller, A.: Concept mapping for Eliciting Verified Personal Ontologies. International Journal of Continuing Engineering Education and Lifelong Learning 14(3), 212–228 (2004)

Corbett, A.T., Bhatnagar, A.: Student Modeling in the ACT Programming Tutor: Adjusting a Procedural Learning Model Model with Declarative Knowledge. In: Jameson, A., Paris, C., Tasso, C. (eds.) User Modeling. Springer, New York (1997)

Czarkowski, M., Kay, J., Potts, S.: Web Framework for Scrutable Adaptation. In: Proceedings of Workshop on Learner Modelling for Reflection, International Conference on Artificial Intelligence in Education, pp. 11–18 (2005)

Dimitrova, V.: StyLE-OLM: Interactive Open Learner Modelling. International Journal of Artificial Intelligence in Education 13(1), 35–78 (2003)

Eyssautier-Bavay, C., Jean-Daubias, S., Pernin, J.-P.: A Model of Learners Profiles Management Process. In: Dimitrova, V., Mizoguchi, R., Du Boulay, B., Graesser, A. (eds.) Artificial Intelligence in Education. IOS Press, Amsterdam (2009)

Gakhal, I., Bull, S.: An Open Learner Model for Trainee Pilots. ALT-J: Research in Learning Technology 16(2), 123–135 (2008)

Gaudioso, E., Montero, M., Talavera, L., Hernandez-del-Olmo, F.: Supporting Teachers in Collaborative Student modeling: A Framework and Implementation. Expert Systems with Applications 36(2), 2260–2265 (2009)

Grigoriadou, M., Tsaganou, G., Cavoura, T.: Dialogue-Based Reflective System for Historical Text Comprehension. In: Proceedings of Workshop on Learner Modelling for Reflection, (Supplemental Proceedings, 5) International Conference on Artificial Intelligence in Education, pp. 238–247 (2003)

Johan, R., Bull, S.: Consultation of Misconceptions Representations by Students in Education-Related Courses. In: Dimitrova, V., Mizoguchi, R., du Boulay, B., Graesser, A. (eds.) Artificial Intelligence in Education. IOS Press, Amsterdam (2009)

Johnson, M., Bull, S.: Belief Exploration in a Multiple-Media Open Learner Model for Basic Harmony. In: Dimitrova, V., Mizoguchi, R., du Boulay, B., Graesser, A. (eds.) Artificial Intelligence in Education. IOS Press, Amsterdam (2009)

Kay, J.: Learner Know Thyself: Student Models to Give Learner Control and Responsibility. In: Halim, Z., Ottomann, T., Razak, Z. (eds.) Int. Conference on Computers in Education, AACE (1997)

Kay, J., Lum, A.: Exploiting Readily Available Web Data for Scrutable Student Models. In: Looi, C.-K., McCalla, G., Bredeweg, B., Breuker, J. (eds.) Artificial Intelligence in Education. IOS Press, Amsterdam (2005)

Kay, J., Reimann, P., Yacef, K.: Mirroring of Group Activity to Support Learning as Participation. In: Luckin, R., Koedinger, K.R., Greer, J. (eds.) Artificial Intelligence in Education. IOS Press, Amsterdam (2007)

Kerly, A., Bull, S.: Children's Interactions with Inspectable and Negotiated Learner Models. In: Woolf, B.P., Aïmeur, E., Nkambou, R., Lajoie, S. (eds.) ITS 2008. LNCS, vol. 5091, pp. 132–141. Springer, Heidelberg (2008)

Kumar, A., Maries, A.: The Effect of Open Student Model on Learning: A Study. In: Luckin, R., Koedinger, K.R., Greer, J. (eds.) Artificial Intelligence in Education. IOS Press, Amsterdam (2007)

Lazarinis, F., Retalis, S.: Analyze Me: Open Learner Model in an Adaptive Web Testing System. International Journal of Artificial Intelligence in Education 17(3), 255–271 (2008)

Lee, S.J.H., Bull, S.: An Open Learner Model to Help Parents Help their Children. Technology, Instruction, Cognition and Learning 6(1), 29–51 (2008)

Linton, F., Schaefer, H.-P.: Recommender Systems for Learning: Building User and Expert Models through Long-Term Observation of Application Use. User Modeling and User-Adapted Interaction 10, 181–207 (2000)

Lloyd, T., Bull, S.: A Haptic Learner Model. International Journal of Continuing Engineering Education and Lifelong Learning 16(1-2), 137–149 (2006)

Mazza, R., Dimitrova, V.: Visualising Student Tracking Data to Support Instructors in Web-Based Distance Education. In: 13th International World Wide Web Conference - Alternate Educational Track (2004)

Mabbott, A., Bull, S.: Alternative Views on Knowledge: Presentation of Open Learner Models. In: Lester, J.C., Vicari, R.M., Paraguaçu, F. (eds.) ITS 2004. LNCS, vol. 3220, pp. 689–698. Springer, Heidelberg (2004)

Mabbott, A., Bull, S.: Student Preferences for Editing, Persuading and Negotiating the Open Learner Model. In: Ikeda, M., Ashley, K., Chan, T.-W. (eds.) ITS 2006. LNCS, vol. 4053, pp. 481–490. Springer, Heidelberg (2006)

Mabbott, A., Bull, S.: Comparing Student-Constructed Open Learner Model Presentations to the Domain. In: Koedinger, K., Luckin, R., Greer, J. (eds.) Artificial Intelligence in Education. IOS Press, Amsterdam (2007)

Mazzola, L., Mazza, R.: Toward Adaptive Presentations of Student Models in eLearning Environments. In: Dimitrova, V., Mizoguchi, R., Du Boulay, B., Graesser, A. (eds.) Artificial Intelligence in Education. IOS Press, Amsterdam (2009)

Mitrovic, A., Martin, B.: Evaluating the Effect of Open Student Models on Self-Assessment. International Journal of Artificial Intelligence in Education 17(2), 121–144 (2007)

Mohanarajah, S., Kemp, R.H., Kemp, E.: Opening a Fuzzy Learner Model. In: Proceedings of Workshop on Learner Modelling for Reflection, International Conference on Artificial Intelligence in Education, pp. 62–71 (2005)

Morales, R., Pain, H., Conlan, T.: Effects of Inspecting Learner Models on Learners' Abilities. In: Moore, J.D., Luckhardt Redfield, C., Johnson, W.L. (eds.) Artificial Intelligence in Education. IOS Press, Amsterdam (2001)

Papanikolaou, K.A., Grigoriadou, M.: Sharing Knowledge and Promoting Reflection through the Learner Model. In: Gutierrez-Santos, S., Mavrikis, M. (eds.) Proceedings of the Intelligent Support for Exploratory Environments Workshop, Third European Conference on Technology-Enhanced Learning, EC-TEL 2008 (2008), http://sunsite.informatik.rwth-aachen.de/Publications/CEUR-WS/Vol-381/paper05.pdf

Papanikolaou, K., Grigoriadou, M., Kornilakis, H., Magoulas, G.D.: Personalising the Interaction in a Web-based Educational Hypermedia System: The Case of INSPIRE. User-Modeling and User-Adapted Interaction 13(3), 213–267 (2003)

Perez-Marin, D., Alfonseca, E., Rodriguez, P., Pascual-Neito, I.: A Study on the Possibility of Automatically Estimating the Confidence Value of Students' Knowledge in Generated Conceptual Models. Journal of Computers 2(5), 17–26 (2007)

Perez-Marin, D., Alfonseca, E., Rodriguez, P., Pascual-Neito, I.: Automatic Generation of Students' Conceptual Models from Answers in Plain Text. In: Conati, C., McCoy, K., Paliouras, G. (eds.) UM 2007. LNCS (LNAI), vol. 4511. Springer, Heidelberg (2007b)

Rueda, U., Larrañaga, M., Ferrero, B., Arruarte, A., Elorriaga, J.A.: Study of Graphical Issues in a Tool for Dynamically Visualising Student Models. In: Proceedings of Workshop on Learner Modelling for Reflection (Supplemental Proceedings, 5), International Conference on Artificial Intelligence in Education (2003)

Shahrour, G., Bull, S.: Does 'Notice' Prompt Noticing? Raising Awareness in Language Learning with an Open Learner Model. In: Neijdl, W., Kay, J., Pu, P., Herder, E. (eds.) AH 2008. LNCS, vol. 5149, pp. 173–182. Springer, Heidelberg (2008)

Tchetagni, J., Nkambou, R., Bourdeau, J.: Explicit Reflection in Prolog-Tutor. International Journal of Artificial Intelligence in Education 17(2), 169–215 (2007)

Upton, K., Kay, J.: Narcissus: Group and Individual Models to Support Small Group Work. In: Houben, G.-J., McCalla, G., Pianesi, F., Zancanaro, M. (eds.) UMAP 2009. LNCS, vol. 5535, pp. 54–65. Springer, Heidelberg (2009)

VanLabeke, N., Brna, P., Morales, R.: Opening Up the Interpretation Process in an Open Learner Model. International Journal of Artificial Intelligence in Education 17(3), 305–338 (2007)

Weber, G., Brusilovsky, P.: ELM-ART: An Adaptive Versatile System for Web-Based Instruction. Int. J. of Artificial Intelligence in Education 12, 351–384 (2001)

Yacef, K.: The Logic-ITA in the Classroom: A Medium Scale Experiment. International Journal of Artificial Intelligence in Education 15(1), 41–62 (2005)

Zapata-Rivera, J.D., Greer, J.E.: Interacting with Inspectable Bayesian Models. International Journal of Artificial Intelligence in Education 14, 127–163 (2004)

Zapata-Rivera, D., Hansen, E., Shute, V.J., Underwood, J.S., Bauer, M.: Evidence-Based Approach to Interacting with open Student Models. International Journal of Artificial Intelligence in Education 17(3), 273–303 (2007)

Chapter 16
Mining Data for Student Models

Ryan S.J.d. Baker

Department of Social Science and Policy Studies, Worcester Polytechnic Institute,
100 Institute Road, Worcester, MA 01609 USA
rsbaker@wpi.edu

Abstract. Data mining methods have in recent years enabled the development of more sophisticated student models which represent and detect a broader range of student behaviors than was previously possible. This chapter summarizes key data mining methods that have supported student modeling efforts, discussing also the specific constructs that have been modeled with the use of educational data mining. We also discuss the relative advantages of educational data mining compared to knowledge engineering, and key upcoming directions that are needed for educational data mining research to reach its full potential.

16.1 Introduction

In recent years, student models within intelligent tutoring systems have expanded to include an impressive breadth of constructs about the individual student (or pairs or groups of students), with increasing levels of precision. Student models increasingly assess not just whether a student knows a skill or concept, but a broad range of affective, meta-cognitive, motivational, and behavioral constructs. These advances in student modeling, discussed in some detail in earlier chapters, have been facilitated by advances in educational data mining methods (Baker and Yacef 2009; Romero and Ventura 2007) that leverage fine-grained data about student behavior and performance. As this data becomes increasingly available at large-scale (cf Borgman et al. 2008; Koedinger et al. in press), there are increasing opportunities for developing increasingly sophisticated and broad-based models of the student using an intelligent tutoring system.

In this chapter, we give a brief overview of educational data mining methodologies, focusing on techniques that have seen specific application in order to enrich and otherwise improve student models. We also discuss the key differences between educational data mining and knowledge engineering approaches to enriching student models, and discuss key steps that would facilitate the more rapid application of educational data mining methods to a broader range of educational software and learner constructs.

Data mining, also called Knowledge Discovery in Databases (KDD), is the field of discovering novel and potentially useful information from large amounts

R. Nkambou et al. (Eds.): Advances in Intelligent Tutoring Systems, SCI 308, pp. 323–337.
springerlink.com

of data (Witten and Frank 1999). On the Educational Data Mining community website, www.educationaldatamining.org, educational data mining (abbreviated as EDM) is defined as: "Educational Data Mining is an emerging discipline, concerned with developing methods for exploring the unique types of data that come from educational settings, and using those methods to better understand students, and the settings which they learn in."

There are multiple taxonomies of the areas of educational data mining, one by Romero and Ventura (2007), and one by Baker (in press), also discussed in (Baker and Yacef 2009). The two taxonomies are quite different, and a full discussion of the differences is out of the scope of this chapter (there is a brief discussion of some of the differences between the taxonomies in (Baker and Yacef 2009)). Within this chapter, we focus on the taxonomy from Baker in press; Baker and Yacef 2009, which categorizes research into educational data mining into the following areas:

- Prediction
  - Classification
  - Regression
  - Density estimation
- Clustering
- Relationship mining
  - Association rule mining
  - Correlation mining
  - Sequential pattern mining
  - Causal data mining
- Distillation of data for human judgment
- Discovery with models

Beyond these categories, there is also a continual presence of knowledge engineering methods (Feigenbaum and McCorduck 1983; Studer et al. 1998) within educational data mining conferences. Knowledge engineering methods utilize human judgment to create a model of a construct of interest, rather than doing so in an automated fashion. We will discuss the similarities and differences between knowledge engineering and data mining, and the relative advantages and disadvantages of each class of method for developing student models, later in this document.

In this chapter, we emphasize the student modeling applications of prediction methods, clustering methods, and methods for the distillation of data for human judgment. Relationship mining is a key class of method within educational data mining research and is a frequently-used method in EDM venues, as summarized in Baker and Yacef 2009, but is typically focused more on data mining to promote discovery for scientific researchers or other end users, rather than on improving student models. One exception to this is the use of relationship mining to improve models of domain structure (e.g. Nkambou et al. 2007); recent research along these lines is discussed in detail in another chapter within this volume. Discovery with models research is generally focused on promoting scientific discovery. Discovery with models research involves student models, but in a different direction from the work discussed in this chapter. When discovery with models research involves student models, it leverages the outputs of student models (among other sources) to promote scientific discovery, rather than using automated discovery to promote the improvement of the student models themselves.

Another key way that data mining has influenced student models is by influencing the internal structure of Bayes Nets and Bayesian Knowledge Tracing; these issues are discussed in another chapter within this volume, and are therefore not discussed in detail in this chapter.

16.2 Prediction Methods

In prediction, the goal is to develop a model which can infer a single aspect of the data (predicted variable) from some combination of other aspects of the data (predictor variables). Prediction requires having labels for the output variable for a limited data set, where a label represents some trusted "ground truth" information about the output variable's value in specific cases. Ground truth labels do not need to be perfect in order to be useful for the development of reliable models through data mining; a data mining approach that is not over-fit can accommodate a moderate degree of noise in the original labels, so long as the labels are not systematically biased. Labels which have noise are sometimes referred to as "bronze-standard" labels. The degree of noise in the original labels can often be assessed by assessing the inter-rater reliability of the labels, frequently with Cohen's Kappa (Cohen 1960).

Broadly, there are three types of prediction: classification, regression, and density estimation. Classification and regression historically have played more prominent roles in educational data mining than density estimation. In classification, the predicted variable is a binary or categorical variable. In regression, the predicted variable is quantitative. In both cases, any type of input data is possible, although some algorithms are not able to handle all types of data.

The range of prediction methods used in educational data mining approximately corresponds to the types of prediction methods used in data mining more broadly; however, the techniques emphasized in educational data mining have varied from those most popular in other domains. In particular, support vector machines (Steinwart and Christmann 2008) and neural networks (Grossberg 1988), popular methods in other domains, have been relatively less emphasized in educational data mining. Contrastingly, linear methods have been relatively more emphasized. It is not necessarily the case that educational data is particularly likely to be linear – in fact, many have argued that educational and learning data frequently have a non-linear character (Marder and Bansal 2009; Teigen 1994). However, the relatively high noise in educational data, combined with the relative expense of labeling data in many cases, may bias in favor of approaches which are less likely to over-fit. Over-fitting is when a model does well on the original training data, as the expense of doing poorly on new data (Hawkins 2004), in this case, data from new students or new intelligent tutor lessons.

Another distinctive feature of educational data mining research is the use of methods involving modeling frameworks drawn from the psychometrics literature (cf. Maris 1995), in combination with machine-learning space-searching techniques (cf. Yu and Liu 2003) (many examples of this also exist in data mining for domain models, discussed in another chapter in this volume). These methods have the benefit of explicitly accounting for meaningful hierarchy and non-independence in data. For instance, it can be important to detect both which students engage in a behavior of interest, and exactly when (or at least in which parts of the interface) an individual student is engaging in that behavior (cf. Baker et al. 2008).

One classification method with prominence both in educational data mining and in data mining within other domains is decision trees (Quinlan 1993), used in a considerable amount of educational data mining research (cf. Baker et al. 2008; Merceron and Yacef 2005; Walonoski and Heffernan 2006). Decision trees are able to handle both quantitative and categorical features in making model assessment, a benefit given the highly heterogenous feature data often generated by tutors (cf. Baker et al. 2006; Beal et al. 2006; Cetintas et al. 2009; D'Mello et al. 2008; Walonoski and Heffernan 2006), and can explicitly control for over-fitting with a "pruning" step (Quinlan 1993).

Educational data mining methods have enabled the construction of student models (or student models components) of a wide number of constructs. For instance, classification and regression methods have been used to develop detectors of gaming the system (Baker et al. 2008; Baker and de Carvalho 2008; Walonoski and Heffernan 2006). These detectors have accurately predicted differences in student learning (Cocea et al. 2009), and have been embedded into intelligent tutoring systems and used to drive adaptive behavior (cf. Baker et al. 2006; Walonoski and Heffernan 2006b). Similarly, classification methods have been used to develop detectors of student affect, including frustration, boredom, anxiety, engaged concentration, joy, and distress (Conati and McLaren 2009; D'Mello et al. 2008). Detectors of affect and emotion have been used to drive automated adaptation to differences in student affect, significantly reducing students' frustration and anxiety (Woolf et al. in press) and increasing the incidence of positive emotion (Chaffar et al. 2009). Classification methods have also been used to develop detectors of off-task behavior (Baker 2007; Cetintas et al. 2009), predicting differences in student learning. Additionally, classification methods have been used to infer low self-efficacy (McQuiggan et al. 2008) and slipping (Baker et al. 2008).

Classification methods have also enabled improvements to Bayesian Modeling of student knowledge, discussed in another chapter in this volume. For instance, (Baker et al. 2008) integrates models of student slipping into Bayesian Knowledge Tracing, leading to more accurate prediction of future student performance within Cognitive Tutors.

16.3 Clustering

In clustering, the goal is to find data points that naturally group together, splitting the full data set into a set of groups, called clusters. Clustering does not require (and does not use) labels of any output variable; the data is clustered based solely on internal similarity, not on any metric of specific interest. If a set of clusters is optimal, within a category, each data point will in general be more similar to the other data points in that cluster than data points in other clusters. The range of clustering methods used in educational data mining approximately corresponds to the types of prediction methods used in data mining more broadly, including algorithms such as k-means (Hartigan and Wong 1979) and Expectation Maximization (EM)-Based Clustering (Bradley et al. 1998), and model frameworks such as Gaussian Mixture Models (Reynolds 2008).

Clustering has been used to develop student models for several types of educational software, including intelligent tutoring systems. In particular, fine-grained models of student behavior at the action-by-action level are clustered in terms of features of the student actions. For instance, Amershi & Conati used clustering on student behavior within an exploratory learning environment, discovering that certain types of reflective behavior and strategic advancement through the learning task were associated with better learning (Amershi and Conati 2009). In addition, Beal and her colleagues applied clustering to study the categories of behavior within an intelligent tutoring system (Beal et al. 2006). Other prominent research has investigated how clustering methods can assist in content recommendation within e-learning (Tang and McCalla 2005; Zaïane 2002).

Clustering is generally most useful when relatively little is known about the categories of interest in the data set, such as in types of learning environment not previously studied with educational data mining methods (e.g. Amershi and Conati 2009) or for new types of learner-computer interaction, or where the categories of interest are unstable, as in content recommendation (e.g. Tang and McCalla 2005; Zaïane 2002). The use of clustering in domains where a considerable amount is already known brings some risk of discovering phenomena that are already known. As work in other areas of EDM goes forward, an increasing amount is known about student behavior across learning environments. One potential future use of clustering, in this situation, would be to use clustering as a second stage in the process of modeling student behavior in a learning system. First, existing detectors could be used to classify known categories of behavior. Then, data points not classified as belonging to any of those known behavior categories could be clustered, in order to search for unknown behaviors. Expectation Maximization (EM)-Based Clustering (Bradley et al. 1998) is likely to be a method of particularly high potential for this, as it can explicitly incorporate already known categories into an initial starting point for clustering.

16.4 Distillation of Data for Human Judgment

One key recent trend facilitating the use of educational data mining methods to improve student models is the advance in methods for distilling data for human

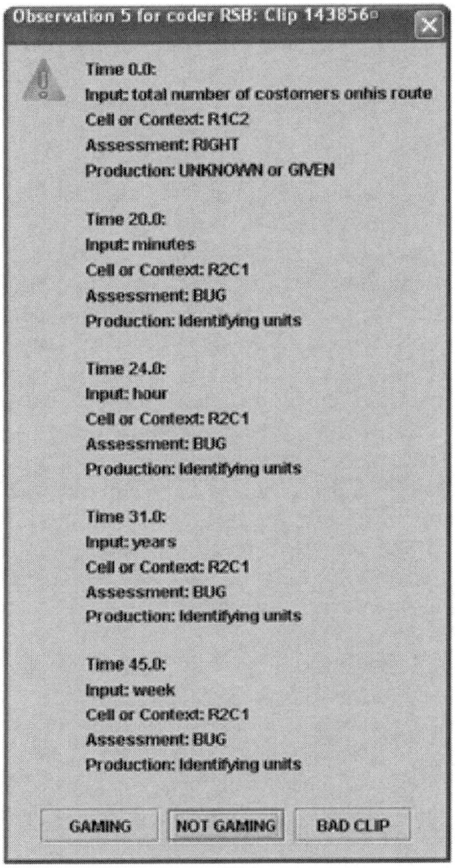

Fig. 16.1 A text replay of student behavior in a Cognitive Tutor (from Baker and de Carvalho 2008)

judgment. In many cases, human beings can make inferences about data, when it is presented appropriately, that are beyond the immediate scope of fully automated data mining methods. The information visualization methods most commonly used within EDM are often different than those most often used for other information visualization problems (cf. Hershkovitz and Nachmias 2008; Kay et al. 2006), owing to the specific structure often present in intelligent tutor data, and the meaning embedded within that structure. For instance, data is meaningfully organized in terms of the structure of the learning material (skills, problems, units, lessons) and the structure of learning settings (students, teachers, collaborative pairs, classes, schools).

Data is distilled for human judgment in educational data mining for two key purposes: classification and identification. One key area of development of data

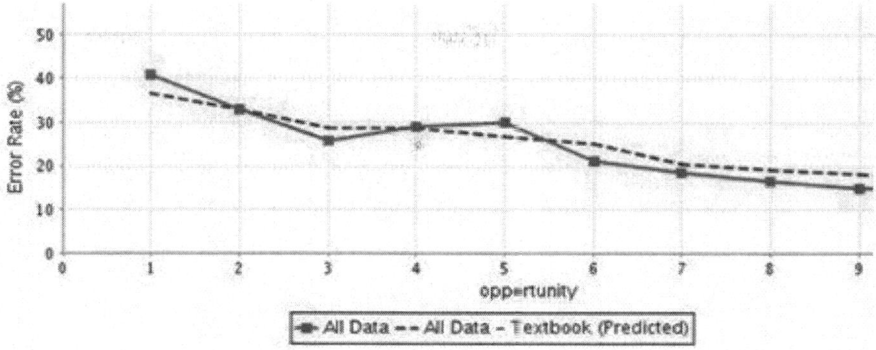

Fig. 16.2 A learning curve of student performance over time in a Cognitive Tutor (from Koedinger et al in press)

distillations supporting classification is the text replay methodology (Baker et al. 2006). An example of a text replay is shown in Figure 16.1. In this case, subsections of a data set are displayed in text format, and labeled by human coders. These labels are then generally used as the basis for the development of a predictor. Text replays are significantly faster than competing methods for labeling, such as quantitative field observations or video coding (Baker et al. 2006; Baker and de Carvelho 2008), and achieve good inter-rater reliability (Baker et al. 2006; Baker et al. in press). Text replays have been used to support the development of prediction models of gaming the system in multiple learning environments (Baker et al. 2006; Baker et al. in press), and to develop models of scientific reasoning skill in inquiry learning environments (Montalvo et al. paper under review; Sao Pedro et al. paper under review). An alternate approach, displaying a re-constructed replay of a student's screen, has also been used to label student data for use in classification (cf. de Vicente et Pain 2002); however, this approach has become less common, as it is significantly slower than text replays (cf. Baker et al. 2006), while not giving more information about student behavior or expression outside the system, unlike methods such as quantitative field observation and video methods.

Identification of learning patterns and learner individual differences from visualizations is a key method for exploring educational data sets. For instance, Hershkovitz and Nachmias's learnograms provide a rich representation of student behavior over time (Hershkovitz and Nachmias 2008). Within the domain of student models, a key use of identification with distilled and visualized data is in inference from learning curves, as shown in Figure 16.2. A great deal can be inferred from learning curves about the character of learning in a domain (Corbett and Anderson 1995; Koedinger et al. in press), as well as about the quality of the domain model. Classic learning curves display the number of opportunities to practice a skill on the X axis, and display performance (such as percent correct or time taken to respond) on the Y axis. A curve with a smooth downward progression that is steep at first and gentler later indicates that successful learning is occurring. A flatter curve, as in Figure 16.2, indicates that learning is occurring, but with significant difficulty. A sudden spike upwards, by contrast, indicates that more than one knowledge component is included in the model (cf. Corbett and Anderson 1995). A flat high curve indicates poor learning of the skill, and a flat low curve indicates that the skill did not need instruction in the first place. An upwards curve indicates the difficulty is increasing too fast. Hence, learning curves are a powerful tool to support quick inference about the character of learning in an educational system, leading to their recent incorporation into tools used by education researchers outside of the educational data mining community (e.g. Koedinger et al. in press).

16.5 Knowledge Engineering and Data Mining

An alternate method for developing student models is knowledge engineering (Feigenbaum and McCorduck 1983; Studer et al 1998). Knowledge engineering approaches develop models that can engage in problem-solving, reasoning, or decision making, making the same decisions that a human expert would; they can do so simply by replicating the decision-making results or by attempting to develop a cognitive model that reasons in the same fashion that a human expert would. As a method, knowledge engineering relies upon human researchers studying the construct of interest, and directly developing – engineering – the model of the construct of interest. The mapping between features of the data set and the construct of interest is directly made by the engineer. As such, knowledge engineering can be contrasted to classification or regression, which use labels generated through expert decision-making but develop the mapping between the features of the data set and the construct of interest through an automated process.

Knowledge engineering is frequently used to develop domain models, as discussed in another chapter in this volume. Within student modeling, knowledge engineering has been a prominent method for modeling sophisticated student behaviors within intelligent tutoring systems, with a focus on gaming the system and help-seeking behaviors. For instance, Beal et al (2006) used knowledge engineering to model gaming the system. Shih et al (2008) used knowledge engineering to develop a mathematical model that could detect self-explanation and appropriate use of bottom-out hints. Buckley et al (2006) used knowledge engineering to assess students' level of systematicity during problem-solving in interactive simulations.

Within student modeling, knowledge engineering is frequently used to develop models of sophisticated student behavior in a hybrid fashion, where knowledge engineering is used to develop the functional form of a mathematical model, and then automated parameter-fitting is used to find (or refine) values for the parameters of that model. For instance, Aleven et al (2004, 2006) developed a model of a range of student help-seeking behaviors in Cognitive Tutors, using knowledge engineering to develop the functional form of a mathematical model, and then automated parameter-fitting to find values for the parameters of that model. Several of the components of Aleven et al's model predicted differences in student learning. In another example, Beck (2005) presented a model of hasty guessing (called disengagement in the original paper, but renamed hasty guessing in later work) in an intelligent tutor for reading, developed using knowledge engineering to develop an item-response theory model, and then using automated parameter-fitting to find values for the parameters of that model. Beck's model successfully predicted differences in student post-test scores. Johns and Woolf (2006) used a similar combination of knowledge engineering and parameter fitting to model gaming the system.

In addition, educational data mining research often also involves some degree of knowledge engineering during the process of generating the data set features to use within classification or regression. During this step of the data mining process, researchers often attempt infer what types of features an expert coder would use – although this trend is diminishing as features are increasingly re-used in creating new data mining models, either from the same research group, or across research groups (cf. Baker et al 2008; Baker et al. 2008; Cetintas et al. 2009; Walonoski and Heffernan 2006).

As can be seen, knowledge engineering and educational data mining have both been used to model gaming the system. Aside from this overlap, the two approaches have been used to model different phenomena, with knowledge engineering methods emphasized in modeling help-seeking while educational data mining methods have been emphasized in modeling affect, self-efficacy, and off-task behavior. It is worth noting that the domains emphasized in educational data mining are often cases where recognition is fairly easy for humans (e.g. it is feasible to tell that a student is bored by looking at him/her – e.g. D'Mello et al. 2007), but where it is difficult to analyze exactly how those decisions are made in terms of features of data available in the log files. In these cases, an automated process that can test large numbers of alternatives can be considerably more time-efficient than attempting to develop such a mapping through pure rational thought.

In terms of accuracy or effectiveness, educational data mining and knowledge engineering have largely not been directly compared. Mostly they have been used to model different phenomena; even when the same phenomena has been modeled with both educational data mining and knowledge engineering methods, it has been modeled in different learning systems. One exception, however, exists in models of gaming the system within Cognitive Tutors. Roll et al. (2005) compared early versions of knowledge engineered and data mined models of gaming the system (e.g. Aleven et al. 2004; Baker et al. 2004). Roll and his colleagues found significant correlation between the predictions made by the two models. However, despite that correlation, they found the data mined model was substantially more successful in predicting human labels of gaming behavior than the knowledge engineered model, suggesting that the data mined model had higher construct validity. The comparison used in this paper was not cross-validated, but the data mined model also performed better in predicting student behavior during cross-validation (e.g. Baker et al. 2004). Beyond this, Roll and colleagues found that both models successfully predicted post-test performance, although in both cases the relationship was unstable (see Cocea et al. 2009 for a discussion of this issue for the data mined model). This suggests that despite the higher construct validity of the data mining model, both models captured the underlying phenomena in qualitatively similar fashions.

This single study, of course, is not conclusive proof that educational data mining achieves higher construct validity than knowledge engineering. In order to investigate this question more thoroughly, it will be necessary to conduct a broader comparison, involving a variety of constructs, and held-out data including new students. One competititon that may produce evidence that can be used to infer the relative accuracy and predictive power of knowledge engineered and data mined methods in educational data will occur in the next year. The 2010 KDD Cup (to be announced at http://www.sigkdd.org/kddcup/index.php) will involve predicting future student correctness in held-out intelligent tutor data from the Pittsburgh Science of Learning Center DataShop (Koedinger et a., in press), and is likely to attract submissions of both types.

16.6 Key Future Directions

Though educational data mining methods have contributed in significant ways to the sophistication of student models, there are two factors that are currently slowing the extension of these improvements to the full range of intelligent tutoring systems and learner characteristics.

One significant limitation is the relatively low degree of investigation into the generalizability of models between or within intelligent tutoring systems. It is becoming increasingly common to use cross-validation at the student level to verify that a model is applicable to new students; however, it remains rare for researchers to validate that a model generalizes across subsets of an intelligent tutoring curricula. In one example of this type of validation, Baker et al (2008) determined that their data-mined model of gaming the system remained accurate within new intelligent tutor lessons drawn from the same overall system and curricula, using cross-validation at the lesson level. However, few other examples exist in the published literature.

Furthermore, the author of this chapter is not aware of any papers that explicitly study whether any models remain accurate when applied to different tutoring systems. Some papers have studied a related issue, whether the model features utilized in a model of a given construct in one tutoring system are effective in a different tutoring system (e.g. Cetintas et al. 2010; Walonoski and Heffernan 2006; Baker et al. 2008). However, these papers largely have restricted themselves to simply noting the common features rather than explicitly studying whether adding these features leads to a more accurate model. Studying the transfer of data-mined models to new intelligent tutors is a highly important area of future work. So long as it is necessary to develop an entirely new model for each new intelligent tutoring system, the process of extending the advances in student modeling made through data mining to all tutoring systems will be slowed considerably. The methods from the data mining subfield of transfer learning (cf. Dai et al. 2007; Daume 2007), transferring data-mined models to new contexts and new sampling distributions, may have a considerable amount to contribute to research in this area. More collaboration between transfer learning and educational data mining researchers would likely benefit both communities (providing a rich set of challenges to transfer learning researchers), as well as benefiting the developers of student models for intelligent tutoring systems.

A second limitation to educational data mining thus far is the lack of tools explicitly designed to support educational data mining research. Data mining tools such as Weka (Witten and Frank 1999) and KEEL (Alcalá-Fdez et al. 2009) support data mining practitioners in utilizing well-known data mining algorithms on their data, with a usable user interface; however, these tools' default user interfaces currently do not support the types of cross-validation that are necessary in educational data to infer generalizability across students. This has led to a considerable amount of EDM research that does not take these issues into account. It is possible to conduct cross-validation across data levels in another tool, RapidMiner (Mierszwa et al. 2006). RapidMiner does not directly support student-level or lesson-level cross-validation, but its "batch cross-validation" functionality makes it possible to conduct student-level or lesson-level cross-validation through predefining student batches outside of the data mining software. Beyond this issue, no data mining tool is currently integrated with tools for the text replay, survey, and quantitative field observation methods increasingly used to label data for using classification or regression, for student models. Integrating data mining tools with data labeling tools and providing support for conducting appropriate validation of generalizability would significantly facilitate research in using data mining to improve student models.

Even without these types of support, research into using data mining to support the development of student models has made significant impacts in recent years. To the degree that researchers address these limitations, the impact of educational data mining can be magnified still further.

16.7 Conclusion

In this chapter, we have discussed how data mining methods have contributed to the development of student models for intelligent tutoring systems. In particular, we have discussed the contribution to student modeling coming from classification methods, regression methods, clustering methods, and methods for the distillation of data for human judgment. Classification, regression, and clustering methods have supported the development of validated models of a variety of complex constructs that have been embedded into increasingly sophisticated student models, enabling broader-based adaptation to individual differences between students than was previously possible. Clustering methods have supported the discovery of how students choose to respond to new types of educational human-computer interactions, enriching student models; classification and regression models have afforded accurate and validated models of a broader range of student behavior. Among other constructs, these methods have supported the development of models of gaming the system, help-seeking, boredom, frustration, confusion, engaged concentration, self-efficacy, scientific reasoning strategies, and off-task behavior. Distillation of data for human judgment has itself facilitated the development of models of this nature, speeding the process of labeling data with reference to difference in student behaviors, in turn speeding the process of creating classification and regression models. In turn, these discoveries have increased the sophistication and richness of student models, covering a broader range of behavior. These richer student models have afforded more broad-based adaptation to student individual differences, significantly improving student learning.

Acknowledgements. The development of this paper was supported by the Pittsburgh Science of Learning Center (National Science Foundation) via grant "Toward a Decade of PSLC Research", award number SBE-0836012, by the National Science Foundation via grant "AMI: ASSISTments Meets Inquiry", award DRL#0733286, and by the U.S. Department of Education via grant "Promoting Robust Understanding of Genetics with a Cognitive Tutor that Integrates Conceptual Learning with Problem Solving", award number #R305A090549. All opinions presented in this documents represent the opinions of the author, and do not necessarily represent these opinions or viewpoints of any of the funders.

References

Alcala-Fdez, J., Sánchez, L., Garcıa, S., del Jesús, M., Ventura, S., Garrell, J., Otero, J., Romero, C., Bacardit, J., Rivas, V., Fernández, J., Herrera, F.: KEEL: A software tool to assess evolutionary algorithms for data mining problems. Soft Computing 13(3), 307–318 (2008)

Aleven, V., McLaren, B.M., Roll, I., Koedinger, K.R.: Toward tutoring help seeking: Applying cognitive modeling to meta-cognitive skills. In: Lester, J.C., Vicari, R.M., Paraguaçu, F. (eds.) ITS 2004. LNCS, vol. 3220, pp. 227–239. Springer, Heidelberg (2004)

Aleven, V., McLaren, B., Roll, I., Koedinger, K.: Toward meta-cognitive tutoring: A model of help seeking with a Cognitive Tutor. International Journal of Artificial Intelligence and Education 16, 101–128 (2006)

Amershi, S., Conati, C.: Combining Unsupervised and Supervised Classification to Build User Models for Exploratory Learning Environments. Journal of Educational Data Mining 1(1), 18–71 (2009)

Arroyo, I., Cooper, D., Burleson, W., Woolf, B.P., Muldner, K., Christopherson, R.: Emotion Sensors Go To School. In: Proceedings of the International Conference on Artificial Intelligence in Education, pp. 17–24 (2009)

Baker, R.S.J.d.: Data Mining for Education. In: McGaw, B., Peterson, P., Baker, E. (eds.) To appear in International Encyclopedia of Education, 3rd edn. Elsevier, Oxford (in press)

Baker, R.S.J.d.: Modeling and Understanding Students' Off-Task Behavior in Intelligent Tutoring Systems. In: Proceedings of ACM CHI, Computer-Human Interaction, pp. 1059–1068 (2007)

Baker, R.S.J.d., de Carvalho, A.M.J.A.: Labeling Student Behavior Faster and More Precisely with Text Replays. In: Proceedings of the 1st International Conference on Educational Data Mining, pp. 38–47 (2008)

Baker, R.S.J. d., Yacef, K.: The State of Educational Data Mining in 2009: A Review and Future Visions. Journal of Educational Data Mining 1(1), 3–17 (2009)

Baker, R.S.J.d., Corbett, A.T., Aleven, V.: More Accurate Student Modeling Through Contextual Estimation of Slip and Guess Probabilities in Bayesian Knowledge Tracing. In: Woolf, B.P., Aïmeur, E., Nkambou, R., Lajoie, S. (eds.) ITS 2008. LNCS, vol. 5091, pp. 406–415. Springer, Heidelberg (2008)

Baker, R.S., Corbett, A.T., Koedinger, K.R.: Detecting Student Misuse of Intelligent Tutoring Systems. In: Lester, J.C., Vicari, R.M., Paraguaçu, F. (eds.) ITS 2004. LNCS, vol. 3220, pp. 531–540. Springer, Heidelberg (2004)

Baker, R.S.J.d., Corbett, A.T., Koedinger, K.R., Evenson, S., Roll, I., Wagner, A.Z., Naim, M., Raspat, J., Baker, D.J., Beck, J.: Adapting to when students game an intelligent tutoring system. In: Ikeda, M., Ashley, K.D., Chan, T.-W. (eds.) ITS 2006. LNCS, vol. 4053, pp. 392–401. Springer, Heidelberg (2006)

Baker, R.S.J.d., Corbett, A.T., Roll, I., Koedinger, K.R.: Developing a Generalizable Detector of When Students Game the System. User Modeling and User-Adapted Interaction 18(3), 287–314 (2008)

Baker, R.S.J.d., Corbett, A.T., Wagner, A.Z.: Human Classification of Low-Fidelity Replays of Student Actions. In: Proceedings of the Educational Data Mining Workshop at the 8th International Conference on Intelligent Tutoring Systems, pp. 29–36 (2006)

Baker, R.S.J.d., Mitrovic, A., Mathews, M.: Detecting Gaming the System in Constraint-Based Tutors. To appear in Proceedings of the 18th International Conference on User Modeling, Adaptation, and Personalization (in press)

Beal, C.R., Qu, L., Lee, H.: Classifying learner engagement through integration of multiple data sources. In: Proceedings of the 21st National Conference on Artificial Intelligence, Boston, pp. 2–8 (2006)

Beck, J.: Engagement tracing: using response times to model student disengagement. In: Proceedings of the 12th International Conference on Artificial Intelligence in Education (AIED 2005), pp. 88–95 (2005)

Borgman, C.L., Abelson, H., Dirks, L., Johnson, R., Koedinger, K.R., Linn, M.C., Lynch, C.A., Oblinger, D.G., Pea, R.D., Salen, K., Smith, M.S., Szalay, A.: Fostering Learning in the Networked World: The Cyberlearning Opportunity and Challenge. In: A 21st Century Agenda for the National Science Foundation, National Science Foundation, Washington, DC (2008)

Bradley, P.S., Fayyad, U., Reina, C.: Scaling EM (expectation maximization) clustering to large databases. Technical Report MSR-TR-98-35 Microsoft Research. Redmond,WA (1998)

Buckley, B.C., Gobert, J., Horwitz, P.: Using log files to track students' model-based inquiry. In: Proceedings of the 7th International Conference on Learning Sciences: 57-63 (2006)

Cetintas, S., Si, L., Xin, Y.P., Hord, C.: Automatic Detection of Off-Task Behaviors in Intelligent Tutoring Systems with Machine Learning Techniques. IEEE Transactions on Learning Technologies (2009)

Chaffar, S., Derbali, L., Frasson, C.: Inducing positive emotional state in Intelligent Tutoring Systems. In: Proceedings of the 14th International Conference on Artificial Intelligence in Education (2009)

Cocea, M., Hershkovitz, A., Baker, R.S.J.d.: The Impact of Off-task and Gaming Behaviors on Learning: Immediate or Aggregate? In: Proceedings of the 14th International Conference on Artificial Intelligence in Education, pp. 507–514 (2009)

Cohen, J.: A coefficient of agreement for nominal scales. Educational and Psychological Measurement 20(1), 37–46 (1960)

Conati, C., Maclaren, H.: Empirically building and evaluating a probabilistic model of user affect. User Modeling and User-Adapted Interaction 19(3), 267–303 (2009)

Corbett, A.T., Anderson, J.R.: Knowledge Tracing: Modeling the Acquisition of Procedural Knowledge. User Modeling and User-Adapted Interaction 4, 253–278 (2007)

Dai, W., Yang, Q., Xue, G., Yu, Y.: Boosting for transfer learning. In: Proceedings of the International Conference on Machine Learning (2007)

Daume, H.: Frustratingly easy domain adaptation. In: the Proceedings of the Association for Computational Linguistics (2007)

de Vicente, A., Pain, H.: Informing the detection of the students' motivational state: an empirical study. In: Cerri, S.A., Gouardéres, G., Paraguaçu, F. (eds.) ITS 2002. LNCS, vol. 2363, pp. 933–943. Springer, Heidelberg (2002)

D'Mello, S.K., Craig, S.D., Witherspoon, A.W., McDaniel, B.T., Graesser, A.C.: Automatic Detection of Learner's Affect from Conversational Cues. User Modeling and User-Adapted Interaction 18(1-2), 45–80 (2008)

D'Mello, S.K., Picard, R., Graesser, A.C.: Towards an Affect Sensitive Auto-Tutor. IEEE Intelligent Systems 22(4), 53–61 (2007)

Feigenbaum, E.A., McCorduck, P.: The fifth generation: Artificial Intelligence and Japan's Computer Challenge to the World. Addison-Wesley, Reading (1983)

Grossberg, S.: Nonlinear neural networks: principles, mechanisms and architectures. Neural Networks 1, 17–61 (1988)

Hartigan, J.A., Wong, M.A.: A K-means clustering algorithm. Applied Statistics 28, 100–108 (1979)

Hawkins, D.M.: The Problem of Overfitting. Journal of Chemical Infomration and Modeling 44, 1–12 (2004)

Hershkovitz, A., Nachmias, R.: Developing a Log-Based Motivation Measuring Tool. In: Proceedings of the First International Conference on Educational Data Mining, pp. 226–233 (2008)

Johns, J., Woolf, B.: A Dynamic Mixture Model to Detect Student Motivation and Proficiency. In: Proceedings of the 21st National Conference on Artificial Intelligence (AAAI 2006), pp. 163–168 (2006)

Kay, J., Maisonneuve, N., Yacef, K., Reimann, P.: The Big Five and Visualisations of Team Work Activity. In: Ikeda, M., Ashley, K.D., Chan, T.-W. (eds.) ITS 2006. LNCS, vol. 4053, pp. 197–206. Springer, Heidelberg (2006)

Koedinger, K.R., Baker, R.S.J.d., Cunningham, K., Skogsholm, A., Leber, B., Stamper, J.: A Data Repository for the EDM community: The PSLC DataShop. In: Romero, C., Ventura, S., Pechenizkiy, M., Baker, R.S.J.d. (eds.) To appear in Handbook of Educational Data Mining. CRC Press, Boca Raton (in press)

Marder, M., Bansal, D.: Flow and diffusion of high-stakes test scores. Proceedings of the National Academy of Sciences of the United States 106(41), 17267–17270 (2009)

Maris, E.: Psychometric Latent Response Models. Psychometrika 60(4), 523–547 (1995)

McQuiggan, S., Mott, B., Lester, J.: Modeling Self-Efficacy in Intelligent Tutoring Systems: An Inductive Approach. User Modeling and User-Adapted Interaction 18, 81–123 (2008)

Merceron, A., Yacef, K.: Educational Data Mining: a Case Study. In: Proceedings of the International Conference on Artificial Intelligence in Education (AIED 2005), pp. 467–474. IOS Press, Amsterdam (2005)

Mierswa, I., Wurst, M., Klinkenberg, R., Scholz, M., Euler, T.: YALE: Rapid Prototyping for Complex Data Mining Tasks. In: Proceedings of the 12th ACM SIGKDD International Conference on Knowledge Discovery and Data Mining (KDD 2006), pp. 935–940 (2006)

Montalvo, O., Baker, R.S.J.d., Sao Pedro, M., Nakama, A., Gobert, J.D.: Recognizing Student Planning with Machine Learning (paper under review)

Nkambou, R., Nguifo, E.M., Couturier, O., Fourier-Vigier, P.: Problem-Solving Knowledge Mining from Users' Actions in an Intelligent Tutoring System. In: Kobti, Z., Wu, D. (eds.) Canadian AI 2007. LNCS (LNAI), vol. 4509, pp. 393–404. Springer, Heidelberg (2007)

Quinlan, J.R.: C4.5: Programs for Machine Learning. Morgan Kaufmann, San Francisco (2008)

Reynolds, D.A.: Gaussian Mixture Models. Encyclopedia of Biometric Recognition. Springer, Heidelberg (2008)

Roll, I., Baker, R.S., Aleven, V., McLaren, B.M., Koedinger, K.R.: Modeling Students' Metacognitive Errors in Two Intelligent Tutoring Systems. In: Ardissono, L., Brna, P., Mitrović, A. (eds.) UM 2005. LNCS (LNAI), vol. 3538, pp. 379–388. Springer, Heidelberg (2005)

Romero, C., Ventura, S.: Educational Data Mining: A Survey from 1995 to 2005. Expert Systems with Applications 33, 125–146 (2007)

Sao Pedro, M., Baker, R.S.J.d., Montalvo, O., Nakama, A., Gobert, J.D.: Using Text Replay Tagging to Produce Detectors of Systematic Experimentation Behavior Patterns (Paper under review)

Shih, B., Koedinger, K.R., Scheines, R.: A response time model for bottom-out hints as worked examples. In: Baker, R.S.J.d., Barnes, T., Beck, J.E. (eds.) Proceedings of the First International Conference on Educational Data Mining, pp. 117–126 (2008)

Steinwart, I., Christmann, A.: Support Vector Machines. Springer, Berlin (2008)

Studer, R., Benjamins, V.R., Fensel, D.: Knowledge engineering: Principles and methods. Data and Knowledge Engineering (DKE) 25(1–2), 161–197 (1998)

Tang, T., McCalla, G.: Smart recommendation for an evolving e-learning system: architecture and experiment. International Journal on E-Learning 4(1), 105–129 (2005)

Teigen, K.H.: Yerkes-Dodson: a law for all seasons. Theory & Psychology 4(4), 525–547 (1994)

Walonoski, J.A., Heffernan, N.T.: Detection and Analysis of Off-Task Gaming Behavior in Intelligent Tutoring Systems. In: Ikeda, M., Ashley, K.D., Chan, T.-W. (eds.) ITS 2006. LNCS, vol. 4053, pp. 382–391. Springer, Heidelberg (2006)

Walonoski, J.A., Heffernan, N.T.: Prevention of Off-Task Gaming Behavior in Intelligent Tutoring Systems. In: Ikeda, M., Ashley, K.D., Chan, T.-W. (eds.) ITS 2006. LNCS, vol. 4053, pp. 722–724. Springer, Heidelberg (2006)

Witten, I.H., Frank, E.: Data mining: Practical Machine Learning Tools and Techniques with Java Implementations. Morgan Kaufmann, San Francisco (1999)

Woolf, B.P., Arroyo, I., Muldner, K., Burleson, W., Cooper, D., Dolan, R., Christopherson, R.M.: The Effect of Motivational Learning Companions on Low-Achieving Students and Students with Learning Disabilities. To appear in the Proceedings of the International Conference on Intelligent Tutoring Systems (in press)

Yu, L., Liu, H.: Feature selection for high-dimensional data: a fast correlation-based filter solution. In: Proceedings of the International Conference on Machine Learning, pp. 856–863 (2003)

Zaïane, O.: Building a recommender agent for e-learning systems. In: Proceedings of the International Conference on Computers in Education, pp. 55–59 (2002)

Chapter 17
Managing Learner's Affective States in Intelligent Tutoring Systems

Claude Frasson and Pierre Chalfoun

HERON Laboratory, University Of Montreal, 2920 Chemin de la Tour, Montreal, H3C 3J7, Canada
{chalfoun,frasson}@iro.umontreal.ca

Abstract. Recent works in Computer Science, Neurosciences, Education, and Psychology have shown that emotions play an important role in learning. Learner's cognitive ability depends on his emotions. We will point out the role of emotions in learning, distinguishing the different types and models of emotions which have been considered until now. We will address an important issue concerning the different means to detect emotions and introduce recent approaches to measure brain activity using Electroencephalograms (EEG). Knowing the influence of emotional events on learning it becomes important to induce specific emotions so that the learner can be in a more adequate state for better learning or memorization. To this end, we will introduce the main components of an emotionally intelligent tutoring system able to recognize, interpret and influence learner's emotions. We will talk about specific virtual agents that can influence learner's emotions to motivate and encourage him and involve a more cooperative work, particularly in narrative learning environments. Pushing further this paradigm, we will present the advantages and perspectives of subliminal learning which intervenes without conscious perception. Finally, we conclude with new directions to emotional learning.

17.1 Emotions and Learning

Learning involves mainly two processes: reasoning and memorizing. Reasoning is developed during cognitive tasks in which a learner tries to solve a problem using deductions or inductions. If a new knowledge (fact, rule) is obtained then the knowledge will be memorized in long term memory. Memorizing is also triggered when a learner tries to remember a previously acquired element using several means such as similarity, imagery, case-based analysis, etc. The two processes work alternatively and can be considered as a "Cartesian" approach of brain functions. Recent researches in neurosciences, education, and psychology have shown that emotions play an important role in learning. People often separate emotions and reason, believing that emotions are an obstacle in rational decision making or

R. Nkambou et al. (Eds.): Advances in Intelligent Tutoring Systems, SCI 308, pp. 339–358.
springerlink.com

reasoning but recent work have shown that in every case the cognitive process of an individual is strongly dependent on his emotions which can drastically influence performance (Damasio 1994; Spering et al. 2005). Numerous students submitted to an examination have been faced to stress and anxiety and by consequence to the "memory blank", a situation in which they are unable to neither retrieve any information nor make any deductions. Generally, negative emotions reduce or block thought processes, slow down the decisions and memory capacity (Idzihowski and Baddeley 1987). Positive emotions provide better conditions for problem solving and improve innovation. Students expressing anxiety or showing hyperactivity will not store knowledge efficiently (Isen 2000; Pekrun 2008). The learning process involves in particular three cognitive processes, namely attention, memorization, and reasoning, with respect to each of which the learner's cognitive ability depends on his emotions.

Attention means focusing. Learning can take place only if the student listens. A necessary condition for successful learning is hence to gain the *attention* of the student (Gagné 1985) depending on the learner's emotions, this first step can be more or less difficult. In fact, strong emotions, particularly if they are negative, disturb any kind of attention and concentration and prevent the learner from focusing on a subject. Moreover, negative emotions lead to difficulties in switching the attention to a new focus (Compton 2000).

Memory is one of the most important concepts in learning. Knowledge memorization represents a principal objective of instruction and is necessary for the major complex cognitive tasks of learning process (Bloom 1994). Memorizing is a process which involves the storage of information as well as the retrieval of knowledge and the following properties (Bower 1992):

- The effectiveness of the memorization process is narrowly linked to someone's emotions: Whereas positive emotions enhance the memory performance in general, unrelated and negative emotions disturb in particular the retrieval process.
- On the other hand, emotions which are related to the content to be memorized help in storing it.
- The retrieval process is improved when the emotional state is closest to the one at the time the desired information to retrieve was memorized.

Reasoning is the subsequent task attached to memory in which the learner is supposed to reason with the acquired knowledge. The process of reasoning enables a learner to perform more complex cognitive tasks – such as comprehension, classification, application of knowledge, analysis, and problem solving – which represent the ultimate goal of the learning process (Bloom 1994):

- Positive emotions improve any kind of reasoning since relations can be made more easily between objects or ideas (Isen 2000). In general, such emotions lead to a more creative, flexible, and divergent thinking process, whereas negative emotions cause a more linear, convergent, and sequential thinking (a step by step process) (Lisetti and Schiano 2000).
- Positive emotions promote efficiency and thoroughness in decision making and problem solving (Isen 2000).

Emotions can be used in the learning content to increase learner's attention and improve his memory capacity. In the following section we examine what types of different emotions can be considered.

17.2 The Different Types of Emotions

In the literature, there is sometimes confusion between the terms emotions, feelings, and affects. Although there is no common agreement on the definitions, it is important to make a distinction (Shouse 2005).

A *feeling* is an internal perception of a situation (or sensation) which is compared with previous experiences. Feelings (and ability to feel) differ from an individual to another as the volume and variety of previous sensations is different. They need an evaluation of a given situation.

An *emotion* is the consequence and display of a feeling. Emotions are comprised in a set of emotional states which can trigger different reactions (heart rate, transpiration, skin conductivity, muscle tension, blood pressure). It is a visible or measurable consequence of a feeling. Emotion is an immediate reaction to a feeling.

An *affect* is a stimulation state, or instinct able to change or provoke affective situations or experiences; affects are unconscious, without individual control. Successive affects will create different feelings according to the time. It is a physiological property of the body to generate intensities of affective situations. An individual can affect or be affected.

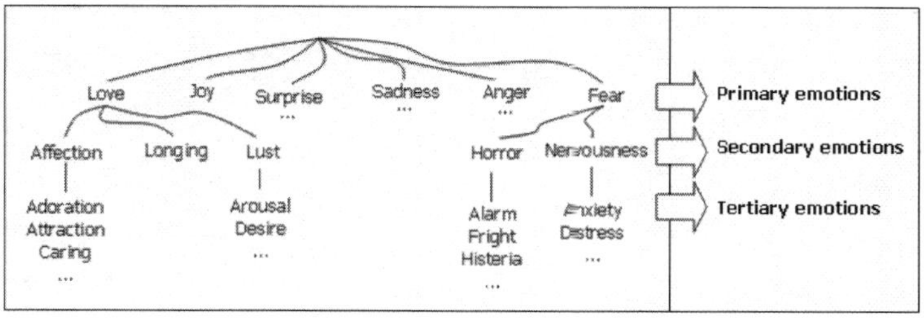

Fig. 17.1 Tree like taxonomy of Philipp Shaver

Thus affects will create feelings, which will be displayed as emotions. For instance, as babies have no previous history of feelings their emotions will be direct expression of the affective stimulations (Shouse 2005). Different models of emotions have been determined and hence different ways to represent them. Commonly, researchers have defined several types of emotions, called basic emotions. The first computational model of emotions was established by Ortony, Clore and Collins (Ortony et al. 1988), known since that time as the OCC model. This model specifies 22 types of emotions: fortunes of others (happy-for, resentment, gloating, pity), well-being (joy, distress), Prospect-Based (satisfaction, fear-confirmed, relief, disappointment), Attribution (pride, shame, admiration, reproach), Attraction (love, hate), Well-Being/Attribution compounds (gratification, remorse, gratitude, anger). This model might be too complex for the development of emotional characters, so Philipp Shaver in (Parrott 2001) asked college students to rate 213 emotional terms on how much emotion they conveyed. This study conducted to the definition of 5 or 6 clusters of basic emotions: *love, joy, surprise, sadness, anger, fear* and each emotion leading to secondary and tertiary emotions.

With the objectives to proceed to a facial recognition of emotions, Ekman (Ekman et al. 2002) distinguished then six different basics emotions: *sadness, happiness, disgust, fear, surprise*, and *anger*. The faces help to moderate and choreograph conversations, and conveyed information. Even if all emotions are not perceived easily and clearly on a human face, we believe that each emotion has a specific facial expression with flagrant or subtle modifications. An example: sad, mournful expressions: use eyebrow depression, eye narrowing, nasal muscle elevation, lip compression, and mouth lowering (turned down at edges).

As emotions can have an impact on the voice, Juslin and Sherer (Juslin and Scherer 2005) has shown how different emotions can change the tone, articulation and intensity of the voice. Elliot (Elliot 1992) built a computing system combining facial expression and voice tones, based on 13 couples of opposed emotions. This system produced emotions which were better recognized by users than emotions expressed by humans.

Table 17.1 Elliot's set of emotions

Joy	Distress
Satisfaction	Disappointment
Liking	Disliking
Happy-for	Resentment
Relief	Fear-confirmed
Gratitude	Anger

Sorry-for	Gloating
Pride	Shame
Gratification	Remorse
Hope	Fear
Admiration	Reproach
Love	Hate
Jealousy	non-Jealousy

However, we must be conscious that on some learner faces, we may not see facial expressions, because of impassive faces or cultural barriers and this represents a limitation of this method. Ochs (Ochs and Frasson 2004) built a system called Case Based Emotioning System (CBE), enabling any user to indicate on a scale how he would feel in a given hypothetic situation. After a period of data acquisition the Case Base Reasoning System was able to make rather accurate predictions using an Emotional Case library.

More recently, researchers considered not only the simplified set of basic emotions (anger, disgust, fear, happiness, sadness and surprise) but also learning-centered affective states such as : anxious, confusion, boredom, contempt, curiosity, eureka, frustration (D'Mello et al. 2009). Transition states from one emotional state to another as well as duration of emotional state are still to be clarified as they depend on individual characteristics.

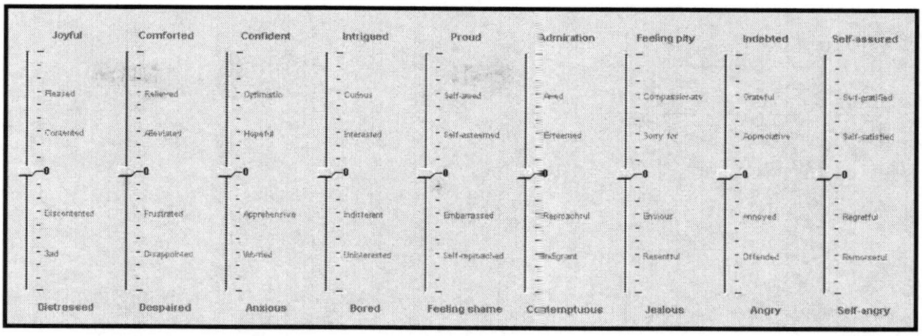

Fig. 17.2 Och and Frasson's Case Based Emotioning system

17.3 Detecting Emotions

Various sensors can be used to detect emotions during a learning session. The following table from (Arroyo et al. 2009) gives an idea of some of them. Sensors are associated with specialized software able determine an emotion or a set of emotions. Several sensors may give a more precise conclusion as to the predicted emotion.

Other physiological sensors have a direct correlation with emotions (Chalfoun and Frasson 2009; Lang 1995). For instance, the following sensors can measure emotion according to two dimensions: *valence* (positive or negative emotion), and *arousal* (intensity of emotion). The GSR sensor (Galvanic Skin Response) allows recording skin conductivity measuring the rate of skin transpiration. RSP sensor (respiration) allows recording variations of respiration. GSR and RESP are positively correlated with arousal. BVP sensor (Blood Volume Pressure) allows recording blood flow variations from which we can extract heart rate (HR). TEMP sensors allow recording temperature variations. BHV, HR, and TEMP are positively correlated with valence.

All these means are in fact indirect consequences of emotional stimuli. Their intensity varies with individuals and can just give an indication on the type of triggered emotion.

Brain interface

In the human brain, each individual neuron communicates with the other by sending tiny electrochemical signals. When millions of neurons are activated, each contributing its small electrical current generating a signal that is strong enough to be detected by an electroencephalogram (EEG) (Cantor 1999; Heraz and Frasson 2009). These devices can be used to detect and measure emotions. Fig. 17.4 represents a learner wearing a pendant EEG able to transmit brainwave activity wireless. The EEG cables are connected to the ears and sensors on the brain.

Table 17.2 Emotional sensors

Sensor	Name	Descriptions
	Postures analysis seat	Detects if the learner is moving back or front to the screen.
	Conductance Bracelet	Measures skin conductivity which in turn has been known to correlate with arousal.
	Facial Expression Sensor	Predicts states such as acknowledgment, interest, reflexion, incertitude.
	Pressure Mouse	Measures learner global pressure using mouse manipulation.
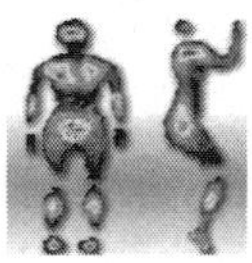	Blood pressure measurement system	Measures blood pressure distribution on the back and under the learner.
	Eye detection	Detects coordinates of eyes and mouth in order to predict facial expression

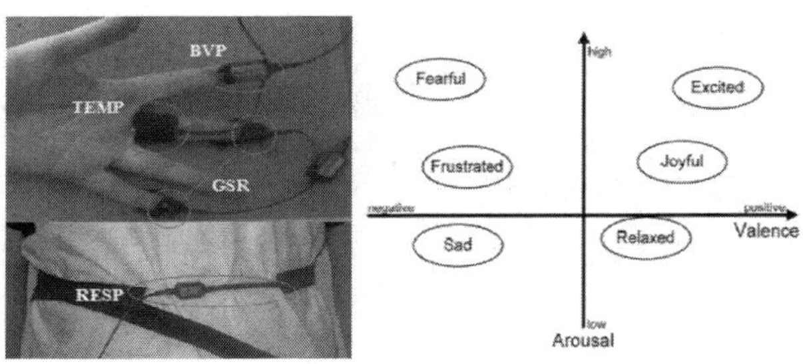

Fig. 17.3 Physiological devices for emotion detection

Commonly, brainwaves are categorized into 4 different frequency bands, or types, known as delta, theta, alpha, beta and gamma waves. Each of these wave types often correlates with different mental states. Fig. 17.5 lists the different frequency bands and their associated mental states, even if we have to note that there

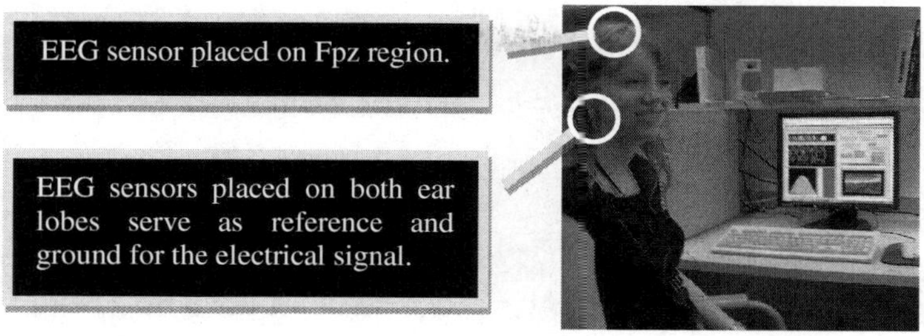

Fig. 17.4 A learner wearing Pendant EEG

Table 17.3 Brainwaves Categories

Bandwith name	Frequency range	Mental states (General characteristics)
Delta (δ)	1-4 Hz	Sleep, repair, complex problem solving
Theta (θ)	4-8 Hz	Creativity, insight, deep states
Alpha (α)	8-12 Hz	Alertness and peacefulness, readiness, meditation
Beta (β)	13-21 Hz	Thinking, focusing, sustained attention
SMR	12-15 Hz	Mental alertness, physical relaxation
High beta	20-32 Hz	Intensity, hyperalertness, anxiety
Gamma(Y)	38-42 Hz	Cognitive processing, learning

is presently no consensus over the use of a fixed threshold to split the frequencies into bands.

Delta frequency band is associated with deep sleep. Theta is dominant during dream sleep, meditation, and creative inspiration. Alpha brainwave is associated with tranquility and relaxation. By closing one's eyes can generate increased alpha brainwaves. Beta frequency band is associated with an alert state of mind and concentration (Demos 2005). Response preparation and inhibition is one of the roles of the cortical sensorimotor beta rhythm (Zhang et al. 2008).

In the present case of pendant EEG, the electrical signal recorded by the EEG is sampled, digitized and filtered to divide it into 4 different frequency bands: Beta, Alpha, Theta and Delta (Fig. 17.5).

Our previous work (Heraz and Frasson 2007; Heraz et al. 2008) indicated that an EEG is a good source of information to detect emotion. Results show that the student's affect (Anger, Boredom, Confusion, Contempt, Curious, Disgust, Eureka, and Frustration) can be accurately detected (82%) from brainwaves By looking on the signal produced at a given time we can guess deduce in which mental state is the learner, how long this state can exist and what event can change it. The

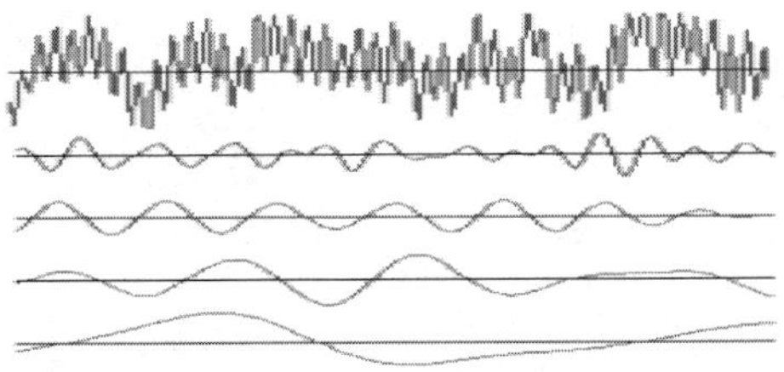

Fig. 17.5 A raw EEG sample and its filtered component frequencies. Respectively (from the top): Recorded EEG signal, Beta, Alpha, Theta and Delta brainwaves

difficulty of this approach is however to remove any noise effect in the outputs and to interpret their variations. The signals are by nature complex, and so, difficult to interpret from an emotional perspective. Measuring the intensity of emotions represents another challenge but this research track is promising and upcoming works should allow discovering new perspectives. By combining all sensors we should understand better not only the types of emotions but also the transitions between emotional states.

17.4 Inducing Emotions

As we mentioned earlier, some emotional states can strengthen knowledge acquisition while other affective situations will slow down or even block cognitive abilities. However, few works in computer science attempted to induce emotions and verify their impact on learning and cognitive capabilities. For instance, at MIT Media Lab, (Picard et al. 2001) used pictures to induce a set of emotions which include happiness, sadness, anger, fear, disgust, surprise, neutrality, platonic love and romantic love. Moreover at affective Social Computing Laboratory, Nasoz et al. (Nasoz et al. 2003) used results of Gross and Levenson (Gross and Levenson 1995) to induce sadness, anger, surprise, fear, frustration, and amusement.

Researchers in psychology have developed a variety of experimental techniques for inducing emotional states aiming to find a relationship between emotions and thought tasks; one of them is the Velten procedure which consists of randomly assigning participants to read a graded set of self-referential statements for example, "I am physically feeling very good today". A variety of other techniques exists including guided imagery (Ahsen 1989) which consists of asking participants to *imagine* themselves in a series of described situations, for example: "You are sitting in a restaurant with a friend and the conversation becomes hilariously funny and you can't stop from laughing". Some other existing techniques are based upon exposing participants to films, music or odors. Gross and Levenson found that 16 video clips could induce really one of the following emotions (amusement, anger, contentment, disgust, fear, neutrality, sadness, and surprise) from a set of the 78 films shown to 494 subjects (Gross and Levenson 1995).

In addition, active listening, empathy, sympathy and venting may be strategies aimed at reducing negative effect (Klein et al. 1999). For instance, by playing a computer game, user indicated on a scale the level of frustration that he experienced with the system. Then, a support agent can offer an affective feedback by sending the user some text that mirrored the level of frustration that the user reported. It has been shown that, with this method some of the user's negative feelings are reduced and the tendency of people to interact with computers is boosted.

Moreover, other researchers have developed systems that are able to control and influence the learner's emotions using empathy. For instance, the "affective companion" adapts to the learner's emotions, by adjusting the difficulty of an exercise (Isen 2000); the "affective tutor", on the other hand, is itself affectively adaptive to user's emotions (Estrada et al. 1994). McQuiggan developed and tested the advantage of affect-based empathetic responses on hints and suggestions (McQuiggan and Lester 2006).

Similarly, Partala focused on human-computer interaction, studying especially the effects of affective interventions using synthetic speech with emotional content (Partala and Surakka 2004). The interventions were given when subjects' problem solving had been interrupted by mouse delays. Compared to no intervention condition, the results showed that the use of positive worded affective intervention has improved the smiling activity. At the same time frowning activity was clearly decreased during affective intervention.

Other researchers have developed hybrid techniques which combine two or more procedures; Mayer et al. used the guided imagery procedure with music procedure to induce four types of emotions, joy, anger, fear, sadness (Mayer et al. 1995). They used guided imagery to occupy the foreground attention and music to emphasize the background. According to Prendinger, human-computer interaction would become more productive by offering help and assistance to confused users (Prendinger and Ishizuka 2005). The authors studied the effect of *empathic* embodied feedback on deliberately frustrated users. These interfaces include guided imagery vignettes, music and images in the context of a job interview simulation and preparation.

In (Chaffar and Frasson 2006), a specific module (ESTEL) was developed to induce an optimal emotional state, which represents a positive state of mind that maximizes learner's performance. Using a Personality Identifier which determines

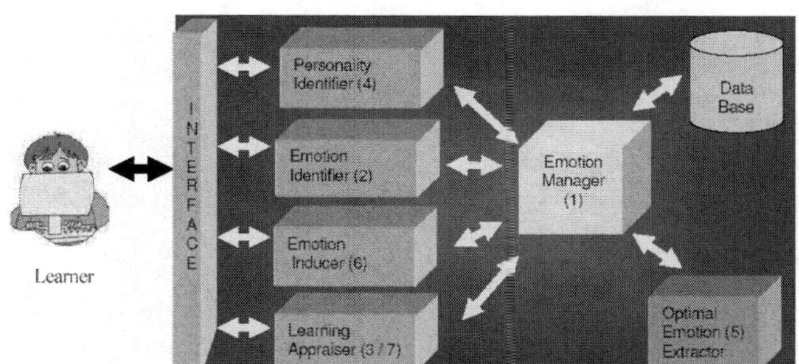

Fig. 17.6 ESTEL's Induction System

his personality (for instance extraversion), the Optimal Emotion Extractor retrieves joy as the optimal emotional state for this personality. The Emotion Inducer will elicit joy in this learner by using a hybrid technique which consists of displaying different interfaces.

The Emotion Inducer is inspired by the study of Mayer et al. (Mayer et al. 1995) that has been done to induce four specific emotions (joy, anger, fear, and sadness). After inducing emotion, the Emotion Manager module will restart the Learning Appraiser module for evaluating learning efficiency.

This architecture shows that managing emotional conditions requires a complete cycle including: identification of emotional state of the learner, identification of adequate emotional state to improve the performances, induction of new emotion, and verification of learning improvement. These modules are also components of emotional Intelligent Tutoring Systems (ITS).

17.5 Emotionally Intelligent Tutoring Systems

Emotional intelligence is the ability to recognize, interpret, and influence someone's emotions (Goleman 2010). When the effect of feeling on thinking is known, this can be used in order to improve someone's cognitive abilities (Salovey et al. 2000). It is hence natural to include emotional intelligence into the learning process. A human tutor is in fact an emotional practitioner in the sense that he can influence learner's emotions with the objective of improving his learning efficiency (Hargreaves 2002). Since interaction with a computer triggers similar responses as if it was with another person a new generation of ITS should be able to influence learner's emotions in a same way as a human tutor.

Motivated by this principle we have introduced the concept of an Emotionally Intelligent Tutoring System (EITS) in (Ochs and Frasson 2004). An EITS is an ITS which includes functional capabilities able to (1) know learner's emotions and, (2) induce emotions to the learner in order to improve his performance. More precisely, an EITS needs to achieve the following conditions:

1. know the current emotional state of the learner,
2. determine the impact of an action on learner' emotional state,
3. identify the most advantageous emotional state of a learner to enhance his performance.

We have developed and tested the two first conditions. The first condition was achieved using nine emotional scales allowing the user to indicate his current emotional state. In order to realize the second condition we used methods derived from Case Based Reasoning (CBR) to create a Case Based Emotioning system (CBE) able to predict the emotional effect of specific actions on the emotional state of a learner. To model the impact of these actions on learner's emotion we have used a directed graph, where the vertices are emotional states and the connecting edges the possible actions which can occur in a learning session (for instance criticism, encouragement, congratulation …). This representation allows the CBE system to assess the emotional state resulting from an action on an initial emotional state. Including this CBE into an ITS allows to generate in a user a specific emotional state.

Knowing which emotion to generate we need however to determine which one will be able to improve learner performance (condition 3 mentioned above). We focus on this point by analyzing the different effects of emotions on learner performance and particularly the emotional conditions for reaching the best performance.

For instance, a computer that flatters a user will generate positive emotions in him. A learner will, hence, experience a variety of emotions upon interacting with an ITS in the same way as in the context of traditional learning, and similarly to the human teacher, a virtual tutor can be viewed as an emotional practitioner able to influence the learner's emotions. Moreover, these emotions will strongly influence his cognitive abilities (Isen 2000). Given these two facts, ITS should, therefore, be able to manage these emotions in a way beneficial for the learning process, e.g., to generate specific emotions in the learner.

By consequence of the above discussion it appears advantageous to include specific capabilities of emotional intelligence into ITS: Learning involves a variety of cognitive processes, and someone's performance therein highly depends on his emotions. An ITS able to manage learner's emotions contains additional human capabilities and is potentially more efficient.

17.6 Affect in Virtual Environments

The intertwined relationship between affect and intelligent tutoring system took a step further in the virtual world. This section will examine the three most active areas of research in emotions and virtual environments: Embodied agents, narrative learning environments or NLE and quite recently a new subliminal teaching technique that has provides promising results regarding performance and students affect.

17.6.1 Embodied Agents

An embodied agent can be defined as a digital, visual representation of an interface, often taking a human form (Cassell 2002). Virtual agents expressing affect started to appear in an ITS as early as 1997 with COSMO (Lester et al. 1997). This pedagogical agent expressed joy and even jubilation when learners achieved successfully a task. Since then, communication through embodied agents within virtual environments in the ITS community has only grown in popularity and complexity. Affective issues such as empathy, self-efficacy and motivation have been implemented in various forms in a very broad range of different virtual environments (four such environments are shown in Fig. 17.7). Because of their strong life-like presence, animated pedagogical agents can capture students' imaginations and play a critical motivational role in keeping them deeply engaged in a learning environment's activities (Lester et al. 1997). Indeed, one of the main goals of an

ITS is to be able to recognize and address the emotional state of the learner and react accordingly through the presence of the pedagogical agent.

The affective tutor is one such system where frustration is detected real-time with a multi-modal approach combining Gaussian affective process classification and Bayesian inference (Kapoor 2007). The Wyang Outpost is another ITS intended for middle school and high school level mathematics learning. We can see Jake and Jane, two virtual agents drawn in flash, mirror emotions to emulate empathy and thus help keep high school children more engaged in the lesson (Arroyo et al. 2004).

Affective tutor Wyang Outpost AutoTutor MOCAS

Fig. 17.7 Four virtual I.T.S. systems employing various forms of embodied agents

AutoTutor is another ITS employing a pedagogical agent that helps students learn by employing a constructivism approach when analyzing their responses using natural language processing. The animated agent uses synthesized speech, gestures and facial expressions to encourage learners to articulate lengthy answers that exhibit deep reasoning, rather than to recite small bits of shallow knowledge (D'Mello et al. 2005). Finally, MOCAS (Motivational and Culturally Aware System) employs the self-determination theory to produce pedagogical agents whose behaviors are closely aligned to learner's motivational and cultural needs (Blanchard and Frasson 2004). MOCAS's autonomy-supportive design and the rule-based methodology adapt its teaching given the cultural backgrounds of its learners. All four presented systems in Fig. 17.7 employ physiological sensors to record affective data in order to express synthetic human emotions through various multimodal channels (i.e. voice, text, gesture). As previously mentioned in this chapter, physiological signals are generally correlated with emotions by associating specific signals, such as skin conductance and heart rate, to valance and/or arousal (Lang 1995). For further reading on the impact of emotional agents on learner's interactions, we highly recommend the recent excellent review from Beale and Creed (Beale and Creed 2009).

17.6.2 Narrative Learning Environments

Narrative has been an important form to transmit knowledge across generations, and is innate to the human nature. Narrative is also a valuable vehicle to structure knowledge and to help us in the process of meaning making. Cognitive psychologists have recognized narrative as relevant to the way we store and make sense of episodic experience, often described as the phenomenon of narrative construction of reality. Due to the explorative and complex nature of narrative, an intelligent learning environment (ILE) based on a narrative approach can promote several kinds of activities for learners:

- co-construction: participate in the construction of a narrative;
- exploration: engage in active exploration of the learning tasks, following a narrative approach and trying to understand and reason about an environment and its elements;
- reflection: engage in consequent analysis of what happened within the learning session.

By applying a narrative approach in the development of ILEs, it is possible to attain an application that may help learners by illustrating phenomena and procedures, and by motivating them to stay engaged and immersed in learning tasks. Additionally narrative learning environments can facilitate activities associated with learning such as role-playing and exploration, reflection, and the sharing of ideas. Fig. 17.8 presents four NLE's that utilize different pedagogical strategies and affect in the context of narration.

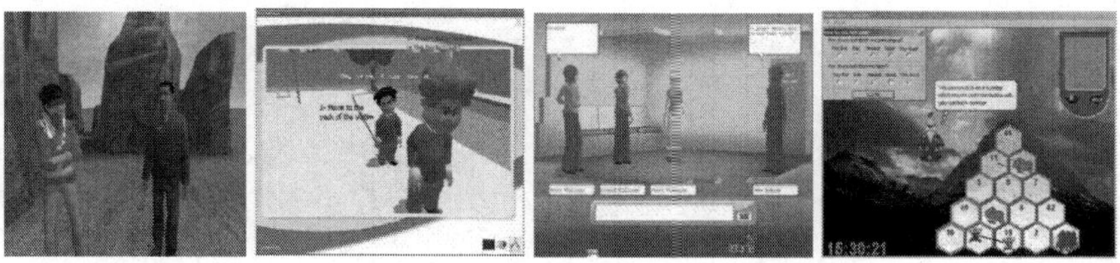

Crystal Island FearNot! EMMA Prime Climb

Fig. 17.8 Four different NLE's each utilizing a different pedagogical approach

Crystal Island is one such narrative-centered learning system used for the teaching of microbiology and genetics (McQuiggan and Lester 2007). The animated agents in Crystal Island are built on empathy (they can express emotions as shown in Fig. 17.8) in order to promote intrinsic motivation in high school students. Overcoming a challenging task provides a student with a personal sense of achievement and a test of her abilities. Indeed, excessively low-challenge periods may cause the student to feel bored, but high-challenge periods may bring about frustration and feelings of hopelessness. FearNot! is a completely different and very interesting interactive NLE developed for education against bullying behavior in schools (Aylett et al. 2005). The autonomous agents in FearNot! implements emotional expressiveness and personality, two important characteristics of synthetic characters, to attain a desirable level of empathy and believability in order to help deal with virtual bullies. The third illustrated NLE, called EMMA, makes a contribution to the issue of what types of automation should be included in interactive narrative environments, and as part of that the issue, of what types of affect should be detected and how. The generation of emotional believable animations based on the detected affective states contributes to the ease and innovative user interface in edrama, which leads to high-level user engagement and enjoyment (Zhang et al. 2007). Prime Climb is another very interesting NLE where Merlin, the virtual tutor, interacts with learners by recognizing multiple user emotions during the interaction with an educational computer game (Conati and Maclaren 2009). The model is based on a probabilistic framework that deals with the high level of uncertainty involved in recognizing a variety of user emotions by combining in a Dynamic Bayesian Network information on both the causes and effects of emotional reactions. The tutor intervenes intelligently by processing information through the complex model in order to help 6th and 7th grade students practice number factorization.

The NLE make extensive use of animated agents to interact with the user. While this is common practice, it is important to not neglect the degree of behavioral realism of these agents for it can have different effects depending on the users. For a complete review on the subject please see (Campbell et al. 2009; Groom et al. 2009).

17.6.3 Subliminal Learning

In recent years, researchers in human-computer interfaces (HCI) as well as in various fields such as Intelligent Tutoring Systems (ITS) have taken advantage of adaptive and customizable HCI to record and analyze emotions (Villon and Lisetti 2006). This is not surprising since emotions, especially motivation and engagement, are widely related in various cognitive tasks as described earlier in this chapter. Learning in virtual worlds has taken a very important part in the HCI community for recent evidence has shown the relevance of using such virtual ITS for affective feedback and adaptation (Blanchard et al. 2007; McQuiggan and Lester 2006). Nevertheless, cognitive learning theories base mostly their intervention on attention to the specified task at hand. Complex information is broken down into pieces to gradually enable the learner to concentrate on one small part of the puzzle at a time. However, a large body of work in neuroscience and other fields lead us to believe that learning simple to complex information can be done without perception or complete awareness to the task at hand (DeVaul et al. 2003; Dijksterhuis and Nordgren 2006; Nunez and Vincente 2004; Watanabe et al. 2001). In fact, the existence of perceptual learning without perception has been neurologically proven and accepted (Del Cul et al. 2007). Furthermore, recent work has put forth the performance increase in performance when using a subliminally teaching Intelligent Tutoring System (Chalfoun and Frasson 2008). Yet, subliminal learning systems are still widely absent in the HCI community. The work by Chalfoun and Frasson focuses on subliminal stimuli in a 3D virtual system to enhance learning (Chalfoun and Frasson 2008).

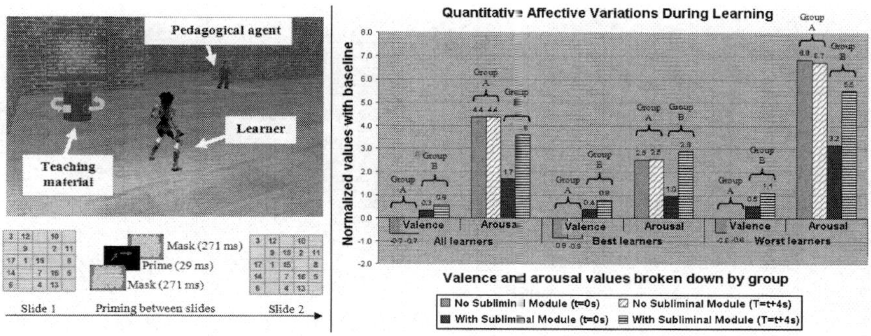

Fig. 17.9 Subliminal module implemented in the immersive version of MOCAS and results

Before going further however, we need to clearly establish the terminology that will be used. Indeed, the simple mention of the word subliminal can lead to discord and confusion. We establish a stimulus as being subliminal when it is sent too fast for a user to consciously report its existence. Conscious perception is well established in neuroscience and its properties are well known. One of those properties is the existence of a visual threshold for conscious access called VT. It is precisely this VT that we establish as being the "line" between conscious and unconscious perception. The technique used to send a given stimuli *below* the VT of awareness is called subliminal projection, as opposed to a paraliminal projection where flashed stimuli is consciously perceived for it is presented at speeds *above* the VT (Del Cul et al. 2007).

The current experiment presented in Fig. 17.9 uses precise and timed subliminal projections in a 3D intelligent tutoring system while monitoring the physiological reactions of the learner. Those visual recording are crucial to remove noise and identify events of special interest. Moreover, we constructed the subliminal cues in a way which would accelerate the learning process by triggering and enhancing an already possessed knowledge without the user's awareness. This step is important for it enables the learner to feel responsible for his own success and hopefully help him stay motivated. The histogram on the right hand side of the figure presents a detailed and precise look at the optimal affective conditions that set the best learners apart. These signal values are normalized by mean-shifting, that is subtracting each signal's value from the signal's baseline mean then dividing the result by the signal's standard deviation. This widely accepted technique enables us to compare learners' results for it solves the problem of extra-individual physiological differences. The figure shows the average affective values for a period of 4 second following every subliminal stimulus. The full brown bars represent the average value of the signal for all subliminal stimuli at the precise moment the stimulus was projected (t=0s, s is for seconds). The horizontal dashed bars represent the same averaged value except that it is computed for the 4 seconds following that projected stimulus (T=t – 4s). Since one was not primed with subliminal stimuli, we placed markers for each learner at the precise moment where subliminal cues would have been projected if these learners would have been taking the course with the subliminal module.

The results shown in Fig. 17.9 are not only statistically significant ($p<0.001$, $\alpha=0.05$) but very important for they enable us to distinguish between the best and worst learners in terms of valence and arousal but also in terms of how much variation is considered optimal for success. In this case, having an average positive valence variation increase of about 0.8 and arousal increase between 2.5 and 2.9 is what our system should be looking for. In fact, we can clearly see at the far right part of the figure that the worst learners, those who made the most mistakes, were the ones who had a negative valence variation. Checking the results with Lang's two dimensional space (Lang 1995) informs us that a negative valence could lead to a negative emotional state and thus not optimal for learning. Since subliminal projections increase valence variations, our system could then detect this negative emotional state and start projecting more stimuli until an optimal state is reached. We demonstrated in (Chalfoun and Frasson 2008) that the subliminal module helped reduce dramatically the number of mistakes made. Fig. 17.9 might have helped explain why in terms of valence and arousal.

17.7 Future Directions

The affective issues addressed here have focused on the important contribution of emotions on the cognitive processes involved in learning such as attention, memorization and reasoning. Indeed, a great body of work has addressed the various approaches to detect and induce emotions with sensors and smart interfaces respectively within an EITS. The same could be said with emotions in virtual environments. Narrative learning environments, simulators and affective agents have shown the importance of empathy and role-playing in keeping learners engaged and motivated throughout the learning session.

It is the authors' belief that resolving future affective issues within an EITS resides within a multidisciplinary approach to affect. Indeed, the potential integration of EEG into the ITS community can greatly enhance mental state detection and adaptation. Indeed, recent work in the ITS field combined neuroscience, psychology and pedagogy by demonstrating that using neurological properties of unconscious cognition can have a positive impact on learner's intuition as well as his self-esteem in problem solving tasks (Chalfoun and Frasson 2010; Jraidi and Frasson 2010). Finally, Chaouachi and al. have shown that learner's affective states can have a direct impact on his engagement level measured by a well established EEG-mental engagement index developed at NASA (Chaouachi et al. 2010).

When looking at research on Artificial Intelligence how can we try to reproduce human behavior if we ignore emotions which in fact sustain knowledge acquisition? The same remark applies to other disciplines as emotions play an important role in their domain. Artificial Intelligence, Education, Neuroscience, Psychology, Medicine are just some of them. Being at the crossroads of these exchanges constitutes a highly exciting experience.

References

Ahsen, A.: Guided imagery: the quest for a science. Part I, II & III. Education 10(1), 2–32 (1989)

Arroyo, I., Beal, C.R., Murray, T., Walles, R., Woolf, B.P.: Web-based Intelligent Multimedia Tutoring for High Stakes Achievement Tests. In: Lester, J.C., Vicari, R.M., Paraguaçu, F. (eds.) ITS 2004. LNCS, vol. 3220, pp. 468–477. Springer, Heidelberg (2004)

Arroyo, I., Cooper, D., Burleson, W., Woolf, B.P., Muldner, K., Christopherson, R.: Emotion Sensors go to School. In: Proceedings of 14th International Conference on Artificial Intelligence in Education, Brighton, UK (2009)

Aylett, R., Louchart, S., Dias, J., Paiva, A.: FearNot! an Experiment in Emergent Narrative. In: Panayiotopoulos, T., Gratch, J., Aylett, R.S., Ballin, D., Olivier, P., Rist, T. (eds.) IVA 2005. LNCS (LNAI), vol. 3661, pp. 305–316. Springer, Heidelberg (2005)

Beale, R., Creed, C.: Affective interaction: How emotional agents affect users. International Journal Of Human-Computer Studies 67, 755–776 (2009)

Blanchard, E., Frasson, C.: An autonomy-oriented system design for enhancement of Learner's motivation in eLearning. In: Lester, J.C., Vicari, R.M., Paraguaçu, F. (eds.) ITS 2004. LNCS, vol. 3220, pp. 325–335. Springer, Heidelberg (2004)

Blanchard, E., Chalfoun, P., Frasson, C.: Towards advanced Learner Modeling: discussions on quasi real-time adaptation with physiological data. In: Proceedings of the 7th IEEE conference on Advanced Learning Technologies, Nigata, Japan (2007)

Bloom, B.S.: Bloom's Taxonomy. University of Chicago Press, Chicago (1994)

Bower, G.: How Might emotions affect learning ?. In: Handbook of emotion and memory (1992)

Campbell, R.H., Grimshaw, M., Green, G.: Relational agents: A critical review. The Open Virtual Reality Journal 11, 7 (2009)

Cantor, D.S.: An overview of quantitative EEG and its applications to neurofeedback. In: Introduction to Quantitative EEG and Neurofeedback. Academic Press, London (1999)

Cassell, J.: Nudge nudge wink wink: elements of face-to-face conversation for embodied conversational agents. In: Embodied Conversational Agents. MIT Press, Cambridge (2002)

Chaffar, S., Frasson, C.: Predicting Learner's Emotional Respons. In: Intelligent Distance Learning Systems. In: Proceedings of Name, Florida, USA (2006)

Chalfoun, P., Frasson, C.: Subliminal Priming Enhances Learning and Performance in a Distant 3D Virtual Intelligent Tutoring System. In: Proceedings the AACE World Conference on E-learning in Corporate, Government, Healthcare, & Higher Education: E-LEARN 2008 (2008)

Chalfoun, P., Frasson, C.: Optimal Affective Conditions for Subconscious Learning in a 3D Intelligent Tutoring System. In: Proceedings the International conference on Human Computer Interactions International, HCII (2009)

Chalfoun, P., Frasson, C.: Showing the positive influence of subliminal cues on learner's performance and intuition: an ERP study. In: Proceedings of the 10th International Conference on Intelligent Tutoring Systems (Poster), Montreal, Canada (2010)

Chalfoun, P., Frasson, C.: Subliminal priming enhances learning in a distant virtual 3D Intelligent Tutoring System. IEEE Multidisciplinary Engineering Education Magazine: Special Issue on Intelligent Tutoring Systems 3(4), 125–130 (2008)

Chaouachi, M., Chalfoun, P., Jraidi, I., Frasson, C.: Affect and Mental Engagement: Towards Adaptability for Intelligent Systems. In: Proceedings the The 23rd International FLAIRS Conference, Florida, USA (2010)

Compton, R.: Ability to disengage attention predicts negative affect. Cognition and Emotion (2000)

Conati, C., Maclaren, H.: Empirically building and evaluating a probabilistic model of user affect. User Modeling and User-Adapted Integration 19(3), 267–303 (2009)

D'Mello, S.K., Craig, S.D., Gholson, B., Franklin, S., Picard, R.W., Graesser, A.C.: Integrating Affect Sensors in an Intelligent Tutoring System. In: Affective Interactions: The Computer in the Affective Loop Workshop at 2005 International Conference on Intelligent User Interfaces, New York, USA (2005)

D'Mello, S.K., Person, N., Lehman, B.: Antededent-Consequent Relationships and Cyclical Patterns between Affective States and Problem Solving Outcomes. In: Proceedings of AIED 2009, Brighton, United Kingdom (2009)

Damasio, A.: Descarte's Error - Emotion, Reason and the Human Brain. Putman Press, New York (1994)

Del Cul, A., Baillet, S., Dehaene, S.: Brain Dynamics Underlying the Nonlinear Threshold for Access to Consciousness. PLoS Biology 5(10), 16 (2007)

Demos, J.: Getting Started With Neuro-Feedback (2005)

DeVaul, R., Pentland, A., Corey, V.: The Memory Glasses: Subliminal vs. Overt Memory Support with Imperfect Information. In: Proceedings the IEEE International Symposium on Wearable Computers (2003)

Dijksterhuis, A., Nordgren, L.F.: A Theory of Unconscious Thought. Perspectives On Psychological Science 1(2), 14 (2006)

Ekman, P., Friesen, W.V., Hager, J.C.: The facial action coding system, 2nd edn. Weidenfeld & Nicolson, London (2002)

Elliot, C.: The Affective Reasonner: A process Model of Emotions in a Multi-Agent System. Northwestern University (1992)

Estrada, C.A., Isen, A.M., Young, M.J.: Positive affect influences creative problem solving and reported source of practice satisfaction in physicians. Motivation and Emotion 18, 285–299 (1994)

Gagné, R.: The conditions of learning, 4th edn. Holt, Rinehart & Winston, New York (1985)

Daniel, G.: Social Intelligence: The New Science of Human Relationships (2010)

Groom, V., Nass, C., Chen, T., Nielsen, A., Scarborough, J.K., Robles, E.: Evaluating the effects of behavioral realism in embodied agents. International Journal Of Human-Computer Studies 6, 7842–7849 (2009)

Gross, J.J., Levenson, R.W.: Emotion elicitation using films. Cognition and Emotion, 987-108 (1995)

Hargreaves, A.: Mixed emotions: teacher's perceptions of their interactions with students. Teaching and teacher education 16, 811–826 (2002)

Heraz, A., Frasson, C.: Predicting the three major dimensions of the learner's emotions from brainwaves. International Journal of Computer Science 31 (2007)

Heraz, A., Daouda, T., Frasson, C.: Decision Tree for Tracking Learner's Emotional State predicted from his electrical brain activity. In: Woolf, B.P., Aïmeur, E., Nkambou, R., Lajoie, S. (eds.) ITS 2008. LNCS, vol. 5091, pp. 822–824. Springer, Heidelberg (2008)

Heraz, A., Frasson, C.: How Do Emotional Stimuli Influence the Learner's Brain Activity? Tracking the brainwave frequency bands Amplitudes. In: Proceedings the International Conference on Agents and Artificial Intelligence (2009)

Idzihowski, C., Baddeley, A.: Fear and performance in novice parachutists. Ergonomics 30, 1463–1474 (1987)

Isen, A.M.: Positive Affect and Decision Making(ed) Handbook of Emotions. Guilford, New York (2000)

Jraidi, I., Frasson, C.: Subliminally Enhancing Self-Esteem: Impact on Learner Performance and Affective State. In: Proceedings of International Conference of Intelligent Tutoring Systems, Pittsburgh, USA (2010)

Juslin, P.N., Scherer, K.R.: Vocal expression of affect In: The New Handbook of Methods in Nonverbal Behavior Research. Oxford University Press, Oxford (2005)

Kapoor, A.: Automatic prediction of frustration. International Journal of Human-Computer Studies 6, 5724–5736 (2007)

Klein, J., Moon, Y., Picard, R.W.: This computer responds to user frustration. In: Proceedings the CHI 1999 extended abstracts on Human factors in computing systems (1999)

Lang, P.J.: The emotion probe. American Psychologist 520(5), 13 (1995)

Lester, J., Voerman, J., Towns, S., Callaway, C.: Cosmo: A Life-like Animated Pedagogical Agent with Deictic Believability. In: Proceedings the Working Notes of the IJCAI Workshop on Animated Interface Agents: Making Them Intelligent (1997)

Lisetti, C., Schiano, D.: Automatic Facial Expression Interpretation: Where Human-Computer Interaction, Artificial Intelligence and Cognitive Science Intersect. Pragmatics and Cognition 81(1) (2000)

Mayer, J., Allen, J., Beauregard, K.: Mood Inductions for Four Specific Moods. Journal of Mental imagery 19, 133–150 (1995)

McQuiggan, S.W., Lester, J.: Learning empathy: a data-driven framework for modeling empathetic companion agents. In: Proceedings the International Conference on Autonomous Agents (2006)

McQuiggan, S.W., Lester, J.: Modeling and Evaluating Empathy in Embodied Companion Agents. International Journal of Human-Computer Studies 65(4), 12 (2007)

Nasoz, F., Lisetti, C.L., Avarez, K., Finkelstein, N.: Emotion Recognition from Physiological Signals for User Modeling of Affect. In: Proceedings of Name, USA (2003)

Nunez, J.P., Vincente, F.D.: Unconscious learning. Conditioning to subliminal visual stimuli. The Spanish Journal of Psychology 7(1), 15 (2004)

Ochs, M., Frasson, C.: Emotionally Intelligent Tutoring System. In: Proceedings the FLAIRS, Florida, USA (2004)

Ortony, A., Clore, G.L., Collins, A.: The cognitive Structure of Emotions. Cambridge University Press, Cambridge (1988)

Parrott, W.: Emotions in Social Psychology. Psychology Press, Philadelphia (2001)

Partala, T., Surakka, V.: The effects of affective interventions in human-computer interaction. Interacting with Computers 16, 295–309 (2004)

Pekrun, R.: The Impact of Emotions on Learning and Achievement: Towards a Theory of Cognitive/Motivational Mediators. Applied Psychology 41(4) (2008)

Picard, R.W., Vyzas, E., Healey, J.: Toward machine emotional intelligence: analysis of affective physiological state. IEEE Transactions Pattern Analysis and Machine Intelligence 23(10), 6 (2001)

Prendinger, H., Ishizuka, M.: The Empathic Companion: A Character-Based Interface That Addresses Users' Affective States. Applied Artificial Intelligence 19(3), 18 (2005)

Salovey, P., Bedell, B., Detweiler, J., Mayer, J.: Current Directions in Emotional Intelligence Research. In: Handbook of Emotions, 2nd edn., Guilford Press, New York (2000)

Shouse, E.: Feeling, Emotion, Affect. Journal of Media and Culture 8(6) (2005)

Spering, M., Wagener, D., Funke, J.: The role of emotions in complex problems solving. Cognitive and Emotion 19 (2005)

Villon, O., Lisetti, C.L.: A User-Modeling Approach to Build User's Psycho-Physiological Maps of Emotions using Bio-Sensors. In: Conati, C., McCoy, K., Paliouras, G. (eds.) UM 2007. LNCS (LNAI), vol. 4511, pp. 319–323. Springer, Heidelberg (2007)

Watanabe, T., Nanez, J.E., Yuka, S.: Perceptual learning without perception. Nature 413, 5 (2001)

Zhang, L., Gillies, M., Barnden, J.A., Hendley, R.J.: An Improvisational AI Agent and Emotionally Expressive Characters. In: Workshop on Narrative Learning Environments, AIED (2007)

Zhang, Y., Chen, Y., Bressler, S.L., Ding, M.: Response preparation and inhibition: the role of the cortical sensorimotor beta rhythm. Neuroscience 156(1), 238–246 (2008)

Part IV
ITS Authoring and Applications

Chapter 18
Building Intelligent Tutoring Systems: An Overview

Roger Nkambou[1], Jacqueline Bourdeau[2], and Valéry Psyché[2]

[1] Université du Quebec à Montréal, 201 Du Président-Kennedy Avenue, PK 4150, Montréal , QC, H2X 3Y7, Canada
[2] Centre de recherche LICEF, Télé-université, Université du Québec à Montréal, 100 Sherbrooke Street W., Montréal, QC, H3T 1J, Canada
nkambou.roger@uqam.ca,
{jacqueline.bourdeau,valery.psyche}@licef.ca

Abstract. This chapter addresses the challenge of building or authoring an Intelligent Tutoring System (ITS), along with the problems that have arisen and been dealt with, and the solutions that have been tested. We begin by clarifying what building an ITS entails, and then position today's systems in the overall historical context of ITS research. The chapter concludes with a series of open questions and an introduction to the other chapters in this part of the book.

18.1 Introduction

Intelligent Tutoring Systems (ITSs) are complex computer programs that manage various heterogeneous types of knowledge, ranging from domain to pedagogical knowledge. Building such a system is thus not an easy task. ITS authors need to be well equipped to face multiple issues related to their building process. In fact, the resources needed to build an ITS come from multiple research fields, including artificial intelligence, the cognitive sciences, education, human-computer interaction and software engineering. This multidisciplinary foundation makes the process of building an ITS a thoroughly challenging task, given that authors may have very different views of the targeted system. Some promote pedagogical accuracy (ensuring that tutoring decision making is based on sound pedagogical principles), while others focus on effective diagnosis of learners' errors (using appropriate knowledge structure and algorithms to interpret learners' decisions correctly). Murray (1999) identified seven different classes of tutoring system, each corresponding to a different author view, conditioned by the author's needs. Murray's study clearly shown that most of the existing authoring systems were designed for building part or all of a specific class of ITSs. Furthermore, there is a lack of methods and standard tools which could ease the authoring process.

Users interested in building ITSs fall into two groups: those with programming skills and those without. While the former can use snippets of code and class

R. Nkambou et al. (Eds.): Advances in Intelligent Tutoring Systems, SCI 308, pp. 361–375.
springerlink.com

libraries (API) without requiring an intuitive user interface, the latter have a need for tools that are easy to use, and which reflect their mental model of the artifact they are building. For example, a non-programmer should be able to integrate a content-planning module without having to program it, or add domain knowledge without having to understand either the knowledge model, the language used for coding this model, or the underlying logic. Two types of systems to help build ITSs were therefore identified (Murray 1999, 2003): shells for programmers and authoring tools for non-programmers. Both provide suitable resources to facilitate the building of ITSs.

In this chapter, the challenge of building or authoring an ITS is addressed, along with the problems that have arisen and been dealt with, and the solutions that have been tested. The chapter begins with a presentation of historical and current work on building ITS shells. Then, an update of Murray's review of authoring systems is given, with an emphasis on other important factors that should be considered in the classification of authoring tools. The chapter ends with a series of open questions and an introduction to the other chapters in this part of the book.

18.2 The Shell-Based Approach

The shell-based approach is well known in artificial intelligence. Since the beginning of expert systems research, there have been many proposals for shells to facilitate their building. A shell is a software development environment containing the basic components for building expert systems. The first experiment on such an approach was done with E-Mycin (Crawford 1987), a general purpose Expert System shell derived from Mycin. E-Mycin was built by removing all domain-dependent knowledge from Mycin, leaving only the inference mechanisms in the system. This allowed the use of these mechanisms with other domain knowledge. Thus, the shell-based approaches focus mainly on the system components but little on the user interface, making shell-based systems very suitable for users with programming skills.

Viewed as a knowledge-base system, an ITS contains general knowledge that governs decision-making in the expert, tutor and student modules. An interesting parallel can thus be drawn with the approach of classical expert systems. This is the analogy underlying a number of shells proposed to facilitate the construction of ITSs. While some of them include a very limited user interface, they are built to be used by ITS developers with some programming skills. They provide code libraries or conceptual frameworks for building parts of an ITS. Some of them focus on user modeling, while others place the emphasis on curriculum planning or content acquisition. As examples, Kobsa & Pohl (1995) developed a user modeling shell named BGP-MS that offers host applications methods for communicating observations regarding the user, and for obtaining information such as the user's presumed knowledge, beliefs and goals. Along these lines, Kay (1995) developed the UM toolkit, a shell for building student models which enable reflection. The student can access his model built with this tool and find answers to questions such as: What does the system know about me? How did it reach these conclusions about me? Paiva and Self (1995) developed TAGUS, a shell for student modeling. TAGUS uses logic to represent the knowledge, reasoning and cognitive strategies of the learner. It provides an interface with services for accessing and updating information about the learner's knowledge state. More recently, Zapata-Rivera and Greer (2004) proposed SModel, a Bayesian student modeling server which provides several services to a group of agents in a CORBA platform.

Multiple other shell projects have been developed, focusing on particular ITS components. For instance, KEPLER (Vivet 1988), an expert system shell for building the domain and tutor modules of an intelligent tutoring system, was used to develop AMALIA, a tutoring system for teaching algebraic manipulations. The PIXIE shell (Sleeman 1987) was proposed to develop the diagnosis and remediation processes within an ITS. SCENT-3 (McCalla and Greer 1988) helped for fine-grain task sequencing. PEPE, a competence-based computational framework, was used for content planning (Wasson 1992).

ITS shells sometimes target the whole system (all components included). FITS (Ikeda and Mizoguchi 1994) is a good example of such a shell. FITS is a domain-independent framework that provides building blocks for student, tutor and domain modeling.

Recently, Stankov et al. (2008) developed a system called the Tutor–Expert System (Tex-Sys). Tex-Sys is an ITS shell that provides a generic module, implementing ITS components that can be used for the deployment of any given content. The content is specified in terms of user and domain knowledge databases. Two versions of Tex-Sys are provided, each dealing with an implementation approach taken by the ITS. DTex-Sys provides the client-server implementation, where the generic components (pedagogical controls) are implemented in a web server. The problem with this shell is that generic components deal only with pedagogical control of the ITS, not learner or domain control. The other version, xTex-Sys, provides another implementation based on a service-oriented architecture, where generic components (including learner and expert modules) are implemented as web services. The main drawback with the Tex-Sys approaches is that there is very little information about the shell's content. What knowledge is stored in it? How is it used? There is no answer to these questions.

A similar approach that targets all ITS components is the shell developed by Goodkovsky (1997). It provides simple component implementation (domain model, expert model, student model) as well as procedural models of the tutor's activity and the tutoring criteria and constraints. The usefulness of this shell can be questioned, however. For example, the domain model it provides is as simple as a set of domain concepts, which is a very limiting knowledge structure for a good tutoring system (see the chapter on domain modeling).

In summary, while the idea of providing ITS developers with a shell that targets the whole system is a very nice one, existing shells tend to focus on the big picture, neglecting the detailed rationale for each component of the system. We believe that approaches that target particular ITS components are more profound. We also feel, given the complexity of the functions of an ITS, that further refinements must be made by considering the development of very specific shells that meet the special needs associated with certain complex and essential ITS functions, as has been done with PIXIE (Sleeman 1987). Thus, complex mechanisms such as cognitive diagnosis (Pelleu et al. 2007) may benefit from special attention that may result in generic implementations to be adapted in different problem-solving contexts. Such a shell would certainly be a welcome addition to the ITS-building toolbox, and would encourage the reuse of predefined small building blocks when developing new ITSs.

18.3 The Authoring Tools Approach: An Update on Murray's Review of ITS Authoring Tools

Authoring tools go beyond the simple shell by providing an additional user interface that allows non-programmers to formalize and visualize their knowledge. The goal is to increase both the accessibility and the affordability of authoring ITSs (Heffernan et al. 2006). After developing and demonstrating powerful systems, ITS research teams are prepared to simplify the ITS building process by developing higher-level tools which do not require programming skills and are therefore accessible to instructional designers and teachers. Decreasing the implementation costs by reducing the time/product ratio is another important target of authoring tools.

Murray, Blessing and Ainsworth (2003) edited a landmark book on the topic of ITS authoring tools. Murray (1999, 2003) has classified existing authoring tools under two categories: pedagogy-oriented and performance-oriented.

18.3.1 Pedagogy-Oriented Authoring Tools

Pedagogy-oriented tools are those that focus on how to sequence and teach relatively canned content. REDEEM (Ainsworth et al., 2003) is an example of the tools in this category. It does not explicitly generate an instructional plan, but allows the production of a representation of instructional expertise, enabling the author to categorize the didactic material, or tutorial page, according to its level of difficulty, its generality and the prerequisites that connect it to other materials. This represents an implicit sequencing of content and learning activities, based on underlying tutoring strategies.

CREAM-Tools (Nkambou et al. 2003) is another example in this category. It provides operations for organizing content in terms of interconnected structures, giving it the characteristics of a pedagogy-oriented authoring tool. Moreover, its organizing capabilities go beyond didactic material (which is what REDEEM is equipped to handle) allowing it to deal with both cognitive (organization of domain knowledge) and pedagogical aspects (organization of learning objectives).

Hypermedia tools such as Interbook (Brusilovsky et al. 1998) and MetaLinks (Murray 2003) also fall into this category. These systems manage the hyperlinks between units of content (both the form and the sequencing of the content). Hyperlinks provided to the learner can be intelligently filtered, sorted and annotated with respect to a model or a learner profile, sometimes based on an ad hoc ontology. The filtering of links can be based on prerequisites, cognitive load, appropriateness of the topic, difficulty, etc.

Other authoring tools which use pedagogy-oriented domain modeling are Eon (Murray 1998), IDE (Russell et al. 1988) and GTE (Van Marcke 1992). Specifically, GTE is a rule-based tool that performs actions according to a given pedagogical goal. Eon, a "one-size-fits-all" authoring tool which provides a full-fledged set of ITS tools, was initially implemented to perform activity streams based on a given instructional goal, in order to provide the author with multiple tutoring strategies. Finally, Eon uses an approach similar to REDEEM's parameterized one that allows the author to generate tutoring strategies from scratch. Several tutors have been implemented using Eon.

It should be noted that these tools are often based on a behavioristic-empiricist approach and tend to produce 'instructivist' tutors (Jonassen and Reeves 1996) with the possible exception of Eon (which is theory-independent). Also, their instructional strategies are fixed (predefined) and they usually do not have ontology-oriented representations.

18.3.2 Performance-Oriented Authoring Tools

Performance-oriented tools are those that focus on providing a rich learning environment in which students can learn skills by practicing them and receiving feedback. RIDES (Munro et al. 1997) is an example of the authoring tools in this category. It is used for the construction of tutors that teach students how to operate devices through simulations. RIDES (for Rapid ITS Development Environment) generates instruction by providing tools for building graphical representations of a device and defining the device's behavior. In the past years, many RIDES tutors have been implemented. A system that adds capabilities to those of RIDES is DIAG (Towne 1997), a tool that simulates equipment faults and guides students through the process of diagnosing and repairing them. DIAG is concerned with the creation of domain knowledge and performs student error diagnosis by providing a mechanism that is applicable to many domains related to diagnosing equipment failure.

CREAM-Tools also belongs to this category, since it allows a connection between skills and the way they are acquired. For example, specific learning materials are linked to specific skills to support their learning. In this way, CREAM-Tools allows automatic generation of instruction and especially of complex learning materials that provide the student with a rich learning environment. When problems or exercises are created using CREAM-Tools, a knowledge structure for student tracking and error diagnosis during the problem-solving phase is also generated.

Other well-known systems in this category include SIMQUEST (Van Joolingen et al. 1997), Demonstr8 (Blessing 1997) and XAIDA (Hsieh et al. 1999). SIMQUEST provides tools for designing and creating dynamic and interactive simulation-based learning environments. Demonstr8 supports the development of model-tracing tutors by inducing production rules from examples. It addresses the ability of non-cognitive scientists to program a model-tracing tutor with limited training. With Demonstr8, an author has available three tools: a palette to create the student interface, a method for creating higher-order declarative representations of these student interface elements; a programming by demonstration method for creation of productions.

XAIDA (for Experimental Advanced Instructional Design Advisor) was originally designed to allow expert maintenance technicians to develop ITSs for maintenance topics. XAIDA relies on an instructional device known as a transaction shell, an instructional procedure applicable to particular instructional objectives of a specific type. XAIDA consists of four sub-tools (transaction shells), each of which uses a different scheme for representing and teaching a specific aspect of maintenance knowledge.

Another interesting system that may also be classified here is the Knowledge Construction Dialog (KCD) tool suite (Jordan et al. 2001), a set of tools that ease the implementation of natural language dialog capabilities within an ITS. The KCD suite was used in building important inputs for ATLAS's components: a plan operator library that is used by ATLAS's dialog manager and planner component (APE) and a semantic grammar for ATLAS's natural language understanding component (CARMEL). ATLAS is the natural language processing component that was used in several intelligent tutoring systems such as ATLAS-Why2 (VanLehn et al. 2002).

18.3.3 Instructional-Design-Oriented Authoring Tools

Some authoring tools have a specific focus on the instructional design (ID) and provide authors with relevant assistance in that process. Even though some of these were included in Murray's classification, we believe they can be considered separately as ID-oriented systems.

Authoring systems in this category include Merrill's ISD-Expert (Merrill 1993), a system that provide rules to guide the ID process. The system suggests the best content structure (an organization of subject matter content) that is consistent with the instructional goals, subject matter knowledge and student profile. Expert CML (Jones and Wipond 1991) and IDE (Russell et al. 1988) are other ID-oriented systems. Using IDE, instructional designers can enter, edit and manipulate their instructional analysis and specifications in the form of complex networks of interrelated notecards (Pirolli and Russell 1991). Smarties (Hayashi et al. 2009) is one of the recent authoring systems within this category. It provides the user with a tool for building learning scenarios by utilizing explicit knowledge related to instructional design theories, which serves as the rational basis for decision making in this context. In the same family, a lighter tool is CIAO, a hypermedia ontology-based authoring assistant that was developed to assist authors of IMS-LD scenarios by providing them with theory-aware services (Bourdeau et al. 2007; Psyché et al. 2005). Another tool in this category is aLFanet (Santos et al. 2003) an authoring system which interprets an IMS-LD schema to develop a pedagogical scenario. It provides scenarios tailored to the particular interests, level of knowledge and experience of the learners.

In fact, an interesting particularity of these recent ID-oriented systems is that they tend more and more to incorporate emerging ID and eLearning standards such as IEEE-LOM (Learning Objects Metadata), SCORM (Sharable Content Object Reference Model), EML (Educational Modelling Language) and IMS-LD (IMS-Learning Design).

ID-oriented systems have a special focus on educational principles, giving first priority to instructional design in the ITS design project. In this perspective, ITSs are perceived as artifacts for the purpose of instruction (Pirolli and Russell, 1991). Thus, their pedagogical value, meaning how well they teach, is very important. Instructional design is a process that can guarantee this pedagogical value. The main question here is how this process can be supported in an ITS development project. We believe that an explicit, formal specification of a shared conceptualization of ID expertise and related theories could be a solution to that issue. Hence, ontology engineering can play a role in this context.

What would be efficient directions to take for the future in the ITS authoring process? It is worth repeating that ITS tutors are usually complex systems involving different dimensions. As in the shell-based approach, some ITS authoring tools focus on only a part of the system, while others consider all components of the system. Also, some are dedicated for tutors in specific learning domains, while others can be used for any domain. To produce generic (covering several areas) or complete tutors (implementing all ITS components) is a challenge. Therefore, when building an ITS, the following principles apply:

- First, approaches involving small building blocks should be preferred in order to reduce the time/product ratio;
- Secondly, increased assistance to authors should be a requirement, because the tools are still complex, although they are becoming more accessible as the teams develop higher-level tools;
- Finally, ontologies should play a role in formalizing the different types of expertise involved in ITS building.

In the ITS community, the trend currently observed is a segmentation: there are more and more specific foci of research on particular aspects of an ITS, such as open learner modeling or educational data mining. This trend increases the chance of seeing local standards emerging. However, this may not occur if knowledge and results are not shared across teams. In other words, those who share the same vision should be given opportunities for dialogue. Again, ontology engineering should help here by providing a framework for engineering ITSs, to facilitate interoperability and shareability between components.

18.4 Recent Approaches in Research and Development

This section highlights and discusses recent approaches in the building of ITSs. First, recent authoring tools are characterized; then, other types of software tools; finally, we discuss Woolf's framework for positioning building tools as they correspond to components and functions of an ITS.

18.4.1 Authoring Tools

Several ITS teams have been attempting to build authoring tools that would allow for sharing of components across ITSs and reduce development costs (Heffernan et al. 2006). The current R-D practice is to develop building tools that are paradigm-specific (Kodaganallur et al. 2005), such as CTAT for model-tracing systems (Aleven et al. 2006) and example-based systems (Aleven et al. 2009); ASPIRE for constraint-based systems (Mitrovic et al. 2009); TuTalk Tool Suite (Jordan et al. 2007) for dialogue-based learning agents; and authoring tools for inquiry-based systems (Murray et al. 2004; Gijlers et al. 2009) and for virtual reality and game-based ITS (Johnson and Valente 2008). Using a "backbone", teams built tools such as the Cognitive Model SDK (Blessing and Gilbert, 2008; Blessing et al. 2009), or developed tools that allow for derivation and variabilization, such as the ASSISTment Builder (Turner et al. 2005; Razzaq et al. 2009; – also see Chapter 20). In other words, recent developments in ITS show a "specialization" by paradigm, discipline of reference and privileged application domains, resulting in a similar specialization in the authoring tools that are derived from them. The paradigm for the CTAT tools and ASSISTment Builder is the ACT* cognitive architecture; their domain of reference is cognitive psychology; the privileged component is the student model; and most of their applications are in math and science. ASPIRE authoring tool has constraint-based modeling as a paradigm, and computer science as a domain of reference. Discovery Learning Environments' paradigm is Discovery Learning; the domain of reference is the sciences, and the applications are mainly in science learning, etc.

Some new web paradigms are becoming good metaphors for collaborative authoring. For example, the open authoring model inspired by Wikipedia seems to be quite appropriate in the ITS context (Aleahmad et al. 2008).

18.4.2 General Software Engineering Tools

Besides the development of authoring tools, another way to facilitate the ITS building process is to view ITSs as software. As such, software engineering (SE) methods and tools can help. Tools provided in this context will be said to be SE-oriented. Various research proposals have been put forward with the aim of integrating different software engineering approaches. One group, the pattern-based approaches, are aimed at providing ITS developers with interesting patterns they can use to build the ITS. A pattern is a generalized solution of a typical problem within a typical context. A thorough analysis of existing ITS development solutions is an important stage in determining such patterns. Devedzic and Harrer (2005) discussed possible architectural patterns that can be found in existing ITSs. Harrer and Martens (2006) described the basis for a pattern language and catalogue for building ITS. Along the same lines, Salah and Zeid (2009) developed PLITS, a pattern language for ITS. PLITS was built from pattern mining by reverse-engineering many existing ITSs. As a proof of the concept, the authors used PLITS to build the Arabic Tutor, a web-based Intelligent Language Tutoring System for teaching a subset of the Arabic language.

Another alternative for building ITSs is to use a multiagent systems (MAS) approach for the basic building infrastructures. In fact, ITSs fulfill all of the conditions to be viewed as multiagent systems: 1) they are made of different interconnected, complex components; 2) they provide multiple, different and complementary services; 3) each of their components is functionally autonomous and equipped with specific knowledge structure and reasoning mechanisms. In this light, many ITSs have been built using agent and multiagent technologies. In particular, Vicari and Gluz (2007) have developed several ITSs to exemplify a set of Agent-Oriented Software Engineering (AOSE) methods derived from ITS research, which defines applicability criteria, design principles and implementation guidelines to be applied in the software analysis, design and development process. The agentification sometimes targets a specific ITS component: the tutor (Mengelle et al. 1998) or the learner (Vassileva et al. 2003). It can also target a specific ITS service (e.g., planning, dialogue management, collaboration) or the whole system (Capuano et al. 2000; Hospers et al. 2003; Nkambou and Kabanza 2001). Even though there are many ITSs that were built using the agent and MAS approach, there is no agent-based framework specially dedicated to facilitating the ITS building process; rather, classic models and tools developed in the MAS community are used. FIPA specifications are well-known basic packages that can be used to support the building of agent-based ITSs. However, programming skills are required in order to use this building approach. Many agent- and MAS-oriented platforms (agent builders) such as JADE (Bellifemine et al. 2008) can be used to ease the development process.

18.4.3 A Framework for ITS Building Tools

Recently, a framework for organizing the necessary building blocks found in authoring systems for building ITS was proposed (Woolf 2008). Four layers were identified, each including specific classes of building blocks (Figure 18.1). The knowledge representation level includes tools for easily representing knowledge. At this level, the user should adopt the right formalism and select the right language or tool to ease the representation process. Level 2 is about the type of domain and student models, level 3 contains tools for implementing teaching knowledge while level 4 comprises those for communication knowledge.

Providing such a framework can be seen as a starting point for developing a real methodology for ITS engineering, where each step may provide guidelines that help the author make the best decision and select the relevant tools to produce the output artifacts of that step. For instance, at level 1, based on some evaluation criteria, such as those proposed in Chapter 2 (section 2.3.1.4) of this book for the evaluation of

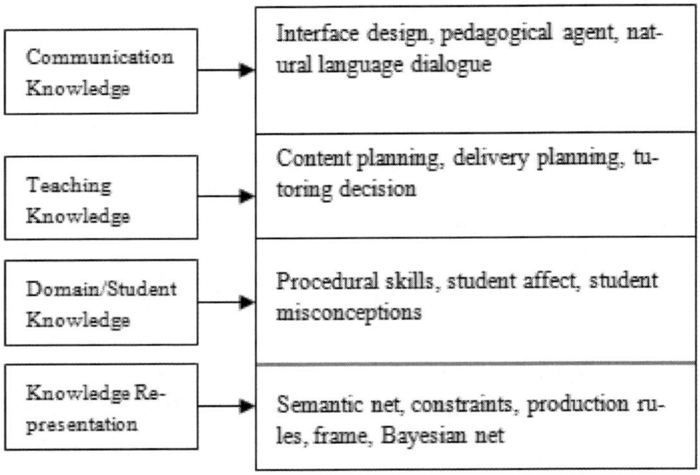

Fig. 18.1 A framework of intelligent tutor building blocks (adapted from Woolf (2008))

knowledge representation languages (expressiveness, inference power, etc), such guidelines may help the user select the right knowledge representation formalism in a given context.

The proposed framework sounds conceptually simple but may be very difficult to implement, due to the heterogeneous nature of the methods and tools that may be found at each level as well as the issue of tool integration at the same level or across levels. For example, how can a tool used for curriculum planning (at level 3) be plugged into a domain model built using CBM (at level 2)? So, even if a decision at the lower level becomes constraints for the upper level, a decision at the latter may still be incompatible with the one at the former. In short, such a framework cannot work without standards established in the community, and we are very far from that in the AIED community.

A simpler approach that should be investigated in the future is to ease the possibility of providing shareable, albeit non-standard, conceptualizations of ITS rationales. In other words, there should be a way for researchers who have the same conceptualization of some fundamental ITS concepts to share it. Fortunately, ontology provides a solution for this. It may allow small teams to clearly and formally define ITS artifacts as they conceive them, and then build their system on that formal conceptualization. As an example, a formal representation of an explicit conceptualization of instructional design and learning theories called OMNIBUS was developed by Bourdeau et al. (2007). OMNIBUS can be used by any ITS developer who shares that conceptualization to build his or her own system. The benefit of initiatives such as OMNIBUS is that they prepare a solid semantic ground on which different tools for building ITSs can be implemented. This shared semantic ground is a guarantee that can ease communication between the different tools. This approach will make it easier to move from the current proprietary, non-shareable solutions for building ITSs to others that are interoperable, reusable and easy to integrate.

18.5 Conclusion: Biodiversity or Tower of Babel? Future or Pipedream?

Multiple solutions have been provided for building ITSs, ranging from programmer-oriented tools to software for non-programmers. However, it is worth noting that there is no standard in the AIED research community to guide this process. In other words, ITS building cannot yet be considered an engineering process, as there are no methods and standard tools available to support it. As a result, after thirty years, existing solutions are still not widely shared in the field, making it difficult to find adequate building blocks and guidance to build an ITS. By comparison, the more recent research field of multi-agent systems is developing in a community where many standard development principles, methods and tools for building MASs emerge. The lack of standards in the ITS community is probably due to the multidisciplinary nature of the field. There are multiple views of the target artifacts and services that an ITS should provide. In this light, an overall reflection on the problem of building ITSs leads us to raise the following questions:

1. Is the authoring bottleneck a 'natural' border to preserve the biodiversity of ITS species? Or is the adoption of standards a necessity for the survival of the species?
2. Is the idea of one-size-fits-all a pipedream, or is it truly the future of ITS research?

Our conception of the human brain may provide answers: in contrast to our former view of the brain as a set of regions with specific functions, our understanding now is that the brain works as a whole, and several regions can perform several functions, together or as substitutes for each other. Reconsidering the architecture of ITSs in that light may provide fruitful insights into how to build them.

The chapters in this part of the book provide the reader with two examples of authoring systems and an example of an ITS. Chapter 19 presents a thorough comparative analysis between CTAT, a well-known authoring tool for cognitive tutors, and ASTUS, a new cognitive tutor authoring tool. Through examples, the chapter addresses many limitations of CTAT and shows how ASTUS copes with these limitations. Chapter 20 is about ASSISTMENT, a suite of web-based tools that help researchers to easily design, build and then compare different ways of teaching students in order to improve their achievement. A randomized controlled experiment conducted using these tools is described. Chapter 21, the last in this part, presents ANDES, one of the most popular ITSs. ANDES is an intelligent homework helper for physics. That is, it replaces students' pencil and paper as they do problem-solving homework. The author presents ANDES' behavior, the development experience, evaluations of its pedagogical effectiveness and recent progress on dissemination/scale-up.

References

Ainsworth, S., Major, N., Grimshaw, S.K., Hayes, M., Underwood, J.D., Williams, B., Wood, D.J.: REDEEM: Simple Intelligent Tutoring Systems From Usable Tools. In: Murray, T., Blessing, S., Ainsworth, S. (eds.) Tools for Advanced Technology Learning Environments. Kluwer Academic Publishers, Amsterdam (2003)

Aleahmad, T., Aleven, V., Kraut, R.: Open Community Authoring of Targeted Worked Example Problems. In: Woolf, B.P., Aïmeur, E., Nkambou, R., Lajoie, S. (eds.) ITS 2008. LNCS, vol. 5091, pp. 216–227. Springer, Heidelberg (2008)

Aleven, V., McLaren, B.M., Sewall, J., Koedinger, K.R.: A New Paradigm for Intelligent Tutoring Systems: Example-Tracing Tutors. International Journal of Artificial Intelligence in Education 19, 105–154 (2009)

Aleven, V., McLaren, B.M., Sewall, J., Koedinger, K.R.: The Cognitive Tutor Authoring Tools (CTAT): Preliminary evaluation of efficiency gains. In: Ikeda, M., Ashley, K.D., Chan, T.-W. (eds.) ITS 2006. LNCS, vol. 4053, pp. 61–70. Springer, Heidelberg (2006)

Bellifemine, F., Caire, G., Poggi, A., Rimassa, G.: JADE: A software framework for developing multi-agent applications. Lessons learned Information & Software Technology 50(1-2), 10–21 (2008)

Blessing, S.B.: A Programming by demonstration authoring tool for model-tracing tutors. The International Journal for Artificial Intelligence in Education 8, 233–261 (1997)

Blessing, S., Gilbert, S.: Evaluating an Authoring Tool for Model-Tracing Intelligent Tutoring Systems. In: Woolf, B.P., Aïmeur, E., Nkambou, R., Lajoie, S. (eds.) ITS 2008. LNCS, vol. 5091, pp. 204–215. Springer, Heidelberg (2008)

Blessing, S.B., Gilbert, S.B., Ourada, S., Ritter, S.: Authoring Model-Tracing Cognitive Tutors. International Journal of Artificial Intelligence in Education 19, 189–210 (2009)

Bourdeau, J., Mizoguchi, R., Hayashi, Y., Psyche, V., Nkambou, R.: When the Domain of the Ontology is Education. In: Proc. of the 4th Conf. on Intelligent, Interactive Learning Objects Repository Networks, I2LOR 2007 (2007)

Brusilovsky, P., Eklund, J., Schwarz, E.: Web-based education for all: A tool for developing adaptive courseware. Computer Networks and ISDN Systems 30(1-7), 291–300 (1998)

Capuano, N., Marsella, M., Salerno, S.: ABITS: An Agent Based Intelligent Tutoring System for Distance Learning. In: Gauthier, G., VanLehn, K., Frasson, C. (eds.) ITS 2000. LNCS, vol. 1839. Springer, Heidelberg (2000)

Crawford, J.: EMYCIN: an expert system shell. Series Technical report, University of Sydney Basser Dept. of Computer Science (1987)

Devedzic, V., Harrer, A.: Software Patterns in ITS Architectures. International Journal of Artificial Intelligence in Education 15(2), 63–94 (2005)

Gijlers, H., Saab, N., Van Joolingen, W.R., De Jong, T., Van Hout-Wolters, B.: Interaction between tool and talk: How instruction and tools support consensus building in collaborative inquiry-learning environments. Journal of Computer Assisted Learning 25, 252–267 (2009)

Goodkovsky, V.A.: Pop Class Intelligent Tutoring Systems: Shell, Toolkit & Design Technology. In: New Media and Telematic Technologies for Education in Eastern European Countries. Twente University Press, Enschede (1997)

Harrer, A., Martens, A.: Towards a Pattern Language for Intelligent Teaching and Training Systems. In: Ikeda, M., Ashley, K.D., Chan, T.-W. (eds.) ITS 2006. LNCS, vol. 4053, pp. 298–307. Springer, Heidelberg (2006)

Hayashi, Y., Bourdeau, J., Mizoguchi, R.: Using Ontological Engineering to Organize Learning/Instructional Theories and Build a Theory-Aware Authoring System. International Journal of Artificial Intelligence in Education 19, 211–252 (2009)

Heffernan, N., Turner, T., Lourenco, A., Macasek, G., Nuzzo-Jones, G., Koedinger, K.: The ASSISTment Builder: Towards an Analysis of Cost-Effectiveness of ITS Creation. In: Proc. of FLAIRS 2006 (2006)

Hsieh, P., Halff, H., Redfield, C.: Four Easy Pieces: Development Systems for Knowledge-Based Generative Instruction. International Journal of Artificial Intelligence in Education (IJAIED) 10, 1–45 (1999)

Hospers, M., Kroezen, A., Nijholt, R., Heylen, D.: Developing a generic agentbased intelligent tutoring system and applying it to nurse education. In: Proceedings of the IEEE International Conference on Advanced Language Technologies, Athens (2003)

Ikeda, M., Mizoguchi, R.: FITS: A Framework for ITS – A Computational Model of Tutoring. J. of AI in Education 5(3), 319–348 (1994)

Johnson, L., Valente, A.: Collaborative Authoring of Serious Games for Language and Culture. In: Proc. of SimTecT 2008 (2008)

Jonassen, D.H., Reeves, T.C.: Learning with technology: Using computers as cognitive tools. In: Jonassen, D.H. (ed.) Handbook of research for educational communications and technology. Macmillan, New York (1996)

Jones, M., Wipond, K.: Intelligent Environments for Curriculum and Course Development. In: Goodyear (ed) Teaching Knowledge and Intelligent Tutoring, Ablex, Norwood (1991)

Jordan, P., Rose, C.P., Vanlehn, K.: Tools for Authoring Tutorial Dialogue Knowledge. In: Proceedings of AI in Education (2001)

Jordan, P.W., Hall, B., Ringenberg, M., Cue, Y., Rosé, C.: Tools for Authoring a Dialogue Agent that Participates in Learning Studies. In: Looi, C.K., McCalla, G., Bredeweg, B.J. (eds.) Proceedings of the 12th Artificial Intelligence In Education. ISO Press, Amsterdam (2007)

Kay, J.: The um toolkit for cooperative user modelling. User Modeling and User-Adapted Interaction 4(3), 149–196 (1995)

Kobsa, A., Pohl, W.: The user modelling shell system BGP-MS. User Modelling and User Adapted Interaction 4(2), 59–106 (1995)

Kodaganallur, V., Rosenthal, D., Weitz, R.: A Comparison of Model-Tracing and Constraint-Based Intelligent Tutoring Paradigms, IJAIED (15) (2005)

Koedinger, K., Corbett, A.: Cognitive Tutors. In: Sawyer, K. (ed.) The Cambridge Handbook of the Learning Sciences. Cambridge University Press, Cambridge (2006)

McCalla, G., Greer, J.: Intelligent advising in problem solving domains: the SCENT-3 architecture. In: Proc. International Conference on Intelligent Tutoring Systems, Montreal, Canada (1988)

Mengelle, T., de Léan, C., Frasson, C.: Teaching and Learning with Intelligent Agents: Actors. In: Goettl, B.P., Halff, H.M., Redfield, C.L., Shute, V.J. (eds.) ITS 1998. LNCS, vol. 1452, pp. 284–293. Springer, Heidelberg (1998)

Merrill, M.D.: An Integrated Model for Automating Instructional Design and Delivery. In: Spector, J.M., Polson, M.C., Muraida, D.J. (eds.) Automating Instructional Design: Concepts and Issues. ETD, Englewood Cliffs (1993)

Mitrovic, A., Martin, B., Suraweera, P., Konstantin, Z., Milik, N., Holland, J.: ASPIRE: An Authoring System and Deployment Environment for Constraint-Based Tutors. International Journal of Artificial Intelligence in Education 19, 155–188 (2009)

Munro, A., Johnson, M.C., Pizzini, Q.A., Surmon, D.S., Towne, D.M., Wogulis, J.L.: Authoring simulation-centered tutors with RIDES. International Journal of Artificial Intelligence in Education 8(3-4), 284–316 (1997)

Murray, T.: Authoring Knowledge Based Tutors: Tools for Content, Instructional Strategy, Student Model, and Interface Design. Journal of the Learning Sciences 7(1), 5–64 (1998)

Murray, T.: Authoring Intelligent Tutoring Systems: An Analysis of the State of the Art. Int'l J. Artificial Intelligence Education 10, 98–129 (1999)

Murray, T.: An overview of intelligent tutoring system authoring tools: Updated analysis of the state of the art. In: Murray, T., Blessing, S., Ainsworth, S.E. (eds.) Tools for Advanced Technology Learning Environments. Kluwer Academic Publishers, Amsterdam (2003)

Murray, T., Blessing, S., Ainsworth, S.: Authoring Tools for Advanced Technology Learning Environment. Kluwer Academic Publishers, Amsterdam (2003)

Murray, T., Woolf, B., Marshall, D.: Lessons Learned from Authoring for Inquiry Learning: A Tale of Authoring Tool Evolution. In: Lester, J.C., Vicari, R.M., Paraguaçu, F. (eds.) ITS 2004. LNCS, vol. 3220, pp. 197–206. Springer, Heidelberg (2004)

Nkambou, R., Frasson, C., Gauthier, G.: CREAM-Tools: An Authoring Environment for Knowledge Engineering in Intelligent Tutoring Systems. In: Murray, T., Blessing, S., Ainsworth, S. (eds.) Tools for Advanced Technology Learning Environments. Kluwer Academic Publishers, Amsterdam (2003)

Nkambou, R., Kabanza, F.: Designing Intelligent Tutoring Systems: A multiagent Planning Approach. ACM SIGCUE Outlook 27(2), 46–60 (2001)

Paiva, A., Self, J.: TAGUS - A User and Learner Modeling Workbench. User Modeling and User-Adapted Interaction 4(3), 197–226 (1995)

Pelleu, J., Nkambou, R., Bourdeau, J.: Explicit Reflexion in Prolog-Tutor. International Journal of Artificial Intelligence in Education 17(2), 169–215 (2007)

Pirolli, P., Russell, D.M.: Instructional design environment: Technology to support design problem solving. Instructional Science 19(2), 121–144 (1991)

Psyché, V., Bourdeau, J., Nkambou, R., Mizoguchi, R.: Making Learning Design Standards Work with an Ontology of Educational Theories. Frontiers in Artificial Intelligence and Applications 125, 539–546 (2005)

Razzaq, L., Patvarczki, J., Almeida, S., Vartak, M., Feng, M., Heffernan, N., Koedinger, K.: The ASSISTment Builder: Supporting the Life Cycle of Tutoring System Creation. IEEE Transaction on Learning Technologies 2(2), 157–166 (2009)

Russell, D., Moran, T., Jordan, D.: The Instructional Design Environment. In: Psotka, J., Massey, L.D., Mutter, S.A. (eds.) IntelligentTutoring Systems, Lessons Learned. Lawrence Erlbaum, Hillsdale (1988)

Salah, D., Zeid, A.: PLITS:A Pattern Language for Intelligent Tutoring Systems. In: Proceedings of the 15th European Conference on Pattern Languauges of Programs (2009)

Santos, O.C., Boticario, J.G., Koper, E.J.R.: aLFanet. Paper presented at the m-ICTE, Badajoz, Spain (2003)

Sleeman, D.: Pixie: a shell for developing intelligent tutoring systems. Artificial Intelligence and Education 1, 239–263 (1987)

Stankov, S., Rosic, M., Zitko, B., Grubisic, A.: TEx-Sys model for building intelligent tutoring systems. Computers & Education 51(3), 1017–1036 (2008)

Towne, D.M.: Approximate reasoning techniques for intelligent diagnostic instruction. International Journal of Artificial Intelligence in Education 8(3-4), 262–283 (1997)

Turner, T.E., Macasek, M.A., Nuzzo-Jones, G., Heffernan, N.T., Koedinger, K.: The Assistment Builder: A Rapid Development Tool for ITS. In: Looi, C.K., McCalla, G., Bredeweg, B.J. (eds.) Proceedings of the 12th Artificial Intelligence In Education. ISO Press, Amsterdam (2005)

Van Joolingen, W.R., King, S., de Jong, T.: The SimQuest authoring system for simulation-based discovery learning. In: du Boulay, B., Mizoguchi, R. (eds.) Artificial intelligence and education: Knowledge and media in learning systems. IOS Press, Amsterdam (1997)

Van Marcke, K.: Instructional Expertise. In: Frasson, C., McCalla, G.I., Gauthier, G. (eds.) ITS 1992. LNCS, vol. 608. Springer, Heidelberg (1992)

VanLehn, K., Jordan, P., Rosé, C.P., et al.: The architecture of Why2-Atlas: A coach for qualitative physics essay writing. In: Cerri, S.A., Gouarderes, G., Paraguacu, F. (eds.) ITS 2002. LNCS, vol. 2363, p. 158. Springer, Heidelberg (2002)

Vassileva, J., McCalla, G.I., Greer, J.E.: Multi-Agent Multi-User Modeling in I-Help. User Model. User-Adapt. Interact 13(1-2), 179–210 (2003)

Vicari, R.M., Gluz, J.G.: An Intelligent Tutoring System (ITS) view on AOSE. International Journal of Agent-Oriented Software Engineering 1(3-4), 295–333 (2007)

Vivet, M.: Knowledge based tutors: towards the design of a shell. International Journal of Educational Research 12(8), 839–850 (1988)

Wasson, B.: PEPE: A computational framework for a content planner. In: Dijkstra, S.A., Krammer, H.P.M., van Merrienboer, J.J.G. (eds.) Instructional Models in Computer-Based Learning Environments. NATO ASI Series F, vol. 104, pp. 153–170 (1992)

Woolf, B.: Building Intelligent Interactive Tutors: Student-centered strategies for revolutionizing e-learning. Morgan Kaufmann, New York (2008)

Zapata-Rivera, J.D., Greer, J.: Inspectable Bayesian student modelling servers in multi-agent tutoring systems. International Journal of Human-Computer Studies archive 61(4), 535–563 (2004)

Chapter 19
Authoring Problem-Solving Tutors: A Comparison between ASTUS and CTAT

Luc Paquette, Jean-François Lebeau, and André Mayers

Computer Science Department, Université de Sherbrooke,
2500 boulevard de l'Université, Sherbrooke, Québec, J1K 2R1, Canada
{luc.paquette,andre.mayers}@USherbrooke.ca

Abstract. ASTUS is an Intelligent Tutoring System (ITS) framework for problem-solving domains. In this chapter we present a study we performed to evaluate the strengths and weaknesses of ASTUS compared to the well-known Cognitive Tutor Authoring Tools (CTAT) framework. To challenge their capacity to handle a comprehensive model of a well-defined task, we built a multi-column subtraction tutor (model and interface) with each framework. We incorporated into the model various pedagogically relevant procedural errors taken from the literature, to see how each framework deals with complex situations where remedial help may be needed. We successfully encoded the model with both frameworks and found situations in which we consider ASTUS to surpass CTAT. Examples of these include: ambiguous steps, errors with multiple (possibly correct) steps, composite errors, and off-path steps. Selected scenarios in the multi-column subtraction domain are presented to illustrate that ASTUS can show a more sophisticated behavior in these situations. ASTUS achieves this by relying on an examinable hierarchical knowledge representation system and a domain-independent MVC-based approach to build the tutors' interface.

19.1 Introduction

Intelligent Tutoring Systems (ITS) that support a learning-by-doing pedagogical strategy are usually developed in line with one of three established approaches: model-tracing tutors (Anderson and Pelletier 1991), constraint-based tutors (Mitrovic et al. 2003) and example-tracing tutors (Aleven et al. in press; Razzaq et al. 2009). All of these fit VanLehn's tutoring framework (VanLehn 2006), in which a model of a task domain is used to evaluate each of the learner's steps (which are themselves driven by mental inferences) as correct or incorrect. Model-tracing tutors such as Cognitive Tutors (Anderson 1995) and Andes (VanLehn 2005) have been proven to be successful in the classroom (Koedinger et al. 1997), but their success is mitigated by their cost (Murray 2003), which is mainly due to the effort needed to develop a generative model of the task domain. Both constraint-based and example-tracing tutors are designed to reduce this cost. The former use an

R. Nkambou et al. (Eds.): Advances in Intelligent Tutoring Systems, SCI 308, pp. 377–405.
springerlink.com

evaluative model involving constraints defined over a set of pedagogically relevant solutions, and the latter, a task-specific evaluative model built from a domain expert's interactions with the learning environment. Example-tracing tutor frameworks can offer tools to generalize the resulting model (Aleven et al. in press; Matsuda 2005), but adding such levels of complexity is an obstacle to their democratization, and still does not make them as flexible and comprehensive as model-tracing tutors. Constraint-based tutors may be particularly effective in handling ill-defined tasks in well-defined domains, such as design-based ones; however, they cannot follow learners as closely as model-tracing tutors, a capacity which is especially interesting for well-defined tasks. To reduce the effort needed to develop model-tracing tutors, one approach is to rely on a more (Cognitive Tutors) or less (Andes) domain-independent framework. In such a context, the knowledge representation system used to build the model is a key part of the framework, and its expressivity and reasoning capacity determine which domains can be modeled and how straightforward it is to model one.

Our work is based on the hypothesis that a more sophisticated knowledge representation system not only widens the range of domains that can be modeled, but also facilitates the testing of varied domain-independent pedagogical strategies, including some that are more elaborate than the ones usually found in model-tracing tutors. In fact, our efforts can be seen as an attempt to achieve an objective similar to that of Heffernan (Heffernan et al. 2008), who expanded Cognitive Tutors with a domain-specific (algebra) pedagogical model in order to provide tutorial dialogs closer to those used by experienced human tutors. Thus, our framework ASTUS is designed with two objectives: to reduce the prohibitive effort usually associated with the development of model-tracing tutors, and to provide the ITS community with a modular framework. In terms of Wenger's classic ITS architecture (Wenger 1987), ASTUS's domain-independent expert and interface modules interpret domain-specific models, and pedagogical and learner model modules can be customized to try out different pedagogical avenues. To achieve this, ASTUS uses a knowledge representation system that can be seen as a middle ground between the Cognitive Tutors production systems and the Andes solution graphs, in which the set of next possible steps is updated before each of the learner's actual steps. This approach, which is online (like Cognitive Tutors) but also top-down (like Andes), was adopted under the hypothesis that the advantages of both can be combined. Designed to model domains from a pedagogical perspective rather than to model the cognitive process used to solve them, ASTUS's knowledge representation system represents procedural knowledge hierarchically, using knowledge components of different grain sizes. Of these components, the examinable ones represent the skills explicitly tutored and the black-box ones model the underlying required abilities. Finally, to ensure that the tutor has complete access to the learning environment interface, as required by Anderson et al. (1995), these knowledge components act as a Model in terms of the Model-View-Controller (MVC) architectural design pattern on which the interface module is based.

In this chapter we present a study that evaluates the effectiveness of ASTUS by comparing it with Cognitive Tutor Authoring Tools (CTAT), a model-tracing tutor framework derived from the Cognitive Tutors (Koedinger 2003). In order to compare the two frameworks, we used each of them to author a tutor for the multi-column subtraction task domain. Each contains a generative model that can be used to trace a correct problem-solving path, but also many different incorrect ones. The objectives of this study are to evaluate whether both frameworks allow us to exhaustively model the multi-column subtraction domain; to identify the features that make each framework more or less suitable for the situations covered by a given set of scenarios; and to show which types of pedagogical behavior are made possible by each framework. The tutors were authored with the sole purpose of comparing the frameworks; we do not plan to experiment with them in a classroom.

In the next sections of this chapter, we describe the two ITS frameworks (CTAT in Section 2, ASTUS in Section 3). For each, we begin by giving an overview of its design and then present its knowledge representation system in detail, with examples from the multi-column subtraction domain. Section 4 presents the methodology of our study, detailing each step that was taken in order to achieve our objectives. For instance, we explain why we deemed the multi-column subtraction domain a good choice to evaluate the strengths and weaknesses of the two frameworks. In Section 5, we discuss the features of the resulting tutors and present scenarios which illustrate how each framework deals with different situations encountered in modeling the multi-column subtraction domain. In particular, we discuss the difference between the frameworks' respective tracing algorithms, their difficulties in modeling specific situations and the types of pedagogical behavior they support. Finally (Section 6), we conclude that authoring a tutor in either framework is a similar task, but that with ASTUS, more attention must be paid in building the task domain model so that it can be fully exploited by the domain-independent pedagogical module. However, without any extra domain-specific effort, the resulting tutor offers elaborate pedagogical behaviors that are usually not supported in problem-solving tutors. The chapter ends with a brief presentation of future work towards a new version of the ASTUS framework.

19.2 The CTAT Framework

CTAT[1] is a freely distributed domain-independent ITS framework that can be used to create tutors for various well-defined task domains (examples include stoichiometry and genetics). The main objective of CTAT is to reduce the amount of work required in authoring these tutors (Aleven et al. 2006). The CTAT framework allows the creation of two different types of tutor: Cognitive Tutors and Example-Tracing Tutors) (Aleven et al. 2006). Cognitive Tutors are based on the ACT-R theory of cognition (Anderson 1993; Anderson and Lebiere 1998) and use a cognitive model of the skills being tutored. To create such a model, expertise in AI programming is required, as the model may be implemented using the Cognitive Tutor Development Kit (TDK) (Anderson and Pelletier 1991) or the Jess rule

[1] http://ctat.pact.cs.cmu.edu

engine[2] (only the latter is distributed along with CTAT). In our study, the CTAT tutor was created as a Cognitive Tutor implemented via the Jess rule engine. In the following sections, when we speak of the CTAT framework, we are referring to Jess-based Cognitive Tutors authored using CTAT.

19.2.1 Knowledge Representation

CTAT tutors, being derived from the TDK-based Cognitive Tutors, use Jess production rules to represent procedural knowledge and Jess facts to represent declarative knowledge. These facts, referred to as Working Memory Elements (WMEs), can be used to represent the task, model the learner's perception of the interface and store temporary results in working memory. The content of a WME is defined by slots that can be filled with primitive data (boolean, integer, string, etc.) or references to other WMEs. For example, in our multi-column subtraction tutor, the WME for a column contains the following slots (the type of each slot is indicated here for clarity's sake; it is not specified in the actual model):

```
WME Column {
  name [string]
  nextColumn [Column]
  previousColumn [Column]
  minuend [Textfield]
  subtrahend [Textfield]
  difference [Textfield]
  hasBeenBorrowedFrom [boolean]
}

WME Textfield {
  name [string]
  value [string]
}

WME DecrementColGoal {
  column [Column]
}
```

Production rules provide a cognitively plausible path to explain a learner's actions, whether they are steps in the interface or mental inferences. Thus, they can exhibit two kinds of behavior (on the RHS): recognizing a step (no more than one per rule) from an event sent by the interface, or updating the content of the working memory (adding, modifying or removing WMEs) to reflect a learner's mental inferences.

```
(defrule AddDrementColumnGoal
 ?problem <- (problem (subgoals $?first ?evaluateGoal $?rest)
 ?evalGoal <- (evalColumnDecrementationGoal (column ?col))
 ?col <- (column (minuend ?minuend &:(neq ?minuend 0))
=>
 (bind ?decColGoal (assert (decrementColGoal (column ?col))))
 (modify ?problem
(subgoals (create$ $?first $?rest ?decColGoal))))
```

The above rule is fired when there is a goal of evaluating the decrementation of a column's minuend for a column with a minuend different from zero. The WME it creates will allow other rules to match, in a chain, including the final one that will result in the action of decrementing the minuend. The following rule is triggered if there is a goal of subtracting the problem's current column, and the step is

[2] http://www.jessrules.com

recognized if the correct value is entered in the column's difference slot, as specified by the *predict-observable-action* statement.

```
(defrule Subtract
 ?problem <- (problem (currentColumn ?column)
 (subgoals $?first ?goal $?rest))
 ?goal <- (subtractColumnGoal (column ?column))
 ?column <- (column (difference ? diff)
 (minuend ?minuend)
 (subtrahend ?subtrahend))
=>
 (bind ?difference (- ?minuend ?subtrahend))
 (predict-observable-action ?diff WRITE-VALUE difference))
```

When a step is executed in the interface, CTAT tries to find a chain of rules that leads to it. A loop is initiated in which the content of the currently available WMEs is compared with the rules' firing conditions in order to find matches. As rules can alter the content of the working memory when they are fired, additional production rules can be fired and the matching process is started over until the rule producing the learner's step is found. If no such rule is found, the step is considered an error. In CTAT, pedagogically relevant errors are modeled using production rules marked as "buggy". Buggy rules, like normal ones, can either match a step or modify the content of the working memory. In both cases, errors are detected after exactly one incorrect step, when the chain of production rules that leads to this step contains a buggy rule.

19.3 The ASTUS Framework

ASTUS, like CTAT, is designed to be a domain-independent ITS framework available to the ITS community.[3] As the main objective of ASTUS is to allow experimentation with varied pedagogical strategies in the context of problem-solving tutors, much work has been focused on its foundations, the domain-independent expert and interface modules. Although we have implemented a basic knowledge-tracing (Corbett and Anderson 1995) algorithm that fits ASTUS's hierarchical knowledge representation and a prototypal pedagogical module as a customizable expert system, we first made sure to support a complete inner loop [**Erreur ! Source du renvoi introuvable.**] that can show ASTUS's potential. The next steps include developing these pedagogical and learner model modules to offer basic services and fully show the benefits of modeling a task domain with ASTUS, as well as developing authoring tools that will make modeling easier.

19.3.1 Knowledge Representation

The knowledge representation system in ASTUS is derived from preliminary work on MIACE (Mayers et al. 2001), a cognitive architecture inspired by ACT-R that

[3] An alpha version, limited to internal usage, has been completed and a beta version designed to be shared is under active development (http://astus.usherbrooke.ca).

proposes original twists useful in the ITS context (Fournier-Viger et al. 2006a; Fournier-Viger et al. 2006b; Najjar et al. 2001). Using ASTUS's knowledge representation system, a task domain's declarative knowledge is divided into semantic (factual) and episodic (autobiographical) components, whereas procedural knowledge is modeled at three different grain sizes. First, *complex procedures* are dynamic plans generating a set of goals (intentions satisfied by procedures), according to an algorithm (e.g., sequence, condition, iteration). For example, in the subtraction tutor, the complex procedure *SubtractWithBorrow* is a partially ordered sequence:

```
SubtractWithBorrow (TopSmallerColumn c) {
 subgoal[1] := BorrowFrom(query{nextColumn(c)})
 subgoal[2] := BorrowInto(c)
 subgoal[3] := GetDifference(c)
 order-constraints {(2, 3)}
}
```

Second, *primitive procedures* represent mastered abilities that correspond to steps in the learning environment. Here is one of the two primitive procedures in the subtraction tutor (the other is *ReplaceTerm*):

```
EnterDifference (Column c, Number dif) {
        c.difference := dif
}
```

Third, *queries* and *rules* represent basic or mastered mental skills, such as pattern-matching and elementary arithmetic. Along with complex procedures, they represent mental inferences. As queries and rules define how procedural knowledge components can access semantic ones, they are described in more detail in the discussion of this below.

A class of problem is associated with a goal (in the subtraction tutor, the *SubtractInColumns* goal) that can be satisfied by different procedures, complex or primitive, some of which may be marked as incorrect to represent pedagogically relevant errors (for instance, in the subtraction tutor, the incorrect procedure *AddInsteadOfSubtract*). As complex procedures specify sub-goals, the resulting graph has a goal as root and a set of primitive procedures as leaves.

Aside from goals, which define a semantic abstraction over procedures, semantic components include *concepts*, *relations*, *functions* and *contexts*, and their corresponding instances, *objects*, *facts*, *mappings* and *environments*. Concepts represent pedagogically relevant abstractions and are defined using both *is-a* relationships and essential features. Functions and relations, respectively, represent single- and multi-valued non-essential associations between objects. For example, the subtraction tutor includes these semantic knowledge components:

```
Concept Column {
  position [integer]
  minuends*[Minuend]
  subtrahends*[Subtrahend]
  difference[integer] (the value is initially unknown)
} *a tuple where the first is the current one.
```

```
Function nextColumn {
 column[Column] (argument)
 next[Column] (image)
}

Concept Term {
 units[integer](0-9)
 tens[integer](0-1)
}

Concept Minuend isA Term

Concept Subtrahend isA Term
```

Contexts reify the subdivisions of the learning environment, which generally correspond to windows in the interface. Examples of multi-context learning environments include "wizards" and pop-up dialog boxes (the subtraction tutor has only one context). Thus an environment contains all the instances (objects, facts, mappings) related to a distinct subtask of the task domain contained in a context. Domain-level (vs. task-level) objects that represent constants (e.g., the integers 0-9 in the subtraction tutor) are part of a global context that is automatically handled by the framework.

The episodic knowledge components are instances of the semantic and procedural knowledge components that represent the learner's solution path. The current episode is a graph that contains procedures in progress, done or undone, next possible primitive procedures and planned goals that have not yet been developed. Goals and procedures are specified with parameters and queries. The values of the former come from the parent component (either a goal or a procedure instance) and the values of the latter come from the current environment, according to domain-independent requests such as "get the unique object representing a given concept" or "get the image of a given function". For example, in the procedure *SubtractWithBorrow* (detailed above) a query fetches the image of the function *nextColumn* for a column specified as a parameter. Queries can also inspect the current episode. For example, in the subtraction tutor, a procedure inspects it to see whether a *BorrowFrom* goal has been satisfied or not.

Rules are used to make relations and functions operational and to classify objects (adding extra is-a relationships). Implemented using Jess, rules help to abstract many of the domain-specific computations that are not relevant in the tutoring process. For example, in the subtraction tutor, the *nextColumn* function is made operational with this rule:

```
(defrule nextColumn
  (Column ?column (position ?p))
  (Column ?next (position ?next:&(= next (+ ?p 1))
  =>
  (Instantiate Function "nextColumn" ("next", ?next)
                         ("column", ?column)))
```

and a column is classified as a TopSmallerColumn with this rule:

```
(defrule TopSmallerColumn
 (Column ?column)
 (CurrentSubtrahend (column ?col) (subtrahend ?subtrahend))
   (CurrentMinuend (column ?col) (minuend ?minuend))
   (test (< (getValue ?minuend) (getValue ?subtrahend)))
   =>
   (Classify ?col "TopSmallerColumn"))
```

Before each of the learner's steps, the episodic graph is developed following each applicable complex procedure's plan, further specified by its arguments, to find the set of next possible primitive procedure. If a step committed by the learner, is included in this set, the graph is updated accordingly, in the other case, the step is added to the off-path steps stack. The graph is not updated while there is a least one step on the stack. Either the tutor or the learner can undo off-path steps, allowing resuming a problem-solving path that can be traced (i.e., not necessarily a correct one). Thus, the episodic components form an interpreted high-level log of the learner's steps, which are stored in a low-level log to allow an exact replay. The latter is required because even if the arguments collected from the learner's interaction in the learning environment match the arguments of a next possible primitive procedure, they may not be exactly the same. For example, the match can be limited to check for a common concept.

19.4 Methodology

In this study, we use incorrect procedural knowledge of the multi-column subtraction domain to add knowledge components to the model and thus yield more data to enrich our comparison. It is uncertain whether being able to give a detailed diagnosis of errors is useful in tutoring (Sleeman et al. 1989), but if it is not supported, it is not possible to conduct studies to evaluate the gains or lack thereof. The methodology we followed comprises six steps. This section describes each step and its relevance to our comparison of the ASTUS and CTAT frameworks.

19.4.1 Choice of the Task Domain

We chose to model multi-column subtraction because it is representative of well-defined tasks in well-defined domains. Indeed, the arithmetic procedure of subtraction is well documented in the literature and is a clear and precise algorithm. Even though it is well-defined, the subtraction domain contains many inferences that must be deduced by the tutor from steps that may occur in a non-strict order. These characteristics give enough complexity to the solving algorithm to produce an interesting model that can be used to compare the features of the CTAT and ASTUS knowledge representation systems.

VanLehn (1990) assembled a list of 121 procedural errors that can occur when a learner subtracts numbers. Incorporating these errors in the tutors adds knowledge components to implement in CTAT and ASTUS, thus introducing new and possibly complex situations for our models. These situations give us insights about the strengths and weaknesses of the two knowledge representation systems.

19.4.2 Framework-Independent Procedural Model

Once we had chosen multi-column subtraction as the task domain for our comparison, we created a procedural model independent of any tutoring framework.

This model is based on a procedural model presented by Resnick (1982) and was used as a reference when we implemented the CTAT and ASTUS tutors. We chose this model as our reference because it can resolve any multi-column subtraction problem, and it is explicit regarding the inferences and steps that must be made and taken by the learner. To make this model more suitable for a tutoring context, we modified it slightly by adding a new way to subtract a column and including more detail in the borrowing section of the algorithm. The borrowing part of the algorithm was not deemed to be a natural extension of the semantic knowledge we assume the learner know (the base-ten numeral system's basis).

More specifically, the first modification made to Resnick's model is a new way to subtract the current column when it does not contain a number in its subtrahend (i.e., equivalent to a zero). In this case, instead of doing a subtraction, the algorithm simply copies the minuend in the difference section below the line. This modification was taken from a set of subtraction rules given by Brown and VanLehn (1982) and is important in our model, since subtraction errors sometimes depend on the presence or absence of a subtrahend.

The second modification is applied to the original model's borrowing algorithm, to more closely capture the semantic knowledge of the domain in a procedure. When borrowing across zeros, our model does not change zeros to nines from right to left. Instead, it finds the first term which is not zero, decrements it and then iterates from left to right, changing zeros to nines. This procedure more closely represents the semantics of the base-ten numeral system: when borrowing one unit of the next column (to the left) is subtracted to add ten units to the current column. In the case of borrowing across a zero ten units are borrowed into and one is borrowed from the column, thus changing the zero to a nine.

19.4.3 Error Modeling

When solving problems using arithmetic procedures, systematic errors can be explained by the application of a faulty method (VanLehn 1990). Hence, for each of the 121 errors, we defined the modifications that are needed to our framework-independent model in order for it to be able to produce this error. Each error was modeled the same way: based on the error's definition, we established the modifications that had to be made to the correct model in order to recreate it. Once a model had been generated for all of the errors, we had to apply modifications to the implementation of both tutors. Also, since all the errors were successfully incorporated in our framework-independent model, we can conclude that a failure to implement an error in the CTAT or ASTUS framework is due to limitations of the framework and not because the error cannot be modeled in our problem-solving procedure.

19.4.4 Subtraction Interface

Like the procedural model, the graphical user interface of the tutors was designed independently of the tutoring framework. It was inspired by the interface of

POSIT (Orey 1990), a subtraction ITS that diagnoses learners' errors while they are solving problems (Orey and Burton 1990) and the traditional pen and paper approach. Our interface imposes as few limitations as possible on how learners interact with the learning environment, in order to allow them to express each of the 121 errors.

The interface we designed (shown in Figure 19.3) allows two types of steps: 1) entering the difference for a column; and 2) replacing terms (minuends or subtrahends) in order to borrow from or borrow into a column. The UI actions used to trigger these steps are similar: the learner clicks on a column's difference input box or on the term he or she wishes to replace, and then enters a value using the numerical pad on the right. Once the "OK" button is pressed on the numerical pad, the problem display on the left is updated accordingly.

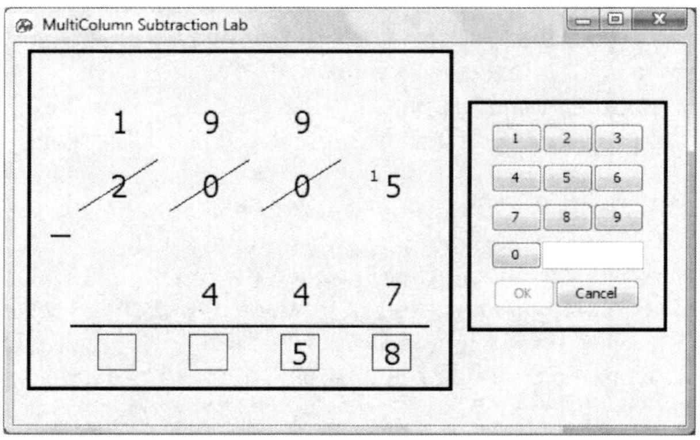

Fig. 19.1 The graphical user interface for the subtraction tutor

19.4.5 Implementation of the Tutors

To implement the CTAT and ASTUS tutors, we began by covering only the knowledge required for a completely correct problem-solving path. As with the framework-independent procedural model, errors should be incorporated without modifying the model already implemented in the tutors.

The model we took from Resnick (1982) accurately describes one way of executing the subtraction procedure, but we had to loosen some of its restrictions to make it applicable in the ITS context, where the model is primarily used to trace the learner's solving path. In particular, the steps that must be executed to subtract two numbers in column can be applied in many different orders while still obtaining correct results, so it is important to consider the multiple ways in which the learner can solve a problem. To be able to correctly trace the learner's solving path, we loosened the borrowing process to remove the constraint on the order of the required steps. For instance, the steps of changing a zero to a nine, decrementing a column and adding ten to a column can be executed in any order. Also, once the current column has been incremented by ten, it can be subtracted even if the borrowing process is not yet finished. Even though we gave learners more freedom, we kept the restrictions on the currently subtracted column so that they must complete all the steps relative to its subtraction before moving to the next one. This restriction has no impact on error modeling and although it is possible to subtract any column as long as subsequent borrows do not affect it, the ordered column sequence from right to left is a key part of the tutored subtraction algorithm.

The model created for our CTAT tutor is organized so that its production rules implicitly replicate a "while loop" in which each of the columns of the problem is subtracted individually, starting from the rightmost and ending with the leftmost. To trace the subtraction of individual columns, our model implements a sub-goal system. The use of such a system is common in examples of tutors given with the CTAT framework and it has the advantage of being flexible with regard to the order of the learner's steps. Sub-goals are normal WMEs that are added to the working memory with the purpose of indicating the steps that are currently possible. In our CTAT model, sub-goals are added to the working memory in the order defined by the framework-independent procedural model. An executed step can thus be accepted if it matches one of the current sub-goals, even if it does not follow the usual ordered sequence of the algorithm. The sub-goal associated with this step is then removed from the working memory to prevent it from being executed again. Once all of the sub-goals for a column have been successfully achieved, the model finds the next column and restarts the process until there is no column left to be subtracted, in which case a sub-goal is added calling for the learner to click on the "Done" button to indicate that he or she considers the problem solved. The "Done" button can also be clicked on when all the columns left to be subtracted have a difference of zero, since these extra zeros are not significant.

ASTUS's model consists of an ordered "for-each" iteration procedure on the columns of the problem. Thus, as in the CTAT model, each column is subtracted individually, from right to left. As each column is subtracted, the episodic graph is developed to obtain the set of primitive procedures that can be executed; the restriction on the order of execution of these primitive procedures is already contained in the complex procedures used to model the column's subtraction. To allow the learner to solve the problem with reasonable freedom, we ensured that the complex procedures contained only those order constraints that are required (borrowing into a column before subtracting it) or pedagogically relevant (subtracting only one column at a time). The problem is considered solved when the "for-each" iteration ends (when the last column of the iteration has been subtracted), or when the "Done" button is pressed (the idea was borrowed from CTAT, but the actual implementation is different because it is handled automatically by the framework), to let the learner finish the problem as soon as all the remaining columns have a difference of zero.

19.4.6 Incorporation of Errors

Once we had completed the basic implementation of the two tutors, we started improving them by incorporating the errors we had modeled. Of the 121 documented errors (VanLehn 1990), we implemented 46. The first 26 errors were modeled by systematically incorporating each error, in the same order we followed in our framework-independent model. After those errors were completed, we had enough experience to evaluate the challenges incorporation of specific errors would pose. Of the 95 remaining errors, we implemented 20 that were more challenging and required modeling behaviors that had not been previously encountered. The remaining 75 errors were not incorporated in the tutors because they were similar to the ones previously implemented and thus did not pose new challenges.

Both frameworks allowed the successful incorporation of the 46 selected errors, but the effort required showed the strengths and weaknesses of the CTAT and ASTUS knowledge representation systems. The "Results" section explains these strengths and weaknesses and the features from which they arise.

19.5 Results

As both frameworks allowed us to produce a comprehensive model of the subtraction domain, we need to show: how each framework's features have influenced the modeling process of the tutors and which type of pedagogical interactions are offered by each of the frameworks.

19.5.1 Modeling Process

The results concerning the modeling process are presented in five sections. The first gives details on how ambiguous steps are managed by the two frameworks. Then we compare how specific error types are modeled and handled; in particular, we discuss errors containing multiple steps and errors containing correct steps. Next we present how each framework allows the reuse of knowledge components. Finally, we examine the coupling between the interface and the model and how this influences the implementation of the tutor.

19.5.1.1 Ambiguities

When tracing a learner's solving path, situations may arise in which different procedures yield the same steps. These steps are qualified as being ambiguous because they can be interpreted in more than one way. Ambiguities may be present in an error-free problem-solving path, but in the subtraction tutors, they result from the inclusion of errors. When we added errors to our tutors, we found three kinds of situations that led to ambiguities. First, two unrelated errors can lead to the same step for specific problems. For instance, the *add-instead-of-sub* (adding the term of a column instead of subtracting) and *smaller-from-larger* (subtracting the smaller number from the larger regardless of which one is the minuend) errors both result in the learner entering 2 as the difference when they are applied to a "5 – 7" column. Second, errors can involve the same behavior but in different application conditions, thus resulting in ambiguities when many sets of conditions are satisfied at the same time. For an example of this situation, consider the errors *diff-0 – N = 0* (writing 0 as the difference when subtracting a number N from 0) and *0 – N = 0-after-borrow* (the same behavior but occurring only when the column has already been borrowed from). Third, ambiguities occur when some or even all of the steps of an incorrect procedure can be considered correct. In this case, the tutor must be able to determine which procedure was executed when an unambiguous step is taken, or give priority to the correct procedure when it is not possible to decide whether the error was committed or not. An example of this situation is the *borrow-skip-equal* error (skipping columns where the minuend and subtrahend are equal during the borrowing process), in which correct steps are omitted rather than incorrect ones being performed. Details of this error are presented in the "Errors containing correct steps" subsection.

Using the model tracing algorithm of CTAT, ambiguities are resolved according to rule priority. When a step is executed, a search is performed to determine which rule path can explain it. Since the execution order of this search is based on the priority of each rule, the rules of the matching path with the highest priorities will be discovered first and will be fired, preventing the discovery of any ambiguities that might have occurred. Additionally, buggy rules always have lower priorities than valid ones, so a step that can be evaluated as correct will be so evaluated even if an incorrect path exists.

ASTUS's hierarchical procedural model allows a tracing algorithm in which an episodic graph containing all the applicable procedures is updated before each of the learner's steps. This feature allows ASTUS to detect and handle ambiguous steps. The graph is used to select the appropriate explanation for previous steps where evidence resolving the ambiguities was found. Evidence includes the execution of a non-ambiguous step or the signal sent by the "Done" button. In the former case, the framework uses the episodic graph to find the nearest common ancestor of the procedure actually executed and the ambiguous one. In the latter case, the graph is searched for a procedure that is completed and that can satisfy the root goal.

In summary, CTAT prevents ambiguous states by always firing the production rules with the highest priority. This method is easy to implement and handles most of the ambiguity in a way that has no negative effects on the tracing process. It is possible to introduce special treatment of ambiguities by adding extra production rules to the model. An example is given in the "Errors containing correct steps" subsection below. CTAT is thus capable of handling any ambiguity, but the drawback is a more complex model containing superfluous knowledge components. With ASTUS, all ambiguities are handled directly by the framework's tracing algorithm, thanks to its top-down hierarchical procedural model and the resulting episodic graph. This approach requires more effort when developing the framework, but authors do not have to worry about ambiguities. Both ASTUS and CTAT must deal with permanent ambiguities that no evidence can resolve, either by using priorities or by relying on the learner model.

19.5.1.2 Errors with Multiple Steps

Some errors require multiple steps from the learner for complete diagnosis. This is the case for errors where the whole problem is solved incorrectly, such as *add-no-carry-instead-of-sub* (the learner executes an addition without carrying instead of

(a)	(b)
45	44
- 25	- 22
60	62

Fig. 19.2 Examples for the add-no-carry-instead-of-sub error. (a) The first column is ambiguous. (b) The error should not be diagnosed

a subtraction). As well as involving multiple steps, these errors can also produce ambiguities. For instance, in the *add-no-carry-instead-of-sub* error described previously, columns where the sum of the minuend and the subtrahend is the same as their difference can exist. An example of this would be the 15 – 15 problem where the sum and the difference of the unit column are both 0 and, because the carry is not performed, there is no way of knowing whether the learner intended to add or subtract the columns (i.e., the ambiguity is permanent).

When it comes to reacting to the learner's errors, CTAT implements a policy of immediate feedback in which there is no delay between the execution of the error and the tutor's feedback. Thus, CTAT's error detection only allows one step to be executed before feedback is given; this poses particular challenges when dealing with errors containing more than one step. For instance, when dealing with errors that are composed of multiple steps that can be executed in different orders, since there is no way of knowing which one the learner will execute first, all of them need to trigger the error diagnosis. Also, because the learner does not have the opportunity to complete all of the steps in the error, CTAT has less evidence that the error it diagnosed is really the one being made. Other challenges come from errors that require an ordered sequence of steps. In those cases, since only one step can be executed before feedback is given, the tutor should diagnose the error when the first step of the sequence is executed. This is difficult when the first steps are ambiguous. Figure 19.2 illustrates such a situation in the case of the *add-no-carry-instead-of-sub* error. If the error diagnosis is made on the basis of the first column alone, the error is not detected in Figure19.2 (a) since the first step is ambiguous. On the other hand, if we allow the error to be diagnosed on any column, it will be detected in Figure 19.2 (b) even if the first column was correctly subtracted[4]. To achieve accurate diagnosis in every situation, this ambiguity must be handled in the procedural knowledge, in a way that is specific to this error. Thus, when an addition is detected, the model needs to iterate on all the columns that were previously processed to see whether the entered results can be interpreted as either the subtraction or the addition of each column. If an ambiguity is found for each of the preceding columns, then the *add-no-carry-instead-of-sub* diagnosis can accurately be given.

[4] Since the learner correctly subtracted the first column we can assume that he/she can subtract (without borrowing) and that the second error is not due to the *add-no-carry-instead-of-sub* error.

Even if we overlook CTAT's immediate feedback policy, it would still be challenging to handle errors with multiple steps because production rules are independent of each other. The lack of an explicit hierarchy in the procedural model has two important consequences: it is 1) difficult to identify that multiple steps are caused by the same occurrence of an error and 2) difficult to evaluate how many steps are caused by an occurrence of an error. These two characteristics prevent the system from delaying feedback until the learner completes all of the steps in an error.

In ASTUS, an error with multiple steps can easily be modeled by creating an incorrect complex procedure. The model's hierarchy allows the creation of an episodic graph in which the primitive procedures caused by a specific occurrence of an error are easily identified. It is thus possible for the tutor to associate multiple steps with a specific error occurrence.

19.5.1.3 Errors Containing Correct Steps

There are situations where an error is not described by the incorrect steps it causes but rather by the correct ones that are skipped when it is made. Being able to handle these errors is a specific case of ambiguity management, and thus the features of each framework that make it possible or challenging are the same ones described previously in the ambiguity subsection. An example of such a situation is the *borrow-skip-equal* error (skipping columns where the minuend and subtrahend are equal during the borrowing process). An example of its execution is shown in Figure 19.3, where all of the executed steps for the first column are correct and the error can only be diagnosed when the learner subtracts the second column and enters zero as its difference. Furthermore, in the situation shown in Figure 19.3 (a), there are four correct steps that must be performed before the next column can be used to diagnose the error. It is essential that the exact sequence of steps is executed before diagnosing the *borrow-skip-equal* error since, as shown in figures 19.3 (b) and 19.3 (c), the exclusion or addition of one could change the diagnosis. In 19.3 (b), five correct steps have been performed and one of them was changing a zero to a nine in a column where the minuend and the subtrahend are equal. Because of this step, we know that the error made is not *borrow-skip-equal*, but probably a slip where the learner forgot one of the columns. On the other hand, Figure 19.3 (c) shows a situation where only three of the four required steps have been executed. This shows why it is mandatory to check for the presence of the exact sequence of steps defined by the error; in this case, removing one of them can completely change the diagnosis. The solving path shown in this figure could be associated with either the *borrow-across-zero* (not borrowing on zeros and skipping to the next column) or the *always-borrow-left* (always borrowing from the leftmost column) errors.

In CTAT, we need to be able to determine when entering the difference of the next column will cause the error to be diagnosed. Since all of the correct steps are already covered by existing production rules, we have to add rules in order to 1) mark all steps that would be performed if the learner was executing the error and 2) iterate through the columns to check that each of these steps was executed. Handling errors containing correct steps is similar to handling errors with multiple

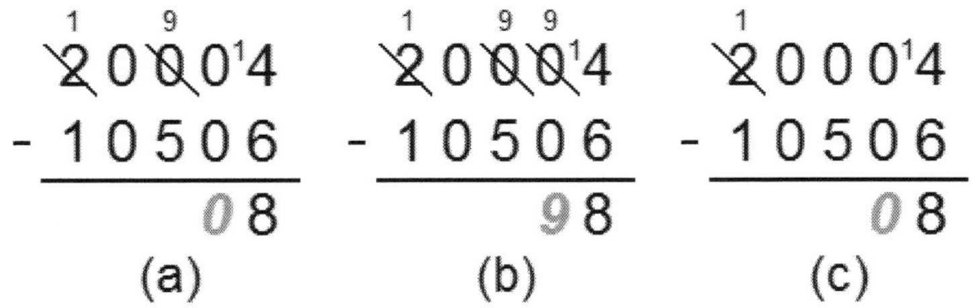

Fig. 19.3 An example of the borrow-skip-equal-error. (a) The error can only be diagnosed when the difference is entered in the second column. (b) When an additional step is present, we cannot evaluate the error as borrow-skip-equal. (c) One of the steps is ng and the error could either be borrow-across-zero or always-borrow-left

steps: the mechanism used to manage the ambiguities must be implemented in the model. Implementing such a mechanism can be complex and tedious: for the *borrow-skip-equal* error, 11 production rules are required to give a correct diagnosis, while only 14 are required to trace an error-free subtraction solving path. Adding ambiguity management in the model for this error requires almost as much effort as implementing the complete model for the basic subtraction tutor. Thus, one advantage of a framework with a knowledge representation system similar to ASTUS's is its built-in approach to manage ambiguities, significantly reducing the effort needed to model complex errors.

Modeling this kind of error in ASTUS does not cause any difficulty: an incorrect complex procedure containing an ordered sequence of sub-goals is created, with a last one implicitly indicating that a part of the correct procedure was skipped. In the case of the *borrow-skip-equal* error, the last sub-goal is to subtract the next column. The main difference is that in ASTUS the error is part of the domain, whereas in CTAT it is recognized using additional rules but not explicitly modeled (e.g., the tutor cannot generate the results of this error to demonstrate it).

19.5.1.4 Reuse of Knowledge Units

Both CTAT and ASTUS allow the reuse of knowledge components in different situations. In this section we give examples of how each of the frameworks reuses knowledge components in the subtraction domain. Both knowledge representation systems allow an author to reuse previously defined procedural and semantic knowledge components to model errors. Reusing existing components reduces the complexity of the resulting model and decreases the effort needed to implement it. An example of how the modeling process can be simplified is taken from the *always-borrow* error, in which the learner systematically borrows before subtracting a column even if it is not necessary. To implement this error, we simply reuse the borrowing procedure that has been previously modeled, by forcing its use on a column that would not require it.

In CTAT, procedural knowledge can be reused with the help of buggy production rules that do not produce a step. These rules can be used to alter the content of the working memory in order to create a rule path containing valid rules ultimately flagged as incorrect. In our model, we use previously defined sub-goals to reproduce existing behaviors in an erroneous context. For instance, with the *always-borrow* error, a buggy rule adds existing sub-goals relative to the borrowing process (i.e., decrementing the next column, adding ten to the current column) in a situation where they are not required. The equivalent can be achieved in ASTUS by defining incorrect complex procedures that use existing goals. Hence, procedures that had been previously defined are used to describe erroneous behaviors. For example, the *always-borrow* error is modeled by an incorrect procedure using the goals of borrowing from and borrowing into a column.

Error composition happens when multiple erroneous behaviors are present at the same time in a solving path. In both CTAT and ASTUS, the reuse of correct procedural knowledge in modeling pedagogically relevant errors allows the recognition of composite errors because the correct procedures that are reused can themselves have erroneous alternatives. The "Pedagogical interactions" section in 5.2 shows an example of a learner displaying both the *always-borrow* (borrowing even if not required) and the *borrow-from-bottom* (borrowing from the subtrahend) behaviors.

In CTAT, production rules test conditions on existing WMEs and perform computations to modify or create new ones. In ASTUS, the complex procedures follow a fixed, domain-independent behavior so that domain-specific conditions and computations are not directly included in them, but added indirectly via queries and rules. For example, the mapping between a column and the number of zeros to its left can be used by both the *borrow-decrementing-to-by-extras* (when borrowing into, increment by 10 minus the number of zeros borrowed across) and *decrement-by-one-plus-zeros* (when borrowing from, decrement by 1 plus the number of zeros borrowed across) errors. In CTAT, this behavior can be emulated with highly prioritized production rules that perform computations and store the result in WMEs. In many cases, including the errors cited above, high priority of the rules is necessary because the computations must be performed before the steps of the borrowing process have been executed, even though the buggy rule can be executed after them. For instance, in the *decrement-by-one-plus-zeros* error, decrementing the minuend of the column that is borrowed from can be executed after the zeros have been changed to nine. It is then crucial that the number of zeros to the left of the current column be counted before the terms are changed. Another example of the reuse of knowledge components in ASTUS is the classification of a column object as having "been borrowed from". Some errors occur only for these columns and the procedures modeling them require the column to be an instance of the "column borrowed from" concept. In CTAT, WMEs can contain slots with a boolean value to emulate classification. In our subtraction model, this is indeed the case. Such a slot must be manually updated in every production rule that could change its value, whereas in ASTUS, classification is automatically retested when an object is modified.

19.5.1.5 The Coupling between the Model and the Interface

In this section we detail how each framework manages its user interface, based on examples taken from our subtraction tutors. The two frameworks have opposite approaches in terms of how they link the user interface and the model. In CTAT, the interface and the model are almost entirely separate: the interface has no access to the content of working memory and the only information concerning the interface that the model receives is through Selection Action Input (SAI) events. A SAI event is sent when a step is performed on the interface, and it contains the element of the interface that triggered the event (selection), the action that has been performed on this element (action) and the value that was entered (input).

Since there is no direct link between the interface and the model, information concerning the problem's current state that is useful for tracing must be stored in WMEs that are updated using the SAI events. For instance, in our subtraction tutor, the working memory contains WMEs representing each of the interface's text fields, to store their value. We must ensure that the values contained in working memory are synchronized with the content shown on the interface by using the data contained in the SAI events.

Another effect of having a weak link between the interface and the model is that the interface's adaptability to the problem data is limited. Since the interface has no access to the content of working memory, it cannot use the problem data to generate its interface elements. Thus, it is difficult to dynamically add interface elements in order to match the problem state. In our subtraction tutor, this means that the interface has a fixed number of columns and a problem cannot contain more or less than that number, as this would require creating (or hiding) the text field dynamically.[5] The severity of this limitation varies according to the particular design of a tutor's interface: for example, it would have been possible to dynamically add columns if we used a "table component"[6] instead of individual text fields for each term.

The ASTUS framework enforces a stronger link between the model of the task domain and the interface (Fortin et al. 2008). Unlike CTAT, ASTUS gives the interface elements access to the instances of the current environment. This is achieved by the use of *views,* scripts that describe the representation of concepts in the interface. This MVC-based approach allows the interface to reflect the content of the current environment at all times. When the effects of a primitive procedure are applied to the environment, the views of the modified objects are notified so that they can update their UI representation accordingly. For example, in our subtraction tutor, when borrowing from or decrementing a minuend, the primitive procedure adds a new minuend to a column object, which view is then notified so that it displays the new minuend and strikes the old one. This approach enables a tutor built with ASTUS to have a flexible interface that adapts itself dynamically. For this reason, the ASTUS subtraction tutor can have the right number of columns to fit any problem.

Each primitive procedure is associated with an *interaction template* that triggers the execution of a step when the required basic UI actions are matched (Fortin et al. 2008). These templates offer a solution to the difficulty which CTAT tutors circumvent by "tutorable" interface elements (for example, a panel

---

[5] It is possible for a production rule to call static Java methods to produce side-effects on the tutor's interface; a sample "Truth tables" tutor uses this technique.

[6] This is the solution used in sample addition and subtraction tutors.

containing multiple combo-boxes and a button). Instead, with our approach, steps can be triggered by multiple interactions on multiple interface elements. The most important benefit of this approach is its capacity to generate the interactions by using the template and the episodic graph to produce a step with complete visual feedback (i.e., mouse moves and clicks). It is then possible to use the demonstration of a step as pedagogical feedback. There is an obvious cost for supporting this form of feedback, but we found out that it also helped us come up with more comprehensive models, as data implicitly defined by the interface must be explicitly encoded to define the templates. In some specific cases, we may implement multiple interactions in a single component if decomposing the step is not pedagogically relevant. For example, the "numerical pad" of our subtraction tutors is a single component.

We believe an MVC approach to the model-interface coupling has many advantages over a weaker one. It allows us to adapt the interface to the content of a particular problem, it prevents synchronization issues between the interface and the environment and it allows sophisticated pedagogical interactions such as demonstration. On the other hand, a weaker link such as the one offered by CTAT is easier to implement and, more importantly, makes it easier to develop tools that can be used to create the tutors' interfaces. These authoring tools are important, since they can greatly reduce the effort required to create a new tutor, but they are always more effective when dealing with rather simple or formatted interfaces.

19.5.2 Pedagogical Interactions

In this section we show how each of the frameworks gives pedagogical feedback. We start by describing the different types of interactions that are supported by at least one framework. We then use four scenarios taken from the subtraction tutors to give examples of how CTAT and ASTUS behave when reacting to learners' steps. Table 19.1 (presented at the end of this section) summarizes how the individual interaction types are supported by each framework.

19.5.2.1 Pedagogical Interaction Types

- Immediate feedback: minimal feedback to indicate whether a step is correct or incorrect. This includes flag feedback: changing the color of an input value to indicate whether it is correct or incorrect.
- Interface highlights: focusing the learner's attention on a specific part of the interface, for example by painting a rectangle over it.
- Next-step hint: giving a hint towards one of the next correct steps.
- Error-specific feedback: giving feedback regarding an error that is contained in the model.
- Off-path error recognition: recognizing certain off-path steps and giving feedback accordingly.
- Demonstration: demonstrating, with full visual feedback, how to perform a step (i.e., the tutor takes control of the mouse and keyboard).

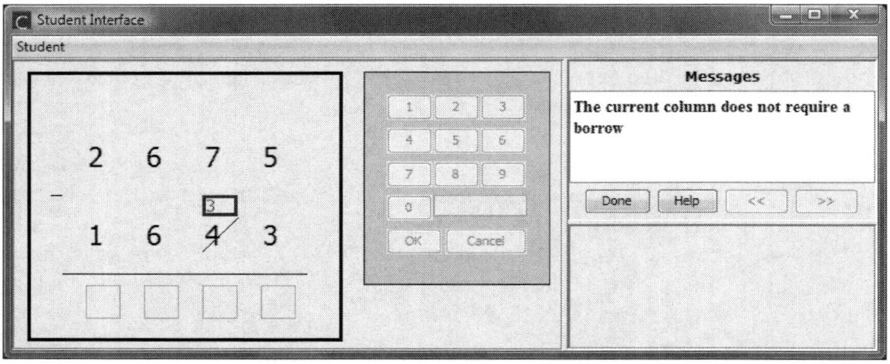

Fig. 19.4 An example of CTAT's reaction to a composite error

19.5.2.2 Scenario 1

The first scenario illustrated in Figure 19.4 shows a case of error composition: solving the problem 2675 – 1643, the learner changes the 4 (subtrahend of the second column from the right) to a 3. This action is caused by the composition of two errors: *always-borrow* (borrowing even if it is not required) and *borrow-from-bottom* (borrowing from the subtrahend). For both frameworks, the recognition of composite errors is made possible by reusing procedural knowledge, but the feedback they produce is different.

In this situation, the CTAT tutor reacts (see Figure 19.4) by: 1) flagging the incorrect value entered by writing it in red; 2) highlighting the replaced term by drawing a blue frame around it; and 3) giving a message related to the error. This message is produced from a template associated with the first buggy rule encountered (*always-borrow*) while the second (*borrow-from-bottom*) is ignored.

In the same situation, the ASTUS tutor reacts by first undoing the learner's step and updating the interface accordingly; and second, displaying a message that indicates which errors have been traced. As shown in Figure 19.5, in cases of error composition, ASTUS's written feedback includes all of the errors found.

19.5.2.3 Scenario 2

The second scenario shows how the tutors interact with the learner when a next-step hint is requested. We show how the learner's solving path affects the chosen hints by illustrating how the tutors react to two hint requests. These requests lead to the same step but at two different points in the solving path of the 2005 – 1017 problem. The first hint request is executed when no step have been yet performed by the learner (Figure 19.6) while the second one is made when the only remaining step is to decrement the minuend (2) from the leftmost column (Figure 19.7).

The method used by CTAT in providing hints is to 1) find the rule chain that produces the next step; 2) generate messages using the templates associated with

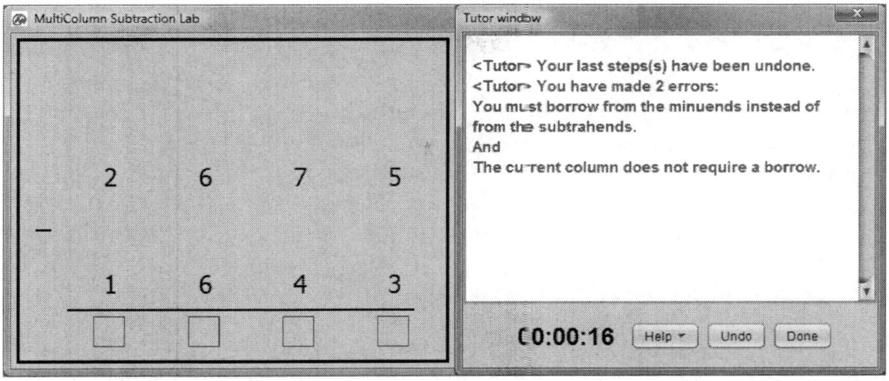

Fig. 19.5 An example of ASTUS's reaction to a composite error

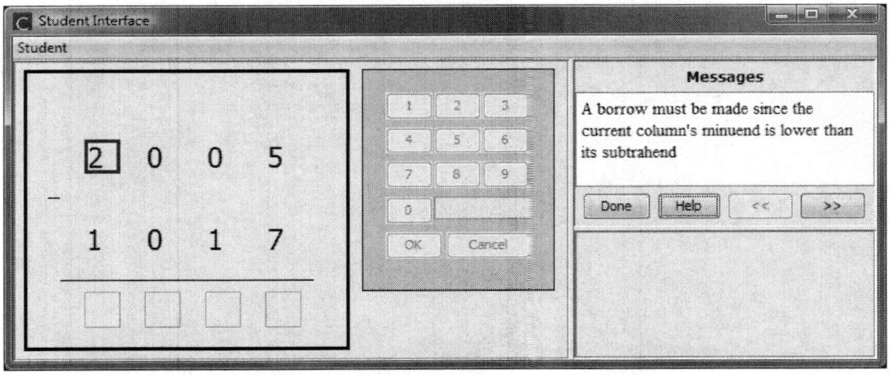

Fig. 19.6 CTAT's feedback when a hint is requested at the beginning of the problem

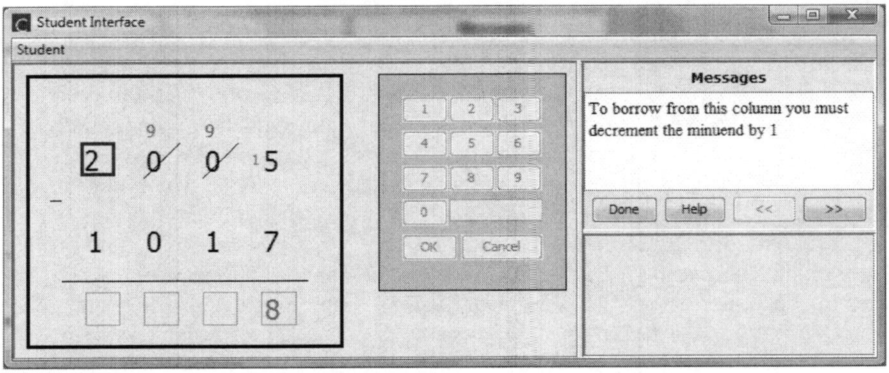

Fig. 19.7 CTAT's feedback when a hint is requested and only the decrementing remains

each of the rules and display them in the same order the rules were fired in (the arrow buttons can be used to navigate the hint list); and 3) highlight the interface component on which the action must be executed. In this scenario, the two hint requests generate different production rule chains and thus the first hint message displayed differs depending on the solving path. For the first request, the hint given concerns the condition required to borrow (Figure 19.6) and, for the second, the hint concerns decrementation of the minuend (Figure 19.7). In both cases, even if the messages displayed are different and refer to different parts of the problem, the highlight is applied to the minuend of the leftmost column. In the case illustrated by Figure 19.6, the highlight should focus the user's attention on the current (rightmost) column but, since the highlight is determined by the SAI event contained in the production rules, the displayed highlight does not correspond to the hint message. Even though highlighting is only supported for rules containing SAI events, we see nothing in the knowledge representation system that would prevents its implementation in rules that modify the working memory.

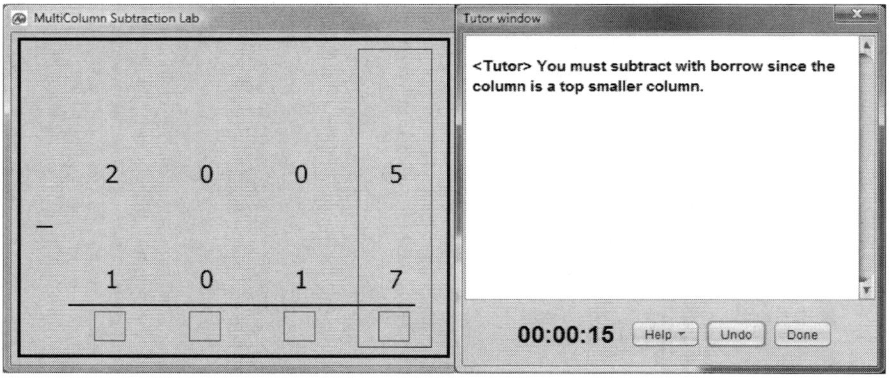

Fig. 19.8 ASTUS's feedback when a hint is requested at the beginning of the problem

In ASTUS, the tutor uses the examinable knowledge components of the model to automatically generate next-step hints.[7] In contrast with CTAT, no message templates are required, although such templates are also supported for situations where customized messages are helpful. Using the hierarchical procedural model, the episodic graph and the learner model, the tutor can evaluate which procedures the learner is most likely to be executing and choose the one on which help should be given. For example, in figure 19.8, the next step is to decrement the minuend of the leftmost column. Since the borrowing process has not been initiated yet, the tutor evaluates that the learner needs help on the conditional procedure which determines whether borrowing is required. A hint is then generated using this procedure's definition. In figure 19.9, the next step is also to decrement the minuend,

[7] Hint generation requires that knowledge components receive meaningful names in all supported languages; internationalization issues may arise.

but the tutor recognizes that the borrowing process has been started and now gives feedback for the sequence procedure used when borrowing across a zero. With ASTUS, hints can be given on partially completed procedures (borrow across zero) while in CTAT, once a rule has been fired, it will never produce a hint again. ASTUS also behaves differently from CTAT in highlighting the interface. The use of views allows the tutor to highlight any object, even if its view is composed of multiple interface components, as is the case for columns. Additionally, ASTUS uses the procedure parameters to determine which components to highlight; highlights are thus supported for every procedure, not only the primitive ones. All objects showing up as arguments that have a view in the current environment may be highlighted, but complex procedures may give special purposes to specific parameters and may use this information to select which ones will be highlighted (e.g., a for-each procedure has a parameter that designates the set to iterate on).

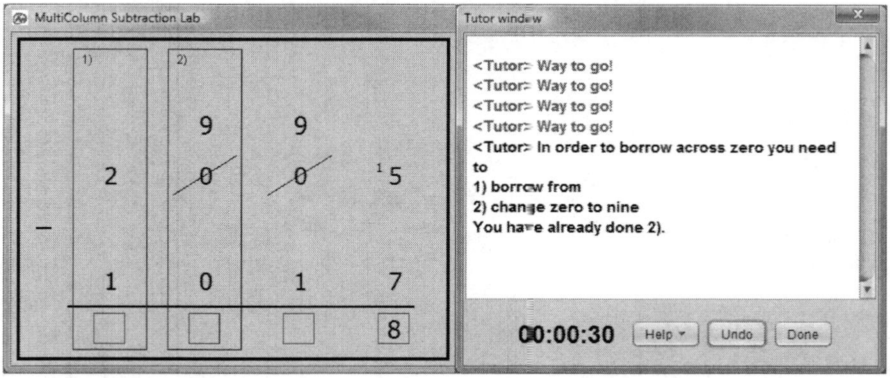

Fig. 19.9 ASTUS's feedback when a hint is requested and only the decrementing remains

19.5.2.4 Scenario 3

As shown in the second scenario, both CTAT and ASTUS produce multiple messages when a hint is requested. These messages are linked to the different procedural components used to produce the next step from the chain of matched rules in CTAT or the episodic graph in ASTUS. The tutor can provide hints at multiple levels, from more abstract (such as subtracting a column) to more specific (such as entering the difference).

Besides being able to vary the hint level between top-down and bottom-up, ASTUS can also offer help on any of the paths leading to a step. When the learner requests help and multiple different steps can be executed, ASTUS can ask for additional information to determine the best hint to provide. For this purpose, ASTUS uses links similar to the "Explain more" feature employed by Andes (VanLehn 2005). Each link represents a goal the learner can choose to request help on. For example, in the "2005 – 1017" problem, if the learner asks for help

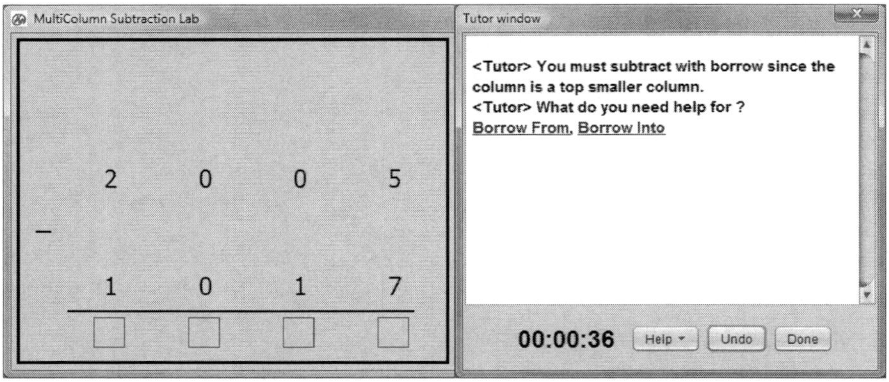

Fig. 19.10 In ASTUS, the tutor asks the learner which goal he/she wishes to be helped with

twice before performing any step, the tutor asks which step he/she intends to do. In this situation, the tutor would propose the "Borrow From" and the "Borrow Into" goals (Fig. 19.10).

We did not find an equivalent mechanism in CTAT's Jess-based cognitive tutors, although example-tracing tutors can give hints for different next steps depending on which interface component currently has the focus. A similar behavior could be implemented, as it would consist of searching for a production rule which results in an action that uses the interface component currently being focused on.

19.5.2.5 Scenario 4

The last scenario we present illustrates feedback automatically generated when off-path steps are recognized by the tutor. In this example, the learner tries to solve the "2675 – 1640" problem by starting with the leftmost column and entering 1 as the difference.

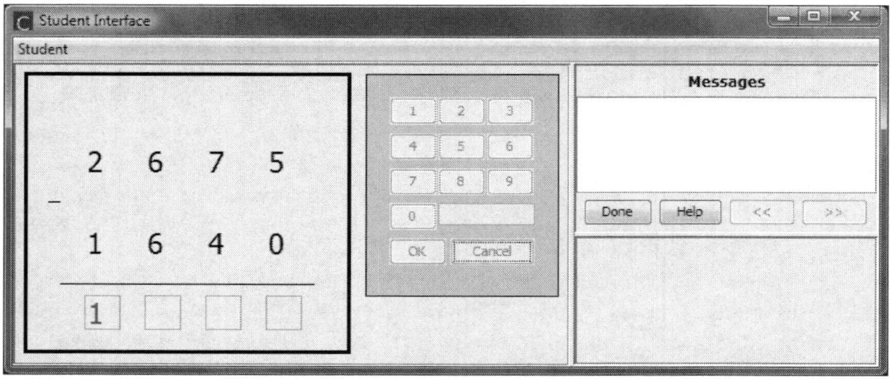

Fig. 19.11 CTAT's feedback for an step executed in a column other than the current one.

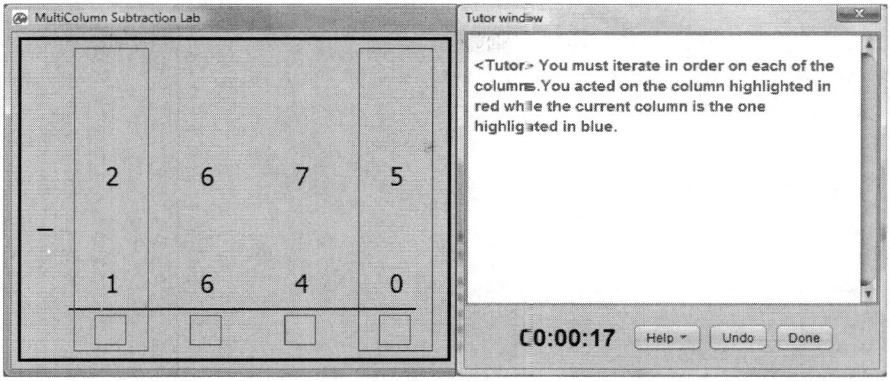

Fig. 19.12 ASTUS's feedback to a step executed in a column other than the current one

Table 19.1 Summary of the supported feedback types

Type of feedback	CTAT	ASTUS
Immediate feedback	Flag feedback Red is used for errors, green for correct steps	Messages in the tutor window ("Way to go!" and "step has been undone")
Interface highlights	Can highlight interface components such as text fields	Can highlight any semantic knowledge viewable on the interface
Next-step hints	Abstract to specific	Abstract to specific Hints on multiple next-steps Next step hint generation
Error-specific feedback	Supported for modeled errors.	Supported for modeled and composite errors.
Off-path step recognition	Not supported	Supported for planning errors such as forgetting an iteration in a "for-each" procedure.
Demonstration	Not supported	Supported

The feedback given by CTAT in this situation is shown in figure 19.11. Since there are no buggy rules that lead to the executed step, CTAT treats it as a non-specific error and only uses flag feedback to indicate that the step is incorrect.

With ASTUS's features, it is possible to implement recognition of errors that go against the scripted behavior of a complex procedure when an off-path step is executed, and to generate error messages accordingly. Figure 19.12 illustrates ASTUS recognizing an error in the iteration on the problem's columns. To make this diagnosis, ASTUS examines the currently available procedures to identify one describing an ordered iteration on a set of columns. It then checks the arguments of the primitive procedure executed by the learner to link one of them to a column on which the iteration is applied. If the identified column is not the one for which steps are currently available, the tutor generates feedback according to a domain-independent predefined template.

19.6 Conclusion

The goal of the study presented in this chapter was to compare the knowledge representation system of the ASTUS framework with a well known production rule-based system such as the CTAT framework. We especially wanted to see whether the two frameworks allow us to model the well-defined task domain of multi-column subtraction in its entirety, what kind of pedagogical interactions they can offer and which situations are more difficult to model in each of them. To achieve our objectives, we compared the modeling process of two subtraction tutors (one in each of the frameworks) into which we incorporated 46 pedagogically relevant errors.

Both CTAT and ASTUS were flexible enough to correctly model the subtraction task domain while still allowing freedom in the order of the learner's steps and being able to model all the errors found in the literature (VanLehn 1990). Even though all errors can be diagnosed by both frameworks, limitations were encountered in incorporating some of them into the models.

Regarding the pedagogical aspect of the tutors, both frameworks can provide immediate feedback, produce hint or error-specific message from templates associated to a procedural knowledge component and highlight elements in the interface. ASTUS has the advantages of being able to: recognize error composition, generate next-step hints or error-specific feedback on off-path steps, offer more sophisticated highlights, demonstrate how a step must be performed on the interface and giving the learners more control over the kind of help they need.

As we incorporated errors into our tutors, we encountered many situations that posed challenges to the modeling process. One challenge that appeared frequently was the management of ambiguous steps. CTAT's solution is simple to implement and ensures that ambiguities are prevented by always selecting the procedures with the highest priority. The author can still implement more complex management by adding production rules to the model. On the other hand, ASTUS handles complex ambiguity cases in its tracing algorithm. Another difficulty that the frameworks must be able to handle is the reuse of knowledge components. In our study, we encountered many situations where previously defined procedural or semantic knowledge components could be reused in order to keep the model as simple as possible. Both CTAT and ASTUS implement mechanisms that allow the author to easily reuse previously defined components. The nature of the problem can also bring difficulties in authoring a tutor In the case of subtraction task domain, the tutor must adapt the interface in function of the number of columns specified for each different problem. With CTAT, it is more difficult because of the lack of a direct link between the model and the interface.

The two frameworks follow different approaches regarding the authoring process of a tutor. CTAT's tracing algorithm and knowledge components are kept simple, as the primary focus has been on decreasing the effort needed to create tutors (Aleven et al. 2006), in order to reduce the time needed by experienced modelers and make the modeling process accessible to non-programmers (Example-Tracing tutors (Koedinger 2003) goes further in this way) . On the other hand, ASTUS is designed to be used by programmers able to manipulate more complex knowledge components. These components allow us to incorporate a number of domain-independent behaviors such as ambiguity management and the generation of next-step hints or error-specific feedback on off-path steps.

With this study, we have shown how ASTUS's knowledge representation system allows us to model complex well-defined task, and that it can be compared to other existing frameworks such as CTAT. There is still much work that can be done to improve our framework. Our efforts have been largely focused on the knowledge representation and our next steps will be to develop other domain-independent modules such as the learner model, the outer loop's task selection and sophisticated domain-independent pedagogical strategies. It would also be interesting for us to do a similar comparative study with a constraint-based framework such as ASPIRE (Mitrovic et al. 2006).This would allow us to evaluate whether our system has advantages over the constraint-based approach and which elements of this approach can be adapted to improve the system we are developing.

Acknowledgments. We would like to thank Vincent Aleven, Ryan S.J.D. Baker and Jean-Pierre Mbungira for their comments and suggestions on our study and its results.

References

Aleven, V., McLaren, B.M., Sewall, J., Koedinger, K.R.: The Cognitive Tutor Authoring Tools (CTAT): Preliminary Evaluation of Efficiency Gains. In: Ikeda, M., Ashley, K.D., Chan, T.-W. (eds.) ITS 2006. LNCS, vol. 4053, pp. 61–70. Springer, Heidelberg (2006a)

Aleven, V., Sewall, J., McLaren, B.M., Koedinger, K.R.: Rapid Authoring of Intelligent Tutors for Real-World and Experimental Use. In: Proceedings of the 6th IEEE International Conference on Advanced Learning Technologies. IEEE Computer Society, Los Alamitos (2006b)

Aleven, V., McLaren, B.M., Sewall, J., Koedinger, K.R.: A New Paradigm for Intelligent Tutoring Systems: Example-Tracing Tutors. International Journal of Artificial Intelligence in Education 19, 105–154 (2009)

Anderson, J.R., Pelletier, R.: A Development System for Model-Tracing Tutors. In: Proceedings of the International Conference of the Learning Sciences, VA, Association for the Advancement of Computing in Education, Charlottesville (1991)

Anderson, J.R.: Rules of the Mind. Lawrence Erlbaum Associates Inc., Hillsdale (1993)

Anderson, J.R., Corbett, A.T., Koedinger, K.R., Pelletier, R.: Cognitive Tutors: Lessons Learned. The Journal of the Learning Sciences 4(2), 167–207 (1995)

Anderson, J.R., Lebiere, C.: The Atomic Components of Thought. Lawrence Erlbaum, Mahwah (1998)

Brown, J.S., VanLehn, K.: Towards a Generative Theory of "Bugs". In: Addition and Subtraction: A Cognitive Perspective. Lawrence Erlbaum, Hillsdale (1982)

Corbett, A.T., Anderson, J.R.: Knowledge Tracing: Modeling the Acquisition of Procedural Knowledge. User Modeling and User-Adapted Interaction 4, 253–278 (1995)

Fortin, M., Lebeau, J.F., Abdessemed, A., Courtemanche, F., Mayers, A.: A Standard Method of Developing User Interfaces for a Generic ITS Framework. In: Woolf, B.P., Aïmeur, E., Nkambou, R., Lajoie, S. (eds.) ITS 2008. LNCS, vol. 5091, pp. 312–322. Springer, Heidelberg (2008)

Fournier-Viger, P., Najjar, M., Mayers, A., Nkambou, R.: A Cognitive and Logic Based Model for Building Glass-Box Learning Objects. Interdisciplinary Journal of Knowledge and Learning Objects 2, 77–94 (2006)

Fournier-Viger, P., Najjar, M., Mayers, A., Nkambou, R.: From Black-box Learning Objects to Glass-Box Learning Objects. In: Ikeda, M., Ashley, K.D., Chan, T.-W. (eds.) ITS 2006. LNCS, vol. 4053, pp. 258–267. Springer, Heidelberg (2006)

Heffernan, N.T., Koedinger, K.R., Razzaq, L.: Expanding the Model-Tracing Architecture: A 3rd Generation Intelligent Tutor for Algebra Symbolization. International Journal of Artificial Intelligence in Education 18(2), 153–178 (2008)

Koedinger, K.R., Aleven, V., Heffernan, N.: Toward a Rapid Development Environment for Cognitive Tutors. In: Proceedings of AI-ED, Artificial Intelligence in Education: Shaping the Future of Learning through Intelligent Technologies, pp. 455–457. IOS Press, Amsterdam (2003)

Koedinger, K.R., Andreson, J.R., Hadley, W.H., Mark, M.A.: Intelligent Tutoring Goes to School in the Big City. International Journal of Artificial Intelligence in Education 8, 30–43 (1997)

Matsuda, N., Cohen, W.W., Koedinger, K.R.: Applying Programming by Demonstration in an Intelligent Authoring Tool for Cognitive Tutors. In: AAAI Workshop on Human Comprehensible Machine Learning, pp. 1–8 (2005)

Mitrovic, A., Koedinger, K.R., Martin, B.: A Comparative Analysis of Cognitive Tutoring and Constraint-Based Modelling. In: Brusilovsky, P., Corbett, A.T., de Rosis, F. (eds.) UM 2003. LNCS, vol. 2702, pp. 313–322. Springer, Heidelberg (2003)

Murray, T.: An Overview of Intelligent Tutoring System Authoring Tools: Updated Analysis of the State of the Art. In: Murray, T., Blessing, S., Ainsworth, S. (eds.) Authoring Tools for Advanced Learning Environments, pp. 491–544. Kluwer Academic Publishers, Dordrecht (2003)

Mayers, A., Lefebvre, B., Frasson, C.: Miace, a Human Cognitive Architecture. SIGCUE Outlook 27(2), 61–77 (2001)

Mitrovic, A., Suraweera, P., Martin, B., Zakharov, K., Milik, N., Holland, J.: Authoring Constraint-based Tutors in ASPIRE. In: Ikeda, M., Ashley, K.D., Chan, T.-W. (eds.) ITS 2006. LNCS, vol. 4053, pp. 41–50. Springer, Heidelberg (2006)

Najjar, M., Mayers, A., Fournier-Viger, P.: A Novel Cognitive Computational Knowledge Model for Virtual Learning Environments. The WSEAS Transactions on Advances in Engineering Education 3(4), 246–255 (2006)

Orey, M.A.: Interface Design for High Bandwidth Diagnosis: The Case of POSIT. Educational Technology 30(9), 43–48 (1990)

Orey, M.A., Burton, J.K.: POSIT: Process Oriented Subtraction-Interface for Tutoring. Journal of Artificial Intelligence in Education 1(2), 77–105 (1990)

Razzaq, L., Patvarczki, J., Almeida, S.F., Vartak, M., Feng, M., Heffernan, N.T., Koedinger, K.R.: The ASSISTment builder: Supporting the life cycle of ITS content creation. IEEE Transactions on Learning Technologies 2(2), 157–166 (2009)

Resnick, L.B.: Syntax and Semantics in Learning to Subtract. In: Moser, J. (ed.) Addition and Subtraction: A Cognitive Perspective. Lawrence Erlbaum, Hillsdale (1982)

Sleeman, D., Kelly, E.A., Martinak, R., Ward, R.D., Moore, J.L.: Studies of Diagnosis and Remediation with High School Algebra Students. Cognitive Science 13, 551–568 (1989)

VanLehn, K.: Mind Bugs: The Origin of Procedural Misconceptions. MIT Press, Cambridge (1990)

VanLehn, K.: The Behavior of Tutoring Systems. International Journal of Artificial Intelligence in Education 16(3), 227–265 (2006)

VanLehn, K., Lynch, C., Schulze, K., Shapiro, J.A. Shelby, R., Taylor, L., Treacy, D., Weinstein, A., Wintersgill, M.: The Andes Physics Tutoring System: Lessons Learned. International Journal of Artificial Intelligence in Education 15(3), 1–47 (2005)

Wenger, E.: Artificial Intelligence and Tutoring Systems: Computational and Cognitive Approaches to the Communication of Knowledge. Morgan Kaufmann Publishers, Los Altos (1987)

Chapter 20
Open Content Authoring Tools

Leena Razzaq[1] and Neil T. Heffernan[2]

[1] Center for Knowledge Communication, 140 Governor's Drive,
University of Massachusetts Amherst, MA 01003-4610
leena@cs.umass.edu
[2] Worcester Polytechnic Institute, Computer Science Department,
Massachusetts, 01609-2280, USA
nth@wpi.edu

Abstract. Education researchers often disagree about the best ways to improve student achievement. The difficulty of designing, conducting, and analyzing experiments means that there is often a dearth of empirical data to support or refute ideas. To design and conduct a simple randomized controlled experiment to compare two different ways of teaching requires a great deal of effort by a teacher or a researcher. The difficulty of conducting such experiments, and then later analyzing the results, may be why so few randomized controlled experiments are conducted in education. One of the goals of the ASSISTment System is to reduce some of those difficulties. We have built web-based tools that allow researchers to easily design, build and then compare different ways to teach children. These tools can administer randomized controlled experiments to large numbers of students. This paper describes these tools and describes a randomized controlled study that was conducted using them.

20.1 Introduction

Similar to many education researchers, mathematics education researchers tend to have heated arguments about what are the best ways to raise student achievement levels. Unfortunately, there is often a dearth of empirical data to support either side. This lack of empirical data is, at least partly, caused by the difficulty of designing, conducting, and analyzing experiments. While not all questions are best settled with randomized controlled experiments, some are. However, to design and conduct a simple randomized controlled experiment to try to compare two different ways of teaching requires a great deal of effort by a teacher or a researcher. Researchers not only must design and author the content for pretests, post-tests and each experimental condition, but they must also randomly assign students to conditions and collect and analyze large amounts of data.

The difficulty of conducting such experiments and then later analyzing the results, are probably some of the reasons that so few randomized experiment are conducted in education. Intelligent tutoring systems can alleviate some of these

R. Nkambou et al. (Eds.): Advances in Intelligent Tutoring Systems, SCI 308, pp. 407–420.
springerlink.com

difficulties such as randomly assigning students, delivering content and collecting data. However, most tutoring systems require computer programming skills to accomplish these tasks and do not make it easy for researchers without programming skills to create new experiments.

One of the goals of the ASSISTment System is to reduce some of those difficulties. The ASSISTment System is a web-based tutoring system that includes web-based authoring tools that allow researchers to easily design, build and then compare different ways to teach mathematics. No computer programming skills are needed. These tools can administer randomized controlled experiments to large numbers of students.

This paper describes the ASSISTment System's tools for education research. We also describe how we designed and conducted a randomized controlled experiment to compare two ways of tutoring eighth grade math.

20.2 The ASSISTment System

The ASSISTment System is joint research conducted at Worcester Polytechnic Institute and Carnegie Mellon University and is funded by grants from the U.S. Department of Education, the National Science Foundation, and the Office of Naval Research. The ASSISTment System's goal is to provide cognitive-based assessment of students while providing tutoring content to students.

The ASSISTment System aims to assist students in learning the different skills needed for the Massachusetts Comprehensive Assessment System (MCAS) test or (other state tests) while at the same time assessing student knowledge to provide teachers with fine-grained assessment of their students; it assists while it assesses. The system assists students in learning different skills through the use of scaffolding questions, hints, and messages for incorrect answers (also known as buggy messages) (Razzaq et al. 2007). Assessment of student performance is provided to teachers through real-time reports based on statistical analysis. Using the web-based ASSISTment System is free and only requires registration on our website; no software need be installed. Our system is primarily used by middle- and high-school teachers throughout Massachusetts who are preparing students for the MCAS tests. Currently, we have over 3000 students and 50 teachers using our system as part of their regular math classes. We have had over 30 teachers use the system to create content.

We are attempting to support the full life cycle of content authoring with the tools available in the ASSISTment System. Teachers can create problems with tutoring, map each question to the skills required to solve them, bundle problems together in sequences that students work on, view reports on students' work and use tools to maintain and refine their content over time.

20.2.1 Structure of an ASSISTment

Koedinger et al. (Koedinger et al. 2004) introduced example-tracing tutors, which mimic cognitive tutors (Anderson et al. 1995) but are limited to the scope of a single problem. The ASSISTment System uses a further simplified example-tracing tutor, called an ASSISTment, where only a linear progression through a problem is supported which makes content creation easier and more accessible to a general audience.

An ASSISTment consists of a single main problem, or what we call the original question. For any given problem, assistance to students is available either in the form of a hint sequence or scaffolding questions. Hints are messages that provide insights and suggestions for solving a specific problem, and each hint sequence ends with a *bottom-out* hint, which gives the student the answer. Scaffolding problems are designed to address specific skills needed to solve the original question. Students must answer each scaffolding question in order to proceed to the next scaffolding question. When students finish all of the scaffolding questions, they may be presented with the original question again to finish the problem. Each scaffolding question also has a hint sequence to help the students answer the question if they need extra help. Additionally, messages called *buggy* messages are provided to students if certain anticipated incorrect answers are selected or entered. For problems without scaffolding, a student will remain in a problem until the problem is answered correctly and can ask for hints, which are presented one at a time. If scaffolding is available, the student will be programmatically advanced to the first scaffolding problems in the event of an incorrect answer on the original question.

Hints, scaffolds, and buggy messages together help create ASSISTments that are structurally simple but can address complex student behavior. The structure and the supporting interface used to build ASSISTments are simple enough so that users with little or no computer science and cognitive psychology background can use it easily. Figure 20.1 shows an ASSISTment being built on the left and what the student sees is shown on the right. Content authors can easily enter question text, hints and buggy messages by clicking on the appropriate field and typing; formatting tools are also provided for easily bolding, italicizing, etc. Images and animations can also be uploaded in any of these fields.

The Builder also enables scaffolding within scaffold questions, although this feature has not often been used in our existing content. In the past, the Builder allowed different lines of scaffolds for different wrong answers but we found that this was seldom used and seemed to complicate the interface causing the tool to be harder to learn. We removed support for different lines of scaffolding for wrong answers but plan to make it available for an expert mode in the future. In creating an environment that is easy for content creators to use, we realize there is a trade-off between ease of use and having a more flexible and complicated ASSISTment structure. However, we think the functionality that we do provide is sufficient for the purposes of most content authors.

20.2.2 Skill Mapping

We assume that students may know certain skills and rather than slowing them down by going through all of the scaffolding first, ASSISTments allow students to try to answer questions without showing every step. This differs from Cognitive

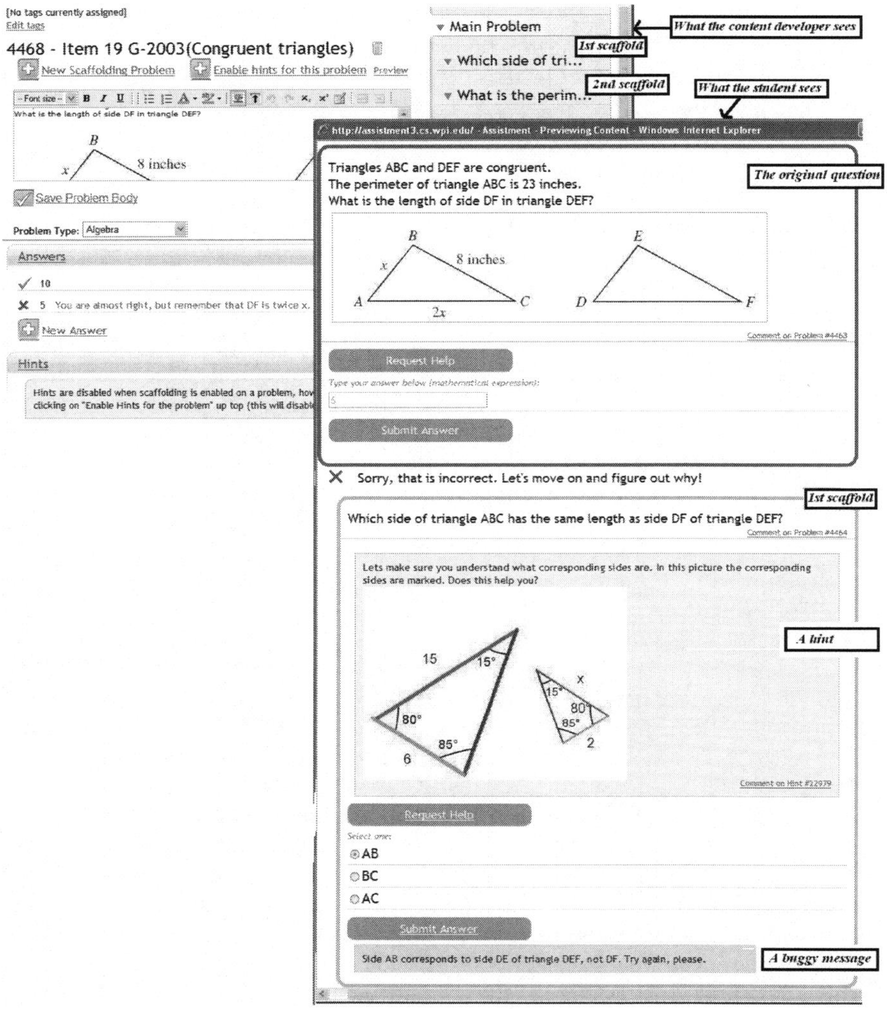

Fig. 20.1 ASSISTment being built on the left and what the student sees is shown on the right

Tutors (Anderson et al. 1995) and Andes (VanLehn et al. 2005) which both ask the students to fill in many different steps in a typical problem. We prefer our scaffolding pattern as it means that students get through items that they know faster and spend more time on items they need help on. It is not unusual for a single Cognitive Tutor Algebra Word problem to take ten minutes to solve, while filling in a table of possibly dozens of sub-steps, including defining a variable, writing an equation, filling in known values, etc. We are sure, in circumstances where the student does not know these skills, that this is very useful. However, if the student already knows most of the steps this may not be pedagogically useful.

The ASSISTment Builder also supports the mapping of knowledge components, which are organized into sets known as transfer models. We use knowledge components to map certain skills to specific problems to indicate that a problem requires knowledge of that skill. Mapping between skills and problems allows our reporting system to track student knowledge over time using longitudinal data analysis techniques (Feng et al. 2006).

In April of 2005, our subject-matter expert helped us to make up knowledge components and tag all of the existing 8th grade MCAS items with these knowledge components in a seven-hour long "coding session". Content authors who are building 8th grade items can then tag their problems in the Builder with one of the knowledge components for 8th grade. Tagging an item with a knowledge component typically takes 2-3 minutes. The cost of building a transfer model can be high initially, but the cost of tagging items is low.

We currently have more than twenty transfer models available in the system with up to 300 knowledge components each. See (Razzaq et al. 2007) for more information about how we constructed our transfer models. Content authors can map skills to problems and scaffolding questions as they are building content. The Builder will automatically map problems to any skills that its scaffolding questions are marked with.

20.2.3 Problem Sequences

Problems can be arranged in problem sequences in the system. The sequence is composed of one or more sections, with each section containing problems or other sections. This recursive structure allows for a rich hierarchy of different types of sections and problems.

The section component, an abstraction for a particular ordering of problems, has been extended to implement our current section types and allows for new types to be added in the future. Currently, our section types include "Linear" (problems or sub-sections are presented in linear order), "Random" (problems or sub-sections are presented in a pseudo-random order), and "Choose Condition" (a single problem or sub-section is selected pseudo-randomly from a list, the others are ignored). The "Choose Condition" section is used to randomly assign students to a condition within an experiment.

The ASSISTment System has a Problem Sequence for every skill needed in sixth, eighth and tenth grade math according to the Massachusetts Frameworks. Teachers can easily choose a Problem Sequence from a list of approved sequences and assign them to their classes. Teachers can also create their own sequences from ASSISTments available in the system or from ASSISTments that they authored.

20.2.4 Running Randomized Controlled Studies

We are interested in using the ASSISTment system to find the best ways to tutor students and being able to easily build problem sequences helps us to run randomized

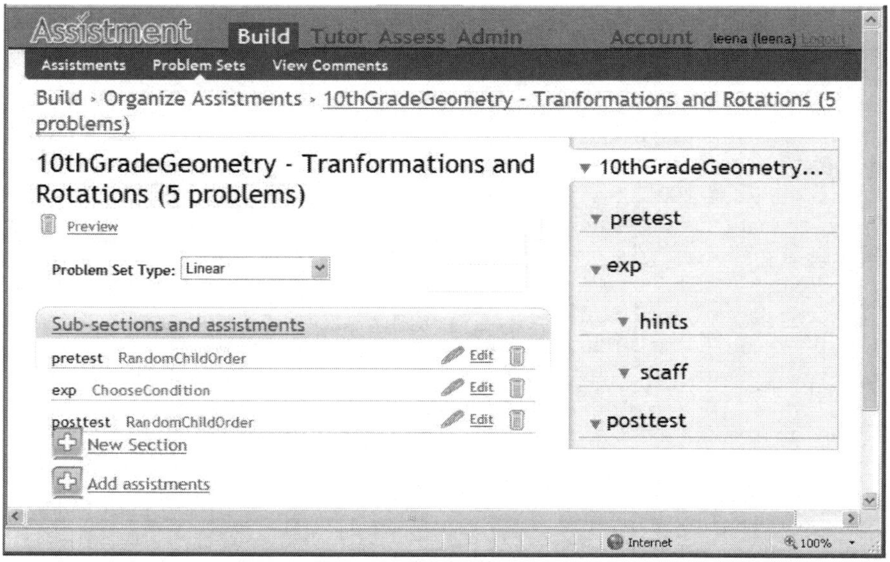

Fig. 20.2 A problem sequence arranged to conduct an experiment

controlled experiments very easily. Figure 20.2 shows a problem sequence that has been arranged to run an experiment that compares giving students scaffolding questions to allowing them to ask for hints. This is similar to an experiment described in (Razzaq and Heffernan 2006).

Three main sections are presented in linear order: a pre-test, experiment and post-test sections. The pre-test and post-test sections are Random sections and will present the questions in a random order. The system allows authors to label an ASSISTment as a “test” question, which means that students will get no feedback on that question. This allows us to compare gains from pre-test to post-test. The experiment section is a Choose Condition section. Within the experiment section there are two conditions and students will randomly be presented with one of them.

We have put together a way for researchers to sign-up for reports on their experiments using a Google Docs Spreadsheet. Researchers can fill out information about their experiment on the spreadsheet, such as the name of the experiment, the purpose, the ID number for the problem sequence and their email addresses. Researchers then will receive an email report every day that students participate in the study by working on the problem set.

The next section describes an experiment conducted with the ASSISTment System.

20.3 Example Randomized Controlled Experiment: Evaluating Educational Content on the Web

Relatively few intelligent tutoring systems (ITS) have become commercially successful even though they have been shown to be capable of producing significant learning gains. This may be due to the high cost of building intelligent tutors and a large part of that cost comes from content creation. After surveying many authoring tools for tutoring content, Murray (1999) roughly estimated that 300 hours of development time was required for one hour of online instruction time. Anderson, Corbett, Koedinger, & Pelletier (1995) suggested that to build the Cognitive Tutor took a ratio of development time to instruction time of at least 200:1 hours. Although the ASSISTment Builder, an authoring tool for example-tracing tutors, produced a smaller ratio of development time to online instruction time of about 40:1 hours (Razzaq et al. 2009), a substantial amount of time is still spent creating content for the ASSISTment System as students (in middle school and high school) often use the system throughout the school year.

The World Wide Web has thousands of pages of educational content, some of them very well written. These web pages often include definitions, worked examples, images and animations. Can we use this wealth of instructional content on the web, saving time and resources spent on content creation? Can students learn from well-designed educational web pages?

The purpose of this study was to determine if students could learn from existing public web pages to help them solve problems in a tutoring system. What percentage of students would show any benefit from having visited a web page? Can we rank the effectiveness of educational web pages? How do these web pages compare to tutored problem solving that is specific to a problem? This section describes two studies that examine the use of educational web pages to help students solve math problems in an interactive learning environment. We found that students could learn from educational web pages and we did not find a significant difference between tutored problem solving and web pages.

20.3.1 Choosing Educational Web Pages

We decided to run this study with eighth graders and chose two topics that are typically covered in middle school to use in our experiment: Pythagorean Theorem and Venn Diagrams. We used Google.com's search engine to find web pages about the two topics on November 24, 2008 and December 15, 2008 and chose two "good" pages for each topic. Figure 20.3 shows the first page of Google's results for "Pythagorean Theorem."

We evaluated 13 pages about Pythagorean Theorem, using our own judgment before finding two that we wanted to use. For instance, we decided not to use the first result, found on Wikipedia.com (http://en.wikipedia.org/wiki/Pythagorean_theorem), because it appeared to be too advanced for eighth graders. We also excluded the state of New York's Regents Exam Prep web page (http://regentsprep.org/regents/Math/fpyth/Pythag.htm) because it had the answer to one of the questions in the problem set used in this study. We looked for colorful engaging web pages that contained background information about the skill as well as examples. We chose PBS's page on the Pythagorean Theorem, (www.pbs.org/wgbh/nova/proof/puzzle/theorem.html), because it was age appropriate and highly ranked by Google. Math Forum's page on the Pythagorean Theorem (mathforum.org/dr.math/faq/faq.pythagorean.html) was excluded because it contained a link to the PBS page.

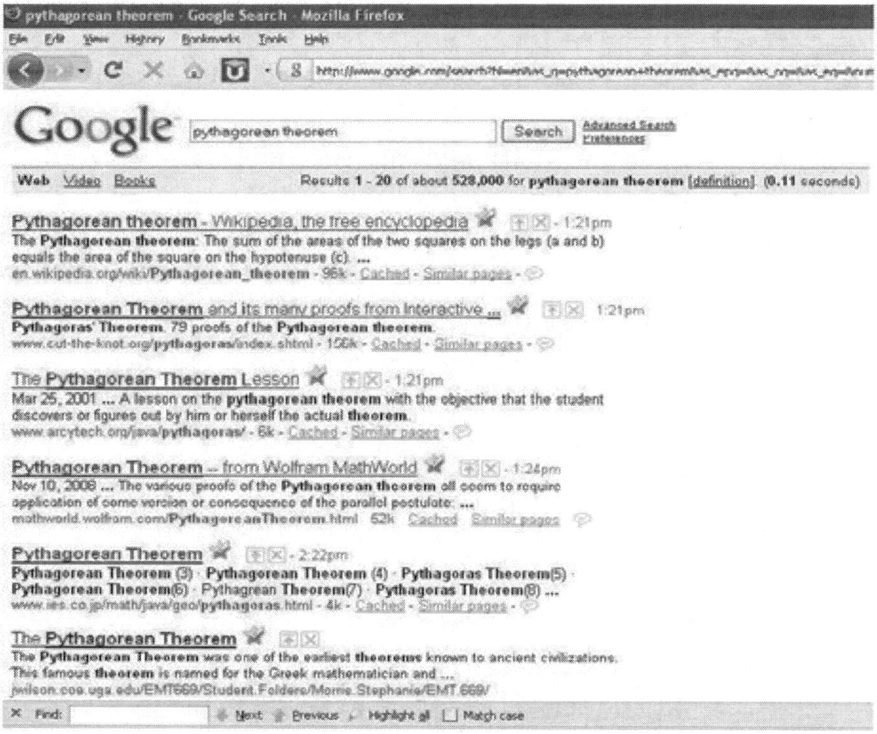

Fig. 20.3 The first page of Google's results for a search on Pythagorean Theorem accessed on November 24, 2008

Originally, we chose a "YouTube" video (http://www.youtube.com/watch?v=0HYHG3fuzvk) because it was the most viewed YouTube video on Pythagorean Theorem and clearly explained the concept. However, the middle school where our study took place blocked YouTube videos on student computers, so we did not use videos in this study. We replaced the video with a web page found at http://www.mathsisfun.com/pythagoras.html, which is shown in figure 20.4.

Similarly, we chose two web pages on the topic of Venn Diagrams. Found at http://regentsprep.org/Regents/math/venn/LVenn.htm and http://www.itl.nist.gov/div897/sqg/dads/HTML/venndiagram.html. In this way, we chose two web pages for each topic in the experiment.

20.3.2 How Does Viewing Educational Web Pages Compare to Tutored Problem Solving?

The purpose of this randomized controlled experiment was to compare tutored problem solving to viewing an educational web page. A previous study showed

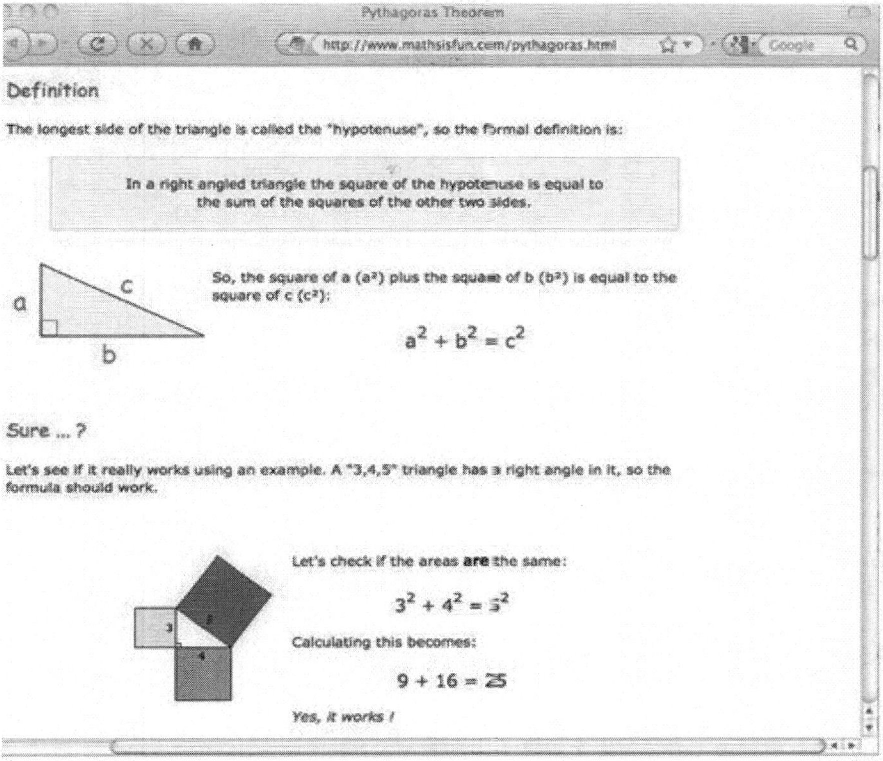

Fig. 20.4 A web page about the Pythagorean Theorem at mathisfun.com

that students could learn from viewing a web page about the skill needed to solve a problem. The ASSISTment System normally uses tutored problem solving to help students solve problems, which requires students to work through problems step-by-step while the system provides hints and feedback on each step. Tutored problem solving has been shown to significantly increase learning especially for students with low math ability [5]. How does this compare to viewing a web page that is more general and does not specifically address how to solve the problem? Will there be a difference between the two conditions based on math ability? If viewing a web page could help students to learn as much or more than tutored problem solving, time and resources spent on content development could be significantly reduced.

20.3.2.1 Experiment Design

This study focused on two topics, Venn Diagrams and Pythagorean Theorem. This experiment used a repeated measures design: students participated in both conditions, web pages and tutored problem solving. The experiment controlled for the order of topics and the order of conditions, which were randomly assigned by

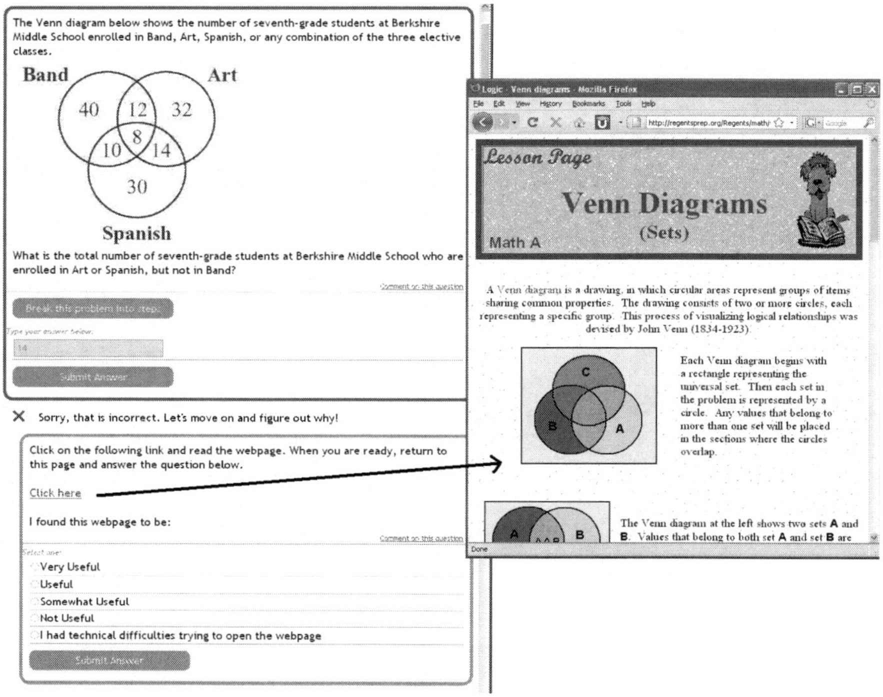

Fig. 20.5 Students were asked to click on a link to display a web page if they needed help

Table 20.1 Students were randomly assigned to one of four groups

	Group 1	Group 2	Group 3	Group 4
First	Venn Diagrams with TPS	Pythagorean Theorem with Web Page	Venn Diagrams with Web Page	Pythagorean Theorem with TPS
Second	Pythagorean Theorem with Web Page	Venn Diagrams with TPS	Pythagorean Theorem with TPS	Venn Diagrams with Web Page

student (see Table 20.1). A pretest and post-test was given before and after each topic, and learning was measured by gain score from pretest to post-test.

Figure 20.6 shows the section that was arranged for Group 3 shown in Table 20.1. This section presents Venn Diagram problems with web pages first and then presents the Pythagorean Theorem problems with tutored problem solving next. The experiment uses four similar sections and randomly assigns students to one of them by using a Choose Condition section.

Fig. 20.6 One of the conditions in the experiment

This study took place during one period of a math enrichment class. Students had already been using the ASSISTment System during this class to help them prepare for the MCAS exam, therefore they were familiar with the system.

20.3.2.2 Results

Seventy-one middle school students (aged 13-14) participated in the study and 66 students completed both problems in both conditions. The Pythagorean Theorem topic was more difficult (mean pretest score = 45%) compared to the Venn Diagram topic (mean pretest score = 58%). The results showed that there was significant learning overall with a mean gain score of 33% [$t(68) = 7.48$, $p < 0.001$] for the Pythagorean Theorem topic and a 19% gain score [$t(67) = 4.18$, $p < 0.001$] for the Venn Diagram topic.

Table 20.2 Pythagorean Theorem, students learned slightly more with tutored problem solving than viewing the web pages

Math Ability	Condition	Mean	Std. Error	95% Confidence Interval	
				Lower Bound	Upper Bound
High	Web	0.348	0.076	0.195	0.500
	TPS	0.444	0.122	0.200	0.688
Low	Web	0.292	0.075	0.142	0.441
	TPS	0.500	0.138	0.223	0.777

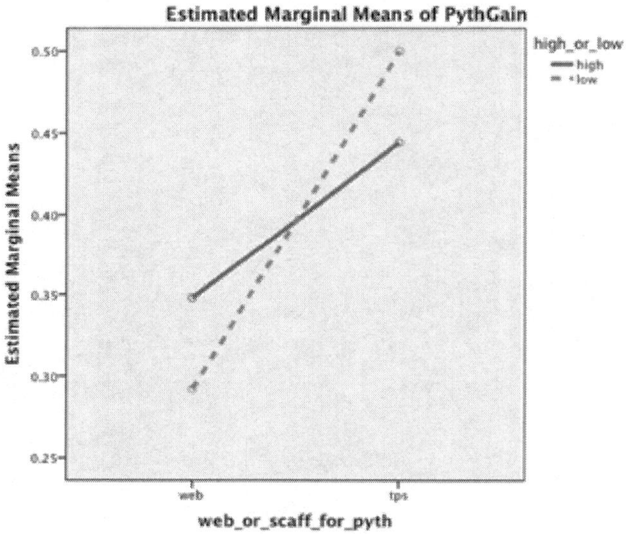

Fig. 20.7 Students learned slightly more from tutored problem solving in the Pythagorean Theorem topic.

The repeated measures analysis showed that the difference between tutored problem solving and web pages was not statistically reliable. Students had taken a practice MCAS test and a median split on the test scores was used to divide students into "low math ability" and "high math ability." No aptitude treatment interaction was found.

For the more difficult topic, Pythagorean Theorem, students learned slightly more with tutored problem solving than viewing the web pages [F(63, 1) = 7.45, p = 0.22]. (See Table 20.2 and Fig. 20.7)

20.3.3 Discussion

This section presented a study to examine the use of educational web pages to help middle school students solve math problems in a tutoring system. If we could show that these web pages were effective, it could encourage tutoring system developers to make more use of the wealth of free educational content available on the web saving time and resources from content development.

The purpose of the study was to determine which method was more effective: tutored problem solving, which was specific to a problem or viewing a web page that was more general. This may be compared to studies such as Ringenberg and VanLehn's (2006) study of worked examples to hints that were specific to a problem. However, the web pages we used tended to have more background information and multiple examples as well as links to other web pages. No significant difference between the two was found in this study. Tutored problem solving appeared to help students slightly more when the topic was more difficult, though not reliably more. Could this result be because the tutored problem solving condition targeted specific problems and how to solve them while the web pages were more focused on the broader principles involved in the topic? We do not know, but believe that there is potential for future work here.

In conclusion, we believe the results of this study show that students can learn from educational web content and developers of ITS should be taking advantage of tutoring content on the web.

20.4 Conclusions

In this paper, we presented the ASSISTment System's tools for designing, conducting and analyzing randomized controlled experiments. No programming skills are needed to use these tools. We also described how we designed and ran an experiment using the ASSISTment System. We were able to easily design a repeated measures experiment that controlled for the order of conditions and the order of topics, which makes a stronger experiment.

We believe these tools will allow researchers to easily design and conduct simple randomized controlled experiments to compare different ways of teaching with interactive learning environments and perhaps settle more questions that educational researchers have.

Acknowledgments. We would like to thank all of the people associated with creating the ASSISTment system listed at www.ASSISTment.org including investigators Kenneth Koedinger and Brian Junker at Carnegie Mellon University. We would also like to acknowledge funding for this project from the U.S. Department of Education, the National Science Foundation, the Office of Naval Research and the Spencer Foundation. This work is also supported by the National Science Foundation under the GK12 program (Grant DGE-0742503) and Grant # 0937060 to the Computing Research Association for the CIFellows Project. Any opinions, findings, and conclusions or recommendations expressed in this material are those of the author(s) and do not necessarily reflect the views of the National Science Foundation or the Computing Research Association.

References

Anderson, J.R., Corbett, A., Koedinger, K., Pelletier, R.: Cognitive Tutors: Lessons Learned. Journal of the Learning Sciences 4(2), 167–207 (1995)

Feng, M., Heffernan, N.T., Koedinger, K.R.: Predicting state test scores better with intelligent tutoring systems: developing metrics to measure assistance required. In: Ikeda, M., Ashley, K., Chan, T. (eds.) ITS 2006. LNCS, vol. 4053, pp. 31–40. Springer, Heidelberg (2006)

Koedinger, K.R., Aleven, V., Heffernan, N.T., McLaren, B., Hockenberry, M.: Opening the door to non-programmers: Authoring intelligent tutor behavior by demonstration. In: Lester, J.C., Vicari, R.M., Paraguaçu, F. (eds.) ITS 2004. LNCS, vol. 3220, pp. 162–174. Springer, Heidelberg (2004)

Murray, T.: Authoring intelligent tutoring systems: An analysis of the state of the art. International Journal of Artificial Intelligence in Education 10, 98–129 (1999)

Razzaq, L., Heffernan, N., Feng, M., Pardos, Z.: Developing Fine-Grained Transfer Models in the ASSISTment System. Journal of Technology, Instruction, Cognition, and Learning 5(3), 289–304 (2007)

Razzaq, L., Heffernan, N.T.: Scaffolding vs. hints in the ASSISTment System. In: Ikeda, M., Ashley, K.D., Chan, T.-W. (eds.) ITS 2006. LNCS, vol. 4053, pp. 635–644. Springer, Heidelberg (2006)

Razzaq, L., Heffernan, N.: To tutor or not to tutor: That is the question. In: Dimitrova, V., Mizoguchi, R., du Boulay, B., Graesser, A. (eds.) Proceedings of the 14th International Conference on Artificial Intelligence in Education, pp. 457–464. Springer, Berlin (2009)

Razzaq, L., Heffernan, N.T., Koedinger, K.R., Feng, M., Nuzzo-Jones, G., Junker, B., et al.: Blending Assessment and Instructional Assistance. In: Nedjah, N., d. Macedo Mourelle, L., Neto Borges, M., Nunesde de Almeida, N. (eds.) Intelligent Educational Machines, pp. 23–49. Springer, Berlin (2007)

Razzaq, L., Parvarczki, J., Almeida, S.F., Vartak, M., Feng, M., Heffernan, N.T., et al.: The ASSISTment builder: Supporting the Life-cycle of ITS Content Creation. IEEE Transactions on Learning Technologies 2(2), 157–166 (2009)

Ringenberg, M.A., VanLehn, K.: Scaffolding Problem Solving with Annotated Worked-Out Examples to Promote Deep Learning. In: Ikeda, M., Ashley, K.D., Chan, T.-W. (eds.) ITS 2006. LNCS, vol. 4053, pp. 625–634. Springer, Heidelberg (2006)

VanLehn, K., Lynch, C., Schulze, K., Shapiro, J.A., Shelby, R., Taylor, L., et al.: The Andes physics tutoring system: Lessons Learned. International Journal of Artificial Intelligence and Education 15(3), 1–47 (2005)

Chapter 21
The Andes Physics Tutoring System: An Experiment in Freedom

Kurt VanLehn, Brett van de Sande, Robert Shelby, and Sophia Gershman

School of Computing, Information, Informatics and Decision Systems Engineering, Arizona State University, 699 S. Mill Avenue, Tempe AZ 85287
{kurt.vanlehn,bvsd}@asu.edu

Abstract. The Andes physics tutoring system is an experiment in student freedom. It allows students to solve a physics problem in virtually any legal way. This means that Andes must recognize an extremely large number of possible steps occurring in an extraordinarily large number of possible orders. Such freedom raises several research questions. (1) How can Andes solve the technical challenge of understanding student's behavior in such a wide-open context? (2) How can Andes give pedagogically useful help and guidance? In particular, how can it guide students who are floundering without curtailing the freedom of students who are not floundering? (3) Will Andes be effective in getting students in real classrooms to learn physics? (4) What does it take to scale up Andes and disseminate it widely? The Andes project, which began in the mid 1990's, has achieved workable solutions to the first three goals: Andes can understand student behavior; It provides pedagogical help similar to that of human experts; Most importantly, Andes causes large, reliable learning gains compared to control classes taught with convention, paper-based instruction. This chapter summarizes the first three results and discusses our progress on the fourth goal, scale-up.

21.1 Introduction

Although problem solving freedom may seem a desirable goal in its own right, it was adopted as the driving goal for the Andes system in the belief that allowing freeform problem solving would facilitate the fourth goal, scale-up. Given a particular physics problem, different instructors will endorse different solutions, and for good reasons. For instance, some instructors prefer mathematically concise solutions in the belief that this makes the physics stand out. Other instructors prefer verbose solutions that make explicit every physics principle's application in the belief that this makes the physics stand out. It may be that both instructors are completely correct, because they are working with different students or they are preparing their students for different subsequent courses. Andes should not take a stand on this issue, but instead should support all instructors' pedagogical

R. Nkambou et al. (Eds.): Advances in Intelligent Tutoring Systems, SCI 308, pp. 421–443.
springerlink.com

practices equally. To put it succinctly, pedagogical equality motivated problem solving freedom.

The plan for this chapter is to first describe how students solve physics problems on Andes, and highlight the freedom they have by comparing Andes' user interface to the user interfaces of other well-known intelligent tutoring systems (ITS). The second section explains how Andes solved the technological problems of scaffolding students despite the freedom they enjoyed. A third section describes evaluations of Andes in the physics classes of the United States Naval Academy. The final section reports progress on scaling up and disseminating Andes.

21.2 The User Interface and Behavior of Andes

Andes' user interface is intended to be as much like pencil and paper as possible. The original version of the Andes user interface is shown in Figure 21.1; a more recently developed one will be discussed later. Students read the problem (top of the upper left window), draw vectors and coordinate axes (bottom of the upper left window), define variables (upper right window) and enter equations (lower right window). These are actions that they do when solving physics problems with pencil and paper.

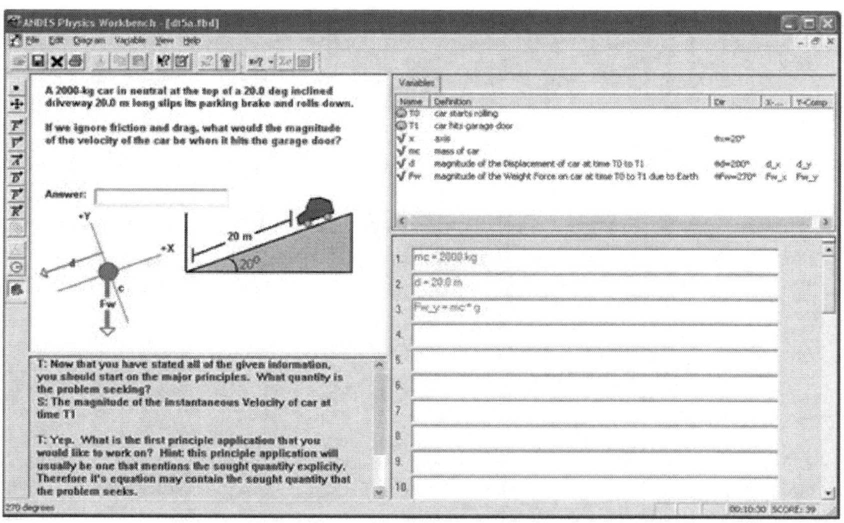

Fig. 21.1 The Andes2 user interface

Although similar to paper and pencil in many ways, Andes' user interface does differ several important ways. First, as soon as an action is done, Andes gives immediate feedback. Entries are colored green if they are correct and red if they are incorrect. This binary indicator is called flag feedback (Anderson et al. 1995). In Figure 21.1, all the entries are green except for equation 3, which is red.

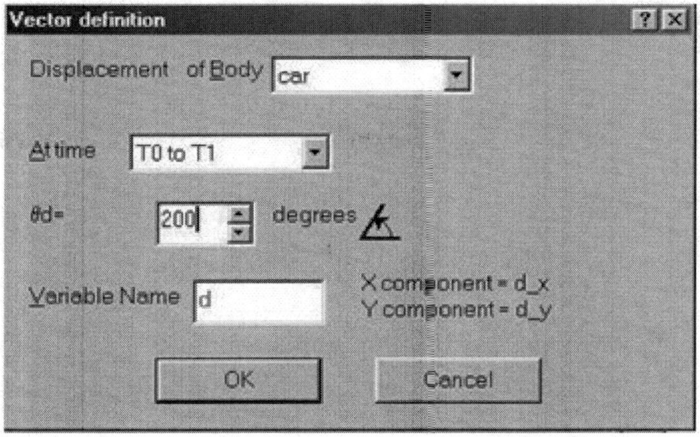

Fig. 21.2 A form for defining a vector

Second, variables must be defined before they can be used in equations. There are two kinds of quantities: vectors and scalars. Vectors are defined by clicking on the symbol for the appropriate vector variable on the tool bar on the left edge of Figure 21.1, then drawing an arrow using the mouse, then filling out a pop-up form like the one in Figure 21.2. Scalar variables are defined by filling out a similar form. Filling out these forms forces students to precisely define the semantics of quantities. Paper does not enforce such precision, so students can easily confuse their variables. For instance, students working on paper often use "v" to stand for all the velocities in a problem. Our collaborating instructors at the United States Naval Academy (USNA) strongly believe that requiring students to define quantities before using them in equations accelerates their learning even though it does restrict their freedom somewhat.

Another violation of freedom is the forms that students fill out when defining quantities. Users must fill in all the blanks before they can close the form. Sophisticated users consider some of the information so obvious that they don't want to fill those blanks in. For instance, when there are only two time points in a problem, there can be only one time interval in a problem, so "duration" must refer to the length of that time interval and "displacement" must refer to the change in location over that time interval. When sophisticated users write such definitions on paper, they often do not use temporal subscripts. They include subscripts only when needed for disambiguation. This was a constant source of irritation to some (but not all) instructors using Andes.

The newest version of the Andes user interface, which is shown in Figure 21.3, does not use forms for defining quantities. Instead, users open a textbox and type their definition into it. Andes matches the user's definition to all definitions possible in this problem, picks the closest match, and confirms it with the user. The text then appears on the interface inside the textbox dragged out by the user. Thus, Andes will accept "Let d be the displacement" and does not insist on the

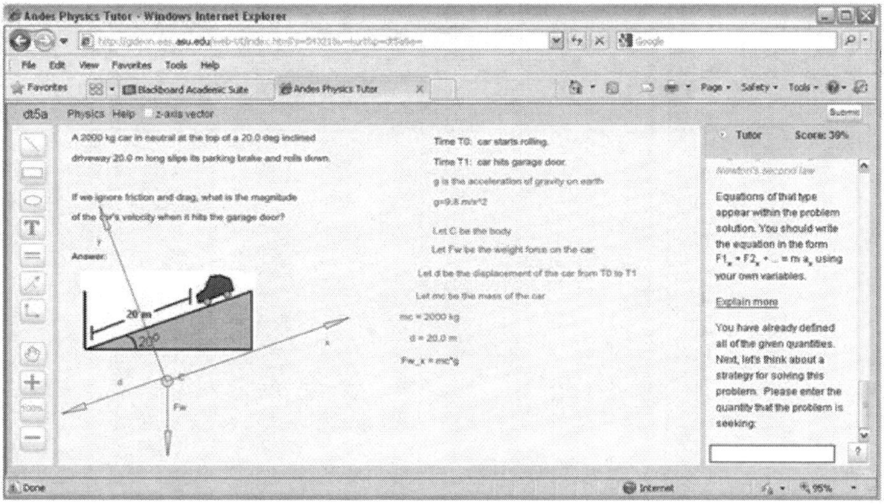

Fig. 21.3 The Andes3 user interface

more complete definition, “Let d be the displacement of the car from T0 to T1.” On the other hand, “Let v be the velocity” is red flagged because it is ambiguous between the instantaneous velocity at T0, the instantaneous velocity at T1 and the average velocity between T0 and T1. This version of Andes (called Andes3) is in beta testing as this chapter is being written. We call it a *freeform* user interface both because it has no forms and because it puts little constraint on the steps and the ordering of steps (except that quantities must be defined before appearing in equations). The new freeform user interface is even closer to paper and pencil.

Andes includes a mathematics package. Students indicate which variable they want to solve for, then Andes tries to solve the system of equations that the student has entered for numerical result for the variable. If Andes succeeds, it enters an equation of the form <variable> = <value>. Although many students routinely use powerful hand calculators and computer-based mathematics packages, such usage requires copying the equations from Andes to their system and back. Andes eliminates this tedious and error-prone copying process. This is one reason that Andes is popular with students. Nonetheless, instructors can turn this feature off.

Andes provides three kinds of help:

Andes pops up an error messages whenever the error is likely to be a slip. That is, the error is probably due to lack of attention rather than lack of knowledge (Norman 1981). Typical slips include using an undefined variable in an equation (which is usually caused by a typographical error) or leaving off the units of a dimensional number. When an error is not recognized as a slip, Andes colors the entry red.

Students can request help on a red entry by selecting it and clicking on a help button. Since the student is essentially asking, “what’s wrong with that?” we call this *What’s Wrong Help*.

If students are not sure what to do next, they can click on a button that will give them a hint. This is called *Next Step Help*.

Notice the students' freedom. Although Andes takes the initiative when it detects a slip, Andes only gives substantive hints only when students ask for them.

What's Wrong Help and Next Step Help generate a hint sequence. The first time the student asks for hints, the first hint in the sequence is printed in the lower left window (Fig. 21.1) or the right window (Fig. 21.3). The student can then ask for another hint, make a new entry or fix an existing red entry. If the student asks for another hint without making any correct entries, then the second hint in the sequence is printed. Most hint sequences have three hints. As an illustration, suppose a student who is solving Figure 21.1 has asked for What's Wrong Help on the incorrect equation *Fw_x = -Fs*cos(20 deg)*. These are the three hints that Andes gives:

Check your trigonometry.

If you are trying to calculate the component of a vector along an axis, here is a general formula that will always work: Let θV be the angle as you move counterclockwise from the horizontal to the vector. Let θx be the rotation of the x-axis from the horizontal. Then: V_x = V*cos(θV-θx) and V_y = V*sin(θV-θx).

Replace cos(20 deg) with sin(20 deg).

This three-hint sequence is typical of many hint sequences. It is composed of a *pointing* hint, a *teaching* hint and a *bottom-out* hint. The pointing hint, "Check your trigonometry," directs the students' attention to the location of the error. If the student knows the appropriate knowledge and the mistake is due to carelessness, then the student should be able to pinpoint and correct the error given such a hint (Hume et al. 1996; Merrill et al. 1992).

The teaching hint, "If you are trying to calculate…," states the relevant piece of knowledge. We try to keep these hints as short as possible, because students tend not to read long hints (Anderson et al. 1995; Nicaud et al. 1999). In other work, we have tried replacing the teaching hints with either multimedia (Albacete and VanLehn,2000a, 2000b) or natural language dialogues (Rose et al. 2002). These more elaborate teaching hints significantly increased learning in laboratory settings, but have not been tried in the field.

The bottom-out hint, "Replace cos(20 deg) with sin(20 deg)," tells the student exactly what to do. Students often abuse these bottom out hints, so Andes uses a scoring function (see below) to discourage such "gaming the system" (Baker et al. 2009).

This completes the description of the Andes user interface. Except for one major restriction (quantities must be defined before using them in equations), students have the freedom to do any step in any order, and to ask for hints at any time. Although Andes does give flag feedback (i.e., red/green) on every step without being asked to do so, students surveys indicate that they like this. If given a choice, they probably would ask for such feedback after every step. In short, with a few exceptions, the Andes user interface gives the user complete freedom to develop solutions.

In contrast to Andes, many ITS constrain student's freedom. For this discussion, let us focus on step-based tutoring systems and ignore answer-based tutoring systems (VanLehn 2006). The difference is that answer-based tutoring systems given feedback and hints on the answer as a whole, whereas step-based tutoring systems decompose the student's entries into steps and give feedback and hints on each step. Most ITS are step-based tutors; Andes is a step-based tutoring system, where a step is a student entry, such as an equation or a quantity definition.

Anyway, many step-based ITS have students enter each step into a specific blank in a form. Only certain steps are allowed in that blank, and the blanks are often labeled. Thus, each blanks represents (reifies) a goal in solving the problem. For instance, the user interface for the SQL Tutor (Mitrovic 2003; Mitrovic and Ohlsson 1999) has blanks for each component of an SQL query. The blanks are labeled (e.g., "Select", "From" or "Where"). Thus, the overall goal of formulating an SQL query is decomposed into several subgoals, and each subgoal is made explicit (reified) as a blank plus its label. This goal reification is part of the scaffolding provided by the SQL tutor and probably contributes to its effectiveness. However, it does reduce the student's freedom. If the SQL tutor's user interface consisted merely of a large text box into which student's entered a whole query, then students would have more freedom.

As a contrasting example, suppose a form-based interface were used for physics problems. Its blanks would have labels such as "Newton's second law applied on the x-axis:" or "Definition of the acceleration of the bowling ball:"

It is an empirical question whether the form-based interface or the freeform interface is more effective, and for which students. Novices may learn faster with a form-based interface whereas intermediates may learn faster with a freeform interface (Singley, 1990). Perhaps transfer from freeform interfaces to paper might be better than transfer from form-based interfaces to paper. The point is merely that the Andes freeform user interface is atypical. Most step-based ITS use form-based interfaces even though they could use a freeform one, and Andes could use a form-based interface even though it uses a freeform one.

21.3 Scaffolding Learning when Students Have Freedom

When a user interface gives students freedom to do any step, students can flounder and they can develop bad problem solving habits. The Andes project developed methods for dealing with both problems and thus scaffolding the students' learning even in the context of a form-free user interface.

When students flounder, most eventually ask for Next Step Help. Two versions of Next Step Help have been developed and evaluated. The first version of Andes (called Andes1) attempted to recognize the student's plan for solving the problem and hint the next step in that plan (Gertner et al. 1998; Gertner and VanLehn 2000). A Bayesian net was used for both plan recognition and next step selection. This approach was used in the 1999 and 2000 experiments. After the 2000 experiment, we evaluated it by randomly selecting episodes where students asked for Next Step Help. For each episode, the student's screen was printed just prior to Andes' first hint, the hard copies were given to 3 physics instructors, and the instructor wrote the sequence of hints they would give at this point. Although the instructors sometimes disagreed on what step to hint next, they did agree in 21 of the 40 episodes. Unfortunately, Andes tended to disagree with the humans. Its hint sequence agreed with the consensual one in only 3 of 21 cases. Thus, Andes1's Next Step Help was not consistent with expert human help.

The problem, in the opinion of instructors, was that students often had no coherent plan for solving the problem. They had generated a few correct entries, often with significant help from Andes, but there appeared to be no pattern in their selection of steps. For these students, the instructors wrote hint sequences that identified the major principle and the first step toward applying that principle. Instead of trying to recognize the student's plan, which is what Andes1 tried to do, the expert instructors hinted the first step in their own plan for solving the problem.

The current version of Next Step Help replicated the instructors' hint sequences. It was based on the assumption that if students are lost and asking for Next Step Help, then they probably have no coherent plan, so they should get a hint on the first relevant step in the instructors' plan for solving the problem. Thus, unless the students have already been introduced to the instructor's plan, Andes walks them through the goal-subgoal-subsubgoal etc chain that leads down to the to-be-hinted step. Figure 21.1 shows a part of such a discussion.

A second problem that can develop with freeform user interfaces is that students can develop bad problem solving habits. Some bad habits are specific to physics, such as plugging in numbers prematurely. (Most instructors prefer that students first generate all the equations needed to solve a problem, and then plug in numbers in order to solve the equations.) Other bad habits involve misuse of the tutoring system itself. For instance, students may "game the system" by clicking rapidly on the hint request button until they get the bottom out hint (e.g. Aleven and Koedinger 2000; Baker et al. 2006).

In order to maintain freedom while discouraging bad problem solving practices, Andes computes and displays a score. Students get points for entering steps correctly without help, and they lose points for gaming and for bad physics practices. The overall score on a problem is continually displayed in the lower right corner. Instructors who use Andes in their classes, and thus can observe students' behavior, report that some students pay a lot of attention to these scores, perhaps even too much.

Note that these scores are not an estimate of the student's mastery of physics. Although Andes1 maintained such a student model, the USNA instructors did not use it for grading. Moreover, they wanted all students to do the same homework problems, so Andes was not allowed to implement mastery learning (Bloom 1984) or macro-adaptation (Shute 1993) by selecting different physics problems for different students. Because there was no use for the model's assessments of student mastery, subsequent versions of Andes did not include such a student model.

21.4 Design and Implementation

This section describes how Andes handles the technical issue of allowing students maximal freedom in entering steps, while still being able to recognize steps and scaffold learning. This section is not an exhaustive description of the design and implementation, as that appears in (VanLehn et al. 2005).

The key to coping with unconstrained physics problem solving is that for each problem, every possible correct step is a mathematical combination of primitive steps. Before the Andes system is distributed, its problem solver solves every problem and stores the set of primitive solution steps in a data structure called a solution graph. There is one solution graph per problem, and it represents all possible solutions allowed by the physics knowledge base in terms of the finest grained steps possible. A typical problem may have hundreds of primitive steps, most of which are equations.

For each problem, Andes also precomputes the problem's *solution point*. For each variable in the solution, the solution point states its value, and there can be hundreds of such variable-value pairs. If the problem has symbolic rather than a numerical givens, then Andes generates "ugly" values for the symbolic givens and computes the solution point in terms of them. For instance, if the problem begins, "A car of mass m slides down a frictionless driveway inclined at θ degrees from the horizontal..." then Andes might substitute 142.56721 kg for m, and 21.8762° for θ. These values are never visible to the student.

Given the solution graph and the solution point, Andes needs to be able to give immediate flag feedback, What's Wrong Help and Next Step Help. The following subsections describe each.

21.4.1 Implementing Immediate Feedback

This section describes how Andes implements immediate feedback. When the student makes an entry, Andes' immediate feedback consists of either coloring it green, coloring it red or popping up an error message such as "Undefined variable."

From an implementation perspective, there are two types of entries that students make when solving a problem: equations and non-equations. Non-equation entries include drawing vectors, drawing coordinate axes and defining scalar variables. A non-equation entry is compared to the non-equation entries in the solution graph. If an exact match is found, the student's entry is colored green; otherwise it is colored red. Thus, a non-equation entry is considered correct if and only if it is a primitive step in at least one solution to the problem.

This simple matching technique won't work for equation entries, because students can enter an equation that is an algebraic combination of primitive equations. Such a composite equation will not be in the solution graph, but it is nonetheless correct. Moreover, an infinite number of correct equations can be generated algebraically from even a single equation (e.g., add 3.1 to both sides).

Thus, there are too many correct equations to precompute and store. This is the price of allowing students' freedom to enter composite equations.

Andes checks equations for correctness using a technique called *color by numbers* (Shapiro 2005). The solution point is substituted into the student's equation. This substitution produces an arithmetic equation. That is, the equation has all numbers and no variables. If the equation balances, with some allowance for round-off errors, then the student's entry is colored green; and red otherwise.

Color-by-numbers is similar to a technique used by tutoring systems ever since Plato (Kane and Sherwood 1980). However, the older technique applies to algebraic *formulas* whereas Andes' technique applies to algebraic *equations*. For instance, suppose a tutoring system asks a fill-in-the-blank question such as "$v1\_x =$ __________" or "The x-component of the initial velocity is __________" and expects an algebraic expression as the answer. It can check the correctness of the student's entry by substituting numbers for variables and simplifying the resulting arithmetic formula. If simplification yields the expected number, then the student's answer is correct. Color-by-numbers essentially does this to both sides of the equation.

Although it was always clear that the solution point could be substituted into an equation and the resulting arithmetic equation could checked for balance, it was not clear what such a check really meant. Joel Shapiro proved that this check was equivalent to finding a algebraic derivation from the primitive equations. Thus, color-by-numbers can be used to replace search-based feedback used by earlier systems (Brna and Caiger 1992; Yibin and Jinxiang 1992) and by Andes1. Andes2 may be the first system to apply the color-by-numbers technique to equations instead of expressions.

21.4.2 Implementing What's Wrong Help

When a red entry is selected and the student clicks on the What's Wrong Help button, Andes applies a large set of error handlers to the entry. Each error handler takes the student's entry as input. If it recognizes the error, it returns a hint sequence and a priority. If several error handlers recognize errors in an incorrect entry, Andes chooses the hint sequence with the highest priority.

Many ITS recognize errors by matching the student's incorrect step to each of the possible correct steps that the student should have made. The match will be partial, because the student's step is incorrect. When the ITS has finished partially matching the student's step to all the correct steps, it selects the best partial match and use it to present a hint sequence. For form-based user interfaces, there are usually only a small number of correct steps that can be entered into a blank in the form, so matching the student's step against each of them is feasible. However, with Andes' freeform approach, there are far too many possible correct steps to match against. Moreover, for equations, the student could be trying to enter a composite equation instead of a primitive one. If Andes were to use the conventional matching approach, it would have to generate all correct composite equations and match them to the student's step. This is clearly infeasible. Thus, Andes had to use a different approach.

Andes recognizes errors by making edits to the student's entry. Each error handler makes different kinds of edits. If the edit generates a correct entry, then the error handler constructs an appropriate hint sequence that refers to the student's entry. For instance, suppose a student who is solving the problem in Figure 21.1 defines a zero-length velocity vector and asserts that it is instantaneous velocity of the car at T1. The velocity is actually non-zero, so this is an error and it turns red. One edit is to change the time specification, so that the vector stands for the instantaneous velocity of the car at T0. Another edit is to change the qualitative length, so that the vector is non-zero. The second edit gets a higher priority, so Andes selects this hint sequence:

Is the car at rest at T1?

Since the car is not at rest at T1, the velocity vector needs to be non-zero.

For equations, error handlers edit the equation then call color-by-numbers on the edited equation. For instance, suppose a student solving the problem of Figure 21.1 enters the incorrect equation *Fw_x =-Fw*cos(20 deg)*. An error handler notes that there is a cosine in the equation, edits the equation to be *Fw_x = -Fw*sin(20 deg)* and submits it to color-by-numbers, which indicates that the edited equation is correct. The error handler returns a high priority hint sequence that starts with "Check your trigonometry."

Often, different error handlers make the same edit, but they have different applicability conditions. For instance, sign errors are common in equations, so there is a general sign-error handler whose hint sequence starts out with "Check your signs." This error handler has a low priority. There is a higher priority error handler that checks for a special kind of sign error. If the term with the sign error is the magnitude of a vector pointing opposite one of the coordinate axes, the hint sequence is (problem-specific text is substituted for the bracketed text):

Think about the direction of the <vector>.

Perhaps you are confusing the MAGNITUDE of the <vector> with its COMPONENT along the <x or y> axis. Because the vector is parallel to the <x or y> axis but in the negative direction, the projection equation is <equation>.

Because the vector is parallel to the <x or y> axis and in the negative direction, replace <magnitude variable> with either -<magnitude variable> or <component variable>.

21.4.3 Next Step Help

Some problems have multiple solutions. For instance, the problem in Figure 21.1 can be solved either with conservation of energy or with Newton's second law. These solutions can share parts, such as defining the given quantities. Thus, a solution graph represents plans for every solution and how they share their parts. For example, Figure 21.4 is part of the information in the solution graph for the problem of Figure 21.1. Although not shown in the figure, there is a branch from step 3 to both step 6 and step 4. This represents that there are alternative solutions. These branches converge later; steps 5 and 6 both point to step 7.

1. Draw the body, which defines the mass variable, mc.
2. Define coordinate axes, rotated 20 degrees
3. Define given quantities
    - 3.1. Draw the displacement, d
    - 3.2. Enter the given value of displacement, d = 20 m
    - 3.3. Enter the given value of mass, mc = 2000 kg
4. Apply translational kinematics and get vf^2 = 2*a*d
    - 4.1. Draw the vector diagram
        - 4.1.1. Draw the initial velocity, vi
        - 4.1.2. Draw the final velocity, vf
        - 4.1.3. Draw the acceleration, a
        - 4.1.4. Draw the displacement, d (shared with step 3.1)
    - 4.2. Apply the fundamental principle: vf_x^2=vi_x^2+2*a_x*d_x
    - 4.3. Project the vectors onto the x-axis
        - 4.3.1. Apply projection: vi_x =0
        - 4.3.2. Apply projection: vf_x=-v2
        - 4.3.3. Apply projection: a_x=-a
        - 4.3.4. Apply projection: d_x= -d
5. Apply Newton's second law and get a= -g*cos(200 deg)
    - 5.1. Draw a free-body diagram
        - 5.1.1. Draw the force of gravity, Fw
        - 5.1.2. Draw the normal force, Fn
        - 5.1.3. Draw the acceleration, a (shared with step 4.1.3)
    - 5.2. Apply the fundamental principle: mc*a_x=Fw_x + Fn_x
    - 5.3. Project the vectors onto the x-axis
        - 5.3.1. Apply projection: Fn_x = 0
        - 5.3.2. Apply projection: Fw_x = Fw*cos(250 deg)
        - 5.3.3. Apply projection: a_x= -a
    - 5.4. Apply the weight law, Fw = mc*g
6. Apply conservation of energy and get vf^2 = 2*g*d*sin(20 deg)
    - 6.1. Define the kinetic energy at time T1: K1
    - 6.2. Define the potential energy due to gravity at T0: P0
    - 6.3. Apply the fundamental principle, P0 = K1
    - 6.4. Define height, h
    - 6.5. Apply the definition of gravitational potential energy: P0 = mc*g*
    - 6.6. Apply trigonometry: h = d*sin(20 deg)
    - 6.7. Draw the final velocity, vf (shared with 4.1.2)
    - 6.8. Apply the definition of kinetic energy, K1 = ½ * mc*vf^2
7. Solve the system of equations for vf and get vf = 11.59 m/s
8. Enter the value of vf into the answer box

Fig. 21.4 The solution graph for the problem of Figure 21.1

The basic job of Next Step Help is to direct the student to some overall problem solving strategy, then guide them through the steps associated with that strategy, taking into account the work the student has already finished. At any time during this process, the student may abort the help and continue solving the problem. The student is free to ignore the advice given and pursue their own problem solving strategy.

Andes first makes sure that the student has "set up" the problem by defining given quantities, bodies, coordinate axes, etc. For instance, if the student has done steps 1 and 3 of Figure 21.4, then Andes will hint step 2.

If the student has already set up the problem, then Andes will help the student select a major principle to apply. For instance, suppose the student has done steps 1, 2, 3 and 6.4, then asks for help. Andes asks, "What quantity is the problem seeking" then "What major principle should you apply to find it?" If the student selects Conservation of Energy, then Andes hints step 6.1, which is the first step of that solution. On the other hand, if the student and Andes have already agreed on a major principle, then Andes will just remind the student of it ("Why don't you continue applying Conservation of Energy to the car.") then hint the first step that has not been done.

Once Andes has selected a step to hint, it generates a hint sequence. It typically consisted of a pointing hint, a teaching hint and a bottom out hint, as illustrated earlier.

In order to make this policy work, Andes must map the student's entries onto steps in the solution graph so that it can mark those steps "done." This is easy for non-equation entries, but it is extremely difficult for equations. For instance, if the student enters *Fw_x = -m*g*sin(20 deg)*, then Andes must somehow figure out that this maps to step 5.3.2 (writing *Fw_x = Fw*cos(250 deg)*) and step 5.4 (writing *Fw = m*g*). Andes1 solved this problem by precomputing all possible algebraic combinations of the steps, but this became infeasible as problems became more complex and numerous. For a long time, this appeared an insurmountable problem.

Fortunately, a solution was found (Shapiro 2005). The algorithm is called *indy check* because its main step is to check the independence of a set of multidimensional vectors. The vectors are the gradients of the solution graph equations and the student's equation. The algorithm computes the gradient of an equation by taking the partial derivative of each variable in the equation, evaluating the resulting expressions at the solution point, and using the resulting numbers as components of the gradient. In order to find a set of solution graph equations that can be combined to form the student's equation, the algorithm finds a set of gradients that can be combined via a weighted sum to form the gradient of the student's equation. The algorithm outputs all such sets.

21.5 Evaluations of Andes at the United States Naval Academy

Andes was evaluated in the USNA's introductory physics class every fall semester from 1999 to 2003. This section describes how the 5 evaluations were conducted and their results.

21.5.1 The Evaluation Method

Andes was used as part of the normal USNA physics course. The course had multiple sections, each taught by a different instructor. Students in all sections took the same final exam and used the same textbook but different instructors assigned different homework problems and gave different hour exams, where hour exams were in-class, hour-long exams given approximately monthly. (Hour exams are often called "quizzes" or "midterms" at other schools.) In sections taught by Andes developers (Profs. Robert Shelby, Donald Treacy and Mary Wintersgill), students did their homework on Andes.

Andes was used as part of the normal USNA physics course. The course had multiple sections, each taught by a different instructor. Students in all sections took the same final exam and used the same textbook but different instructors assigned different homework problems and gave different hour exams, where hour exams were in-class, hour-long exams given approximately monthly. (Hour exams are often called "quizzes" or "midterms" at other schools.) In sections taught by Andes developers (Profs. Robert Shelby, Donald Treacy and Mary Wintersgill), students did their homework on Andes.

Each year, the Andes instructors recruited some of their colleagues' sections as Controls. Students in the Control sections did the same hour exams as students in the Andes section. Control sections did homework problems that were similar but not identical to the ones solved by Andes students.

The USNA requires all its 2nd-year students to take physics, so the students in the experiment could be considered a random sample of USNA sophomores. USNA students are academically well prepared (84% scored above 600 on the MSAT in 2010). They are highly motivated because all students receive 100% scholarships, and if they graduate, then they receive a commission in the US Navy or Marines. On the other hand, they often do have enough time to do all their coursework because they have many required non-academic activities as well.

Both Andes and Control instructors motivated their students to do their homework by scoring it on "effort displayed" rather than competence or correctness. Andes instructors had student submit their Andes solutions and used the Andes score as an "effort" score . On the other hand, the Control instructors initially marked the homework carefully in order to stress that the students should write proper derivations, including drawing coordinate systems, vectors, etc. Later in the semester, the Control instructors grade homework more lightly with short comments like "Draw FBD", "Axes" or "Fill out F=MA by components" in order to continue the emphasis on proper derivations. In some classes, instructors gave a weekly quiz (not an hour exam) consisting of one of the problems from the preceding homework assignment. This encouraged students to both do the assignments carefully and to study the solutions that the instructor handed out.

The same final exams were given to all students in all sections. The final exams comprised approximately 50 multiple choice problems to be solved in 3 hours. The hour exams had approximately 4 problems to be solved in 1 hour. Thus, the final exam questions tended to be less complex (3 or 4 minutes each) than the hour exam questions (15 minutes each). On the final exam, students just entered the answer, while on the hour exams, students showed all their work to derive an answer. On the hour exams, students were scored on (a) their vector drawing, (b) whether they defined variables and/or used standard, legible variable names and used them in preference to numbers in equations, (c) correctness of the equations written, (d) the correctness of the answer. Students were informed of the grading rubric well in advance.

The instructors believe that the open-response, complex problems of the hour exams were a more valid assessment of students' competence, but the multiple-choice final exam were necessary for practical reasons. The hour exam results will be reported first.

21.5.2 Hour Exams Results

It is important to check that the prior knowledge and skill of the Andes students was approximately the same as the prior knowledge and skill of the Control subjects. Students were not assigned randomly to condition, although it could be argued that they were assigned randomly to sections. Students were not given an appropriate pre-test. Thus, the standard methods for insuring equal prior competence were not used. However, over many years of experience, instructors have found that the students' physics grades are strongly correlated with their overall college grade point average (GPA) and their college major. Thus, one method of checking the equality of prior competence is to check that the two conditions have equal distributions GPAs and majors. A t-test showed no reliable differences in GPA in any of the years for which GPA data were available. In order to check for equivalent distribution of majors, the majors were classified as either Engineering, Science or Other. A 3x2 Chi-squared test confirmed that the two conditions had the equivalent distribution of majors in every year for which data were available. Thus, it seems likely that the Andes and Control group students had similar initial competence in physics.

Table 21.1 shows the hour exam results for all 5 years. It presents the mean score (out of 100) over all problems on one or more exams per year. In all years, the Andes students scored reliably higher than the Control students with moderately high effect sizes.[1] The 1999 evaluation had an effect size that was somewhat lower, probably because Andes had few physics problems and some bugs, thus discouraging students from using it. It should probably not be considered representative of Andes' effects, and will be excluded from other analyses in this section.

In order to calculate an overall effect size, it was necessary to normalize the scores across years so that they could be combined. The raw exam scores for each exam were converted to z-scores, and then the data points from years 2000 through 2003 were aggregated. The mean Andes scores was 0.22 (standard deviation: 0.95) and the mean score of Control students was –0.37 (0.96). The difference was reliable ($p < .0001$) and the effect size was 0.61.

[1] Effect size was defined as (AndesMean – ControlMean) / ControlStandardDeviation.

Table 21.1 Results from hour exams evaluations

Year	**1999**	**2000**	**2001**	**2002**	**2003**
Number of Andes students	173	140	129	93	93
Number of control students	162	135	44	53	44
Andes mean (stan. dev)	73.7 (13.0)	70.0 (13.6)	71.8 (14.3)	68.2 (13.4)	71.5 (14.2)
Control mean (stan. dev)	70.4 (15.6)	57.1 (19.0)	64.4 (13.1)	62.1 (13.7)	61.7 (16.3)
P(Andes = Control)	0.036	< .0001	.003	0.005	0.0005
Effect size	0.21	0.92	0.52	0.44	0.60

It is often the case that educational software helps some types of students more than other types. For instance, it is often found that students with high prior knowledge learn equally well from the experimental and control instruction, but students with low prior knowledge learn more from the experimental instruction than from the control instruction. In order to check for such aptitude-treatment interaction, we compared linear regression models of GPA vs. hour exam scores. We used the aggregated z-scores for the exams over years 2000 through 2002, because we lacked GPAs of student in the 2003 Control condition. Figure 21.5 shows the results. The regression lines of the two conditions are nearly parallel, indicating that there was little aptitude-treatment interaction. In a separate analysis, we found that the benefits due to Andes vs. Control did not depend on the student's major (VanLehn et al. 2005): Science, Engineering and Other majors all benefited equally.

The USNA physics instructors believed that the point of solving physics problems is not to get the right answers but to understand the reasoning involved, so they used a grading rubric for the hour exams that scored the students' work in addition to their answers. In particular, 4 subscores were defined (weights in the total score are shown in parentheses):

- *Drawings:* Did the student draw the appropriate vectors, coordinate systems and bodies? (30%)
- *Variable definitions:* Did the student use provide definitions for variables or use standard variable names? (20%)
- *Equations:* Did the student display major principle applications by writing their equations without algebraic substitutions and otherwise using symbolic equations correctly? (40%)
- *Answers:* Did the student calculate the correct numerical answer with proper units? (10%)

Andes was designed to increase student conceptual understanding, so we would expect it to have more impact on the more conceptual subscores, namely

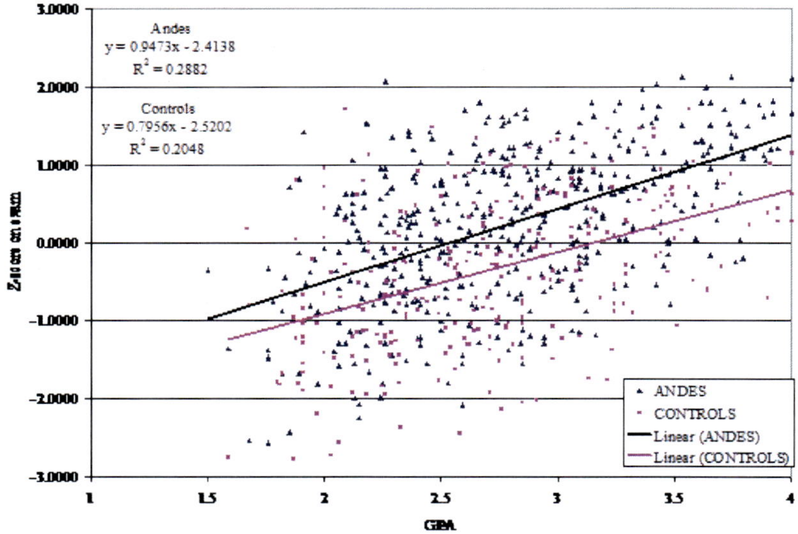

Fig. 21.5 Aptitude-treatment interaction: GPA (x-axis) vs. hour-exam score (y-axis)

Table 21.2 Hour exam subscore effect sizes (and p-values for t-tests)

	2000	**2002a**	**2002b**	**2003**	**Average**
Drawings	1.82 (<.001)	0.49 (.003)	0.83 (<.001)	1.72 (<.001)	1.21
Variable definitions	0.88 (<.001)	0.42 (.009)	0.36 (.026)	1.11 (<.001)	0.69
Equations	0.20 (.136)	0.12 (.475)	0.30 (.073)	-0.17 (.350)	0.11
Answers	-0.10 (.461)	-0.09 (.585)	0.06 (.727)	-0.20 (.154)	-0.08

the first 3. Table 21.2 shows the effect sizes, with p-values from two-tailed t-tests shown in parentheses. Results are not available for 2001. Two hour exams are available for 2002, so their results are shown separately.

There is a clear pattern: The skills that Andes addressed most directly were the ones on which the Andes students scored higher than the Control students. For two subscores, Drawing and Variable definitions, the Andes students scored significantly higher than the Control students in every year. These are the problem solving practices that Andes requires students to follow, because it requires quantities to be defined before they are used in equations. Although these requirements restrict the students' freedom, the instructors' insistence that Andes impose this restriction paid off in increased scores.

The third subscore, Equations, assesses a practice that is not required but it is encouraged by Andes' scoring policy. The effect sizes here were moderate and not statistically reliable. Perhaps Andes should have curtailed freedom even more and made proper usage of symbolic equations a requirement.

The Answers subscore was the same for both groups of students for all years even though the Andes students produced better drawings and variable definitions on those tests. This suggests that the probability of getting a correct answer depends strongly on other skills, such as algebraic manipulation, that are not measured by the more conceptual subscores and not emphasized by Andes. On the other hand, the tied Answer subscores suggest that the Andes students' use of the equation solving tool did not seem to hurt their algebraic manipulation on the hour exams.

In summary of the hour exam results, Andes students tended to learn more than Control students, with an overall effect size of 0.61. Andes was equally beneficial for high-GPA and low-GPA students and for different majors. Breaking down the hour exam scores into subscores showed that Andes students scored higher than Control students on drawing and variable usage, which are two fundamental conceptual skills emphasized by Andes. Andes does not teach algebraic manipulation skills and only encourages proper use of symbolic equations, so it had little apparent advantage over paper for teaching these skills.

21.5.3 Final Exam Scores

Although the hour exams used complex, open-response problems instead of multiple choice problems, and this allowed instructors to use a grading rubric that valued conceptually clear derivations, the final exams were scored on the basis of the student's final answers, and most of the final answers were multiple choices, as are many high stakes assessments. Although the answer-only format probably reduces the validity of the final exams, students studied very hard for them, and we were interested in whether the benefits of Andes would still be visible after such intensive studying.

The final exams covered the whole course, but Andes did not. However, Andes' coverage steadily increased over the years. In 2003, Andes covered 70% of the homework problems in the course. This section reports an analysis of the 2003 final exam data.

First, we need to check that the Andes sections have a representative sample of the whole course population. As mentioned earlier, GPA and major seem to be the best measures of prior knowledge and skill. The Andes students' mean GPA was 2.92 (SD = 0.58), which was marginally significantly higher ($p = 0.0662$) than the non-Andes students' mean GPA of 2.80 (SD = 0.55). Moreover, the distribution of majors among the Andes students was statistically different from the distribution of majors among the non-Andes students ($p < .0001$, 3x2 Chi-squared test, where majors were aggregated into 3 types: engineering, science and other). In particular, there were relatively more engineers than in the Andes sections than in the non-Andes sections. Thus, it appears that the Andes students were *not* representative of the whole population. Thus, we had to use statistical techniques to factor out effects of the higher prior knowledge of the Andes students.

For each group of majors, we regressed the final exam scores of all students in the course against the students' GPAs. Of the 931 students, we discarded scores from 19 students with unclassifiable majors or extremely low scores. This yielded three statistically reliable 2-parameter linear models, one for each type of major. Each model expresses the relationship between general competence and the final exam score. For each student, we subtracted the exam score predicted by the linear model from the student's actual score. This residual score represents how much better or worse this student scored compared to the score predicted solely on the basis of their GPA and their major. That is, the residual score factors out the students' general competence. The logic is the same as that used with an ANCOVA, with GPA and major serving as covariates instead of pre-test scores.

Using these residual scores, we evaluated Andes' impact on students in each of the 3 groups of majors. As Table 21.3 indicates, the residual scores of the engineering and science majors were not statistically different with Andes than with paper homework. However, the other majors did learn more with Andes than with paper homework ($p<.016$; effect size = 0.5). Over all students, the Andes students mean residual score was higher than the mean residual score of the non-Andes students (effect size = 0.25; $p=0.028$).

Although the overall effect of Andes was positive and reliable, Andes had smaller benefits on the learning of the engineering and science majors for two plausible reasons. (1) The engineering majors were concurrently taking a course on Statics, which has very similar content to the physics courses. This dilutes the effect of Andes, since it affected only their physics homework and not their Statics homework. (2) Both the science and engineering majors probably took advanced physics courses in high school, and thus may have developed considerable skill at obtaining correct answers without showing conceptually explicit derivations. This would explain why the Andes engineering and science majors did better than the controls on the hour exams, which were scored conceptually, but they did not do better than the non-Andes students on the final exam, which was not scored conceptually.

Thus, Andes students overall learned significantly more than non-Andes students. The overall effect size was somewhat smaller for the final exam (0.25) than the hour exams (0.61). This may be partially due to the fact that roughly 30% of the final exam addressed material not covered by Andes. It may also be partially due to the format of the final exam. The final exam had students enter only their answers, whereas the hour exams had students show their work, which allowed graders to assess their conceptual understanding more directly.

Although the overall effect of Andes was positive and reliable, Andes had smaller benefits on the learning of the engineering and science majors for two plausible reasons. (1) The engineering majors were concurrently taking a course on Statics, which has very similar content to the physics courses. This dilutes the effect of Andes, since Andes affected only their physics homework and not their Statics homework. (2) Both the science and engineering majors probably took advanced physics courses in high school, and thus may have developed considerable

Table 21.3 Residual scores on the 2003 final exam

	Engineers	**Scientists**	**Others**	**All**
Number of Andes students	55	9	25	89
Number of non-Andes students	278	142	403	823
Andes students mean (stand. dev.)	0.74 (5.51)	1.03 (3.12)	2.91 (6.41)	1.38 (5.65)
Non-Andes students mean (s.d.)	0.00 (5.39)	0.00 (5.79)	0.00 (5.64)	0.00 (5.58)
p(Andes=non-Andes)	0.357	0.621	0.013	0.028
Effect size	0.223	0.177	0.520	0.25

skill at obtaining correct answers without showing conceptually explicit derivations. This would explain why the Andes engineering and science majors did better than the controls on the hour exams, which were scored conceptually, but they did not do better than the non-Andes students on the final exam, which was not scored conceptually.

Thus, Andes students overall learned significantly more than non-Andes students. The overall effect size was somewhat smaller for the final exam (0.25) than the hour exams (0.61). This may be partially due to the fact that roughly 30% of the final exam addressed material not covered by Andes. It may also be partially due to the format of the final exam. The final exam had students enter only their answers, whereas the hour exams had students show their work, which allowed graders to assess their conceptual understanding more directly. There could be other explanations as well.

21.5.4 Discussion of the Evaluations

This section summarizes the evaluation results and compares them to those from the Koedinger et al. (1997) study that tested a combination of an intelligent tutoring system (PAT) and a novel curriculum (PUMP). A revised version of this combination is now distributed by Carnegie Learning as the Algebra I Cognitive Tutor (www.carnegielearning.com). There are only a few in-school, semester-long, controlled studies of intelligent tutoring systems in the peer-reviewed literature, and the widely-cited Koedinger et al. (1997) study is arguably the benchmark study against which others should be compared.

Using experimenter designed tests of conceptual understanding, Koedinger et al. (1997) found effect sizes of 1.2 and 0.7. The Andes hour exams were intended to measure both conceptual understanding and algebraic manipulation skills. When the hour exam scores were broken down into conceptual and algebraic components, Andes students scored significantly higher on the conceptual components that Andes addressed (Diagrams: effect size 1.21; Variables: effect size 0.69). In this respect, the results from the two studies are remarkably similar.

Standard tests may be less sensitive to conceptual understanding due to their final-answer format and different coverage. If so, one would expect them to be less sensitive to the benefits of tutoring. Indeed, the Koedinger et al., (1997) evaluation found smaller effects when using multiple-choice standardized tests: 0.3 for each of two tests. These results are comparable to our results for the multiple-choice final exam, where Andes students scored higher than non-Andes students with an effect size of 0.25. The 0.05 difference in effect size may be due to the facts that 30% of the homework problems are not yet covered by Andes.

Thus, the Andes evaluations and the Koedinger et al. (1997) evaluations have remarkably similar effect sizes: 1.2 and 0.7 on experimenter-designed tests and 0.3 on standard tests.

The Andes evaluations differed from the Koedinger et al. (1997) evaluation in a crucial way. The Andes evaluations manipulated only the way that students did their homework—on Andes vs. on paper. The evaluation of the Pittsburgh Algebra Tutor (PAT) was also an evaluation of a new curriculum, developed by the Pittsburgh Urban Mathematics Project (PUMP), which focused on analysis of real world situations. It is not clear how much gain was due to the tutoring system and how much was due to the reform of the curriculum. In our evaluation, the curriculum was not reformed. Indeed, the Andes students and the Control students were in the same course and used the same textbook. The gains in our evaluation are a better measure of the power of intelligent tutoring systems per se. This is extremely good news.

21.6 Progress toward Scaling up Andes

This chapter has made three of its four points so far. First, it discussed how Andes allows more freedom in problem solving than most ITS. Second, it discussed the challenges of scaffolding students in such a freeform user interface and presented Andes' solutions. Third, it discussed the evaluation results, which showed that Andes was unequivocally beneficial in a long-term, controlled evaluation of learning.

The last topic is to report the progress on scaling up Andes. This has proved to be the most difficult goal of all. Briefly put, we have been "selling" Andes for many years by presenting talks and booths at Physics Education Conferences, networking with instructors, and making countless personal contacts and visits. At this writing, Andes2 has a small set of instructors who use it every semester in their classes. However, tens of instructors have tried it in their classes for one semester and dropped it, and hundreds of instructors have worked Andes problems for several hours then decided not to adopt it in their classes. In many causes, we have interviewed the non-adopters and elicited their views on Andes' strengths and weaknesses.

First the good news. The non-adopters approved of Andes main design goal, which was to provide students with as much freedom as paper and pencil. Even the few restrictions that Andes imposes on freedom (e.g., immediate flag feedback; defining quantities before using them in equations) met with general approval. Moreover, non-adopters were mostly satisfied with the approximately 500 Andes physics problems available even though textbooks have about 2000 problems. Our ability to rapidly add new problems to the system was only occasionally used.

Non-adopter's major reason given for abandoning Andes was that Andes2 could only be run on Windows and it had to be installed. This meant that Mac users had to use an emulator, and that users of public machines (e.g., in school's computer labs) had to either reinstall Andes at each use or have tech support pre-install it on the public machines.

The second major reason given for abandoning Andes was that it was not commercially supported software. Potential adopter were worried that user support might disappear. Nothing we could say would dislodge the concept that software companies are more stable than research groups, even a research group supported by large, long-term NSF center (the Pittsburgh Center for the Learning Sciences).

Lastly, the Andes server logs indicate a large number of casual users who try Andes but give up soon. We have spoken to only a few of these non-adopters. Their major barrier was the complex user interface of Andes2.

Given this clear feedback from non-adopters, Andes3 was developed and is about to be launched.

To allay the first concern (Windows only; installation necessary), Andes3 is a client-server system where the client is written in Javascipt using Ajax techniques. That is, any student whose machine has a browser that is connected to the internet can run Andes by clicking on a URL that points to an Andes problem page. The page downloads like any other HTML resource (i.e., it passes through most machine and cluster firewalls). The page displays an unsolved Andes problem, and waits for the user to click on one of Andes' buttons. When the student makes an entry (e.g., by clicking on a text-entry tool, typing, and pressing the Enter key), the raw entry is sent to the Andes server. The server sends back a small change to the page being displayed so that an entire new page doesn't need to downloaded. This is the same technology used by Google Maps and other recently developed client-server software. The Andes3 client has been tested on a large number of combinations of browsers and platforms. The Andes3 server has been load tested with up to 300 (simulated) users concurrently. Andes3 appears ready for prime time.

To allay the second concern (no apparent long-term support), Andes3 has been configured to run under several physics learning management systems that have provided trusted service for years. We are currently working with WebAssign (www.webassign.com) and Lon-Capa (www.lon-capa.org) as well as our long-time collaborators at the Open Learning Initiative (www.cmu.edu/oli). With its new architecture, Andes problems can be mixed in with physics problems supported by more conventional software (e.g., students enter a final answer and submit it to server for grading). Because these organizations have reputations for providing good service to their users, Andes3 is undergoing testing by them. Moreover, all its code has been made open source, so that these organizations can take over support if anything should happen to us.

To allay the third concern (a complex user interface to learn), the Andes3 user interface was designed to mimic the Microsoft PowerPoint user interface, which many physics instructors and students are already familiar with. This was trickier than it seems, because similarity needs to extend down to the details of selecting, dragging, resizing, deleting, double clicking, etc. For instance, what should happen when a user clicks on the arrow-drawing tool, then simply clicks on the canvas instead of dragging out an arrow? It turns out that PowerPoint draws a down-ward pointing arrow about an inch long. So Andes3 should too, as there are likely to be some users who would expect that.

In short, even though Andes2 has been used successfully in many year-long physics courses by instructors who did not participate in its development, and evaluations have shown that it raises students exam grades by 0.61 standard deviations, a few supposedly minor issues appear to have thwarted its dissemination. It will be interesting to see if removing them allows Andes to become widely used.

Acknowledgements. This research was supported by the Cognitive Sciences Program of the Office of Naval Research under grants N00019-03-1-0017 and ONR N00014-96-1-0260, and by the NSF under grant 0354420 to the Pittsburgh Science of Learning Center. We gratefully acknowledge the Andes Alumni: Patricia Albacete, Cristina Conati, Ellen Dugan, Abigail Gertner, Collin Lynch, Zhendong Niu, Charles Murray, Joel Shapiro, Kay Schulze, Stephanie Siler, Linwood Taylor, Donald Treacy, Anders Weinstein and Mary Wintersgill. Some text, tables and figures were copied from the *International Journal of Artificial Intelligence and Education*, 15(3), K. VanLehn et al., "The Andes physics tutoring system: Lessons learned," pp. 147-204, copyright 2005, with permission from IOS Press.

References

Albacete, P.L., VanLehn, K.: The Conceptual Helper: An intelligent tutoring system for teaching fundamental physics concepts. In: Gauthier, G., VanLehn, K., Frasson, C. (eds.) ITS 2000. LNCS, vol. 1839, p. 564. Springer, Heidelberg (2000a)

Albacete, P.L., VanLehn, K.: Evaluation the effectiveness of a cognitive tutor for fundamental physics concepts. In: Gleitman, L.R., Joshi, A.K. (eds.) Proceedings of the Twenty-Second Annual Conference of the Cognitive Science Society. Erlbaum, Mahwah (2000b)

Aleven, V., Koedinger, K.R.: Limitations of student control: Do students know when they need help? In: Gauthier, G., VanLehn, K., Frasson, C. (eds.) ITS 2000. LNCS, vol. 1839, p. 292. Springer, Heidelberg (2000)

Anderson, J.R., Corbett, A.T., Koedinger, K., Pelletier, R.: Cognitive Tutors: Lessons Learned. The Journal of the Learning Sciences 4(2), 167–207 (1995)

Baker, R.S., Corbett, A., Koedinger, K.R., Evenson, S., Roll, I., Wagner, A.Z., et al.: Adapting to when students game an intelligent tutoring system. In: Frasson, C., Gauthier, G., McCalla, G.I. (eds.) ITS 2006. LNCS, vol. 4053, pp. 392–401. Springer, Heidelberg (2006)

Baker, R.S., de Carvalho, A.M., Raspat, J., Corbett, A., Koedinger, K.R.: Educational software features that encourage or discourage "gaming the system". In: Proceedings of the 14th International Conference on Artificial Intelligence in Education. IOS Press, Amsterdam (2009)

Bloom, B.S.: The 2 sigma problem: The search for methods of group instruction as effective as one-to-one tutoring. Educational Researcher 13, 4–16 (1984)

Brna, P., Caiger, A.: The application of cognitive diagnosis to the quantitative analysis of simple electrical circuits. In: Frasson, C., Gauthier, G., McCalla, G.I. (eds.) ITS 1992. LNCS, vol. 608. Springer, Heidelberg (1992)

Gertner, A., Conati, C., VanLehn, K.: Procedural help in Andes: Generating hints using a Bayesian network student model. In: Proceedings of the 15th national Conference on Artificial Intelligence (1998)

Gertner, A.S., VanLehn, K.: Andes: A coached problem solving environment for physics. In: Gautheier, G., Frasson, C., VanLehn, K. (eds.) ITS 2000. LNCS, vol. 1839, p. 133. Springer, Heidelberg (2000)

Hume, G., Michael, J., Rovick, A., Evens, M.: Hinting as a tactic in one-on-one tutoring. Journal of the Learning Sciences 5(1), 23–49 (1996)

Kane, D., Sherwood, B.: A computer-based course in classical mechanics. Computers and Education 4, 15–36 (1980)

Koedinger, K.R., Anderson, J.R., Hadley, W.H., Mark, M.A.: Intelligent tutoring goes to school in the big city. International Journal of Artificial Intelligence in Education 8(1), 30–43 (1997)

Merrill, D.C., Reiser, B.J., Ranney, M., Trafton, J.G.: Effective tutoring techniques: A comparison of human tutors and intelligent tutoring systems. The Journal of the Learning Sciences 2(3), 277–306 (1992)

Mitrovic, A.: An intelligent SQL Tutor on the Web. International Journal of Artificial Intelligence and Education 13(2-4), 171–195 (2003)

Mitrovic, A., Ohlsson, S.: Evaluation of a constraint-based tutor for a database language. International Journal of Artificial Intelligence and Education 10, 238–256 (1999)

Nicaud, J.F., Bouhineau, D., Varlet, C., Nguyen-Xuan, A.: Towards a product for teaching formal algebra. In: Lajoie, S.P., Vivet, M. (eds.) Artificial Intelligence in Education. IOS Press, Amsterdam (1999)

Norman, D.A.: Categorization of action slips. Psychological Review 88(1), 1–15 (1981)

Rose, C.P., Roque, A., Bhembe, D., VanLehn, K.: A hybrid language understanding approach for robust selection of tutoring goals. In: Cerri, S.A., Gouardéres, G., Paraguaçu, F. (eds.) ITS 2002. LNCS, vol. 2363, pp. 552–561. Springer, Heidelberg (2002)

Shapiro, J.A.: Algebra subsystem for an intelligent tutoring system. International Journal of Artificial Intelligence in Education 15(3), 205–228 (2005)

Shute, V.J.: A macroadaptive approach to tutoring. Journal of Artificial Intelligence in Education 4(1), 61–93 (1993)

Singley, M.K.: The reification of goal structures in a calculus tutor: Effects on problem solving performance. Interactive Learning Environments 1, 102–123 (1990)

VanLehn, K.: The behavior of tutoring systems. International Journal of Artificial Intelligence and Education 16, 227–265 (2006)

VanLehn, K., Lynch, C., Schultz, K., Shapiro, J.A., Shelby, R.H., Taylor, L., et al.: The Andes physics tutoring system: Lessons learned. International Journal of Artificial Intelligence and Education 15(3), 147–204 (2005)

Yibin, M., Jinxiang, L.: Intelligent tutoring system for symbolic calculation. In: Frasson, C., McCalla, G.I., Gauthier, G. (eds.) ITS 1992. LNCS, vol. 608. Springer, Heidelberg (1992)

Part V
Social, Cultural and Privacy Issues

Chapter 22
Computer Supported Collaborative Learning and Intelligent Tutoring Systems

Pierre Tchounikine[1], Nikol Rummel[2], and Bruce M. McLaren[3,4]

[1] Université Joseph Fourier - BP 53 - 38041 Grenoble Cedex 9, France
[2] Ruhr-Universität Bochum, 150 Universitätsstraße, 44801 Bochum, Germany, 0234 32201
[3] Carnegie Mellon University, 5000 Forbes Avenue, Pittsburgh, PA 15213, US
[4] Germany Research Center for Artificial Intelligence (DFKI), Stuhlsatzenhausweg 3, D-66123 Saarbruecken, Germany
pierre.tchounikine@imag.fr,
rummel@psychologie.uni-freiburg.de,
bmclaren@cs.cmu.edu

Abstract. In this chapter we discuss how recent advances in the field of Computer Supported Collaborative Learning (CSCL) have created the opportunity for new synergies between CSCL and ITS research. Three "hot" CSCL research topics are used as examples: analyzing individual's and group's interactions, providing students with adaptive intelligent support, and providing students with adaptive technological means.

22.1 Introduction

The field of Computer Supported Collaborative Learning (CSCL) focuses on how students learn by collaborating and how this collaboration can be supported by technology. Research in collaborative learning has shown, in general, that collaboration can increase group performance and individual learning outcomes. However, an educational setting with collaboration is not, on its own, sufficient for learning to occur (see Slavin 1996, for a review). CSCL research has shown that it is difficult to clearly define the interaction between the initial conditions of collaboration (e.g., the composition of the group or the type of task the group is engaged in) and learning outcomes (Dillenbourg et al. 1996). Moreover, collaboration leads to positive outcomes only when students engage in knowledge-generative interactions such as giving explanations, and engaging in argumentation, negotiation, conflict resolution or mutual regulation (Dillenbourg and Jermann 2007). At the same time, several potentially problematic issues must be avoided, such as unequal engagement or social loafing. Generally, it has been discovered that the occurrence of knowledge-generative interactions is not a given: such interactions do not necessarily emerge spontaneously (Cohen 1994; Salomon and Globerson 1989). Researchers attempting to understand how to

R. Nkambou et al. (Eds.): Advances in Intelligent Tutoring Systems, SCI 308, pp. 447–463.
springerlink.com

foster collaborative learning have thus focused on how best to promote fruitful interactions among collaborative learners.

A well studied and documented approach to supporting collaboration is scripting with *macro-scripts*. The purpose of CSCL macro-scripts and associated technology is to introduce structure and constraints that guide collaborative interactions among distant students or co-present students whose action or interaction is (at least partially) mediated by a computer-based system. A CSCL script typically describes the task to be achieved by students and defines how the task is to be divided into subtasks, the sequencing of these subtasks, the role of each student, sets constraints or rules for the interaction, and prescribes the features or tools of the computer-based system to be used by the students (Fischer et al. 2007; Dillenbourg and Tchounikine 2007; Kollar et al. 2006; Rummel and Spada 2005a and 2005b; Tsovaltzi et al. 2010). Implemented in this manner scripts, thus, provide predefined, fixed assistance to learners as they collaborate. Other support methods can similarly be classified as providing fixed assistance; for instance, giving students declarative instruction on how to collaborate before they collaborate (e.g., Saab et al. 2007) or providing students with examples of good collaboration (e.g., Rummel and Spada 2005b). Scripts have proven to be effective in promoting fruitful interactions and student learning, but the design of scripts follows a "razor's edge" between useful guidance and control of student activities: if the scaffolding they provide is too weak, it will not produce the expected interactions; if it is too strong or irrelevant, it can lead to sterile interactions (Dillenbourg 2002).

Research in CSCL scripts has initially focused on how to design settings (i.e., define the task, the script structure, the content of hints provided to the individuals or the group, or the technology provided to support students' actions) whose properties are likely to prompt students to engage in particular activities and interactions. A recent development in CSCL is to also take into account students' *effective enactment* of the script. A rationale for this is that students' enactment of the script is influenced by many parameters (individual differences, group phenomena) that lay outside of the control of the designer, and, consequently, the enactment may differ from expectations (Tchounikine 2008). In particular, students mostly focus on solving the task (and not on collaborating or interacting as hoped) and it may thus be required to adapt the setting or provide adapted feedback in order to enhance collaboration. Moreover, students may have different and dynamically changing needs and thus require individualized support (Diziol et al. 2010).

An important objective of current research is thus to dynamically adapt to the conditions of students' interactions as they unfold. Addressing this objective requires on-the-fly assessment of students' activities (how they interact, which steps they take at solving the task, how they use the technology) as a basis for adaptation decisions (McLaren et al., in press). It also requires modelling how the system should react, that is, in which way it should adapt its support to individual and collective needs. Intelligent Tutoring Systems (ITS) are traditionally designed to provide user adapted support. Leveraging ITS approaches could thus be a promising direction for achieving adaptive support for CSCL.

In the remainder of this chapter we will present research efforts that address issues related to adaptation and illustrate potential synergies between CSCL and ITS research:

Analyzing individual's and group's interactions (interaction analysis). In section 2 we explore how one can analyze student interactions as they communicate and collaborate with one another, as a basis for providing adaptive guidance and feedback.

Providing students with adaptive intelligent hints. In section 3 we explore how ITS technology and CSCL scripts can be combined to adaptively tailor support and feedback to students' needs.

Providing students with adaptive technological means. In section 4 we discuss why technological frameworks should be flexible and adapt to students' effective activity in order to continuously provide technological support that maintains the targeted pedagogical conditions. We outline the requirements for adaptive technological platforms.

22.2 Interaction Analysis

In CSCL, *interaction analysis* refers to analyzing what students are doing as they communicate and collaborate with one another at a relatively fine-grain level. This is a basis for, in the context of CSCL scripts, adapting the script, adapting the technological framework or providing individual or collective hints. More generally, interaction analysis can be a basis to guide collaborative behavior in general, including within non-scripted settings.

As an example of an interaction analysis approach, past work of McLaren and colleagues (McLaren, et al. in press) involved developing techniques to analyze the interactions and communication of students as they debate, in an online environment, thorny ethical and social issues such as "Should experiments be performed on animals?" These collaborative e-discussions occur in the context of a shared graphical workspace, such as shown in Figure 22.1. Students make contributions by dragging and dropping shapes with different semantics (e.g. a "claim", "argument", "counter-argument"), filling the shapes with text containing their contributions to the discussion (e.g., "I don't agree with John's claim, because …"), and linking the shapes to the contribution of other students with labeled links, such as "opposes" or "supports."

The reason for interaction analysis in the system that McLaren et al. worked with – called ARGUNAUT – was to provide feedback to a teacher so that he or she can help students stay on topic, elicit contributions from all members of the collaboration groups, suggest the use of supported claims and arguments, and generally guide the students toward fruitful discussion and collaboration (Hever et al. 2007; De Groot et al. 2007). Since the burden of moderating multiple, simultaneous e-discussions – that is, supporting many groups of students at the same time – is too difficult for individual teachers, ARGUNAUT attempts to summarize

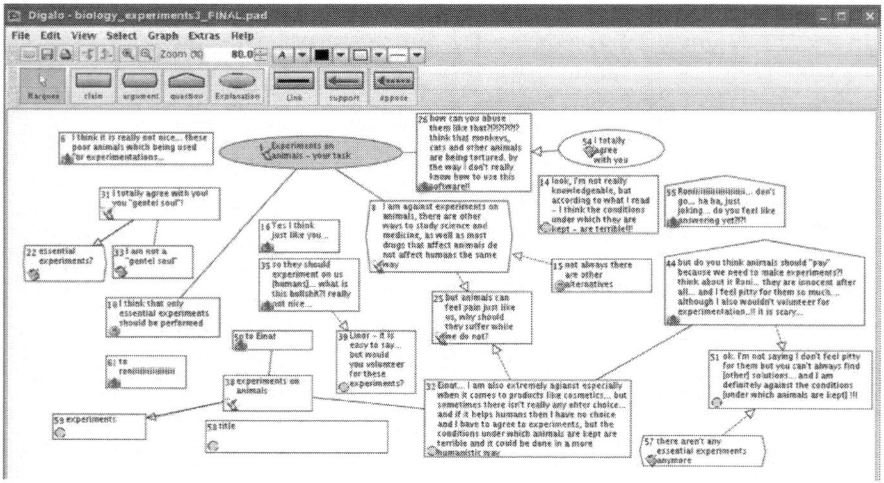

Fig. 22.1 An e-discussion in ARGUNAUT

the students' discussions and alert teachers to critical aspects and events in the e-discussions. In other educational settings, this type of analysis could be the basis for automated feedback to students or script adaptation.

The ARGUNAUT system provides the teachers with online, automated feedback regarding important aspects and characteristics of each discussion, explicitly focusing attention on events or situations that may require the teacher's intervention or support. The key idea is to analyze student contributions and e-discussions using machine learning (Witten and Frank 2005), shallow text processing (Rosé et al. 2008), and case-based graph matching (McLaren 2003). ARGUNAUT leverages (a) the structure of the argument graphs, (b) the textual contributions of the students, and (c) the temporal sequence of those contributions. It also analyzes and classifies, using machine-learned classifiers, student interactions at three levels: (1) the shape-level (i.e., individual contributions), (2) the paired-shape level (i.e., contributions by two students, linked together in the graphical representation), and (3) the cluster level (i.e., linked groups of contributions, consisting of 2 or more shapes). The classifiers that are created at each of these levels help to draw attention to activities of students that are positive, such as responding to one another's arguments, or negative, such as going off topic. Through extensive experimentation, six of the shape and paired-shape classifiers achieved at least satisfactory results (i.e., Kappa values above 0.6), while five of the cluster-level classifiers have achieved such results (for details, see McLaren, et al. in press and Wegerif et al. 2009). An example of the teacher's view of the results of one of the cluster classifiers – Argument + Evaluation – is shown in Figure 22.2.

In the following we review other projects that use machine-learning, language processing or data mining techniques.

Rosé and colleagues have done similar interaction analysis research to the work described above; in fact, they developed the text analysis tool, TagHelper (Rosé et al. 2008), also used in the ARGUNAUT. They analyzed a corpus of 1,250 coded

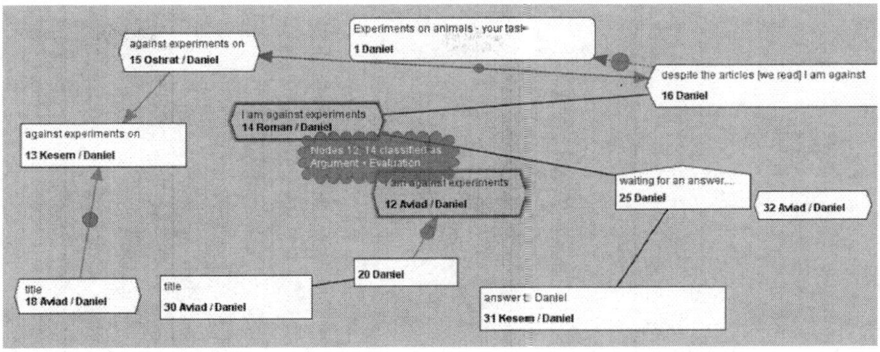

Fig. 22.2 Argument graph display of Argument + Evaluation clusters, with one matching cluster highlighted

text segments along multiple dimensions of argumentation to derive machine-learned classifiers, some of which achieved acceptable Kappa values of 0.7. In their work, Rosé and colleagues originally intended to assist coders in analyzing protocols after the fact; but, more recently, they have focused on providing real-time support to dyads collaborating on a problem-solving task (Kumar et al. 2007). Similar to the ARGUNAUT system, they perform online analysis of textual communication data, in their case, chat data. In contrast to ARGUNAUT, but more like traditional ITS, their analysis results are displayed directly to students, rather than to teachers. An empirical study (Kumar et al. 2007) showed significant learning benefits in terms of analytical knowledge and conceptual understanding when the adaptive support was provided.

Goodman and colleagues (2005) have also applied a machine-learning approach to the problem of interaction analysis. Within the EPSILON system peer groups work together on a problem in the domain of object modeling techniques (OMT). Their collaboration takes place within a shared whiteboard in which diagrams (e.g. class diagrams) are constructed. Peers communicate via a text chat with a sentence opener interface (an interface in which the beginning of sentences are provided as options to the student, e.g., "We have learned …"); an agenda tool supports task management. The system evaluates aspects concerning domain (e.g., domain knowledge of peers), task (e.g., progress in solving the task) and possible problems in the collaboration process (e.g. unanswered questions). The sentence opener interface plays a critical role in interaction analysis; it is used to automatically classify each chat contribution as a dialogue act. The dialogue acts, in turn, are used as features for machine-learning analyses, bypassing the complicated task of natural language (or text) processing tackled by McLaren and colleagues and Rosé and colleagues. The Goodman et al. (2005) analysis approach is similar to ITS: Relevant results are displayed immediately to the peers via meters, while feedback is provided by means of an artificial peer agent that verbally interacts with the participants.

Soller's work, also in the context of EPSILON, was concerned with the analysis of chat conversations that accompanied activities in a problem-solving environment (Soller 2004). The chat tool was enhanced by a sentence-opener interface to structure users' communication (the same one used by Goodman et al.) and to make interaction analysis feasible. The analysis of EPSILON aimed at identifying episodes in which students communicated their knowledge to their peers ("knowledge sharing episodes") and episodes in which they failed to do so ("knowledge sharing breakdowns") using annotated data provided to Hidden Markov Models (HMMs). In a second step, she investigated reasons for knowledge sharing breakdowns using multidimensional scaling (MDS) and clustering techniques. In this way Soller identified different patterns (clusters) of successful knowledge sharing and knowledge sharing breakdowns. Soller's approach analyzes *sequences* of actions, as opposed to sub-graphs within an argument map that are sometimes sequential, sometimes parallel (such as in the ARGUNAUT system described above). The *textual* content of contributions is not analyzed in Soller's system; it relies exclusively on dialogue acts, which are trivially inferred from the selected sentence openers.

Another approach to interaction analysis is that of Ravi and Kim (2007) who analyzed threads in a technical discussion board in order to call an instructor's attention to discussions that contain unanswered questions. They developed two machine-learned classifiers (linear SVMs), one for detecting questions (QC), the second for detecting answers (AC), and achieved accuracies of 88% (QC) and 73% (AC). They also employed shallow language features, similar to the McLaren et al and Rosé et al. approaches, although somewhat more elaborated ones (e.g., tri- and quadro-grams). Furthermore they implemented a rule-based thread profiler that assigns one of four typical profiles to threads, with accuracies varying between 70% and 93%. In follow-up work, Kim and colleagues (2008) describe PedaBot, a threaded discussion system that scaffolds students' discussions by retrieving messages from past technical discussions (e.g., Operating Systems) that are possibly relevant to the current context. The work by Kim and colleagues does not specifically take into account structure and temporal features to support its classification approach, as, for instance, the ARGUNAUT system does.

Jeong (2005) has developed a software tool called the Discussion Analysis Tool (DAT) that uses sequential analysis to capture and model sequences of speech acts. DAT models an online threaded conversation as a network of transitional probabilities, called a transitional state diagram, building the diagram from prelabeled data. For example, in a particular diagram a "challenge" act might occur with a probability of 0.52 after an "argument" is made, while an "explanation" follows an "argument" with a (surprising) probability of 0.08. DAT has been used to, for instance, evaluate the interactions that are most likely to promote critical thinking and the effects of supportive language (e.g., "I agree," or "ask" questions) on subsequent group interactions. DAT can also be used to evaluate whether threaded conversations deviate from a norm; it creates a z-score matrix to show probabilities that were significantly higher or lower than expected in one state diagram compared to another. But Jeong's tool is intended more as a post-hoc analysis tool, rather than a tool for supporting online and "live" analysis, as in a traditional ITS system. Also, Jeong's system does no language analysis; it depends on human post-hoc coding (or real-time labeling) to identify the individual acts.

Other interaction analysis work has focused on identifying frequently occurring patterns. For instance, sequential pattern mining algorithms (Agrawal and Srikant 1995) has been used to try to find salient patterns of student collaboration. Such algorithms have, for instance, been used to account for both language and contextual attributes – to a reasonable degree of accuracy – in classifying email messages (Carvalho and Cohen 2005). Kay et al. (2006) used a variant of the Generalized Sequential Pattern Algorithm (Srikant and Agrawal 1996) to identify common interaction patterns in a source repository and Wiki log data from student software development projects.

The pattern that emerges from the interaction analysis research within CSCL is a movement away from computer tools used almost exclusively to facilitate collaborative communication to computer tools used to *analyze* that communication. A key difference to ITS work is that the nexus of analysis in CSCL systems is the interaction between the collaborating students, rather than between software tutor and human tutor. Furthermore, the communication between collaborators is typically much more varied and unpredictable than the interaction between a software tutor (which is typically consistent and somewhat predictable) and a human tutee. A similarity to ITS work, on the other hand, is that student actions in CSCL systems are now increasingly analyzed for the purpose of providing feedback, guidance, and scaffolding for learning.

22.3 Providing Adaptive Intelligent Hints

Providing adaptive intelligent hints consists of, first, monitoring and evaluating students' collaboration (on the basis of run-time data, similar to interaction analysis, as described in the previous section) and, second, automatically responding to the particular needs of the collaborating participants with context-specific hints and/or scaffolding.

A typical way that student interactions have been assessed is to compare the difference between those interactions and a model of optimal collaboration (a kind of model tracing approach) or by checking constraints (constraint-based approach).

For instance, one way of assessing the quality of student interactions is by tracking student dialogue patterns. This is commonly accomplished by asking students to indicate the type of contribution that they would like to make before they make it. For example, students may select a sentence starter like "We need to work together on this..." to begin their utterance, an approach used in several projects described earlier (Goodman et al. 2005; Soller 2004). Based on the starters that students select, the system can make inferences about what students are saying and use these inferences to provide feedback (Tedesco 2003). However, as students often do not accurately label their utterances, the inferences that the system makes can be inaccurate. Thus, automated dialogue assessment solutions have been developed. So far, this technology has only been used successfully in limited ways, such as classifying the topic of conversation (Kumar et al. 2007), characterizing general argumentation patterns (McLaren et al, in press; discussed above), or assessing student accuracy when they use sentence starters (Israel and Aiken

2007). Some researchers try to circumvent the problems of assessing dialogue by relying on simple metrics like participation to trigger feedback. For instance, these systems evaluate the amount or length of contributions collaborators make to a shared workspace or to a dialogue and support the interaction by directly encouraging non-contributors to participate more (Constantino-Gonzalez et al. 2003). Unfortunately, the same assessment metrics cannot be used to give students feedback on how to participate, which may ultimately be more valuable. Some other works propose to facilitate tutor's perception of collaboration issues such as student's organization breakdowns by proposing students with specifically design organization tools, whose usage makes learners' organization more easily detectable and analyzable (Moguel et al. 2010).

Another approach to providing adaptive support for collaborative learning may be more promising – and leverages ideas from ITS. In individual learning settings, adaptive ITS feedback has been shown to be quite successful at improving student learning in a variety of domains such as physics (VanLehn et al. 2005), mathematics (Koedinger et al. 1997) and reading (Beck et al. 2004). Further, it has been argued that using existing ITS problem-solving models for individual learning to provide interaction support in collaborative learning settings may be fruitful (Walker et al. 2009b). Some adaptive collaboration systems already partially capitalize on the ITS approach. For example, when students submit a group solution in COLLECT-UML (Baghaei et al. 2007), the system evaluates the solution using a constraint-based model, and provides feedback on the quality of the solution. Occasionally, this problem-solving support even leverages student talk rather than only student action: When CycleTalk (Kumar et al. 2007) detects problem-relevant topics in student conversation, it engages the collaborating students in a tutorial dialogue, asking them to answer questions that concern these aspects. This tutorial dialogue often yields increased interaction between the collaboration partners. Both COLLECT-UML and CycleTalk implemented support in a collaborative setting by extending an existing *individual* learning system.

In two other projects, the problem-solving models of an existing intelligent tutoring system were used to provide *interaction support* (Diziol et al. 2010). Here collaborative extensions to the Cognitive Tutor Algebra (CTA), a widely-used tutoring system for mathematics instruction on the high-school level (Koedinger et al. 1997) were developed. The CTA covers different aspects of algebra learning such as linear equations and inequalities. To provide adaptive tutoring, the CTA evaluates the student's problem-solving actions by comparing them to a cognitive model of successful student performance, represented using a set of production rules. If an error is detected, the CTA immediately marks it as incorrect and provides context-sensitive feedback. In one project (Rummel et al. 2010), two students worked on the same computer, and the CTA had a joint model of the dyad's problem-solving. The output of this problem-solving model served as input for an interaction model that assessed students' learning behaviour. If ineffective learning strategies, such as trial and error and hint abuse behaviours, were detected, this triggered adaptive support prompting fruitful collaboration. In a second project (Walker et al. 2009b) a peer tutoring scenario was involved: one student tutored another student, while the students worked on separate computers. The system integrated the CTA problem-solving model with a model of good tutoring in order to provide the peer tutor with support.

There are several other design options for providing interaction support along the lines of ITS. Instead of modelling either one student (the peer tutor) or two students together, another option is to model the interaction partners separately. For instance, the COMET system developed by Suebnukarn and Haddawy (2006) assesses the individual expertise of participants and encourages the collaborators to share their complementary knowledge with others. Similarly a system for adaptive collaboration support could intervene if it detects undesirable asymmetries in students' skill acquisition: if a particularly difficult problem-solving step is mainly solved by one interaction partner, this might indicate that the other student has not yet acquired the relevant skills and could benefit from adaptive collaboration support.

Furthermore, as argued in (Walker et al. 2009a), leveraging existing problem-solving models can facilitate the comparison of different types of adaptive collaboration support, and thus help to gain information on the conditions of optimal assistance. While the evaluation of the two adaptive collaborative extensions to the CTA revealed an impact on student interaction, the improved interaction did not yield the differences in student learning outcome that have been found in other studies (e.g. Baghaei et al. 2007). This indicates that the need for further research on how to optimize collaboration support for particular interaction conditions. This optimization can be considered an instantiation of a more general assistance dilemma (Koedinger and Aleven 2007), where in order to discover how to best deliver assistance, one must manipulate the amount, type and timing of help provided to students. For example, we so far only have limited knowledge on how to time support most effectively. It is still an open question whether it is best to provide adaptive collaboration support immediately, or whether it might sometimes be beneficial to withhold it. Mathan and Koedinger (2005) investigated this research question in an individual learning setting, and discovered that the two timing options served different goals. Immediate feedback ensured that students did not get stuck in problem-solving and thus was more immediately effective and efficient. Delayed feedback enabled students to practice their monitoring skills and consequently yielded improved learning transfer. Similarly, immediate feedback to collaboration may improve the current interaction, while delayed feedback may increase students' collaboration skills and thus promote future interactions (Kapur 2008). Thus, the type of feedback may have to be adapted to the goal of the instruction. On a practical level, the accelerated development of adaptive collaboration support conditions based on existing intelligent tutoring technology may help us to increase our knowledge of optimal collaboration assistance, as it enables us to more rapidly implement and compare different design options concerning the amount, type and timing of support.

In summary, while there are preliminary promising results that hint toward the effectiveness of adaptive collaboration support, clear conditions and guidelines have not yet emerged on how to best deliver adaptive assistance. Leveraging existing problem-solving models can facilitate the implementation of adaptive collaboration support, but this work is still in its early stages. A clear next step is to investigate different types of adaptive collaboration support in more detail to increase our knowledge of when and why adaptive collaboration support is effective.

22.4 Providing Students with Adaptive Technological Means

Providing adaptive technological means consists of, on the basis of monitoring and evaluating students' activity (cf. preceding sections), adapting accordingly the technological means they are presented with.

With respect to CSCL support, the role of the computer-based system is two-fold: (1) provide the necessary technological means and (2) participate to the scripting objective, i.e., supporting and constraining.

Let us consider a setting where students must collaboratively build a graphical model of some physics phenomenon, and anchor its design in explicit argumentation. First, the role of the computer is to provide the *technological means* for students' activities. For instance, students can be proposed with a chat to communicate synchronously and a shared whiteboard to collaboratively edit their model. Second, the computer might be used as a way to provide *support and constraints*. Proposing students with a basic chat allows argumentation to occur, but does not foster it. Differently, proposing students with a dedicated e-discussion tool such as ARGUNAUT (cf. section 2) introduces some support (and constraints). Similarly, a whiteboard allows students to build a model. A dedicated modeling tool introducing specific physics notions and relations (as ARGUNAUT introduces specific communication constructs) will here again introduce some support (and constraints).

More generally, the computer-based system can participate in structuring and constraining the sequences of activities or the way students engage in individual and collective activities by specifying roles, introducing specific dataflow or workflow, specific tools (e.g., modeling tools proposing carefully defined epistemic primitives) or communication functionality that impact students' interaction (e.g., imposing sentence-openers or turn-taking structures). As a matter of fact, technology is not neutral: although not necessarily explicit, a given design always denotes some usage expectations, and influences users' activity.

Seen from the perspective of usage, macro-scripts create socio-technical settings. Technology impacts the script enactment, but this impact is not necessarily the one that is expected, in particular because of the uncertainties of how students will perceive and use the technology. Thus, it is important to take into consideration not only the script and the technology as considered by designers, but also the phenomena related to the effective use of technology.

A general difficulty with designing technology to support human activity is that designers have limited control over how their designs will be enacted. Goodyear (2001) emphasizes the fact that teachers set tasks and students interpret the specifications of the task. Their subsequent activity is related to their interpretation of the task (which may change from student to student and from the teacher's wills), and also to other dimensions (e.g., students' motivations or perception of the setting) that evolve over time, and are interrelated within systemic relations. Students' perception and enactment of CSCL scripts and their use of the provided technological means are intrinsically situated. The students' activity that will emerge from the confrontation of the students with the task and the technological setting is thus subject to different contingences. As a consequence, its details are unpredictable (Tchounikine 2008).

Macro-scripts, as a particular kind of pedagogical method, are also intrinsically related to open issues that cannot be fully defined or predicted. It is not possible to exhaustively list and consider all of the pedagogical parameters of a macro-script situation. As a consequence, when monitoring the script as it is enacted by the students, the teacher often has to manage unexpected events (originating from inside or outside the script), manage requests from the students that will lead him/her to consider the script's or technology's modifications, or use a pedagogical opportunity that appears (Dillenbourg and Tchounikine 2007). For example, teachers may want or need to modify, at run-time, decisions taken when tuning the script: change the groups because a student drops out of the course or because two conflicting leaders emerge from a group and this becomes problematic; postpone some deadlines in order to deal with external or internal reasons (network failure, bad appreciation of the task difficulty, etc.); change the script structure (change the order of phases, add or remove a phase, merge some tasks, change the argumentation tool because students face problems with it); etc. (Dillenbourg and Tchounikine 2007). Here again, this creates uncertainties related to macro-scripts' enactment, to be taken into account when studying the technological dimensions.

By definition, macro-scripts provide learners with some flexibility, i.e., let them decide on their own part of how they will follow the script. Many experiments reported in the literature show that learners use this flexibility, e.g., in context, decompose tasks into subtasks and refine their division of labor, adopt sub-strategies or alternative ways of using the technological means they have been offered: they involve self-organization activities: part of the organization is externally set by the script, and part is related to emergent features of learners' enactment of the script at run-time (Tchounikine 2007).

Phenomena related to the perception by students of the task and the technological setting, and their use of technology, is a general issue of educational technology. It is however of particular importance in open or semi-open settings in CSCL within which learning is supposed to originate from student's process, interactions and initiatives: the focus is not on the task output, but on the process and its characteristics.

From the point of view of computer-based system design, the fact that scripts enactment is difficult to anticipate can be an argument in favor of keeping the technological support very generic. This is the principle underlying platforms such as generic learning management systems. Students are presented with generic platforms that offer students a list of autonomous technological means (chat, forum, whiteboard, etc.) they can choose to use as they want. These means are not thought to as a *milieu* proposing resources and means that have been designed and articulated according to the details of the students' hypothesized activity. Although largely adopted (if anything else, because of its simplicity), this approach does not address the issue of using the technology as a way to provide support and/or constraints in line with the objectives of the script and the expected students' knowledge-construction processes. It also makes it difficult to provide precise scaffolding.

Recent CSCL works has attempted to understand how to design and implement platforms addressing the following system of constraints: (1) provide students with the functionality necessary to achieve the tasks proposed by the script, (2) participate in the objective of structuring students' collaboration by reifying constraints/support related to the script's pedagogical objective, and (3) be sufficiently adaptive to be coherent with students activity if the actual interaction pattern differs from expectations, or if some unpredictable events arise. The system must be sufficiently flexible not to over-constrain students' activity whilst keeping the script's *raison d'être* and remaining coherent with the pedagogical objectives.

In order to both (1) orchestrate activities and manage the workflow and (2) be able to adapt at run-time to requests from students or teachers whilst adhering to the learning objectives constraints, a powerful evolution for CSCL is to address the technological setting as a *script engine*: the technological settings (tools, interfaces, etc.) must be the result of careful *a priori* decisions (when the script is launched) and then run-time decisions, during its enactment. Figure 22.3 (Tchounikine 2008) presents a general theoretical architecture of such an (intelligent) flexible adaptive activity framework.

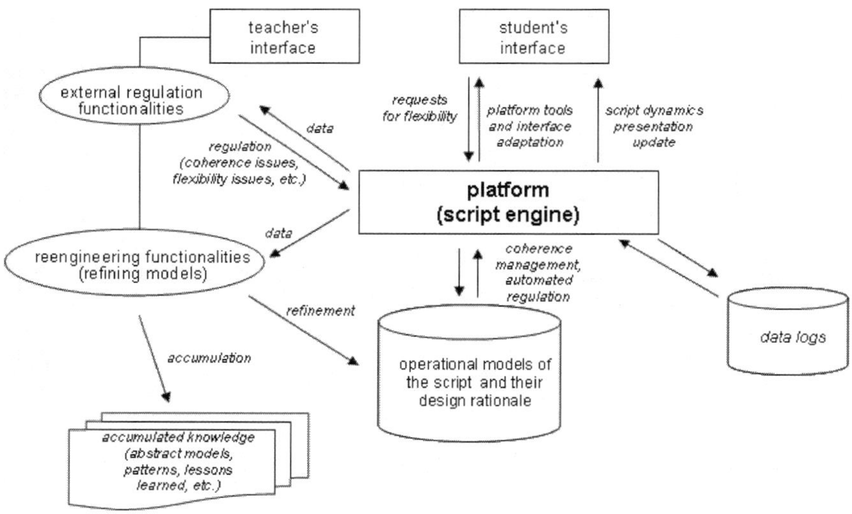

Fig. 22.3 A general theoretical architecture

Putting into practice this approach requires computer-tractable models of the script. Different efforts have been recently dedicated to this such as defining scripts' general components and mechanisms (Kobbe et al. 2007) and exploring different modeling languages that can be used to represent such issues (finite automata, statecharts, activity diagrams, Petri nets, etc.; see Harrer and Malzahn (2006) for a review). As an example of how flexibility can be introduced, Haake and Pfister (2007) propose to describe scripts (roles, possible sequences of actions, etc.) as a finite state automaton, a formalism that allows complex control structures. The script can then be deployed on a platform that is compliant with this formalism. Within such an approach, the platform runs the script and provides access to functionalities or data according to the automata. The script can be modified at any time via its specification, without requiring any hand-modification of the platform, which provides a certain form of flexibility.

Adapting the script or the platform at run-time requires modeling not only the script structure (phases, division of work, etc.) but also the underlying design rationale. For this purpose, Dillenbourg and Tchounikine (2007) propose to dissociate intrinsic constraints (script's core mechanisms, which set up the limits of flexibility, i.e., what cannot be accepted in order for the script to keep its raison d'être) and extrinsic constraints (constraints bound to contextual factors, which define the space for flexibility, i.e., the space within which a script should be modifiable by teachers and/or students because the related decisions result from arbitrary or practical choices).

Another important dimension of such architecture is to support accumulation of knowledge by, for instance, supporting data analysis and recurrent-patterns identification, which will help in iteratively refining scripts, and in progressing in the understanding of script enactment. Here again, Machine Learning techniques appear promising.

Implementing such an innovative approach to CSCL technological settings thus pushes one to consider issues that are similar to those tackled in ITS: creating operational languages that denote the different models and systems of constraints; understanding students' activity by interpreting data and logs (cf. section 2), and dynamically (intelligently) reacting to adapt the script and/or the technological framework or to scaffold individual and groups (cf. section 3).

22.5 Discussion

Historically, in CSCL systems, the computer has been used as a mediating mechanism between learners and teachers or other learners. Whereas Intelligent Tutoring Systems address issues such as the analysis and understanding of learners' activity and production, problem solving or interaction control, classical CSCL systems have not addressed these issues at all. Whereas ITS research has, since its inception, leveraged Artificial Intelligence techniques, CSCL research has instead focused on HCI issues related to providing students with a good experience of communicating with their fellow students (Tchounikine et al. 2009). However, times have changed. In this chapter, we have shown that at least some current CSCL research is exploring issues of adaptivity, automated analysis, and feedback. These changes bring the field of CSCL closer to techniques of ITS research.

In this chapter we have disentangled interaction analysis, providing students with adaptive hints and adapting the technological framework. This is indeed an analytical presentation. Interaction analysis is a basis for adaptivity in general. Adapting the support provided by individual or collective hints or provided by the technology serves the same objective (maintaining positive conditions that enhance interactions) and may be powerfully connected. Moreover, the examples we have raised are but approaches. For instance, some other works address adaptivity by identifying useful *adaptation patterns* to be embedded in systems for adaptive collaboration scripting, i.e., adaptation processes that can be initiated by the system when specific conditions are identified during script enactment (Karakostas and Demetriadis 2009).

One of the reasons of CSCL rapid development in basic settings (i.e., out of research experiences) is the fact many CSCL projects are based on simple, stable, well disseminated, and almost freely available technologies. CS difficulties raised by CSCL (e.g., HCI issues) are less binary and non-contingent problems than ITS issues such as AI issues related to building learners' models or solving problems. In fact, historically, CSCL research focused on education-psychology issues rather than on CS support and, in particular, adaptive support. The CS dimension was often limited to communication devices and simple interfaces. Advanced technologies can however powerfully support and enhance communication and collaboration. Addressing adaptivity will indeed conduct to face difficult issues as ITS does since its early ages. This is a promising perspective for research.

Acknoledgements. The content of this article is based on the authors' works with colleagues and doctoral students. In particular, without the support and contributions of Dejana Diziol, Erin Walker, Oliver Scheuer and Ken Koedinger this chapter would not have been possible.

References

Agrawal, R., Srikant, R.: Mining sequential patterns. In: Proceedings of the 11th International Conference on Data Engineering (ICDE 1995), Taipei, Taiwan, pp. 3–14 (1995)

Baghaei, N., Mitrovic, T., Irwin, W.: Supporting collaborative learning and problem solving in a constraint-based CSCL environment for UML class diagrams. International Journal of Computer- Supported Collaborative Learning 2(2-3), 159–190 (2007)

Beck, J.E., Mostow, J., Bey, J.: Can automated questions scaffold children's reading comprehension? In: Lester, J.C., Vicari, R.M., Paraguacu, F. (eds.) ITS 2004. LNCS, vol. 3220, pp. 478–490. Springer, Heidelberg (2004)

Carvalho, V.R., Cohen, W.W.: On the collective classification of email "speech acts". In: Proceedings of the 28th Annual International ACM SIGIR Conference on Research and Development in Information Retrieval, pp. 345-352. ACM Press, New York (2005)

Cohen, E.G.: Restructuring the Classroom: Conditions for Productive Small Groups. Review of Educational Research 64(1), 1–35 (1994)

Constantino-Gonzalez, M.A., Suthers, D., Escamilla de los Santos, J.: Coaching web-based collaborative learning based on problem solution differences and participation. Artificial Intelligence in Education 13(2-4), 263–299 (2003)

De Groot, R., Drachman, R., Hever, R., Schwarz, B.B., Hoppe, U., Harrer, A., De Laat, M., Wegerif, R., McLaren, B.M., Baurens, B.: Computer Supported Moderation of E-Discussions: the ARGUNAUT Approach. In: Proceedings of the Conference on Computer Supported Collaborative Learning (CSCL 2007), vol. 8, pp. 165–167 (2007)

Dillenbourg, P.: Over-scripting CSCL: The risks of blending collaborative learning with instructional design. In: Kirschner, P.A. (ed.) Three worlds of CSCL. Can we support CSCL, Heerlen, Open Universiteit Nederland, pp. 61–91 (2002)

Dillenbourg, P., Jermann, J.: Designing integrative scripts. In: Scripting Computer-Supported Collaborative Learning - Cognitive, Computational, and Educational Perspectives. Computer-Supported Collaborative Learning Series, pp. 275–301. Springer, Heidelberg (2007)

Dillenbourg, P., Tchounikine, P.: Flexibility in macro-scripts for CSCL. Journal of Computer Assisted Learning 23(1), 1–13 (2007)

Dillenbourg, P., Baker, H.P.M., Blaye, A., O'Malley, C.: The Evolution of Research on Collaborative Learning. In: Learning in Humans and Machines: Towards an Interdisciplinary Learning Science. Elsevier/Pergamon, Oxford (1996)

Diziol, D., Walker, E., Rummel, N., Koedinger, K.: Using intelligent tutor technology to implement adaptive support for student collaboration. Educational Psychology Review (2010), doi:10.1007/s10648-009-9116-9

Fischer, F., Kollar, I., Mandl, H., Haake, J.M. (eds.): Scripting computer-supported communication of knowledge – cognitive, computational, and educational perspectives. Springer, Heidelberg

Goodman, B., Linton, F., Gaimari, R., Hitzeman, J., Ross, H., Zarrella, G.: Using Dialogue Features to Predict Trouble During Collaborative Learning. User Modeling and User-Adapted Interaction 15, 85–134 (2005)

Goodyear, P.: Effective networked learning in higher education: notes and guidelines. Final Report to JCALT: Networked Learning in Higher Education Project (3) (2001), `http://csalt.lancs.ac.uk/jisc/` (Retrieved from December 2, 2006)

Haake, J., Pfister, H.-R.: Flexible scripting in net-based learning groups. In: Fischer, F., Kollar, I., Mandl, H., Haake, J.M. (eds.) Scripting computer-supported cooperative learning. Cognitive, computational, and educational perspectives. Springer, Heidelberg (2007)

Harrer, A., Malzahn, N.: Bridging the gap –towards a graphical modelling language for learning designs and collaboration scripts of various granularities. In: International Conference on Advanced Learning Technologies, Kerkrade, The Netherlands, pp. 296–300 (2006)

Hever, R., De Groot, R., De Laat, M., Harrer, A., Hoppe, U., McLaren, B.M., Scheuer, O.: Combining Structural, Process-oriented and Textual Elements to Generate Alerts for Graphical E-Discussions. In: Proceedings of the Conference on Computer Supported Collaborative Learning (CSCL 2007), vol. 8, pp. 285–288 (2007)

Israel, J., Aiken, R.: Supporting collaborative learning with an intelligent web-based system. International Journal of Artificial Intelligence and Education 17(1), 3–40 (2007)

Jeong, A.: A Guide to Analyzing Message-Response Sequences and Group Interaction Patterns in Computer-Mediated Communication. Distance Education 26(3), 367–383 (2005)

Kapur, M.: Productive failure. Cognition and Instruction 26(3), 379–424 (2008)

Karakostas, A., Demetriadis, S.: Adaptation patterns in systems for scripted collaboration. In: O'Malley, C., Suthers, D., Reimann, P., Dimitracopoulou, A. (eds.) Proceedings of the 9th international Conference on Computer Supported Collaborative Learning. Computer Support for Collaborative Learning. International Society of the Learning Sciences, vol. 1, pp. 477–481 (2009)

Kay, J., Maisonneuve, N., Yacef, K., Zaïane, O.: Mining Patterns of Events in Students' Teamwork Data. In: Proceedings of the Workshop on Educational Data Mining at the 8th International Conference on Intelligent Tutoring Systems, pp. 45–52 (2006)

Kim, J., Shaw, E., Ravi, S., Tavano, E., Arromratana, A., Sarda, P.: Scaffolding On-Line Discussions with Past Discussions: An Analysis and Pilot Study of PedaBot. In: Woolf, B.P., Aïmeur, E., Nkambou, R., Lajoie, S. (eds.) ITS 2008. LNCS, vol. 5091, pp. 343–352. Springer, Heidelberg (2008)

Kobbe, L., Weinberger, A., Dillenbourg, P., Harrer, A., Hämäläinen, R., Häkkinen, P., Fischer, F.: Specifying Computer-Supported Collaboration Scripts. International Journal of Computer-Supported Collaborative Learning 2(2-3), 211–224 (2007)

Koedinger, K.R., Aleven, V.: Exploring the assistance dilemma in experiments with Cognitive Tutors. Educational Psychology Review 19(3), 239–264 (2007)

Koedinger, K.R., Anderson, J.R., Hadley, W.H., Mark, M.A.: Intelligent tutoring goes to school in the big city. International Journal of Artificial Intelligence in Education 8, 30–43 (1997)

Kollar, I., Fischer, F., Hesse, F.W.: Computer-supported collaboration scripts—a conceptual analysis. Educational Review 18(2), 159–185 (2006)

Kumar, R., Rosé, C.P., Wang, Y.C., Joshi, M., Robinson, A.: Tutorial dialogue as adaptive collaborative learning support. In: Luckin, R., Koedinger, K.R., Greer, J. (eds.) Proceedings of Artificial Intelligence in Education, pp. 383–390. IOS Press, Amsterdam (2007)

Mathan, S.A., Koedinger, K.R.: Fostering the Intelligent Novice: Learning from errors with metacognitive tutoring. Educational Psychologist 40(4), 257–265 (2005)

McLaren, B.M.: Extensionally Defining Principles and Cases in Ethics: An AI Model. Artificial Intelligence 150, 145–181 (2003)

McLaren, B.M., Scheuer, O., Mikšátko, J.: Supporting Collaborative Learning and e-Discussions Using Artificial Intelligence Techniques. International Journal of Artificial Intelligence in Education, IJAIED (in press)

Moguel, P., Tchounikine, P., Tricot, A.: Supporting Learners' Self-Organization: an Exploratory Study. To appear in the Proceeding of the International Conference on Intelligent Tutoring Systems ITS 2010 (2010)

Ravi, S., Kim, J.: Profiling Student Interactions in Threaded Discussions with Speech Act Classifiers. In: Luckin, R., Koedinger, K.R., Greer, J. (eds.) Proceedings of the 13th International Conference on Artificial Intelligence in Education, AIED 2007, pp. 357–364. IOS Press, Amsterdam (2007)

Rosé, C., Wang, Y.-C., Cui, Y., Arguello, J., Stegmann, K., Weinberger, A., Fischer, F.: Analyzing Collaborative Learning Processes Automatically: Exploiting the Advances of Computational Linguistics in CSCL. International Journal of Computer-Supported Collaborative Learning (IJCSCL) 3(3) (2008)

Rummel, N., Spada, H.: Instructional support for collaboration in desktop videoconference settings: How it can be achieved and assessed. In: Bromme, R., Hesse, F.W., Spada, H. (eds.) Barriers and biases in computer-mediated knowledge communication and how they may be overcome, pp. 59–88. Springer, New York (2005a)

Rummel, N., Spada, H.: Learning to collaborate: An instructional approach to promoting collaborative problem-solving in computer-mediated settings. Journal of the Learning Sciences 14(2), 201–241 (2005b)

Rummel, N., Diziol, D., Spada, H.: Collaborative learning with the Cognitive Tutor Algebra. An experimental classroom study (2010) (Manuscript under revision)

Saab, N., van Joolingen, W.R., van Hout-Wolters, B.H.A.M.: Supporting communication in a collaborative discovery learning environment: The effect of instruction. Instructional Science 35(1), 73–98 (2007)

Salomon, G., Globerson, T.: When Teams do Not Function the Way They Ought To. International Journal of Educational Research 13, 89–100 (1989)

Slavin, R.E.: Research on cooperative learning and achievement: What we know, what we need to know. Contemporary Educational Psychology 21(1), 43–69 (1996)

Soller, A.: Computational Modeling and Analysis of Knowledge Sharing in Collaborative Distance Learning. User Modeling and User-Adapted Interaction: The Journal of Personalization Research 14, 351–381 (2004)

Srikant, R., Agrawal, R.: Mining sequential patterns: Generalizations and performance improvements. In: Apers, P.M.G., Bouzeghoub, M., Gardarin, G. (eds.) EDBT 1996. LNCS, vol. 1057, pp. 3–17. Springer, Heidelberg (1996)

Suebnukarn, S., Haddawy, P.: Modeling individual and collaborative problem-solving in medical problem-based learning. User Modeling and User-Adapted Interaction 16(3-4), 211–248 (2006)

Tchounikine, P.: Directions to acknowledge learners' self organization in CSCL macro-scripts. In: Haake, J.M., Ochoa, S.F., Cechich, A. (eds.) CRIWG 2007. LNCS, vol. 4715, pp. 247–254. Springer, Heidelberg (2007)

Tchounikine, P.: Operationalizing macro-scripts in CSCL technological settings. International Journal of Computer-Supported Collaborative Learning 3(2), 193–233 (2008)

Tchounikine, P., Mørch, A.I., Bannon, L.: A computer science perspective on TEL research. In: Balacheff, N., Ludvigsen, S., de Jong, T., Lazonder, A., Barnes, S. (eds.) Technology-Enhanced Learning – Principles and Product. Springer, Heidelberg (2009)

Tedesco, P.: MArCo: Building an artificial conflict mediator to support group planning interactions. International Journal of Artificial Intelligence in Education 13, 117–155 (2003)

Tsovaltzi, D., Rummel, N., McLaren, B.M., Pinkwart, N., Scheuer, O., Harrer, A., Braun, I.: Extending a virtual chemistry laboratory with a collaboration script to promote conceptual learning. International Journal of Technology Enhanced Learning 2(1-2), 91–110 (2010)

VanLehn, K., Lynch, C., Schulze, K., Shapiro, J.A., Shelby, R., Taylor, L., et al.: The Andes physics tutoring system: Lessons learned. International Journal of Artificial Intelligence and Education 15(3), 147–204 (2005)

Walker, E., Rummel, N., Koedinger, K.: CTRL: A research framework for providing adaptive collaborative learning support. User Modeling and User-Adapted Interaction: The Journal of Personalization Research (UMUAI) 19(5), 387–431 (2009a)

Walker, E., Rummel, N., Koedinger, K.: Integrating collaboration and intelligent tutoring data in evaluation of a reciprocal peer tutoring environment. Research and Practice in Technology Enhanced Learning 4(3), 221–251 (2009b)

Wegerif, R., McLaren, B.M., Chamrada, M., Scheuer, O., Mansour, N., Mikšátko, J., Williams, M.: Exploring Creative Thinking in Graphically Mediated Synchronous Dialogues. Computers & Education (2009) doi:10.1016/j.compedu.2009.10.015

Witten, I.H., Frank, E.: Data Mining: Practical Machine Learning Tools and Techniques, 2nd edn. Morgan Kaufmann, San Francisco (2005)

Chapter 23
Preserving Learners' Privacy

Esma Aïmeur and Hicham Hage

Département d'Informatique, Université de Montréal, Pavillon André-Aisenstadt,
CP 6128 succ. Centre-Ville, Montréal QC, H3C 3J7, Canada
{aimeur,hagehich}@iro.umontreal.ca

Abstract. E-learning systems have made considerable progress within the last few years. Nonetheless, the issue of learner privacy has been practically ignored. Existing E-learning standards offer some provisions for privacy and the security of E-learning systems offers some privacy protection. Privacy preserving E-learning solutions fall short and still require further development. Additionally, the advent of E-learning 2.0 introduced a whole new set of challenges with regards to privacy preservation. In this chapter we introduce E-learning systems security and privacy preserving approaches, challenges they still face, as well as the challenges brought forth by E-learning 2.0.

23.1 Introduction

When E-learning first emerged, it consisted solely of text, like a book on a screen, and was ineffective and unpopular with learners. Today, E-learning has become richer with multimedia content and more interactive. With E-learning, education is shifting from being Tutor Centered, where the tutor is the center and has access to the resources, and becoming more Learner Centered (Mccombs and Vakili 2005), where the student is the center and the focus of the learning process and has access to a multitude of resources. Although learner centered education is not a novel idea, E-learning, whether by using an LMS (Learning Management System) or an ITS (Intelligent Tutoring Systems) (Brooks et al. 2006; Woolf, 2008) are major contributors to the development and advancement of learner centered education. Indeed, one of the main drives behind E-learning is to personalize the learning experience to the individual learner. As such, in order to tailor the learning experience, E-learning systems take into consideration various factors including the learner's level of knowledge, reasoning method, preferred learning style, cultural background, even the learner's emotional state (Blanchard et al. 2009; Conati 2002; Dolog et al. 2004).

In order to provide such a level of personalization, E-learning systems collect large amounts of information about the learner, information that could be misused, and therefore violating his/her privacy. There are many reasons why learners might need to keep private different parts of their profile, and existing research

R. Nkambou et al. (Eds.): Advances in Intelligent Tutoring Systems, SCI 308, pp. 465–483.
springerlink.com

(Aïmeur et al. 2007; Anwar & Greer 2009; Hage & Aïmeur 2009a) indicates that learners have a preference for privacy in E-learning and tend to perform better. Existing E-learning standards offer some provisions for privacy and the security aspects of E-learning systems do offer some privacy protection; nonetheless it remains unsatisfactory on several levels. On the other hand, privacy preserving E-learning solutions, such as (Aïmeur et al. 2007) and (Anwar and Greer 2009) do satisfy the privacy constraint, but come at a price. Moreover, these solutions are not adequate for E-learning 2.0 and PLEs (Personal Learning Evnironment) (van Harmelen 2006). In particular, with the availability of the numerous tools which are available to the learners, tools that are external to the E-learning system and out of its control, it becomes difficult to protect the learners' information and privacy, which represents a new set of challenges. This chapter highlights the importance of security in E-learning systems, and corroborate the need for privacy. Moreover, it details some approaches to privacy preserving E-learning, their shortcomings and the challenges that still lay ahead. This chapter also provides an introduction to E-learning 2.0 and the new challenges it brings with regards to learner privacy.

The chapter is organized as follows: the next section provides an overview of security in E-learning systems and provides an overview of some common existing threats. The next section introduces privacy preserving E-learning, why privacy is important, some approaches to insure privacy and the challenges to be solved. The next section provides an introduction to E-learning 2.0 and highlights the new challenges it raises with regards to preserving learner privacy, and the last section concludes the chapter.

23.2 Security of E-Learning Systems

Security is an important aspect of E-learning. Indeed, most (if not all) of the E-learning systems and Intelligent Tutoring Systems store information about the learner, and use an underlying layer of communication between the client computer (where the learner is working) and the server (where the application is actually running). In this section we first introduce some notions about security, and then we highlight some underlying threats that need to be considered, from a security point of view.

23.2.1 Pillars of Security

Information security, (in this case the learners' information) is based on three pillars: *Confidentiality*, *Integrity*, and *Availability*. Maintaining the Confidentiality of the information involves protecting the information from unwarranted disclosure, and making sure that only the users with the proper privileges have access to that information. In other words, the user can only access the information he is permitted to. On one hand, the confidentiality of the information is considered during the transfer of the data between the client and the server. Indeed, with the availability

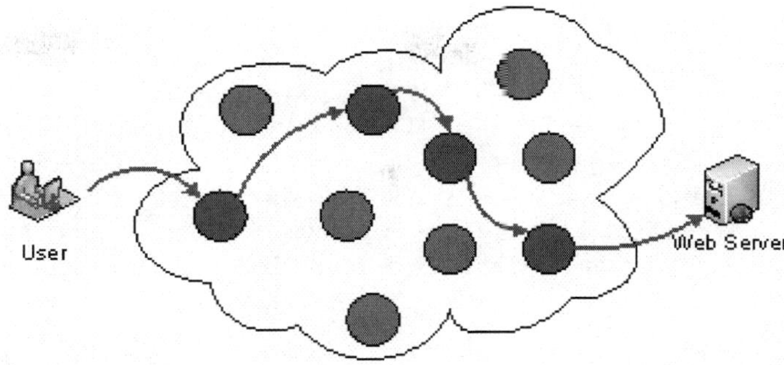

Fig. 23.1 information route from user's PC to web server

of high bandwidth and the speed of the internet connection we tend to forget that in order to reach a certain website, the connection goes through several connection points. Indeed, running a simple "*tracert*" to the website we are trying to reach displays the detailed information about the route taken by any information exchanged between the user's PC and the web server hosting that certain web page. Fig. 23.1 highlights such a route, where the circles in the cloud illustrate possible connection points.

Consequently, imagine a learner sending his login information, or even uploading his homework to the e-learning system: the data could be intercepted and used maliciously by another learner. Similarly, imagine the learner requesting his grade report for the e-learning system: that information could be viewed by an unauthorized person while being sent from the server to the learner. On the other hand, the confidentiality of the information is also considered while it is being stored within the system. Indeed, imagine that any person with access to the registrar's office of your academic institution can also access and view your academic record. Regardless whether you have a good or bad academic record, this is inacceptable. Similarly, the confidentiality of the information stored within the E-learning systems should be guarded, and only the persons with the proper access privileges might have access to that information.

The second pillar is Integrity, which enforces the validity and authenticity of the data. In other words, ensuring information integrity protects the data from any tampering or modifications from unauthorized users. To begin with, the integrity of the information is considered during the transfer of the data between the client and the server. Indeed, consider taking a learner taking an online quiz. The answers to the quiz's questions are sent through the same route described earlier. Without any integrity verification mechanisms, to insure that the data was not modified through the transmission, a malicious user can intercept the answers of the learner and modify them before they reach the e-learning system, successfully tampering with the learner's score. Additionally, the integrity of the information is also considered while it is being stored within the system. Indeed, again in this case, without the proper mechanisms to protect the data integrity, a malicious user with access to the e-learning system could tamper with the information (increasing or decreasing a test score for instance) unnoticed.

The third pillar is Availability, which relates to the availability of the e-learning system. Indeed, such systems must be available at all time, and provisions must be considered to implement and ensure this availability. One might be tempted to think: why is this crucial? Well, consider a learner without access to the system the day homework is due. Not only will the learner not have access to any necessary learning resources available through the e-learning system, but he will not be able to submit his homework. Moreover, consider a learner performing an online quiz. Even if the system was not available for only a few minutes, it is still precious time lost, not to mention the stress and the emotional pressure caused to the learner.

23.2.2 Security Threats

This section highlights *some* of the existing security threats. Actually, what we list here is the tip of the iceberg, and is intended to raise the awareness that when on the internet, we are not as safe and secure as we think we are.

23.2.2.1 SQL Injection

SQL Injection exploits security vulnerabilities at the database level of the system. Such vulnerabilities occur when the user input (data provided by the user) is not properly filtered, allowing the user input to contain executable SQL code. For instance, consider an authentication system that asks the user to provide a user name and a password, and uses the following query to validate the user's credentials:

```
SELECT * FROM user_table
WHERE      user_name = provided_user_name
AND        user_password = provided_password".
```

If the user enters valid values for the variables *provided_user_name* and *provided_password*, the query will work just fine and as expected. Nonetheless, if a malicious user provides the following user name: *"abcd OR 1=1 --"*, the *WHERE* clause of the query becomes: *"WHERE user_name = abcd OR 1=1 -- AND user_password = provided_password"*. In this case, regardless of the password provided by the malicious user, since the "--" is a comment in SQL, the database system will ignore anything that comes after it. Consequently, the query will always return the entire users list from the *user_table* due to the *"OR 1=1"* in the query.

Although the example portrayed here is fairly simple, malicious users using the SQL Injection attack can formulate far more complex queries and do a large amount of damage. Indeed, just to cite a couple of recent events, in August of 2009, the BBC published a story about a US citizen allegedly stealing 130 million credit card numbers using an SQL injection attack (BBC 2009b). More recently, in December of 2009, the New York Times reported on a hacker who accessed, using an SQL Injection attack, the RockYou (rockyou.com) database where he found unencrypted login information for more than 32 million user accounts (O'Dell 2009).

The consequences of a successful SQL Injection attack on an e-learning system are numerous: the attacker could have access to the tutor's resources (upcoming exams or homework, grade books, etc.) or the learner's resource (homework, reports, learning resources, etc.).

23.2.2.2 Cross Site Request Forgery

Cross-Site Request Forgery (CSRF) is an injection type attack, where a malicious web site causes the user's browser to perform unwanted actions on a trusted site. Specifically, the malicious website would try to inject *malicious* requests to the trusted website. For example, consider a user that is logged in to his banking website to pay his bills, while at the same time browsing the malicious website. The malicious website could send a request to the banking site, asking for a money transfer to a specific account held by the attacker Specifically, the malicious website could post an image that links to the website banking site instead, using the following link for example:

"http://mybank.com/transfer?from=account&amount=1000&to=malicious"

It is important to note that such attacks are difficult: the attacker must first gather different information about the targeted site, and the targeted user. Moreover, in order for this attack to happen, the user must be simultaneously have a valid session opened on the targeted site, and be connected to the site of origin of the attack. Nonetheless, these vulnerabilities are real and could have a devastating effect. In this report (Zeller and Felten 2008), a professor from Princeton and his graduate student report on successful CSRF attacks against several popular websites, including ING Direct (ingdirect.com), where they were able to transfer funds out of users' accounts.

A CSRF attack can be used to manipulate the E-learning system into releasing, modifying or even deleting sensitive information. For instance, a learner could manipulate the E-learning system into modifying the grade book in order to increase his own grades.

23.2.2.3 Denial of Service

A *denial-of-service* attack (DoS attack) or *distributed denial-of-service* attack (DDoS attack) is an attempt to overload a computer's resources in order to render it unable to process legitimate users' requests. It is generally conducted against web servers, saturating them with *fake* requests, making them unable to process genuine users' requests. One common method of attack involves overwhelming the target machine by saturating it with *fake* communications requests, such that it cannot respond to legitimate request, or responds so slowly as to be rendered effectively unavailable. A distributed denial of service attack (DDoS) occurs when multiple systems collaborate to flood the resources of the targeted system. Often,

DDoS attacks are conducted using *zombie* machines, computers that were compromised and are now being controlled by the attacker.

In July 2009, South Korea witnessed one of its the largest cyber attacks. DDoS attacks were used to crash the websites of dozens of government offices and banks among others (Lee 2009). Additionally, in August of 2009, Twitter and Facebook were the victims of similar attacks. While Twitter was taken offline for a while by the attacks, Facebook's service was reduced (BBC 2009a). Such attacks are quite common and usually used for extortion purposes (Messmer 2010).

Such an attack could also affect the E-learning systems in various ways: slowing down the system during an exam, or even completely crippling the E-learning system effectively disrupting any learning activity.

23.3 Privacy Preserving E-Learning

One of the main advantages of E-learning and Intelligent Tutoring Systems is their adaptability to the learner's specific needs and preferences. Nonetheless, to do so, these systems collect large amounts of information about the learner, information that could be misused, and therefore violating his *privacy*, which is the claim of individuals to determine what information about themselves is known to others, as well as when and how it is used (Westin 1967).

Although the security of E-learning system is imperative to preserve privacy, it is not enough. Indeed, security will protect learners' information against unwarranted access, but not against abuse from authorized access. Specifically, the insuring the *Integrity* and *Confidentiality* of the learner's information does protect the learner's data (and consequently his privacy) from unauthorised access, nonetheless, E-learning systems gather large amounts of information about the learners, information that is readily made available for the tutor, or even E-learning platform system administrator.

Specifically, *privacy* is nearly absent in current E-learning systems. Only primitive forms of privacy are offered in some platforms, for instance not allowing tutor access to certain information such as auto-evaluations performed by the learners. Nonetheless, the tutor has access to virtually all the remaining information including, but not limited to, who the students are, what parts of the course they referred to, how many times and for how long, as well as all the messages in the forums, and all the information about the quizzes and tests the learner took in his course. While learners' privacy is largely ignored within E-learning, it remains an important aspect for learners.

23.3.1 Why Privacy Preserving E-Learning

Other than the case of Head-in-the-sand privacy (by which the learner wants to keep secret his ignorance even from himself), learners might need to keep private different parts of their profile for *personal*, or *competitive* reasons. In the **Competitive** context, the learner requires his privacy due to competitive considerations. For example, consider a prominent politician taking a course to increase his knowledge in a certain domain of interest to the electors. Other than for protecting himself from any prejudice from the part of the tutor, he has the right and interest in keeping this fact hidden, and his performance results private, from public knowledge and scrutiny, especially from his opponents. As another example, consider a company that uses E-learning for employee training purposes. If competitors have knowledge of the training and the performance of the employees, it could seriously affect the competitiveness of the company and its reputation, especially if the employees performed poorly. On the other hand, in the **Personal** context, the learner requires his privacy due to personal considerations. For example, he may wish to protect himself from a biased tutor. The bias of the tutor might stem from prejudice or stereotyping, based on a previous encounter with the learner, or even from personal reasons. Another reason a learner would prefer to keep his privacy is the increased pressure and stress due to performance anxiety; a learner might feel more comfortable and relaxed knowing the tutor will not know how he performed in the test.

Indeed, existing research demonstrates the effect of emotions on learning (Zins et al. 2007): positive emotions improve the performance whereas negative emotions hinder the thought processes. Additionally, studies are conducted to evaluate the impact of various factors on the learner's emotional state. The purpose of these studies is to avoid situations which create negative emotions, while motivating the occurrence of situations which create positive emotions.

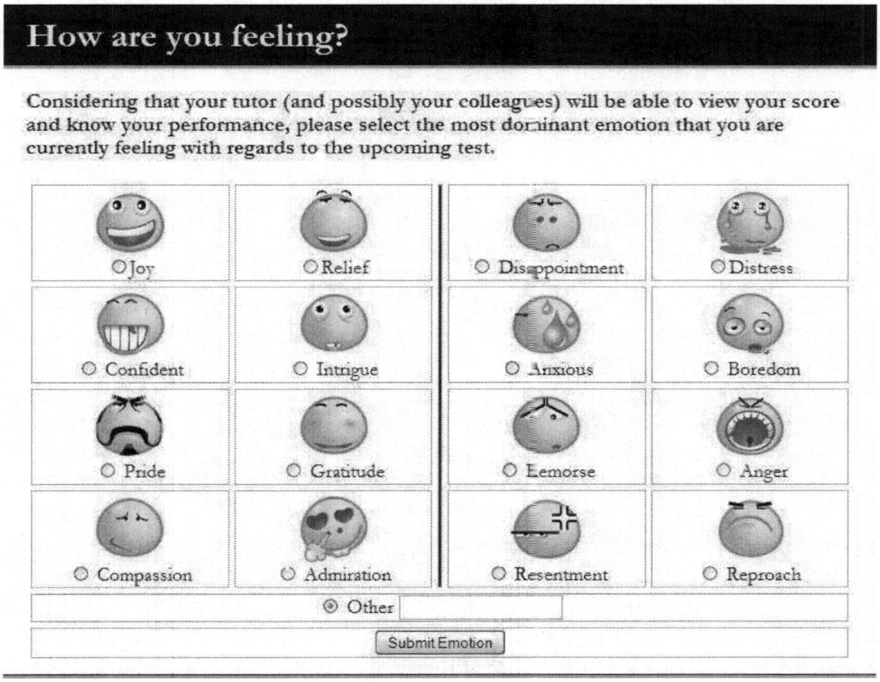

Fig. 23.2 Capturing the participant's most dominant emotion (Hage and Aïmeur 2009b)

In tandem, learners have conveyed a clear preference to privacy in E-learning systems (Aïmeur et al. 2007), and reported being more comfortable engaging in course related forums in privacy preserving mode (Anwar and Greer 2009). Additionally, in a recent study (Hage and Aïmeur 2009a), we investigated the impact of privacy on the leaner's emotions, and whether privacy had a positive or negative impact on learners. Specifically, in this the study, we attempt to determine, in the context of a web-based assessment, whether privacy would have a positive effect (effectively reducing stress and helping learners perform better), or a negative effect (learners would become reckless and careless about their grades). There was a total of 77 participants in the experiment which consisted of two IQ tests, one performed in a traditional none private environment, and the other in a privacy preserving environment. In order to preserve the privacy of the participants, a *random id* (*rid*) was created to and used instead of the actual participants' identifier. Consequently, the participants were informed that all their actions within the privacy preserving environment are recorded using the *rid* which cannot be linked back to them. Nonetheless, in order for us to be able to evaluate the impact of privacy, we had to *deceive* the participants and maintain an actual link between the participants profile and his *rid*.

Moreover, the participants' most dominant emotion was recorded before and after each test (Fig. 23.2) in order to determine the effect of their emotions on the score as well as their attitudes towards their performance in the tests.

In summary, on average, participants performed better in the privacy preserving test (higher score and lower average response time). Additionally, the effect of the negative emotions on the performance of the participant was lower in the privacy preserving environment. In details, the participants were separated into two groups: the first group was composed of the participants who reported a *positive* emotion prior to the test, and the second was composed of the participants who reported a *negative* emotion prior to the test. We then compared the averages of each group. The group which reported a positive emotion, on average, performed better on both tests (with and without privacy). Hence, privacy preserving E-learning is not just necessary to protect learners' information, but can also enhance the learning experience. Consequently there were some proposed approaches.

23.3.2 Existing Approaches

E-learning systems use information about a learner in order to adapt the learning activity and the interactions of the E-learning system. Such information is referred to as the learner profile or learner model. Many E-learning systems use their own internal representation of the learner model. Nonetheless, there are several standards and specifications to represent the learner model, including the IEEE LTSC Personal and Private Information draft standard (LTSC) and the IMS Learner Information Package (IMS). Although these specifications contain some attributes and means that may uphold learner privacy, the detailed specification is still missing. Moreover, the learner involvement in deciding which information is private or not is not enabled (Jerman-Blazic and Klobucar 2005), consequently the learner has no control over which parts of his information is private, and which is public.

On the other hand, there were concerns raised with regards to security. There exists literature, such as (Franz et al. 2006), on how to achieve basic security requirements: *confidentiality, integrity* and *access control*. The security of existing E-learning systems (such as Blackboard, WebCT, or Atutor) does provide a certain level of privacy. As such, **integrity** guarantees that the data is not maliciously or accidentally tampered with or modified: for example, when the learner submits his test, he requires the guarantee that his test answers are not modified after his submission. Moreover, **confidentiality** assures that the data and information is kept secret and private and is disclosed only to the authorized person(s): for example, test scores must be accessible only to the appropriate tutor. The confidentiality of the information is considered at two different stages: while it is being transmitted to/from the E-learning system, and when it is stored within the E-learning system. In the first case, the data can be encrypted such that only the appropriate receiver can read the data. In the second case, **access control** mechanisms can be employed to restrict access to the data. Access control cannot totally guarantee the privacy of the learner: first of all, it does not protect against a *super user* with full access privileges. Moreover, none of the previously mentioned security mechanisms can be used to observe the *core* of the definition of privacy, in such that the learner has no control on what information about him is being gathered by the E-learning system and how it is used. Although Privacy Policies have been provided for this purpose (Yee and Korba 2003), they cannot restrict unwanted access to the data.

Consequently, other approaches were proposed. (Anwar and Greer 2008) proposes a privacy mechanism based on identities. In particular, a learner can have different identities, or personas, that he could use within the different parts of the E-learning system. As long as the learner does not divulge his real identity, and the personas he is using are not linked to each other, or to the learner in question, his anonymity is insured, thus protecting his privacy. Another approach proposed in (Aïmeur et al. 2007) starts by proposing 4 different levels of privacy: *No Privacy*, *Soft Privacy*, *Hard Privacy* and *Full Privacy*. Each level of privacy protects different aspects of the learner's profile. Another dimension that is also considered, which is independent of the learner's personal data, is the tracking of learners within a course. Indeed, learners' activities within the system could be tracked, and a dossier could be built, even though their information within the system are protected. Hence, in addition to the privacy levels, (Aïmeur et al. 2007) also introduces 4 tracking level: *Strong Tracking*, *Average Tracking*, *Weak Tracking* and *No Tracking*. Each level of tracking reduces the amount of trace left by the learner within the E-learning system. In order to satisfy these various privacy and tracking levels, (Aïmeur et al. 2008) proposes *Anonymous Credentials for E-learning Systems* (ACES), a complete set of protocols, relying mainly on blind digital signatures and anonymous credentials, to preserve the learners' privacy.

23.3.3 Challenges

The previous sections presented why the need for privacy in E-learning, and highlighted several approaches to achieve that goal. Yet these existing approaches do have their weaknesses, and this section details some of the major common shortcomings of the existing solutions, shortcomings that need to be addressed in order to have an effective privacy preserving E-learning.

One such weakness is the *overhead* produced by the privacy preserving mechanisms. Indeed, regardless of the chosen solution, preserving the privacy of the learner creates a computational and an operational overhead. Indeed, the cryptographic protocols used to protect the learner's privacy require significant amounts computational resources from the server hosting the E-learning platform, amounts that would grow with the increasing number of learners using the system. On the other hand, in order to preserve their own privacy, learners are required to perform additional operations. Indeed, privacy does come with a price, and learners are required to participate in the management of their information, whether by maintaining their identities, or their own anonymous credentials. Note that the higher the level of privacy required by the learner, the more complicated the privacy preserving approach will become, which implies a higher incurred overhead.

Another shortcoming of privacy is its impact on personalization. Indeed, most of the information gathered on learners within the E-learning system is used in order to personalize the learning experience, capitalizing on the learner's strength, while targeting his weaknesses, and thus tailoring the learning experience according to the learner's learning needs and style. Consequently, the personalization of the learning will be impacted by the fewer available information about the learner (due to his privacy preferences). Indeed, privacy and personalization are like two opposite forces *pulling* the learner's information: the first is pulling to hide it, whereas the second is pulling to gather more of it, in order to better personalize the content. It is a big challenge to find the middle ground such as to satisfy both the privacy and personalization needs.

Another aspect of privacy that requires further investigation is its *impact* on the learner. Indeed, although thus far, the existing research tends to demonstrate that privacy has a positive impact, to the best of our knowledge there were no studies conducted to evaluate the long impact of privacy on the learners. This lack of certitude, whether privacy has a positive or negative impact, is another weakness of privacy in E-learning. Indeed, privacy *might* provide this false sense of security: knowing that as long as you do enough to get the average and pass, no one will know. Consequently, learners might lose their motivation to perform, and they could become more nonchalant, or indifferent to the learning.

The challenges raised in this section relate to privacy preserving solution for E-learning systems. Nonetheless, the advent of what is commonly referred to as E-learning 2.0 raises a new set of challenges with respect to protecting learner privacy.

23.4 Privacy and E-Learning 2.0

E-learning 2.0 does not refer to a new class of LMS (Learning Management Systems) or a new educational technology. Rather it is a natural consequence of changes in how tutors and learners perceive learning in general. Indeed, in recent

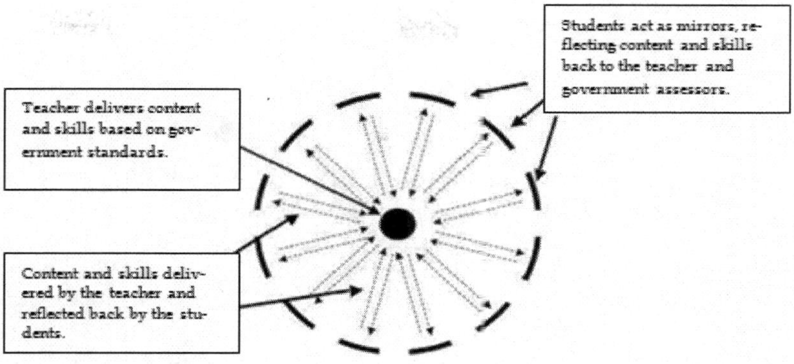

Fig. 23.3 Tutor-centered education (Webilus 2008)

years, education has been shifting from being *tutor-centered*, to being *learner-centered*. In tutor-centered education (Fig. 23.3), the tutor is the active participant in the educational process and learners are considered as passive receptacles of knowledge. Tutor-centered education is a *one size fits all* approach.

On the other hand, in learner-centered education (Fig. 23.4), the learners have access to a variety of knowledge sources and the tutor places more emphasis on what learners can contribute to the educational encounter.

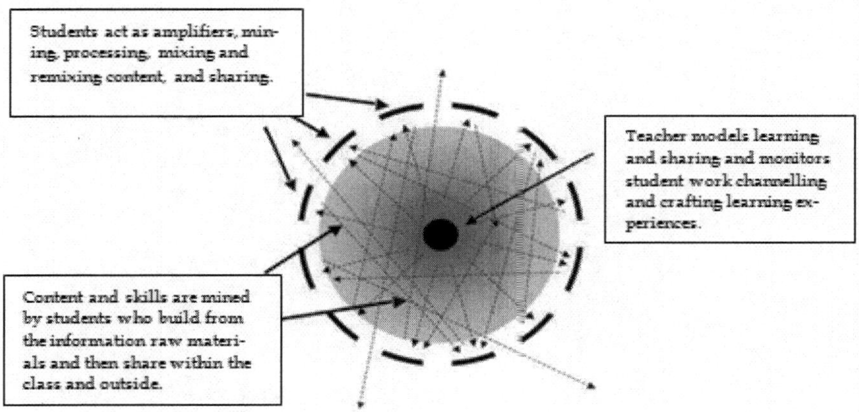

Fig. 23.4 Learner-centered education (Webilus 2008)

It is important to note that *E-learning 2.0 is not a consequence of Web 2.0*. Indeed, both share the same basic concept where the user/learner is not only a spectator and a simple consumer of information, but rather an active participant in the creation of such information. As such, one can view *Web 2.0* tools and technologies as a *natural recourse* to achieve *learner-centered* education.

We will start by defining Web 2.0, then highlight how Web 2.0 is used in E-learning, and then describe the impact on privacy and the challenges.

23.4.1 Web 2.0

Although the term Web 2.0 suggests a new version of the World Wide Web, it does not refer to an update or any technical specifications, but rather to changes in the ways software developers and end-users perceive and use the web. Indeed, the term Web 2.0 refers to a perceived second generation of web-based communities and hosted services (such as blogs, Wikis, etc.) which aim to facilitate creativity, and to promote collaboration and sharing between users.

In short, the following point summarizes the difference between Web 1.0 and Web 2.0: publishing vs. participation. Specifically, in Web 1.0 (*publishing*) the content is controlled by the publisher, and the users are just the recipient of the information. Whereas in Web 2.0 (*participation*) the users are no longer passive recipients of information, but are active participants in the creation of such information, participating in Wikis, tagging, rating, sharing, and/or referring websites. A recently published report (Lenhart et al. 2007) indicates that 64% of online teenagers in the US, ages 12 to 17, engage in at least one type of content creation.

There are three pillars to Web 2.0: the Social Web, Service Oriented Architecture (SOA) and Rich Internet Application (RIA).

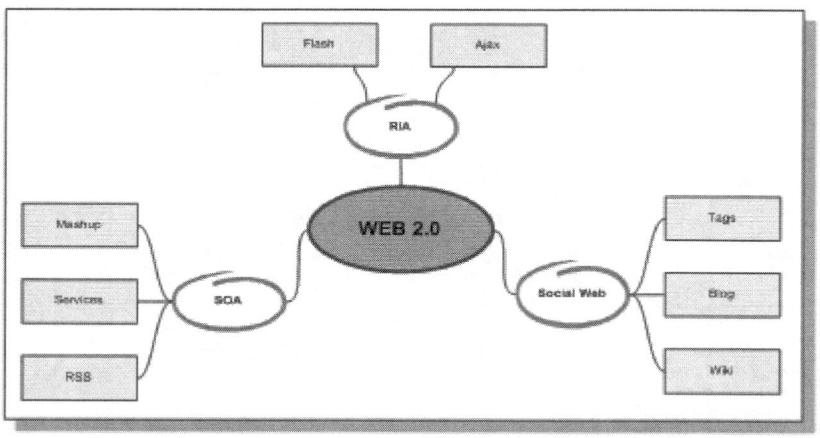

Fig. 23.5 The three pillars of Web 2.0, adapted from (Webilus 2008)

The **Social Web** refers to the "social interactions" between the users of the web, and the resulting virtual "social groups". It allows users to share their writings, videos, photos, and more with their friends, family, colleagues, or the public at large. For instance, the Social Web includes simple publishing through a blog or a wiki. As such, in the case of the blog the owner of the blog and his *faithful* readers can become a social circle where the readers can comment on the blog posts, or each other's comments. Similarly, with the Wiki, the users who regularly visit, contribute to, or maintain the Wiki become a virtual social community centered on the Wiki.

The main drive behind the Social Web is collaboration and the harnessing of collective knowledge. Common features that exist in the Social Web, such as *tagging*, *rating*, *comments* and *recommendations*, exploit and share the knowledge and experiences of the users. As an example, we will consider social bookmarking sites, such as delicious.com or StumbleUpon.com. Such sites enable users to bookmark their favourite web sites, recommend and share these bookmarks with other users, or a community of friends.

Rich Internet applications (RIAs) are web applications that provide functionalities and interactions similar to desktop applications. Typically, RIAs are delivered through browser add-ons or directly through the webpage using for instance Ajax or Macromedia Flash. To illustrate RIAs, consider for instance Google documents (http://docs.google.com) which provides a decently complete set of tools to create and manage documents, spreadsheets, presentations and even forms. The whole set of tools is web based, that is accessible through the browser.

On the other hand, there is a multitude of web pages that illustrate the use of RIA, including web-based virtual computers, such as G.ho.st (http://g.ho.st/). Such environments provide a virtual computer environment, accessible online using any browser, which provides the functionalities and tools or a regular computer, including disk space (5 Gbytes in the case of G.oh.st), a media player, and even an office suite to create, and store documents spreadsheets and presentations.

Service Oriented Architecture (SOA) is an architectural style where the main goal is to relax the dependencies between various components and to achieve loose coupling. Specifically, a *service* is a task performed by the *service provider* to achieve a desired end result for a *service consumer*. Consequently, a service-oriented architecture is a collection of services (service providers and consumers), where these services communicate with each other. Such communication could be just simple data passing or it could involve two or more services coordinating to perform a certain activity. Note that the service provider can also be a service consumer. The flexibility and interoperability of SOA and web services has lead to a new type of web applications called Mashup. Specifically, a mashup describes a Web application that combines multiple services and/or data sources into one single application.

*23.4.2 Web2.0 and E-Learn*ing

This section highlights some examples of "Web 2.0" tools and websites designed for, and used in learning. For instance, a webcast consists of distributing media content over the using streaming media technology. A webcast may be distributed live or on demand. In essence, webcasting is "broadcasting" over the internet. A simple example of webcasting is a TV station that simultaneously streams over the internet the show being broadcasted on TV. On the other hand, a podcast is a series of media content made available via syndication, such as RSS. Dedicated software applications, known as podcatchers automatically identify and retrieve new available media files. The utility of webcasts and podcast in E-learning is very clear: tutors can either webcast their lectures live to students, or the lectures could be made available on demand or through a podcast. Note that a lecture can consist of various media, such as audio only, a slide presentation with audio, a recording of the tutor, etc.

Currently, webcasting and podcasting are being used in several universities worldwide (Shim et al. 2007). It is important to note that webcasting and podcasting are not just used by virtual universities, but also as a complement to lectures in *traditional* classrooms, for instance, Berkely makes publicly available webcasts of several courses (available at http://webcast.berkeley.edu), consisting of either an audio recording of the tutor's lecture, a video recording of the tutor giving his lecture, or a slide presentation of the lecture with the explanations of the tutor.

Alternatively, wikis are websites that generally allow visitors at large to modify their content. Nonetheless, wikis generally can support authentication, such that certain members can modify only certain pages. This feature is important since it enable the use of wikis in group work assignments. Wikis offer the possibility of central access for all the users or limited user groups, which makes it an ideal choice for running projects, drafting documentations and other group work. As such, wikis are used to promote team work and collaboration between students (Raitman et al. 2005). Alternatively, wikis can also be employed by tutors to collaborate on creating learning content. For instance, wikiversity.org offers tutors the chance to collaborate and create freely available learning resources, where currently, on the English site of wikiversity, there are more than 10,000 pages available, covering various topics.

Similarly, *SuTree.com* and *eduSLIDE.net* offer both learners and tutors access to a variety of learning resources. Specifically, SuTree.com offers a variety of how-to videos, ranging from learning how to whistle, to following a complete course watching MIT lectures. eduSLIDE allows tutors to create lessons (presentations) and group them into courses, making these courses available for learners.

Additionally, many existing "web 2.0" pages and tools can help learners during the learning process. For instance, *Footnote.com* allows students to access primary source documents and photos, and to easily create and post online history reports. Moreover, *VoiceThread.com* can be used by both tutors (to create lessons) and learners (for homework purposes) to upload pictures and create an audio narrative to go along with them. *VisualThesaurus.com* offers, as its name indicates, a visual thesaurus. Specifically the lookup word is presented in the center of the graph, and edges connect the lookup word with its synonyms. A color code is used on the edge connecting the word to its synonyms to indicate whether the synonym is a noun, verb, adjective or an adverb. Moreover, the edge connecting the lookup word with its antonym is presented differently. *Wayfaring.com* is a mashup that uses Google maps to list podcasts and webcasts from about 68 universities worldwide. *wePapers.com* allows users to share academic papers, ranging from research papers, tutorials, lectures, to tests and exams. Moreover, users can comment, and even ask questions to the community about these papers. Another useful browser add-on is Diigo (http://www.diigo.com/). Diigo provides learners with the ability to highlight specific parts of webpages, add sticky notes and comments (private or public) to the highlighted sections or the whole page, and learners can share the highlights and notes with their Diigo social network.

Moreover, existing systems rely on the learners' collaboration and social network to enhance the learning experience. For instance, *Knowledge Sea II* (Brusilovsky et al. 2005) treats research papers as regular pedagogical resources, allowing users to annotate and review these resources, and using the annotations to perform the recommendations. Comtella (Vassileva 2004) is another academic system that uses P2P (peer to peer) technology to enable students to share research papers. In addition, Comtella employs a *reputation* scheme (Mao et al. 2007) to motivate and award the students. On the other hand, SHAREK (Hage and Aïmeur 2008) is designed to enable learners to attach external learning resources to the tutor defined learning content within the E-learning system.

The proliferation of tools and websites such as listed earlier has led to the concept of Personal Learning Environment (PLE) (van Harmelen 2006). PLE is a combination of tools and processes, whether formal or informal, which learners use to gather information, reflect on it and work with it. The appeal of PLE for learners relies in the fact that they can choose the tools that best suit their preferences. An interesting representation we came once across compares a Learning Management System (LMS) and a Personal Learning Environment (PLE) using the following analogy: an LMS is similar to a Swiss army knife containing a set of tools, some of which you might never used. On the other hand, a PLE is like having a box containing the tools you use, but most importantly tools that you chose and prefer. Indeed, although it might be more practical to fit a large set of tools into your pocket (Swiss army knife analogy), having only the specialized tools that you are comfortable with does have it advantages.

Many PLE advocates portray an LMS as being inflexible and used to control the learning and the learner, whereas a PLE is portrayed as easy to use, personalized, and liberated. In short, LMS is equivalent to controlling how you learn, whereas PLE corresponds to giving you control over how you learn. Although controlled and passive learning reduces self reliance and causes loss of curiosity and creativity, an uncontrolled education would create a shortage of certified labor and would introduce unqualified people into the labor pool. Currently, this is where E-learning stands today (Fig. 23.6).

The Tutor delivers the learning content to the learner through the LMS. On the other hand, the learner has access to the controlled environment provided by the LMS as well as a PLE containing the set of his favorite tools and resources, which are external to the LMS. As such, the leaner can *freely* perform the learning activity, relying on the content and tools provided through the LMS, and on external *uncontrolled* resources through the PLE. In addition, the learner has access to both his personal social network (outside the LMS), and a peer network through the LMS. Note that some peers can also be part of the learner's external social network. In such a scenario, the tutor *controls* the curriculum (which courses and topics the learner must complete), and he can *validate* the learner's knowledge through assessments. On the other hand, the learner has the freedom to choose *how* to complete the learning activities: whether by solely using the content and

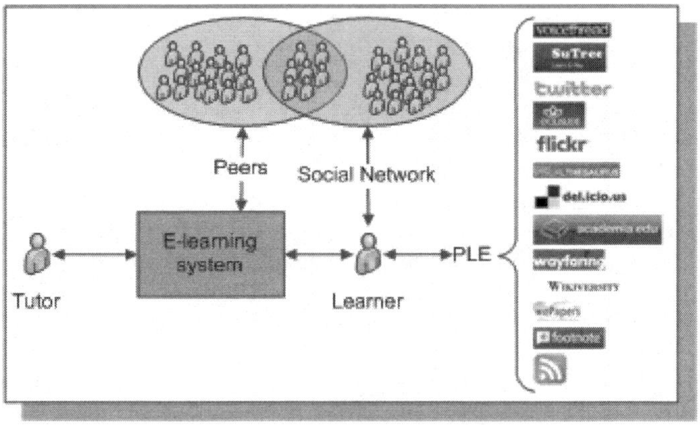

Fig. 23.6 Using LMS and PLE for education

tools provided through the LMS, by relying completely on his PLE, or a combination of both. In the last case, the LMS can be actually viewed simply as another component of the PLE.

Most of the PLE components are external to the E-learning system. Consequently, the educational institutions using an E-learning system, as well as the entity developing it, have no control over these components. This raises an enormous challenge with regards to ensuring and maintaining the learners' privacy.

23.4.3 Impact on Privacy and the Challenges

The challenges brought forth by the decentralized nature of PLEs with regards to learners' privacy are numerous. The first major concern is with regards to the security of these components. Indeed, as detailed in the section "*Pillars of security*", the Confidentiality and Integrity of the learner's information are imperative to protect his privacy.

Specifically, consider a learner using *delicious.com*, a social bookmarking website, to organize a list of websites and resources he is using to help prepare a report. If the confidentiality of that information is compromised, other learners – probably his classmates – might use the same resources in their report, effectively reducing his chances at a better grade. Similarly, the integrity of that information is also important: corrupting or altering these bookmarks would cause delays in preparing the report. Moreover, consider a learner using *zotero.org* to capture and organize references for a research paper he is working on. Again, the confidentiality, as well as the integrity of the information, is important.

The same need for confidentiality and integrity (not to mention availability) is necessary for any these components that are available for learning, and assuming that most do provide an acceptable level of security, would that be enough to protect the learners' privacy!? The answer is a resounding NO! Indeed, as with E-learning systems, the security is important factor to preserve privacy, but is not enough. Specifically, in this case it is harder to protect the learners' privacy, because first, some of his information will be replicated across the various components he is using as part of his PLE. Indeed, most of these systems require registration, and ask for personal information, such as name, sex, email, etc. Having that information replicated all over the internet increases the risk of having parts of it, if not all, disclosed to unwarranted parties. Moreover, when using such systems, learners are providing large amount of information about themselves, and yet, are *blindly* trusting these systems not to disclose their information (Conti 2008). The use of solutions such as OpenID (openid.net) does help reduce the spread the learner's information across various systems, but it does not guarantee the *privacy* of the learner. Indeed, not all the components used for E-learning do support systems such as OpenID, and even when such support is available, the learner is not in control of his information, and he can still be easily tracked through the various systems.

23.5 Conclusion

Today, E-learning offers rich multimedia content and is more interactive. Moreover, E-learning is very flexible: students can choose instructor-led or self-study courses and they can select from a variety of learning tools that best fit their style. Indeed, one of the main advantages of E-learning is its adaptability to the learner's specific needs and preferences. Nonetheless, to do so, the E-learning systems collect large amounts of information about the learner information that could be misused, and therefore violating his *privacy*. The security of E-learning systems offers is imperative to safeguard the information stored within the system, and is essential to preserve privacy. Nonetheless, security alone is not enough and various solutions for privacy preserving E-learning were proposed, some relying on identities, others on anonymous credentials. Although these solutions are technically sound, they do fall short: they introduce a computational and operational overhead, influence the personalization of E-learning systems, and its effect (whether positive or negative) on learners attitudes is still not entirely determined. Preliminary research attributes a positive effect to privacy on learners, but this is a point that requires further investigation. Consequently, privacy preserving E-learning must be able to balance privacy, with access to learners' necessary information required to personalize the learning content and experience, while reducing the overhead incurred by privacy preserving mechanism. Alternatively, the advent E-learning 2.0 and the widespread use of PLEs introduced a new set of challenges that need to be addressed to ensure learner privacy. Indeed, learners regularly use and access resources external to the E-learning system or classroom. These resources are not controlled by the educational institution, and consequently are harder to supervise, increasing the risk to learner's privacy. Moreover, since most of these external resources require learners to register, their personal information will be redundantly duplicated, increasing the risk of unwanted disclosure of that information. Although not everybody will embrace our wish for privacy, as many would agree, we consider privacy to be a fundamental human right: it is not negotiable! Since learning is as important, further research to concord privacy and E-learning is imperative.

References

Aïmeur, E., Hage, H., Mani Onana, F.S.: A Framework for Privacy-Preserving E-learning. In: Joint Itrust and PST conferences on Privacy, Trust Management and Security (IFIPTM 2007), Moncton (2007)

Aïmeur, E., Hage, H., Mani Onana, F.S.: Anonymous Credentials for Privacy-Preserving E-learning. In: The Montreal Conference on eTechnologies 2008, MCETECH 2008, Montreal (2008)

Anwar, M., Greer, J.: Role and Relationship-Based Identity Management for Private yet Accountable E-Learning. In: Joint iTrust and PST Conferences on Privacy, Trust Management and Security, IFIPTM 2008, Trondheim (2008)

Anwar, M., Greer, J.: Implementing Role-and Relationship-based Identity Management in E-learning Environments. In: 14th International Conference on Artificial Intelligence in Education, AIED 2009, Brighton, pp. 608–610 (2009)

BBC, Hackers hit Twitter and Facebook. BBC News (2009a), `http://news.bbc.co.uk/2/hi/8188201.stm` (Retrieved)

BBC, US man 'stole 130m card numbers' BBC News (2009b), `http://news.bbc.co.uk/2/hi/americas/8206305.stm` (Retrieved)

Blanchard, E., Roy, M., Lajoie, S., Frasson, C.: An evaluation of sociocultural data for predicting attitudinal tendencies. In: 14th International Conference on Artificial Intelligence in Education, AIED 2009, Brighton, pp. 399–406 (2009)

Brooks, C.A., Greer, J.E., Melis, E., Ullrich, C.: Combining ITS and eLearning Technologies: Opportunities and Challenges. In: Ikeda, M., Ashley, K.D., Chan, T.-W. (eds.) ITS 2006. LNCS, vol. 4053, pp. 278–287. Springer, Heidelberg (2006)

Brusilovsky, P., Farzan, R., Jae-wook, A.: Comprehensive personalized information access in an educational digital library. In: Proceedings of the 5th ACM/IEEE-CS Joint Conference on Digital Libraries, JCDL 2005 (2005)

Conati, C.: Probabilistic assessment of user's emotions in educational games. Journal of Applied Artificial Intelligence 16(7-8), 555–575 (2002)

Conti, G.: Googling Security: How Much Does Google Know About You? Addison-Wesley Professional, Reading (2008)

Dolog, P., Henze, N., Nejdl, W., Sintek, M.: Personalization in Distributed eLearning Environments. In: 13th World Wide Web Conference, New York, pp. 170–179 (2004)

Franz, E., Wahrig, H., Boettcher, A., Borcea-Pfitzmann, K.: Access Control in a Privacy-Aware eLearning Environment. In: International Conference on Availability, Reliability and Security, ARES 2006, Vienna, pp. 879–886 (2006)

Hage, H., Aïmeur, E.: Harnessing learner's collective intelligence: a Web2.0 approach to E-learning. In: Woolf, B.P., Aïmeur, E., Nkambou, R., Lajoie, S. (eds.) ITS 2008. LNCS, vol. 5091, pp. 438–447. Springer, Heidelberg (2008)

Hage, H., Aïmeur, E.: The impact of privacy on learners in the context of a web-based test. In: 14th International Conference on Artificial Intelligence in Education, AIED 2009, Brighton, pp. 65–72 (2009a)

Hage, H., Aïmeur, E.: The impact of privacy on learners in the context of a web-based test. In: 14th International Conference on Artificial Intelligence in Education, AIED 2009, Brighton (2009b)

IMS, IMS Global Learning Consortium, http://www.imsproject.org/ (Retrieved, September 2007)

Jerman-Blazic, B., Klobucar, T.: Privacy provision in e-learning standardized systems: status and improvements. Computer Standards & Interfaces 27(6), 561–578 (2005)

Lee, J.: Cyberattack rocks South Korea. GlobalPost (2009), http://www.globalpost.com/dispatch/south-korea/090710/cyberattacks (Retrieved)

Lenhart, A., Madden, M., Macgill, A.R., Smith, A.: Teens and Social Media (2007), http://www.pewinternet.org/PPF/r/230/report_display.aspU (Retrieved, January 2008)

LTSC Learning Technologies Standardization Committee (LTSC), http://www.ieeeltsc.org/ (Retrieved, September 2007)

Mao, Y., Vassileva, J., Grassmann, W.: A System Dynamics Approach to Study Virtual Communities. In: 40th Annual Hawaii International Conference on System Sciences, HICSS 2007, Waikoloa, Hawaii, p. 178a (2007)

Mccombs, B.L., Vakili, D.: A Learner-Centered Framework for E-Learning. Teachers College Record 107(8), 1582–1600 (2005)

Messmer, E.: DDoS attacks, network hacks rampant in oil and gas industry, other infrastructure sectors Network World (2010), http://www.networkworld.com/news/2010/012710-ddos-oil-gas.html (Retrieved)

O'Dell, J.: RockYou Hacker: 30% of Sites Store Plain Text Passwords The New York Times (2009), http://www.nytimes.com/external/readwriteweb/2009/12/16/16readwriteweb-rockyou-hacker-30-of-sites-store-plain-text-13200.html (Retrieved)

Raitman, R., Augar, N., Zhou, W.: Employing Wikis for Online Collaboration in the E-Learning Environment: Case Study. In: 3rd International Conference on Information Technology and Applications (ICITA 2005), Sydney, pp. 142–146 (2005)

Shim, J.P., Shropshire, J., Park, S., Harris, H., Campbell, N.: Podcasting for e-learning, communication, and delivery. Industrial Management & Data Systems 107(4), 587–600 (2007)

van Harmelen, M.: Personal Learning Environments. In: 6th International Conference on Advanced Learning Technologies (ICALT 2006), Kerkrade, pp. 815–816 (2006)

Vassileva, J.: Harnessing P2P Power in the Classroom. In: Lester, J.C., Vicari, R.M., Paraguaçu, F. (eds.) ITS 2004. LNCS, vol. 3220, pp. 305–314. Springer, Heidelberg (2004)

Webilus (2008), http://webilus.com/toutes-les-images (Retrieved, January 2008)

Westin, A.: Privacy and Freedom. Atheneum, New York (1967)

Woolf, B.P.: Building Intelligent Interactive Tutors, Student-Centered Strategies for Revolutionizing E-Learning. Elsevier & Morgan Kaufmann (2008)

Yee, G., Korba, L.: The Negotiation of Privacy Policies in Distance Education. In: IRMA International Conference, Philadelphia (2003)

Zeller, W., Felten, E.W.: Cross-Site Request Forgeries Exploitation and Prevention (2008)

Zins, J.E., Bloodworth, M.R., Weissberg, R.P., Walberg, H.J.: The Scientific Base Linking Social and Emotional Learning to School Success. Journal of Educational and Psychological Consultation 17(2), 191–210 (2007)

Chapter 24
Infusing Cultural Awareness into Intelligent Tutoring Systems for a Globalized World

Emmanuel G. Blanchard[1] and Amy Ogan[2]

[1] Department of Educational and Counselling Psychology, McGill University, 3700 McTavish Street, Montreal, QC, H3A 1Y2, Canada
emmanuel.blanchard@mcgill.ca
[2] Human-Computer Interaction Institute, Carnegie Mellon University, 5000 Forbes Avenue Pittsburgh, PA 15213-3891, USA
aeo@andrew.cmu.edu

Abstract. In a global economy, with increasing immigration and cross-cultural interaction, the impact of culture in educational settings cannot be ignored. The impact is two-fold: students from diverse cultural backgrounds will be using the same educational technologies, and intercultural competence will become an increasingly important domain of instruction. In response, this chapter introduces what it means to adapt Intelligent Tutoring Systems for users with diverse cultural backgrounds, and how Intelligent Tutoring Systems can be used to support instruction in culture. We then discuss the major research issues involved in modifying Intelligent Tutoring Systems in support of these efforts. To provide insight into the current landscape of the field, we briefly outline several recent research achievements. In conclusion, we highlight significant current and future issues that arise in the integration of cultural concerns and educational technology.

24.1 Introduction to Culturally-Aware Tutoring Systems

24.1.1 Culture and Educational Technology

In the not so distant past, everyday interactions were generally between people from the same geographical region, who shared a common ground in social norms and expectations. Consequently, issues of miscommunication or friction resulting from cultural differences were not a concern for technologists. The era of globalization has seen an increase in immigration and communication between people from diverse cultures. For instance, by 2031, it is likely that one third of Canadian citizens will belong to a minority group, and an equivalent portion of the population is expected to have a mother tongue other than French or English[1]. Similarly

[1] http://www.statcan.gc.ca/pub/91-551-x/91-551-x2010001-eng.pdf

R. Nkambou et al. (Eds.): Advances in Intelligent Tutoring Systems, SCI 308, pp. 485–505.
springerlink.com

the United States is projected to become a "nation of minorities" by 2050[2]. These two national examples highlight the growing need for attention to diversity in technological solutions. Only recently, however, has culture been recognized as an important consideration in software design. "A *'culturally-aware system'* refers to any system where culture-related information has/had some impact on its design, runtime or internal processes, structures, and/or objectives" (Blanchard et al. 2006). There is a growing community of researchers who investigate cultural issues with regards to educational technology.

Education is an area where the impact of cultural diversity has been identified as critical. UNESCO (2007)0 has highlighted several objectives for intercultural education:

1. To respect "*the cultural identity of the learner through the provision of culturally appropriate and responsive quality education for all*",
2. To provide "*every learner with the cultural knowledge, attitudes and skills necessary to achieve active and full participation in society*",
3. To provide "*all learners with cultural knowledge, attitudes and skills that enable them to contribute to respect, understanding and solidarity among individuals, ethnic, social, cultural and religious groups and nations*".

These guidelines provide technologists with a focus for the development of future educational technologies, those that incorporate two important facets of what is termed *cultural intelligence* (Earley and Mosakowski 2004). In the frame of this paper, this intelligence encompasses an ability to understand another's actions and thoughts with regards to his/her cultural specifics, to undertake actions in order to optimize positive as well as limit negative interactions with foreigners, and to develop culturally-informed perceptions of a socio-cultural environment (Blanchard et al. 2006). In line with UNESCO Guideline 1, technologies should respect the culture of the user (i.e., make culturally-intelligent adaptations) by providing examples and utilizing communication schemas that are more familiar to the user. Not only will this provide a more user-centered experience, but may also make learning more efficient. Such systems could also lead to solutions that better ensure students from minority groups have an equal opportunity to learn. Of interest are intelligent tutoring systems (ITS), which are especially suited to adapt to the needs of individual students. Second, following UNESCO Guidelines 2 and 3, educational technology should also strive to provide opportunities to increase the learner's cultural intelligence. While ITS have been very successful in well-defined domains such as algebra or physics, they hold great potential to be adapted to support learning in the domain of intercultural competence. In the following sections, we show how ITS can be used to effectively introduce cultural knowledge, skills, and attitudes, and how ITS can contribute to the larger effort in culturally-aware educational technologies.

[2] http://www.census.gov/Press-Release/www/releases/archives/population/012496.html

24.1.2 Culturally-Adaptive Systems

As with developing any instructional materials, the startup costs of developing a tutoring system are fairly high (Aleven et al. 2006). However, a major advantage of ITS is their ability to be deployed on a wide scale. Furthermore, there is great potential in taking existing systems and using them in new communities and cultures. The majority of ITS research to date has been done in America and Europe, but there are growing possibilities for educational technologies worldwide. Such a globalization objective increases the need for culturally adaptive systems.

Using culturally-relevant educational practices and resources rather than generic ones is supported by a large body of literature. For instance, Biggs (2001) explained that Hong Kong students performed better when pedagogy was targeted towards Hong Kong cultural norms and references. Conversely, students performed worse when Commonwealth-wide educational practices were employed. It is well established that human communication, in all forms, is highly subject to cultural influence (Bonvillain 2008). Indeed, what is a highly relevant communication practice in a given cultural context may have negative effects in another. This reciprocally impacts the Human-Computer Interaction (HCI) methods used in ITS since designers are likely to unconsciously employ cultural schemas in their systems.

Finally, in ITS, adaptation is traditionally achieved by modeling a student profile in order to deduce the tutoring approach that will maximize learning. It is to be noted that many data sources that are currently used in student modeling are known to vary from a cultural group to another. This includes cognitive processing (Nisbett and Norenzayan 2002) and affective management (Mesquita et al. 1997). In the case of a globally-deployed ITS, it is critically important to embed cultural adaptation mechanisms that prevent a system's decision-making from being grounded on an incomplete or false model of the learner.

24.1.3 Developing Intercultural Competence

Cultural understanding is important in many contexts, from language classrooms to business negotiations or service abroad (Landis et al. 2003). Training programs designed to teach these skills have evolved substantially over the past six decades. The earliest examples began to emerge after World War II, when international travel and collaboration became more prevalent in business and government affairs. Typically, the goal of such programs is to induce changes in knowledge, skills, and/or attitudes (Mendenhall et al. 2004). Knowledge includes basic facts about a new culture, such as common values and beliefs, preferences for physical contact, and typical eating and drinking patterns. Skills usually refer to the learner's ability to interact with someone from the new culture, including communicating their desires and interpreting the behaviors of others. Finally, attitudes have to do with basic beliefs a learner has about people of a different culture and whether a positive, neutral, or negative disposition exists towards them. This should eventually lead them towards better integration in cultural contexts, e.g., by exhibiting fewer stereotypes and misconceptions, and employing better communication (Savicki 2008). As the need for these programs became more evident, scientific interest in creating theories of intercultural growth, identifying underlying cognitive processes, and demonstrating their effectiveness also grew.

Surprisingly, very little of this work leverages state-of-the-art educational technology. Intelligent tutoring technology, while not traditionally aimed at such ill-defined domains, could provide a substantial benefit to cultural instruction.

24.1.4 Intersection of Intercultural Competence and Adaptation

Although we describe two different approaches for integrating cultural issues into intelligent tutoring technology, culture is such a pervasive concept that these two approaches are inextricably intertwined. It would be less efficient, and perhaps ineffective to teach about a particular culture the same way with students from different cultural backgrounds: interpretation of historical events, stereotypes and misconceptions, and shared knowledge may vary from one group to another. For example, France and Japan share little common history - their respective spheres of influence were distinct and they had minimal dealings on the international stage. Currently, there is a generally positive image of Japan among the French population. It is usually seen as a peaceful country with a rich and distinct culture. However, when discussing Japan in China, a wealth of stereotypes exist based on their shared history that do not necessarily reflect the modern Japanese reality. An ITS developed for teaching Japanese history or culture would benefit from considering these factors and adapting to them.

It is equally important that designers of educational technology advance their own cultural skills so that they can more readily detect cultural assumptions they make during the course of development. This will help designers to predict cultural variations and discover potential solutions for future targeted cultural groups. For instance, an American designer gaining additional understanding about Chinese culture might find such knowledge helpful in adaptations for other Asian cultural groups, as many share Confucian influences and have collectivist social orientations. In general, increased cultural understanding may enable designers to think about how the interface or interaction scaffolding might contain cultural elements that should be adapted to different contexts.

In the rest of this chapter, we lay out the issues involved in cultural modeling and applications to educational technology, current achievements in the area of culturally-aware tutoring systems (CATS), and finally, a discussion of open questions and future directions of the field.

24.2 The Cultural Domain: An Overview of Common Theoretical Approaches

This section briefly discusses general conceptions of culture and surveys related approaches employed in diverse fields of research. We hope to provide readers and future CATS developers with an opportunity to develop a theory-based understanding of the cultural domain. The variety of paradigms and orientations also makes readers aware that their conceptual choices are likely to have strong implications when designing CATS.

24.2.1 Defining the Cultural Domain

Before discussing the nature of culture itself, it is essential to disambiguate this concept from the notion of *cultural group*. Cultural groups are a coherent and stable ensemble of individuals to which a culture can be associated (Blanchard et al. 2010). The notion of cultural group is frequently simplified in the frame of large human groups such as countries or religions. However, numerous sub-groups do not fall under this definition, but are useful for explaining individuals' behavioral and cognitive characteristics. Hence, any group of individuals coherent enough to develop a specific set of such characteristics can be considered a cultural group. This includes, for example, businesses, communities of interest or practices (e.g., sport fans and carpenters, respectively; for a more complete overview, see (Lave and Wenger 1991)). This provides additional opportunities for the development of CATS. Most individuals today are subject to multiple cultural influences, sometimes referred to as layers of cultural identity (Rehm 2010; Reinecke et al. 2010), which further complicates the development of cultural user models.

With this clarification, a discussion on the nature of culture can be started. A very large number of definitions for culture have been proposed, which are often strongly influenced by the interests of a particular discipline. In cross-cultural psychology, Kashima (2000) states that two schools of thought exist that define culture either as "*a process of production and reproduction of meanings in particular actors' concrete practices or actions or activities in particular contexts in time and space*", or as "*a relatively stable system of shared meanings, a repository of meaningful symbols, which provides structure to experience*".

Cooper and Denner (1998) present several theoretical approaches to the study of culture in human and social sciences. Their focus varies in order to consider, among other things, (a) core cultural ideas and the key role of shared social values in shaping individuals' cognitive, affective and social processes, (b) the interpretation of individuals' characteristics with regards to their surrounding social and material context, (c) the consequences of differences in social position among cultural subgroups in historical and cultural context, or (d) how individuals develop and claim membership in specific socio-cultural group(s) and its implications for intergroup relations.

It should be noted that in most of the approaches presented above, the notion of culture is only considered in terms of cognitive and behavioral implications. However, this approach may be too restrictive from the viewpoint of CATS development. For instance, other domains, such as archaeology or anthropology, heavily consider cultural artifacts. Such information is highly relevant when designing a virtual cultural environment or when looking for concrete examples to include as pedagogical resources.

24.2.2 Distinguishing Universalisms and Cultural Specifics

The two main approaches in cultural studies consist of identifying either universalisms or group specifics. Universalisms are genuine characteristics of human beings and as such, are supposedly shared by a wide cluster of cultural groups (if not all). Group specifics are characteristics specific to cultural groups in that they are understood or endorsed by an important portion of insiders and unknown or considered external by outsiders. Discussing universalisms or group specifics is equivalent to eliciting cultural aspects that unite all human groups, versus those that distinguish each of them. Oversimplification is a key concern when addressing group specifics: a given characteristic of a cultural group is rarely (if not never) shared by all its members (Scharifian 2003). In order to discriminate between cultural groups, scholars frequently suggest attaching a pool of common characteristics to a cultural group (Scharifian 2003) rather than referring to a unique one.

Universalisms have been posited in many aspects of human life including facial expressions of emotions (Ekman 1972), motivation (Ryan and Deci 2000), and politeness (Brown and Levinson 1987) to cite but a few. Cultural specifics are similarly reported along many dimensions, including cognitive (e.g., core cultural ideas, interpretations, beliefs), behavioral (e.g., body language, rituals, good practices), and physical (e.g., artifacts) (see Blanchard et al. 2010). Although frequently presented as universals, empirical research has demonstrated group specifics in such aspects of human life as basic emotions (Mesquita et al. 1997), frequency of personality profiles (Allik and McCrae 2004), basic wellbeing needs (Hofstede 1984), and cognitive processing (Nisbett and Norenzayan 2002).

Some approaches include both universalism and group-specific considerations. *System of values* is a practical approach to describing cultures that emerged decades ago. It consists of identifying universal dimensions of the major orientations of cultural groups (their behavioral and cognitive tendencies) in order to develop group-specific models, thus providing an easy method for cross-cultural comparisons and assessments, and for potentially explaining cultural specifics. At present, the most popular system of values results from the analysis of a cross national survey of more than 100,000 people by Hofstede (2001; 2010). It characterizes more than 70 country cultures by computing their numeric scores for the following five dimensions: a) power distance (PDI: "*the extent to which the less powerful members of organizations and institutions (like the family) accept and expect that power is distributed unequally*"), b) individualism/collectivism (IDV: "*the degree to which individuals are integrated into groups*"), c) masculinity/femininity (MAS: "*the distribution of roles between the genders*"), d) uncertainty avoidance (UAI: "*a society's tolerance for uncertainty and ambiguity*"), and e) long term orientation (LTO: a more recently added dimension referring to a general interest for "*virtue regardless of truth*"). Table 24.1 presents scores of Hofstede's dimensions for a limited set of nations.

Table 24.1 Hofstede's scores for six different nations, taken from (Hofstede 2010).

	PDI	IDV	MAS	UAI	LTO
Brazil	69	38	49	76	65
Canada	39	80	52	48	23
India	77	48	56	40	61
Japan	54	46	95	92	80
U. K.	35	89	66	35	25
USA	40	91	62	46	29

Numerous studies have used Hofstede's framework in different contexts and disciplines (see Kirkman et al. 2006). However Hofstede's method of analysis has been strongly criticized (McSweeney 2002), and competing systems of values have emerged. Of particular note is GLOBE (House et al. 2004), which has garnered considerable attention in recent years and has the advantage of proposing dimensions that discuss cultural issues at both group and individual levels.

However the *system of values* paradigm is not to be considered a perfect solution. Indeed, many researchers continue to argue about its insufficient consideration of individual variations within cultural groups, as well as risks of oversimplification that may result in cultural stereotyping (McSweeney 2002). Furthermore *systems of values* are mainly developed in business-related research fields. Consequently, dimensions that have been identified may not be well adapted for cross-cultural research in other domains (Blanchard 2009; Stewart and Chakraborty 2010). Indeed, a lack of education-focused studies, especially those based on student sampling, can be easily identified in Kirkman's listing of Hofstede's related studies (Kirkman et al. 2006).

In the next section we discuss how incorporating cultural awareness into ITS could impact different aspects of design, development and operation.

24.3 Cultural Implications for ITS Architectures

Culture in curriculum modules. Cultural implications in domain modeling and the design of a curriculum exist at many levels. For systems intending to teach intercultural competence, culture is the domain matter to be modeled and transmitted to learners. It is immediately obvious that capturing and representing the full richness of a culture is infeasible. Even modeling a reasonable representation of a culture, without falling into oversimplifications and stereotypes, is a complex question that merits exploration and research. For instance, cultural elements (references, artifacts, processes) carry multiple interpretations even within a cultural group (due to varying personal experiences of its members and the existence of sub-cultures, among many factors). Reliably reporting this heterogeneity in educational systems requires the development of methods to evaluate sources of cultural information and the quality of their implementation, as well as presentation that captures this complexity.

Additionally, cultural groups frequently develop specific approaches to a domain in terms of representation and pedagogy, among others. Even hard sciences like mathematics carry cultural specifics in representation, such as different symbols and theorem names (Melis et al. 2010). Hence, there is a need for curriculum modules where culturally-concurrent models of a domain could be merged. New ways of retrieving cultural information and representing it in an ITS knowledge base could be imagined. Learning objects that carry the learning content with information on how it varies with cultural specifics, as well as links to culturally-relevant resources, could be organized within cultural knowledge repositories. In addition to variations in knowledge representation, domain knowledge and pedagogical methods are frequently intertwined in modern curriculum modules. Thus, infusing culture into such modules should also enable the domain knowledge to be employed with culturally specific strategies in educational activities. Techniques to retrieve and use information related to the cultural variability of the content are needed. Such considerations have additional implications for ITS tutor modules.

Culture in tutor modules. Cultural specifics are also likely to affect the relationship between the learner and the tutor module of an ITS. For example, some cultural groups greatly value individual behaviors, whereas collective behaviors such as groupwork and consensus-building are of primary importance in others. Similarly, there are cultural contexts where constructive criticism is welcome and fostered by educators, whereas in others, this speech style is perceived negatively as a lack of respect. Such examples call for research on cultural adaptation of tutoring strategies. This includes the pedagogical agents who are assigned different roles in ITS (e.g., tutor, learning companion) and who are increasingly grounded on cognitive models, sometimes with additional affective features. As discussed previously, both of these domains have been shown to be culturally variable. In order for these agents to behave and think in a more realistic manner, integrating cultural considerations into existing agent architectures must be explored.

As mentioned previously, the existence of subcultures must also be taken into account: certain groups of students may hold certain attitudes that can contradict core cultural values of their national cultural group. To be more effective, and even in some cases to be taken seriously as learning tools, tutors may need to express these attitudes. For instance, armies obviously endorse and reflect some of their national core values. However, all armies of the world similarly seek to develop the collaborative skills of their soldiers, their ability to rely on their comrades, and their feeling of trust in the chain of command. Thus, tutors frequently have to consider different layers of cultural influence to better meet the specific needs and values of their targeted group of people.

Culture and the student model. Modeling cultural aspects of the learner naturally leads to controversy, given the well known risks of oversimplifications and stereotyping. Individuals are influenced by multiple cultural sources; elements such as personal history and experiences may lead to the conclusion that each student is unique and should be considered as such. However, the existence of cultural influences on aspects of students models such as affect (Mesquita et al. 1997) or cognitive processing (Nisbett and Norenzayan 2002) is well documented and can be leveraged in the development of a more accurate student model. Furthermore, legitimate concerns emerge about the development of "*systems that care*" without cultural considerations: ignoring cultural diversity is likely to lead to problems such as ethnocentrism (Lévi-Strauss 1952) in educational practices as well as in information taught to learners. This has the potential to negatively impact not only their perception of the system but also their ability to apply what they might have learned in the real globalized world.

In response, cultural student modeling should not be solely about reporting a set of demographics. Instead, this line of research raises new modeling objectives in order to reconcile the multi-layered nature of cultural identity. This is no more an intractable problem than other challenges ITS researchers have already embraced in their quest towards developing such caring systems. Similar to Self's vision of the future of student modeling (Self 1988), cultural student modeling is about identifying, defining, and developing realistic principles and technologies. Though such a model will promote ITS objectives, it should not be expected that we can obtain a perfect and total representation of the cultural profile of the learner.

Culture and Graphical User Interfaces. First, there are strong assumptions that culture impacts the development and the appraisal of GUIs, whether consciously or unconsciously. Indeed research frequently posits relations between cultural context and specific guidelines endorsed by graphic designers (Marcus and Gould 2000). Symbolism, for example, is culturally variable: meanings associated with symbols used to illustrate concepts, such as icons, avatars and marks, have been shown to greatly differ among cultural groups (Clemmensen 2010). This results in cultural variations both in immediate perception and at the cognitive level. Consequently, it is important to develop a strong knowledge of graphical interpretations in different cultural contexts. It is readily apparent that culturally-adaptive interfaces have a central role to play in globalized ITS

Secondly, according to situated learning research, virtual environments that allow students to immerse themselves in different socio-cultural contexts should provide good opportunities for cultural learning. Such virtual environments can also be augmented by embedding additional resources that can provide explanations about cultural specifics. The experience can be further enriched by populating the virtual environment with embodied agents that represent the local population. Embodied agents are an emerging technology strongly aimed at fostering human-computer interaction.

One main interest of embodied pedagogical agents lies in their ability to transmit more than just verbal or written information: they can perform body gestures and postures. Body language being a well-known example of cultural variations (Bonvillain 2008), it is likely that information transmitted using body language features that are unknown to the learner will either be misunderstood or completely ignored. At the same time, mastering the body language of another culture is an extremely useful intercultural skill. This promotes the development of embodied agents with cultural abilities in conjunction with ITS support.

However, with these opportunities come new challenges. For instance, Baylor's study (Baylor and Kim 2004) reported variation in perceived competence of pedagogical agents according to their ethnicity. This is likely not a universalism, but rather the expression of stereotypes and unconscious mental preconditioning within a cultural group. This effect demonstrates that cultural interpretations occur across various dimensions of computer-assisted learning activities. On a positive note, research suggests that showing more virtual experts with characteristics from minorities could help to reverse negative mental programming (Yee and Bailenson 2006).

Culture and educational technologies. Computer-mediated education needs to consider that information technology and related devices are perceived differently from one cultural context to another (Riley et al. 2009). This could be an expression of the Digital Divide, since having fewer technological experiences can lead users to be more or less demanding towards systems. For instance, computer literacy varies from one cultural context to another, along with availability of computers and internet services. It is legitimate to imagine that students with more videogame literacy are more likely to appraise a serious game according to their experience with state-of-the-art game technologies, and more easily master videogames paradigms incorporated into serious games.

Variations in appraisal of educational technology can also be due to deeper issues such as common practices, attitudes, and popular stereotypes. For instance, the notion of privacy varies across cultural groups (PRIVACY 2010), which in some groups may lead them to develop a negative appraisal towards systems requesting personal information. Scholars thus need to remain aware that evaluation of an educational technology performed in a given cultural context is not necessarily universal and may not be repeatable in a different context. Educational data mining techniques could help to clarify the relation between specific cultural groups and educational technology. Once cultural effects are identified, remediation techniques could be imagined to normalize results or help predict the acceptance of a technology in another cultural context.

24.4 Current Achievements in Culturally-Aware Technology

24.4.1 General Cultural Frameworks for Educational Technology

In the past few years, several frameworks have been proposed for infusing culture into information technology. Each framework has a slightly different focus and for the most part, they are not designed especially for educational issues. Still, computer-assisted educational technology is a natural target of these generic approaches to culture. Some of these initiatives are reported in the next paragraphs.

Birukou and colleagues have proposed the Implicit Culture Framework (ICF) (Birukou et al. 2010) to formalize how cultural knowledge transfers between community members or between communities within multi-agent systems. Their framework consists of definitions of culture-related concepts and phenomena (e.g., *cultural theory, implicit cultural relation, strong and weak cultures, transmission…*), a meta-model, and several algorithms to apply in a multi-agent system context. A general architecture is provided as a guideline for developing systems that support their framework. The ICF could be adapted in order to enculturate educational technology. For instance, an ICF-based system could be designed to create hybrid human-virtual agents communities. Cultural knowledge from the learners could then be transferred to the virtual agents through ICF methodology, thus forming a community that shares cultural values and practices, enabling more efficient interaction.

Based on a more formal analysis of the domain, Blanchard and colleagues have used heavyweight ontology-engineering techniques to propose the Upper Ontology of Culture (UOC) (Blanchard et al. 2010) The UOC is a neutral, theory-driven, and interdisciplinary conceptualization of the cultural domain including cognitive, affective, behavioral, contextual, and physical dimensions. It aims to provide guidelines for the development of culturally-aware applications, the consistent computerization of cultural data and their interoperability, as well as the development of culture-driven automatic reasoning processes. It proposes theory-driven ontological definitions for important concepts of cultural and culture-related domains (i.e., it describes their internal structure and properties) such as *culture* (see Fig. 24.1), *enculturated agent, cultural profile, cognitive information, affective phenomena, notions of collective and individual cultural cognition*. All of these concepts are interrelated, and these interrelations are also explicitly described in the UOC. For instance, to be called "enculturated", an agent needs to have a cultural profile that depends on (a) cultural experiences that provide some domain knowledge, (b) internalized cultures it consciously or unconsciously endorses, thus affecting its behaviors, cognition and affective processes, and (c) group specifics such as phenotype attributes. Allard and colleagues (2010) have followed a similar heavyweight ontological approach to analyse the specific problem of cultural interference of a mother tongue while learning a second language. They have ontologically defined concepts related to language learning and to cultural influences that may arise in such a context.

Reinecke and her colleagues also adopted ontology engineering techniques and proposed the General User Modelling Ontology (GUMO) (Reinecke 2007) a lightweight ontological approach focusing on user modelling. GUMO follows a pragmatic perspective by identifying several features that could complement conventional location information. These features include *e.g., country of current residence, former residence(s), the nationality of both parents, mother tongue, second languages, the main reading/writing direction, age, the most familiar form of instruction in education, political orientation/social structure, and religion.* GUMO is embedded in MOCCA, a web-based to-do list tool that allows users to manage their tasks online (Reinecke et al. 2010). The system was developed as a test-bed for culturally-adaptive interface technology. An evaluation of MOCCA revealed its ability to predict interfaces that would suit users' preferences through the use of adaptation rules based on the previously-mentioned cultural features.

Finally, whereas relations between culture and other human factors of interest for educational purposes (e.g., affect and cognitive processing) have been approached in the UOC, specific models have also been developed for such objectives. For instance, Nazir and his colleagues have proposed a model that merges cultural, affective, and personality features (Nazir et al. 2009). This model refers to Hofstede's system of value (Hofstede 2001) for its cultural component, the PSI

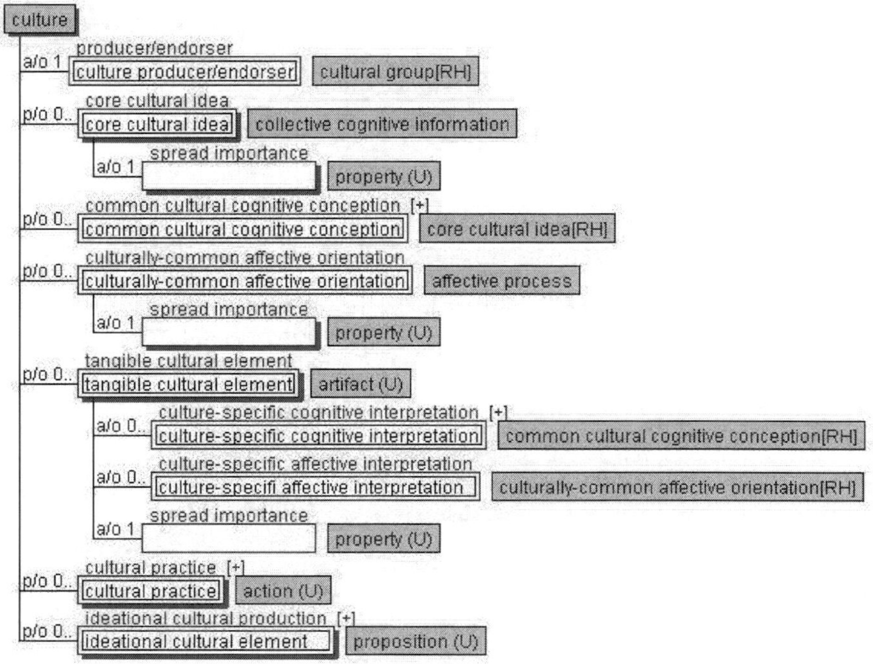

Fig. 24.1 The ontological conceptualization of the "culture" concept in the UOC; 'p/o' refers to part-of links, and 'a/o' refers to attribute links. See (Blanchard et al. 2010) for an in-depth explanation of this conceptualization and the ecology of concepts into which it is integrated.

model (Doerner 2003) for affect, and the Big Five model (McCrae and John 1992) for its personality component. Aylett and her colleagues (2009) reported that integrating such a model into embodied agents populating a role-playing environment could increase users' intercultural empathy. While initially designed for synthesizing realistic enculturated agents, Nazir et al.'s model could be easily adapted to produce advanced user models.

24.4.2 Realtime Cultural Adaptation for Educational Technology

As mentioned earlier, developing realtime cultural adaptation is not a prerogative solely of educational technologies, but rather has implications for information systems in general. Thus, several of the initiatives presented below focus on general cultural issues in HCI that remain of interest in educational technologies.

Cultural adaptation processes have been proposed to improve the adaptive capabilities of classic GUIs. For instance, the MOCCA system (Reinecke et al. 2010), introduced in the latter section, automatically adapted its interface to users' preferences according to a cultural user model that goes beyond location information. The Motivational and Culturally-Aware System (MOCAS) (Blanchard 2009) adopted a different approach for its cultural adaptation process: dynamically computing learner's cultural memberships by determining whether the pedagogical attitudes and results of this learner are in line with those of other members of his or her supposed cultural groups. Computed cultural memberships then influence the selection of culturally-relevant multimedia resources and pedagogical strategies. The overall fitness of the selected adaptations is determined according to the learner's resulting pedagogical assessments and then used to update the learner's cultural memberships, and so on. Finally, Melis and colleagues (2010) deduced from an empirical intercultural analysis of the ActiveMath platform that greater consideration of students' language, and regional specifics would solve cultural misunderstandings that arose when using their platform. They consequently described enculturation solutions for presenting the system and its learning material in a more appropriate manner, for adapting mathematical notations and names according to learners' specifics, and for selecting and sequencing learning objects and scenarios in order to match learners' contextual reality.

Perhaps the most promising development in the area of culturally-adaptive technology is Embodied Enculturated Communication Agents (EECA) (Rehm 2010). This concept initially stemmed from Embodied Communication Agents (ECA), agents able to communicate with users through the genesis of gestures and postures of their virtual body. However, body language is known to differ greatly from one cultural group to another. In developing diverse agents, cultural considerations must be considered. EECAs are currently among the most difficult cognitive agents to implement (Rehm 2010). They require mastery and coherent merging of several issues such as the genesis of realistic 3D behaviors and communication styles, the computerization and integration of cultural competences, and the consideration of cognitive (and potentially affective) implications.

In the past few years, a tremendous number of projects with very different research focii have emerged in this area. Among notable initiatives, Huang and his colleagues (2009) have proposed the Generic Embodied Conversational Agent (GECA) framework in order to speed the development of EECA. Using GECA, only a module describing verbal and non-verbal communication specifics of a targeted group has to be developed in order to provide cultural intelligence to an ECA. This concept was showcased in an application where an EECA played the role of a culturally-intelligent tour guide. Endrass and her colleagues proposed a system where EECAs were attributed culturally-marked communication styles with varying usages of pauses and overlapping speech (Endrass 2010). They found that, even if the fantasy language EECAs used to communicate with each other was unknown to human observers, they perceived the agents as having a western or Asian orientation depending on their communication style. Furthermore, in a preliminary evaluation, observers reported to prefer agents with a communication style similar to their own.

Promoting positive perception of embodied agents can thus be addressed in part through developing agents with the capability to address users' cultural communication specifics. But universalisms in communication are also worth consideration. For example, Brown & Levinson's theory of universal politeness (1987) has been applied to EECAs (Johnson et al. 2005), and Miller and colleagues recently proposed a formalized computational model for agents (Miller et al. 2010) that further facilitates the integration of this politeness theory.

Since Embodied Pedagogical Agents (EPA) (Rickel and Johnson 1997) are ECAs with additional ITS capabilities, it is reasonable to posit that improving their ability to efficiently communicate would improve the quality of their relation with learners and their overall efficiency. Several EPAs with cultural models have already been implemented, mainly for intercultural competence instruction (Johnson 2007; Kim et al.). They are discussed in the next section.

24.4.3 Adaptive Systems for Cultural Instruction

Over the past several decades, as technology-based training has increased in popularity, there has been some history of using it to support intercultural interactions. This training has varied in its goals along with the learning objectives identified for cultural training. There tend to be two directions taken for instruction. In the first, students learn through the use of cultural artifacts, whether they are authentic documents such as films or commercials from the culture, or depictions of monuments and high culture. Carmen Sandiego[3] is an example of a game dating back to 1985 that introduces players to a wide array of cultural knowledge through artifacts, as they chase "bad guys" through different worldwide destinations. Delving more deeply into a specific cultural environment was *A la rencontre de Philippe* (Furstenberg et al. 2001), a game in which students play a French journalist using cultural knowledge to interact with the environment through branching storylines. Student journalists were tasked with helping a broken-hearted French man find a new apartment after being dumped by his girlfriend.

In the second type of instruction, immersive virtual environments use embodied conversational agents (as described in section 4.2) to let students practice intercultural interactions. A very early example of a virtual environment for culture learning is text-based multi-user domains (MUDs). In these environments, there are no artificially intelligent agents. Instead, language students interact with each other online in an imaginary world where they can test their language skills with others and practice interacting in culturally influenced ways (Bruckman 1995; Falsetti and Schweitzer 1995). Modern virtual environment systems often utilize a preexisting immersive technology (e.g., Second Life, Unreal Tournament Engine) to create a simulated representation of another culture complete with architectural features and ambient sounds. One such system that moves towards agents with behaviors based on cultural models is Second China, an island in Second Life that is designed to mimic cultural and visual aspects of China (Henderson et al. 2008). Embedded scenarios within this world deliver important cultural experiences that players can interact with or observe. These scenarios are facilitated by embodied agents located in the environment, which assume culturally appropriate roles. These agents tend to play out scripted interactions that range from tai chi demonstrations to a receptionist who will answer questions.

---

[3] http://en.wikipedia.org/wiki/Carmen_Sandiego

We now have the ability to combine these training systems with artificial intelligence-based scaffolding such as ITS. It is not clear that ITS approaches, most often used in domains like algebra or physics, will translate directly to an ill-defined domain like culture. Ogan, Aleven, and Jones (2010) describe how ITS principles might be adapted for learning in this domain. These principles can be used to develop interactive systems that help students examine cultural artifacts such as feature films or commercials. These systems cover cultural knowledge, analysis of cultural values and behaviors, and may have also focus on developing perspective-taking skills. One such system is ICCAT, which requires students to make culturally aware predictions of events in French films and includes a tutored online discussion component (Ogan et al. 2010).

There also exist a small number of 3D virtual environments that teach intercultural competence with the support of ITS. They typically integrate a set of embodied conversational agents who are imbued with a more complete model of cultural behavior. Interaction with these agents facilitates the practice of communicative skills in the new culture, from making appropriate gestures of greeting to conversing in culturally appropriate ways. These systems cover a range of cultures (e.g., Spanish, Chinese, Iraqi, Dari, Pashto, and French), and exist for various training purposes, ranging from language classrooms to military or business contexts. Two such systems are the Tactical Culture and Language Training System (TCLTS), developed by Johnson et al. (2007; see Fig. 24.2), and BiLAT, developed by Hill et al.

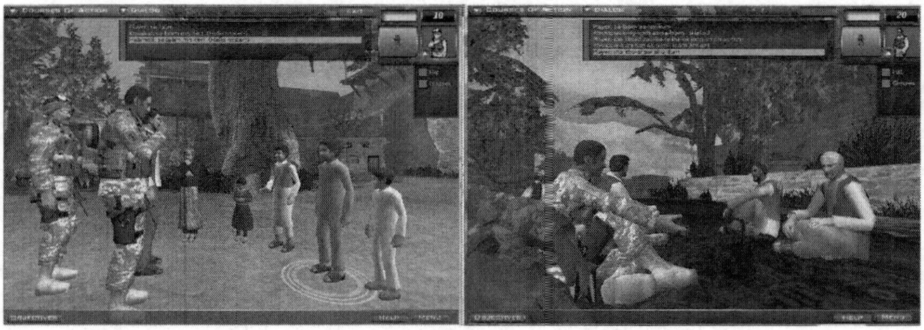

Fig. 24.2 Two illustrations of the TLCTS interface (Tactical Dari version). Courtesy of © Alelo Inc.

In TCLTS, students move through a virtual town to solve missions while making culturally-appropriate gestures, acting in culturally-appropriate ways, and practicing the target language with the aid of voice recognition software. BiLAT focuses in particular on cross-cultural negotiation skills in one-on-one meetings with agents representing the target culture. Systems can also be developed that go beyond national cultures. For instance, Rothwell suggests using culturally-aware educational technology for strengthening a cross-institution and cross-nation culture of nuclear safety (Rothwell 2010).

These interactive systems benefit from adapting the classic approaches of ITS to scaffold and support learning. For example, in a virtual environment, all of the students' communicative actions can be linked to a detailed representation of learning objectives which is managed by an ITS (Lane et al. 2007). Such an ITS coaching component can provide guidance and feedback during face-to-face meetings with a virtual character from the target culture (Lane et al. 2008) or provide an after-action review following each meeting. These systems may also integrate other learning activities, such as multimedia resources, quizzes, and part-task training exercises (e.g., Second China). Ogan and Lane (2010) describe six such systems in greater detail, although a number of them have yet to incorporate an ITS with a student model.

Many of these systems might be considered in their early stages of development or deployment. Therefore, one avenue for future research is in the evaluation of such systems for student learning, compared to either typical classroom approaches or the gold standard of one-on-one human tutoring of such skills. These systems represent a growing trend recognizing the power of immersive virtual environments for teaching social, interpersonal, and cultural domains.

24.5 Discussion

Bridging cultural and educational issues raises several concerns that naturally transfer to and evolve in the context of educational technologies. First, as in many other domains, ethics is a central concern in intercultural education. Culture is tightly coupled to foundational aspects of an individuals' identity. Inadequate representation of the group and the individual's cultural specifics may have a durable negative impact on learners. For instance, imagine the case of a system developed by a western company to teach a scientific domain through the use of EECAs and culturally-adapted resources. Should its Asian cultural adaptation be perceived by Asian students as oversimplifying their cultural specifics, this could cause the perception that westerners, in general, have little or no understanding of their culture, and perhaps do not care about it. Culturally-aware educational technologies could also be diverted from their original objective (i.e. promoting intercultural awareness and consideration of learners' cultural specifics) to serve less respectable goals such as propaganda. While these concerns may appear to go beyond the control researchers have of the technology they develop, procedures could be deployed at an international level to provide systems with certificates of compliance to international standards such as UNESCO guidelines (2007), or well known ISO and IEEE standards.

In fact, validating the quality of culturally-aware educational technology is a complex and difficult question. It is well-known in cultural research that an analysis of a culture by an outsider may be biased by preconceptions, even if anthropological methods such as participant observation (DeWalt et al. 1998) are designed to mitigate this threat. As enculturated individuals, course authors, as well as system designers and evaluators, are prone to such cultural bias: they may simply not consider possible interpretations or categories of behaviors because they are not aware of their existence. A classic solution to this issue is that members of the target culture be part of the development team, which still raises the problem of ensuring their objectivity - they could dislike a particular modelling of their own culture which does not fit their idealized view. Indeed, perceptions of a culture by both insiders and outsiders of the cultural group bring useful information at different levels in the development of culturally-aware educational technology. This highlights the importance of taking into account the context of use of cultural data when validating CATS. For instance, grounding adaptation on misconceptions and stereotypes is surely a situation to be avoided at all cost. Knowing common misconceptions and stereotypes of outsiders towards a specific cultural group would support the deployment of corrective processes in systems aiming at developing intercultural competence. A taxonomy that describes the origin and nature of cultural data (e.g., stereotypes, facts, misconceptions) could be developed to inform when it is relevant to use each kind of information, thus providing a first step towards metrics for assessing the quality of cultural information.

Learning cultural content implies going beyond the question of what has been learned in order to also consider how the cultural knowledge has been integrated. Indeed, in many of the stages of intercultural development, students may express negative attitudes towards the other culture and feel superior about their own (Bennett 1993). It is easy to accidentally support negative attitudes towards another culture, perhaps through teaching stereotypes, by letting students becoming overconfident about their cultural skills, or through neglecting to address attitudes like openness as part of the learning objectives. Even if students do learn cultural knowledge, these negative attitudes may be very detrimental to future intercultural communication.

Finally, researchers should understand that legal implications frequently illustrate varying core cultural ideas and have to be clarified in the course of internationally deploying educational technology. For instance, countries vary in their legal acceptance of community-based statistical evaluations. Countries like the United States promote them to address the considerations of minorities, while others more strictly restrict their use, such as France, which follows a national principle of "republican equality" where sub-community membership is downplayed in public life.

24.6 Conclusion

This chapter attempts to show that approaching the cultural domain is a highly complex objective where risks of oversimplification are high. Teaching culture in an inadequate manner may have a durable negative impact on learners' intercultural competence. Furthermore, grounding adaptive processes on cultural misconceptions could have negative effects on learners' perception of the system. Whether culture should be considered in educational technology appears to be a legitimate question at first sight. However, not considering the cultural dimension would lead to ethnocentric approaches that are disrespectful and potentially harmful for efficient and correct learning. Thus, researchers and designers working in the area of CATS have to be especially conscious compared to other disciplines due to the responsibilities and risks that their role implies.

Nonetheless, current achievements demonstrate that CATS have significant ability to enhance students' cultural awareness, to correct their cultural misconceptions, and to provide more respectful and efficient HCI and teaching approaches. It is this great potential that merits continued investigation into culturally-aware educational technology.

References

Aleven, V., McLaren, B., Sewall, J., Koedinger, K.: The Cognitive Tutor authoring tools (CTAT): Preliminary evaluation of efficiency gains. In: Ikeda, M., Ashley, K.D., Chan, T.-W. (eds.) ITS 2006. LNCS, vol. 4053, pp. 61–70. Springer, Heidelberg (2006)

Allard, D., Bourdeau, J., Mizoguchi, R.: Addressing and modeling knowledge of cross-linguistic influence and related cultural factors using Computer-Assisted Language Learning (CALL). In: Blanchard, E.G., Allard, D. (eds.) Handbook of research on culturally-aware information technology: Perspectives and models. IGI Global, Hershey (2010)

Allik, J., McCrae, R.R.: Towards a geography of personality traits: Patterns of profiles across 36 cultures. J. Cross-Cult. Psychol. 35, 13–28 (2004)

Aylett, R., Paiva, A., Vannini, N., Enz, S., André, E.: But that was in another country: Agents and intercultural empathy. In: Proc. Int. Conf. Autonomous Agents and Multi Agents Systems, pp. 329–336 (2009)

Baylor, A.L., Kim, Y.: Pedagogical agent design: The impact of agent realism, gender, ethnicity, and instructional role. In: Lester, J.C., Vicari, R.M., Paraguaçu, F. (eds.) ITS 2004. LNCS, vol. 3220, pp. 592–603. Springer, Heidelberg (2004)

Bennett, M.J.: Towards ethnorelativism: A developmental model of intercultural sensitivity. In: Paige, R.M. (ed.) Education for the intercultural experience, pp. 27–71. Intercultural Press, Yarmouth (1993)

Biggs, J.B.: Teaching across cultures. In: Salili, F., Chiu, C., Hong, Y. (eds.) Student motivation: The culture and context of learning. Kluwer Academic/Plenum Publisher, New York (2001)

Birukou, A., Blanzieri, E., Giorgini, P.: Implicit culture framework for behaviour transfer. In: Blanchard, E.G., Allard, D. (eds.) Handbook of research on culturally-aware information technology: Perspectives and models. IGI Global, Hershey (2010)

Blanchard, E.G.: Adaptation-oriented culturally-aware tutoring systems: When adaptive instructional technologies meet intercultural education. In: Song, H., Kidd, T. (eds.) Handbook of research on human performance and instructional technology, Information Science Reference, Hershey, PA (2009)

Blanchard, E.G., Mizoguchi, R., Lajoie, S.P.: Structuring the cultural domain with an upper ontology of culture. In: Blanchard, E.G., Allard, D. (eds.) Handbook of research on culturally-aware information technology: Perspectives and models. IGI Global, Hershey PA (2010)

Blanchard, E.G., Roy, M., Lajoie, S.P., Frasson, C.: An evaluation of sociocultural data for predicting attitudinal tendencies. In: Proc. 14th Int. Conf. Artif. Intell. Educ., pp. 399–406 (2009)

Bonvillain, N.: Language, culture, and communication: The meaning of messages, 5th edn. Pearson Prentice Hall, Upper Saddle River (2008)

Brown, P., Levinson, S.: Politeness: Some universals in language usage. Cambridge University Press, Cambridge (1987)

Bruckman, A.: The MediaMOO Project: Constructionism and professional community. Convergence 1(1) (Spring 1995)

Clemmensen, T.: A framework for thinking about the maturity of cultural usability. In: Blanchard, E.G., Allard, D. (eds.) Handbook of research on culturally-aware information technology: Perspectives and models. IGI Global, Hershey PA (2010)

Cooper, C.R., Denner, J.: Theories linking culture and psychology: Universal and community-specific processes. Annual Rev. Psychol. 49, 559–584 (1998)

DeWalt, K.M., DeWalt, B.R., Wayland, C.B.: Participant observation. In: Bernard, H.R. (ed.) Handbook of methods in cultural anthropology. AltaMira Press, Walnut Creek (1998)

Doerner, D.: The mathematics of emotions. In: Proc. 5th Int. Conf. Cogn. Modeling, pp. 75–79 (2003)

Earley, C.P., Mosakowski, E.: Cultural intelligence. Harvard Business Review 82(10), 139–146 (2004)

Ekman, P.: Universals and cultural differences in facial expressions of emotion. In: Cole, J. (ed.) Nebraska Symposium on Motivation, vol. 19. University of Nebraska Press, Lincoln (1972)

Endrass, B., Rehm, M., André, E.: Towards culturally-aware virtual agents systems. In: Blanchard, E.G., Allard, D. (eds.) Handbook of research on culturally-aware information technology: Perspectives and models. IGI Global, Hershey PA (2010)

Falsetti, J., Schweitzer, E.: SchMOOze University: A MOO for ESL/EFL students. In: Warschauer, M. (ed.) Virtual connections: On-line activities and projects for networking language learners. Second Language Teaching and Curriculum Center, pp. 231–232. University of Hawai'i, Honolulu (1995)

Furstenberg, G., Levet, S., English, K., Maillet, K.: Giving a virtual voice to the silent language of culture: The Cultura project. Language learning & technology 5(1), 55–102 (2001)

Henderson, J., Fishwick, P., Fresh, E., Futterknecht, E., Hamilton, B.D.: Immersive learning simulation environment for chinese culture. In: Proc. Interservice/Industry Training, Simulation, and Ed. Conf., I/ITSEC, paper 8334 (2008)

Hofstede, G.: The cultural relativity of the quality of life concept. Acad. Manag. Rev. 9(3), 389–398 (1984)

Hofstede, G.: Culture's consequences: Comparing values, behaviors, institutions, and organizations across nations, 2nd edn. Sage, London (2001)

Hofstede, G.: Scores of Hofstede's cultural dimensions (2010), http://www.geert-hofstede.com/hofstede_dimensions.php (Accessed, February 18, 2010)

House, R.J., Hanges, P.J., Javidan, M., Dorfman, P., Gupta, V.: Culture, leadership and organizations: The GLOBE study of 62 societies. Sage, Thousands Oaks (2004)

Huang, H.-H., Cerekovic, A., Pandzic, I.S., Nakano, Y., Nishida, T.: Toward a multi-culture adaptive virtual tour guid agent with a modular approach. Int. J. AI & Society 24(3), 225–235 (2009)

Johnson, W.L.: Serious use of a serious game for language learning. In: Proc. Art. Intell. in Ed (2007)

Johnson, W.L., Mayer, R.E., André, E., Rehm, M.: Cross-cultural evaluation of politeness in tactics for pedagogical agents. In: Looi, C., McCalla, G., Bredeweg, B., Breuker, J. (eds.) Proc. Art. Intell. in Ed., pp. 298–305. IOS Press, Amsterdam (2005)

Kashima, Y.: Conception of culture and person for psychology. J. Cross. Cult. Psychol. 31(1), 14–32 (2000)

Kim, J., Hill, R.W., Durlach, P., Lane, H.C., Forbell, E., Core, M., Marsella, S., Pynadath, D., Hart, J.: BiLAT: A game-based environment for practicing negotiation in a cultural context. To appear in Int. J. of Art. Intell. in Ed (in press)

Kirkman, B.L., Lowe, K.B., Gibson, C.B.: A quarter century of culture's consequences: a review of empirical research incorporating Hofstede's cultural values framework. J. Int. Bus. Stud. 37, 285–320 (2006)

Landis, D., Bennett, J., Bennett, M.J.: Handbook of intercultural training, 3rd edn. Sage Publications, Inc., Thousand Oaks (2003)

Lane, H.C., Core, M.G., Gomboc, D., Karnavat, A., Rosenberg, M.: Intelligent tutoring for interpersonal and intercultural skills. In: Proc. Interservice/Industry Training, Simulation, and Education Conference, paper 1514 (2007)

Lane, H.C., Hays, M.J., Core, M.G., Gomboc, D., Forbell, E., Auerbach, D., Rosenberg, M.: Coaching intercultural communication in a serious game. In: Proc. 16th International Conf. on Computers in Ed., pp. 35–42 (2008)

Lave, J., Wenger, E.: Situated learning: Legitimate peripheral participation. Cambridge University Press, New York (1991)

Lévi-Strauss, C.: Race et histoire. UNESCO/Denoël (1952)

Marcus, A., Gould, E.W.: Crosscurrents: Cultural dimensions and global web user-interface design. Interactions 7(4), 32–46 (2000)

McCrae, R.R., John, O.P.: An introduction to the five-factor model and its applications. J. Personality 60, 175–215 (1992)

McSweeney, B.: Hofstede's model of national cultural differences and their consequences: A triumph of faith – a failure of analysis. J. Hum. Relat. 55(1), 89–118 (2002)

Melis, E., Goguadze, G., Libbrecht, P., Ullrich, C.: Culturally aware mathematics education technology. In: Blanchard, E.G., Allard, D. (eds.) Handbook of research on culturally-aware information technology: Perspectives and models. IGI Global, Hershey (2010)

Mendenhall, M.E., Stahl, G.K., Ehnert, I., Oddau, G., Osland, J.S., Kühlmann: Evaluation studies of cross-cultural training programs: A review of the literature from 1988 to 2000. In: Landis, et al. (eds.), pp. 129–144 (2004)

Mesquita, B., Frijda, N.H., Scherer, K.R.: Culture and emotion. In: Dasen, P., Saraswathi, T.S. (eds.) Handbook of cross-cultural psychology. Basic processes and pevelopmental psychology, vol. 2, pp. 255–297. Allyn & Bacon, Boston (1997)

Miller, C.A., Ott, T., Wu, P., Vakili, V.: Politeness and etiquette modeling: Beyond perception to behavior. In: Blanchard, E.G., Allard, D. (eds.) Handbook of research on culturally-aware information technology: Perspectives and models. IGI Global, Hershey (2010)

Nazir, A., Enz, S., Lim, M.Y., Aylett, R., Cawsey, A.: Culture–personality based affective model. Int. J. AI & Society 24(3), 281–293 (2009)

Nisbett, R.E., Norenzayan, A.: Culture and cognition. In: Medin, D., Pashler, H. (eds.) Stevens' Handbook of Experimental Psychology, 3rd edn. Memory and Cognitive Processes, vol. 2. John Wiley & Sons, New York (2002)

Ogan, A., Lane, H.C.: Virtual learning environments for culture and intercultural competence. In: Blanchard, E., Allard, D. (eds.) Handbook of Research on Culturally-Aware Information Technology: Perspectives and Models. IGI Global, Hershey (2010)

Ogan, A., Aleven, V., Jones, C.: Advancing development of intercultural competence through supporting predictions in narrative video. International J. of Artificial Intelligence in Education (2010)

PRIVACY, Privacy - Stanford Encyclopaedia of Philosophy (2010), http://plato.stanford.edu/entries/privacy/ (Accessed February 18, 2010)

Rehm, M.: Developing enculturated agents: Pitfalls and strategies. In: Blanchard, E.G., Allard, D. (eds.) Handbook of research on culturally-aware information technology: Perspectives and models. IGI Global, Hershey (2010)

Reinecke, K.: Cultural Adaptivity for the Semantic Web. In: Proc. 6th Int. Semantic Web Conf. (2007)

Reinecke, K., Schenkel, S., Bernstein, A.: Modeling a user's culture. In: Blanchard, E.G., Allard, D. (eds.) Handbook of research on culturally-aware information technology: Perspectives and models. IGI Global, Hershey PA (2010)

Rickel, J., Johnson, W.L.: Integrating pedagogical capabilities in a virtual environment agent. In: Proc. 1st Int. Conf. Autonomous Agents, pp. 30–38 (1997)

Riley, C., Buckner, K., Johnson, G., Benyon, D.: Culture & biometrics: Regional differences in the perception of biometric authentication technologies. Int. J. AI & Society 24(3), 295–306 (2009)

Rothwell, S.L.: Information technology and the development of a global safety culture: A nuclear perspective. In: Blanchard, E.G., Allard, D. (eds.) Handbook of research on culturally-aware information technology: Perspectives and models. IGI Global, Hershey PA (2010)

Ryan, R.M., Deci, E.L.: Self-determination theory and the facilitation of intrinsic motivation, social development, and well-being. Am. Psychol. 55, 68–78 (2000)

Savicki, V.: Developing intercultural competence and transformation: Theory, research, and application in international education. Stylus, Sterling (2008)

Scharifian, F.: On cultural conceptualizations. J. Cogn. Cult. 3(3), 187–207 (2003); Scherer, K.R.: What are emotions? And how they can be defined. Social Science Information 44(4), 695–729 (2005)

Self, J.: Bypassing the intractable problem of student modelling. In: Proc. 1st Int. Conf. Intelligent Tutoring Systems, pp. 18–24 (1988)

Stewart, O., Chakraborty, J.: Culturally determined preferences: Automatic speech recognition (ASR) systems versus live help. In: Blanchard, E.G., Allard, D. (eds.) Handbook of research on culturally-aware information technology: Perspectives and models. IGI Global, Hershey (2010)

UNESCO, UNESCO Guidelines on Intercultural Education (2007), http://unesdoc.unesco.org/images/0014/001478/147878e.pdf (Accessed February 18, 2010)

Yee, N., Bailenson, J.N.: Walk A Mile in Digital Shoes: The Impact of Embodied Perspective-Taking on The Reduction of Negative Stereotyping in Immersive Virtual Environments. In: Proc. of PRESENCE 2006: The 9th Annual International Workshop on Presence (2006)

Index